Insights into Mobile Mult
Communications

Signal Processing and its Applications

Insights into Mobile Multimedia Communications

Edited by

D. R. BULL
C. N. CANAGARAJAH
and
A. R. NIX

Centre for Communications Research
University of Bristol, UK

Academic Press

San Diego • London • Boston • New York
Sydney • Tokyo • Toronto

Academic Press Limited
24–28 Oval Road, London NW1 7DX, UK
http://www.hbuk.co.uk/ap/

Academic Press
525 B Street, Suite 1900, San Diego, CA 92101-4495, USA
http://www.apnet.com

ISBN 0-12-140310-6

A catalogue record for this book is available from the British Library

Typeset by Keyword Typesetting Services Ltd.
Printed and bound by CPI Antony Rowe,
Transferred to digital printing 2006
99 00 01 02 03 04 MP 9 8 7 6 5 4 3 2 1

Contents

The plate section appears between pp. 614 and 615.

Series Preface

Signal processing applications are now widespread. Relatively cheap consumer products through to the more expensive military and industrial systems extensively exploit this technology. This spread was initiated in the 1960s by introduction of cheap digital technology to implement signal processing algorithms in real-time for some applications. Since that time semiconductor technology has developed rapidly to support the spread. In parallel, an ever increasing body of mathematical theory is being used to develop signal processing algorithms. The basic mathematical foundations, however, have been known and well understood for some time.

Signal Processing and its Applications addresses the entire breadth and depth of the subject with texts that cover the theory, technology and applications of signal processing in its widest sense. This is reflected in the composition of the Editorial Board, who have interests in:

(i)   Theory – The physics of the application and the mathematics to model the system;
(ii) Implementation – VLSI/ASIC design, computer architecture, numerical methods, systems design methodology, and CAE;
(iii) Applications – Speech, sonar, radar, seismic, medical, communications (both audio and video), guidance, navigation, remote sensing, imaging, survey, archiving, non-destructive and non-intrusive testing, and personal entertainment.

Signal Processing and its Applications will typically be of most interest to postgraduate students, academics, and practising engineers who work in the field and develop signal processing applications. Some texts may also be of interest to final year undergraduates.

Richard C. Green
Department of Technology,
Metropolitan Police Service, London, UK

Preface

Personal multimedia communications is a major growth area in engineering today with groups in industry and academia across the world working to develop future generations of mobile radio systems, compression algorithms and coding tools. However, before multimedia applications can be truly successful in mobile applications, many technologically demanding problems must be resolved. For example, in order to provide the user with a suitable end to end service for multimedia applications, integrated systems, high performance interfaces and robust and flexible source coding tools need to be developed. All of these issues are currently being addressed in standards bodies ranging from ETSI SMG-2, UMTS and BRAN to ISO MPEG-4 and ITU-H.263. In mobile applications the bandwidth requirements are likely to be dominated by real time video. Even with today's most aggressive compression algorithms, the needs for most video applications far exceed the bandwidth available with existing mobile systems. Although the bandwidth available to the user will increase with the emergence of future GSM extensions, UMTS and local services based on wireless LANs, it is clear that advances in radio technology and air interface design must go hand in hand with innovations in source and channel coding.

The motivation for this book originated in 1995/96 after the 2nd International Mobile Multimedia Communications Conference (MoMuC-2) was held in the UK at the University of Bristol. It was then that the editors (together with Dr Colin I'Anson of HP Laboratories) identified the need for a text which brought together the fields of advanced communication technologies and multimedia coding in a single volume. It was clear that, due to the breadth and evolving nature of this field, such a book could not be written by a single author or even a small group of authors. Multimedia communications research is truly interdisciplinary, drawing together technologies from areas as diverse as signal processing, RF engineering, communication systems, propagation, source and channel coding, network design, VLSI circuit design, computing, human-computer interaction (HCI) and others. In the light of this it was evident that the broad community would be best served by an integrated text comprising chapters from international experts in the various disciplines which together make up this field. The purpose of this book is thus to present an integrated view of advanced radio systems, network architectures and source coding which demonstrates the state of the art, as well as future directions, in this rapidly evolving subject area.

The book is divided into eight sections which together cover the main building blocks in a mobile multimedia communication system. Our aim was not only to cover each area in its own right but also to show how these integrate to form a system solution, while emphasising their interactions and dependencies. The contents of each of these sections are summarised below.

Part 1 provides an introduction to the area and sets the scene for the remaining sections. Here the authors review all the building blocks, highlighting key issues and interdependencies.

Parts 2 and 3 focus on systems aspects of current and proposed future technology and services. An overview of the functionality and capabilities of existing systems is given in Part 2. GMS, JDC and the emerging PMR standard, TETRA, are all covered and compared, as is the H.324 multimedia standard. Part 3 looks further into the future with chapters on the potential of GSM extensions, UMTS and wireless LAN systems. The important debate on the role of the internet in future mobile multimedia systems is also highlighted.

Parts 4–6 of the book cover source coding. Part 4 concentrates on speech and audio. Here we see contributions not only on low bit rate speech coding, but also on higher bit rate techniques appropriate for the enhanced quality needed for audio (e.g. stereo music) transmission. The areas of image, video and multimedia coding are covered in Part 5. The importance of efficient compression algorithms is emphasised, especially for low bit rate applications. Standards such as MPEG-4, together with advanced methods based on wavelets, segmentation and scalability, are all covered. The section also examines the integration of, and trade-offs between, speech and video coding in a multimedia framework. Quality of service in a wireless environment is not simply dictated by available bandwidth and compression tools but also by the nature of the channel and its error characteristics. Video data, in particular, is sensitive to error propagation. This area is explored in Part 6, where a variety of error resilient image and video coding methods are presented.

The bandwidths needed to deliver acceptable multimedia quality of service can only be provided if efficient air interfaces and modulation schemes are available. Part 7 presents contributions which exemplify current research in antenna technology, CDMA and equalisation methods. Both radio system design and propagation aspects are covered. Part 8 continues the theme of Part 7 in the area of wireless local area network technology. WLANs are emerging as a powerful alternative to wired infrastructures in both commercial and consumer applications and are likely to offer the earliest opportunities for true wireless multimedia applications at acceptable quality. This section overviews current research in this area with emphasis on the recently finalised ETSI HIPERLAN standard.

The book ends with a vision of the future multimedia technology. This chapter argues the potential of multimedia communication systems and outlines their requirements while at the same time warning of the dangers of proprietary solutions and closed standards.

We hope that this book will be of interest to a broad audience, including academic researchers, practising engineers and engineering managers in industry. In such a rapidly evolving area, it is not possible to answer all the questions, or perhaps even to pose all the questions. However, we believe that the book will act as a repository of ideas, design examples and advanced techniques which will stimulate the reader. If he or she emerges with a greater awareness of the key issues, both technical and political, and of the relationships between the building blocks, then the main aim of the book will have been achieved.

ACKNOWLEDGEMENTS

We would like to express gratitude to all our colleagues (staff and students) in the Centre for Communications Research at the University of Bristol for creating an environment from which a book such as this could emerge. In particular we thank Joe McGeehan for his leadership and inspiration, Mark Beach for his enthusiasm and support for this project, and Nina Bunton for her assistance throughout. We would also like to thank Colin I'Anson of HP Labs (Bristol) for bringing MoMuC-2 (the conference from which the idea for this book emerged) to Bristol and for his Chairmanship of the event.

We also owe a debt to all the contributors who have shown a great deal of patience with us while we pulled this project together. Not least we would like to thank Bridget Shine from Academic Press for her faith in this project and for her patience when our deadlines became mobile!

D. R. Bull
C. N. Canagarajah
A. R. Nix
Bristol, March 1998

List of Contributors

Adachi, F., NTT Mobile Communications, Network Inc., R&D Department, I-2356 Take, Yokosuka-Shi, Kanagawa 238-03, Japan

Araki, M., Radio Systems Laboratory, NTT Wireless Systems Labs, 1-1 Hikarinooka, Yokosuka-Shi, Kanagawa-Ken, 239 Japan

Arroyo-Fernández, B., European Commission DG XIII-B4, BU-9 4.90, Ave de Beaulieu 9, B-1160, Brussels, Belgium

Athanasiadou, G., Centre for Communications Research, University of Bristol, Merchant Venturers Building, Woodland Road, Bristol BS8 1UB, UK

Barani, B., European Commission DG XIII-B4, BU-9 4.90, Ave de Beaulieu 9, B-1160, Brussels, Belgium

Barberis, S., Wireless Information Network Lab, Rutgers University, PO Box 909 Piscataway, NJ 08855, USA

Bassil, J., Centre for Information Systems, Applied Research and Technology, MLB1 PP14, BT Laboratories, Martlesham Heath, Ipswich IP5 3RE, Suffolk, UK

Beach, M. A., Centre for Communications Research, University of Bristol, Merchant Venturers Building, Woodland Road, Bristol BS8 1UB, UK

Black, A. W., Centre for Satellite Engineering Research, University of Surrey, Guildford GU2 5XH, UK

Broderson, R., Electrical Engineering and Computer Science Department, University of California, Berkeley, CA 94720, USA

Bull, D. R., Centre for Communications Research, University of Bristol, Merchant Venturers Building, Woodland Road, Bristol BS8 1UB, UK

Canagarajah, C. N., Centre for Communications Research, University of Bristol, Merchant Venturers Building, Woodland Road, Bristol BS8 1UB, UK

Cheetham, B., Department of Electrical Engineering and Electronics, University of Liverpool, PO Box 147, Liverpool L69 3BX, UK

Chen, L., Computer Science Department, Royal Holloway College, London University, Egham, Surrey TW20 0EX, UK

Chung-How, J. T., Centre for Communications Research, University of Bristol, Merchant Venturers Building, Woodland Road, Bristol BS8 1UB, UK

Czerepiński, P., Centre for Communications Research, University of Bristol, Merchant Venturers Building, Woodland Road, Bristol BS8 1UB, UK

Ebrahimi, T., Signal Processing Laboratory, CH-1015 Lausanne, Switzerland

Egger, O., Signal Processing Laboratory, CH-1015 Lausanne, Switzerland

Eryurtlu, F., Centre for Satellite Engineering Research, University of Surrey, Guildford GU2 5XH, UK

Evci, C., Alcatel Telecom, Radiocommunications Department, 5 Rue Noel Pons, 92734 Nanterre, Paris, France

Färber, N., Universität Erlangen-Nürnberg, Lehrstuhl für Nachrichtentechnik, Caurstrasse 7, 91058 Erlangen, Germany

Fitton, M. P., Centre for Communications Research, University of Bristol, Merchant Venturers Building, Woodland Road, Bristol BS8 1UB, UK

Girod, B., Universität Erlangen-Nürnberg, Lehrstuhl für Nachrichtentechnik, Caurstrasse 7, 91058 Erlangen, Germany

Hailes S., Department of Computer Science, University College London, London

Hanzo, L., Department of Electronics and Computer Science, Southampton University, Southampton SO17 1BJ, UK

Hardman, V., Department of Computer Science, University College London, London, UK

Higashi, A., NTT Mobile Communications Network Inc., R&D Department, I-2356 Take, Yokosuka-Shi, Kanagawa 238-03, Japan

Hirade, K., NTT Mobile Communications Network Inc., R&D Department, 3-5 Hikari-no-oka, Yokosuka, Kanagawa 239-8536, Japan

Holtzman, J., Wireless Information Network Lab, Rutgers University, PO Box 909 Piscataway, NJ 08855, USA

I'Anson, C., Mobile Communications Department, Hewlett Packard Laboratories, Filton Road, Stoke Gifford, Bristol BS12 6QZ, UK

Ikonomou, D., European Commission DG XIII-B4, BU-9 4.90, Ave de Beaulieu 9, B-1160, Brussels

Kelliher, J., Simoco PO Box 24, St Andrews Road, Cambridge CB4 1DP, UK

Kenington, P., Wireless Systems International Ltd, Clifton Heights, Triangle West, Bristol BS8 1EJ, UK

Kingsbury, N. G., Department of Engineering, Cambridge University, Trumpington Street, Cambridge CB2 1PZ, UK

Kondoz, A. M., Centre for Satellite Engineering Research, University of Surrey, Guildford GU2 5XH, UK

Kudumakis, P., Department of Electronic and Electrical Engineering, Kings College London, Strand, London WC2R 2LS, UK

Kunt, M., Signal Processing Laboratory, CH-1015 Lausanne, Switzerland

Magrath, A. J., Department of Electronic and Electrical Engineering, Kings College London, Strand, London WC2R 2LS, UK

McGeehan, J. P., Centre for Communications Research, University of Bristol, Merchant Venturers Building, Woodland Road, Bristol BS8 1UB, UK

Mandayam, N., Wireless Information Network Lab, Rutgers University, PO Box 909 Piscataway, NJ 08855, USA

Millar, I., Network Technology Department, Hewlett Packard Laboratories, Filton Road, Stoke Gifford, Bristol BS12 6QZ, UK

Mitchell, C. J., Information Security Group, Royal Holloway College, London University, Egham, Surrey TW20 0EX, UK

Munro, A. M., Centre for Communications Research, University of Bristol, Merchant Venturers Building, Woodland Road, Bristol BS8 1UB, UK

Murata, H., Department of Electronics, Kyoto University, Kyoto 606-01, Japan

Nix, A., Centre for Communications Research, University of Bristol, Merchant Venturers Building, Woodland Road, Bristol BS8 1UB, UK

Ohno, K., NTT Mobile Communications Network Inc., R&D Department, I-2356 Take, Yokosuka-Shi, Kanagawa 238-03, Japan

Parris, C., Ensigma Ltd, Turing House, Chepstow NP6 5PB, UK

Pereira, J., European Commission DG XIII-B4, BU-9 4.90, Ave de Beaulieu 9, B-1160, Brussels, Belgium

Perry, R., Mobile Communications Department, Hewlett Packard Laboratories, Filton Road, Stoke Gifford, Bristol BS12 6QZ, UK

Proudler, G., Mobile Communications Department, Hewlett Packard Laboratories, Filton Road, Stoke Gifford, Bristol BS12 6QZ, UK

Ramstad, T., Department of Telecommunications, Norwegian Institute of Technology, O S Bragstads plass 2B, N-7034 Trondheim, Norway

Raychaudhuri, D., NEC USA, C&C Research Laboratories, 4 Independence Way, Princeton, NJ 08540, USA

Redmill, D. W., Centre for Communications Research, University of Bristol, Merchant Venturers Building, Woodland Road, Bristol BS8 1UB, UK

Reusens, E., Signal Processing Laboratory, CH-1015 Lausanne, Switzerland

Richards, B., Electrical Engineering and Computer Science Department, University of California, Berkeley, CA 94720, USA

Sadka, A. H., Centre for Satellite Engineering Research, University of Surrey, Guildford GU2 5XH, UK

Salami, R. A., Department of Electronics and Computer Science, Southampton University, Southampton SO17 1BJ, UK

Sandler, M. B., Department of Electronic and Electrical Engineering, Kings College London, Strand, London WC2R 2LS, UK

Sawahashi, M., NTT Mobile Communications Network Inc., R&D Department, I-2356 Take, Yokosuka-Shi, Kanagawa 238-03, Japan

da Silva, J. S., European Commission DG XIII-B4, BU-9 4.90, Ave de Beaulieu 9, B-1160, Brussels, Belgium

Simmonds, C. M., Centre for Communications Research, University of Bristol, Merchant Venturers Building, Woodland Road, Bristol BS8 1UB, UK

Steele, R., Department of Electronics and Computer Science, Southampton University, Southampton SO17 1BJ, UK

Steinbach, E., Universität Erlangen-Nürnberg, Lehrstuhl für Nachrichtentechnik, Caurstrasse 7, 91058 Erlangen, Germany

Streit, J., Department of Electronics and Computer Science, Southampton University, Southampton SO17 1BJ, UK

Talluri, R., DSP R&D Centre, Texas Instruments Inc., 13510 North Central Expressway, MS 446 Dallas, Texas 75243, USA

Thillainathan, S., Centre for Communications Research, University of Bristol, Merchant Venturers Building, Woodland Road, Bristol BS8 1UB, UK

Toh, C. K., School of Electrical and Computer Engineering, Georgia Institute of Technology, Atlanta, GA 30332-0250, USA

Tominaga, H., Department of Electronics and Communication Enginering, Waseda University, 3-4-1 Ohkubo, Shinjuko-ku, Tokyo 169, Japan

Tsoulos, G., Centre for Communications Research, University of Bristol, Merchant Venturers Building, Woodland Road, Bristol BS8 1UB, UK

Umehira, M., Radio Systems Laboratory, NTT Wireless Systems Labs, 1-1 Hikarinooka, Yokosuka-Shi, Kanagawa-Ken, 239 Japan

Webb, W., Department of Electronics and Computer Science, Southampton University, Southampton SO17 1BJ, UK

Whybray, M., BT Laboratories, Martlesham Heath, Ipswich, Suffolk, UK

Wilkinson, J., Sony Broadcast and Professional Europe, Jays Close, Viables, Basingstoke, Hants RG22 4SB, UK

Wilkinson, T. A., Mobile Communications Department, Hewlett Packard Laboratories, Filton Road, Stoke Gifford, Bristol BS12 6QZ, UK

Williams, J., Department of Electronics and Computer Science, Southampton University, Southampton SO17 1BJ, UK

Williams, S., Network Technology Department, Hewlett Packard Laboratories, Filton Road, Stoke Gifford, Bristol BS12 6QZ, UK

Williamson, M., Centre for Communications Research, University of Bristol, Merchant Venturers Building, Woodland Road, Bristol BS8 1UB, UK

Woodard, J. P., Department of Electronics and Computer Science, Southampton University, Southampton SO17 1BJ, UK

Yabusaki, M., NTT DoCoMo, 1-2356 Take, Yokosuka-Shi, Kanagawa 238-03, Japan

Yoshida, S., Department of Electronics, Kyoto University, Kyoto 606-01, Japan

Part 1

Research Trends and Technical Developments

1

Mobile Multimedia Communications – Research Trends and Technical Developments

C. I'Anson, H. Tominaga, T. Wilkinson, M. Yabusaki
D. R. Bull, A. R. Nix and C. N. Canagarajah

1.1. INTRODUCTION

The market for mobile multimedia is currently in its infancy and technology providers, operators, and users alike wait for a pervasive market place to emerge. Many would claim that a single *killer application*, once known, will enable the market to blossom. Although partially true, this is undoubtedly a simplistic view, which ignores the enormous efforts that must be invested to first resolve the outstanding technical difficulties.

Insights into Mobile Multimedia Communication
ISBN 0-12-140310-6

Application level architectural considerations and standards issues must be resolved to ensure a pervasive and open service. Network level architectures must also be addressed and the air interface must have the capacity and flexibility to deliver real-time multimedia at an acceptable quality of service level. It is equally clear that source coding is no longer simply a matter of optimising rate-distortion characteristics. Instead it must also consider the integration of data, speech, audio and video information while offering a robust, scalable, low latency and interactive service.

This chapter presents an overview of the current state of the art and indicates future directions in wireless multimedia. Our aim is to give the reader a greater insight into the problems faced and the methods with which they may be overcome. Section 1.2 reviews existing services and applications and section 1.3 extends this view into emerging third generation systems. The issues associated with source coding for wireless applications are considered in section 1.4 and section 1.5 concludes with an overview of experimental demonstration systems.

1.2. SERVICES AND APPLICATIONS

As the challenges of wireless multimedia are resolved, different types of user, service and application will become feasible. These include:

- emergency services for remote consultation or scene of crime work;
- virtual universities for remote learning;
- the media industry for remote access to and submission of news reports;
- the security industry for telesurveillance;
- healthcare experts for remote diagnosis and training;
- the public for video telephony and multimedia information distribution in the home;
- business for conferencing.

The basic GSM data services of short message service, fax and data are soon to be enhanced and improved within ETSI to a point where GSM can offer a reliable cellular data service which can be used for a variety of purposes (see Chapters 2 and 10). However, a number of significant challenges still exist, which are not part of the basic GSM service. For example, limited data rates and slow PSTN modem interworking have been identified as critical issues that must be addressed in the near future.

The recently standardised High Speed Circuit Switched Data (HSCSD) standard allows for the combination of multiple time slots within GSM (see Chapter 10). By using up to four time slots in each direction (uplink *and* downlink), the channels can be multiplexed together to offer a *raw* data rate of up to 64 kb/s (38.4 kb/s *user* data rate, or up to 153.6 kb/s with data compression). A second standard, the General Packet Radio Service (GPRS), is based on the transportation and routing of packetised data, reducing greatly the time spent setting up and taking down modem connections.

With digital cellular systems like GSM and PDC there is a basic multimedia capability for voice and modem speed data. The data services are not perfect but their limitations can be resolved by good application and network protocol design. Given the success of the fixed-wire Internet, the demand to access the World Wide Web via wireless is likely to be strong. The concept of real-time multimedia access to the Internet is addressed and discussed as part of Chapter 8.

GRPS is only one of the data services proposed for implementation in the next few years. Another more long-term issue is wireless interfaces for Ethernet and ATM networks (see section 1.3.1). Initially, packing ATM cells for transmission over radio appears to be a simple task, but complex issues such as the framing of cells on the air interface in the best possible manner immediately appear. Chapter 9 provides an overview of ATM wireless LANs and discusses specific issues relating to the PHY, MAC and DATA LINK layers.

For any personal communications system to be trusted, a wide range of security services must be provided. Already GSM supports authentication and confidentiality services over the air interface, but new multimedia capable third generation systems present new challenges (see Chapter 11).

1.3. THIRD GENERATION SCHEMES AND WIRELESS LANs

Third generation radio schemes offer the promise of providing a usable infrastructure for true mobile multimedia communications.

1.3.1. Wireless LANs

Current indoor radio standards such as IEEE802.11 offer data rates of 1–2 Mb/s and, with the emergence of systems such as HIPERLAN (up to 23 Mb/s), high-quality mobile multimedia applications are now becoming a reality. HIPERLAN type I has been designed to support services such as real-time video and this will be enhanced by HIPERLAN II which will be modified to act as a radio tail for ATM-based applications (see Chapters 9, 36, 37, 40). Although HIPERLAN can support very high instantaneous data rates, operation at reduced bit rates and the ability to withstand transmission errors are still highly desirable. In contrast, current in-building cordless telephone systems such as DECT operate at far lower bit rates (multiples of 32 kb/s). These will however increase with the emergence of UMTS in the coming years.

Low-cost localised communication is possible with the IRDA infra-red standard. Infra-red provides an economic solution for walk-up *ad hoc* networking. Many of the difficult addressing problems associated with conventional LANs do not have to be resolved since the user can merely "point and shoot" at the target terminal or peripheral device. The short range of the system can also be used to advantage in avoiding difficult network management problems. Although HIPERLAN offers similar user speeds to IRDA and a much wider coverage area, this is at the cost of more complex technology in the radio and infrastructure management.

Looking further into the future, the use of higher-frequency bands (17 GHz, 40 GHz and 60 GHz) are being considered for high-speed wireless LAN connections (see Chapters 36 and 39). These frequencies have the advantage of offering more bandwidth, but the disadvantage of limited range and a greater reliance on line-of-sight communications.

There is a large number of European funded research projects currently investigating the feasibility of high-speed wireless LAN technology. Many of these projects are summarised and discussed in Chapters 10 and 36. The subject of mobility for a wireless ATM LAN is also discussed in Chapter 40.

1.3.2. Spread Spectrum

The use of spread spectrum techniques will be key to the success of many third generation cellular systems and is already present in a number of second generation standards. While both Frequency Hopping (FH) and Direct Sequence (DS) solutions are being investigated, it is the DS solution that has received the most attention (see Chapters 29 and 33). The trend with DS is towards much higher chipping rates than those originally proposed in the North American IS95 standard. These higher chipping rates offer a better opportunity for diversity processing, lower signal strength variations and improved power control.

Another key advantage of DS-CDMA is the exploitation of soft handover. The degree of soft handover provision is an important aspect in network planning, and while it can significantly improve performance, it also places a greater strain on the infrastructure. DS-CDMA also has the ability to support multiple variable bit rate sources in a flexible way, and this is very desirable in future multimedia networks (see the use of orthogonal codes in Chapter 33 and the integration of voice and data in Chapter 34). This is one area where DS-CDMA has an advantage over its competitors. Variable bit rates cannot so easily be supported with TDMA or FDMA solutions.

Finally, although FH solutions have not been studied in such detail, they still offer a number of significant benefits. Capacity estimates for future DS- and FH-based cellular networks are given in Chapter 29.

1.3.3. Advanced Antenna Solutions

Adaptive antennas will undoubtedly form a vital part of third generation cellular systems. A general introduction to the use of adaptive antennas in cellular base stations is given in Chapter 31. Adaptive antenna technology offers a number of tangible advantages. For example, on the downlink, the system can confine the radiated energy associated with a mobile to a small volume, thus limiting interference and increasing capacity. Also, on the uplink, the system can null out interfering and multi-path components to further increase performance.

The integration of adaptive antennas with DS-CDMA offers a powerful solution to the problems of wireless multimedia. This combination is considered in detail in Chapter 31.

For high-speed indoor LAN applications, the use of switched, or sectorised antennas has also been considered as an alternative to equalisation or multi-carrier techniques. These issues are considered in Chapters 36 and 39.

1.3.4. Equalisation and Flexible Radio Techniques

Multi-path counter measures are required to deliver the necessary high transmission rates for multimedia traffic, in a mobile TDMA environment. Equalisation has been a popular choice for systems ranging from GSM and IS54 to HIPERLAN and the need for efficient high bit-rate equalisation techniques is likely to increase as the demand for wireless multimedia grows.

The equaliser requires considerable processing power, and hence careful examination of the complexity performance trade-offs for these components is an important aspect of system design. This trade-off is investigated in Chapter 32, where a number of new architectures and algorithms are proposed.

Advances in amplifier linearisation techniques, combined with reduced transmit powers, have enabled the adoption of linear modulation schemes in a number of new systems. This technology, combined with the idea of a software reconfigurable multi-standard radio, is redefining the manner in which future mobile handsets will be designed. The concepts and challenges surrounding the creation of a software radio are addressed in Chapter 35.

1.4. SOURCE CODING

1.4.1. Audio and Speech

In source coding, voice and audio are usually processed separately. This is because different coding schemes provide different benefits for speech and non-speech sources. In both cases the drive is to achieve acceptable fidelity while reducing the bit rate, minimising the delay and improving the robustness to transmission errors. Power consumption is a key concern in mobile applications.

Chapter 12 reviews the speech codecs used in existing mobile and cellular systems and compares their performance and implementation complexity. Code Excited Linear Prediction (CELP) represents the state of the art in speech coding for near transparent coding of clean speech at bit rates down to 4 kb/s. In recent years there has also been significant interest in low bit-rate techniques based on sinusoidal modelling and waveform interpolation. These are reviewed in Chapter 13.

The demand for higher quality (see Chapters 13 and 17) has been satisfied by the use of wideband speech. This is assigned the band from 50 Hz to 7 kHz and sampled at 16 kHz for subsequent digital processing. The very low-frequency components increase the voice naturalness, whereas the added high frequencies make the speech sound crisper and more intelligible when compared to narrowband speech. A wideband coding algorithm must, therefore, accurately code the perceptually important lower-frequency components, and yet retain enough of the higher-frequency information such that the perceived richness and fidelity of the original speech is preserved.

The choice of speech codec and modem must be based on a number of criteria including: equipment complexity, power consumption, spectral efficiency, robustness against channel errors, co-channel and adjacent channel interference and cell-dependent propagation phenomena. Researchers are now looking at a more systems-oriented approach to optimise these complexity performance trade-offs (see Chapters 14 and 15). The focus must be on the bigger system picture in order to show the value of each design decision rather than on the individual merits of one element of a system.

Scalable coding and error resilience will become important when operating in varying radio and network environments. In Chapter 16, the significance of a scalable audio coding scheme for mobile applications is addressed. The near future is likely to witness the emergence of adaptive mobile speech communicators that can adapt their parameters to rapidly changing propagation environments and maintain a near-optimum combination of transceiver parameters in various scenarios. Another major consideration for network and multimedia-based applications is coding delay, i.e. the amount of buffering that the encoder requires such that the correlations present in the waveform can be exploited for economical digital representation. The encoder buffering necessary for parameter analysis in linear-prediction-based coders can be several tens of milliseconds. Such delays can have important ramifications for speech coders used in networks such as the ISDN where the one way encoder–decoder delay is a very important criterion for service quality. Delay may necessitate the use of echo cancellation, which, in some applications, is detrimental to non-voice services. Therefore a codec for mobile applications should not only optimise the traditional quality and bit-rate measures but also the delay and complexity to be viable for emerging multimedia applications.

1.4.2. Video

The emergence of communication systems based on UMTS, HIPERLAN and ATM and the associated source coding activities in academia, industry and standards bodies such as ISO MPEG-4 will move us closer to the provision of mobile multimedia, TV and interactive services for a range of terminal types at various bit rates. The key aims must be to provide advanced features such as scalability (Chapter 23), error robustness and content manipulation (Chapters 18, 19) together with a bitstream syntax which facilitates codec reconfigurability. Although most standards currently adopt a hybrid motion compensated DCT approach to video coding, it is evident that alternative approaches can offer benefits especially at low bit rates (see Chapter 18). Examples include wavelet methods (Chapter 21) and segmentation methods (Chapter 22).

Future terminals will therefore need to offer flexible operation and incorporate video codecs capable of optimising their performance dependent on available bandwidth. A number of example scenarios can be envisaged: (i) interoperability between conventional broadcast, wireless LAN, and cellular radio systems to ensure information transfer at the optimum quality according to environment, and to enable the decoding of a high data rate source by equipment with limited capability, (ii) in remote security or surveillance applications, different data rates could be employed according to information content and (iii) in mobile terminals linked to connectionless networks, performance will need to be optimised according to available bitstream content in the context of packet loss or contention.

A further important issue associated with the transmission of image information over noisy and multi-path fading channels is that of error resilience. Conventional error correction schemes incur an overhead, which is in direct opposition to the goal of the compression process. Alternative methods can be developed which reduce the propagation of errors and conceal rather than correct them. These issues are discussed in Chapters 25 to 28.

Future wireless multimedia systems will also have to cope with information transmitted across a heterogeneous network. Bit-rate variations, error characteristics and transcoding arising from communication across differing networks using differing protocols can generate complex interactions which significantly influence the quality of the received video. For example, broadcast TV information may be transcoded for transmission across wireless networks such as HIPERLAN or UMTS. In this context, the development of efficient and accurate propagation modelling tools is essential to understand these interactions and to support codec design and performance evaluation (see for example the tools used in Chapter 40).

Another key systems aspect is the need to operate speech and video over the same bandlimited channel. This represents a challenge to design cooperative voice and speech codecs that can make the very best use of the available bandwidth (see Chapters 20 and 24).

1.5. MULTIMEDIA DEMONSTRATORS

Examples of projects embracing the key architectural aspects of future mobile multimedia networks include InfoPad (Chapter 7), WAND (Chapters 10 and 37), MEDIAN (Chapters 10 and 37), WINHOME (Chapter 37) and Walkstation. InfoPad utilises a high compute-power, server-based architecture with minimum local processing in the mobile units. The mobiles are optimised for power consumption and, since they are analogous to the X-terminal or network computer, they need only a screen for information viewing, an input device and a communications port. A number of air interfaces have been considered prior to the adoption of more aggressive CMOS technology, where CDMA radios and the use of mmWave bands are a natural choice.

Walkstation has a longer history with roots in the Columbia Student Electronic Notebook. The key element is the mobility architecture which depends on a mobile Internet router. This implies, but does not necessitate, the mobile client to be a computer in its own right. Hence the mobile can be based on any portable PC platform. In contrast to InfoPad, the intelligence for an application resides in the user's client device and any server it may choose to use. Central to Walkstation is the concentration of the mobility functions in a combined network router and radio unit that is carried by the mobile user. This type of platform with its standard IP communications base can then benefit from related work in the field of real-time multi-media traffic on the Internet (see Chapter 8 for details).

The WAND and MEDIAN projects are part of the European Union ACTS programme and both are aimed at developing high-speed wireless LAN demonstrators. WAND is based on the 5 GHz HIPERLAN type II standard, whereas MEDIAN is addressing the use of the 60 GHz band to provide data rates up to 155 Mb/s.

The EC ESPRIT project WINHOME is another good example of how wireless local area network technology can be exploited in the home to deliver various classes of multimedia entertainment and information services. WINHOME employs HIPERLAN I technology to distribute a mix of multimedia traffic ranging from low-bit-rate videophone and Internet services to multiple MPEG-2 encoded video channels and high-speed Internet via a cable or satellite set top box.

1.6. CONCLUSIONS

It is clear that significant effort and money is being expended across the world with the aim of making mobile multimedia communications a cost-effective reality. Although much of this work is focused on fundamental technology and its integration into sub-system modules, it is becoming increasingly clear that the performance of an entire system is not simply related to the sum of its parts. Service providers, researchers and standards bodies are all now beginning to see the big picture and the complex interdependencies between different technology components are gradually emerging.

To enable mobile multimedia services to become a reality, the challenge is to combine the areas of application software design, wireless networks and protocols, radio architecture and interfacing, wireless propagation, audio and video coding, source compression and channel coding. It is only by optimising at a system level that a truly optimum and efficient solution to the challenges of mobile multimedia will emerge.

Part 2

Multimedia Services on Digital Cellular Networks

Since the launch of the second generation cellular schemes there has been clear migration from fixed platform services to mobile network. In particular GSM has become a world-wide success and is recognised to play an important part in future mobile multimedia systems. Mobility is considered to be essential in future applications and there has been a lot of activity in developing standards for third generation telecommunication systems. Furthermore, the widespread use and low cost services offered by the Internet has resulted in developing Mobile IP standards for interworking with the Digital Cellular systems. The ITU/IMC VoIP Forum and the ETSI TIPHON projects are developing standards for mobile multimedia over IP.

In Chapter 2 the author shows the main benefits of GSM and its features for mobile multimedia. It is becoming clear that GSM will become the future network for mobile multimedia applications. Chapter 3 presents the historical progress and some technological aspects of the Japanese Digital Cellular Radio services. The author shows the mobile revolution since the launch of the cellular mobile radio in 1979. The author also looks at some of the challenges for the future and shows how the cellular and cordless systems will be integrated to provide an Intelligent Universal mobile telecommunications network. Chapter 4 looks at TETRA which is the ETST digital wireless PMR standard developed to integrate voice and data services. The author shows that TETRA represents the most flexible environment for mobile multimedia on the European stage and the only standard at present which attempts to provide a mobile communications network for data services. In Chapter 5 a mobile multimedia system is developed based on the ITU H.324 family of standards. This standard defines a terminal standard for multimedia applications over the PSTN and a suite of algorithms for speech and video compression, and the necessary multiplex and control protocols. The authors investigate the error characteristics of the mobile network on video compression and propose error resilient techniques suitable for mobile multimedia applications.

2

Wireless Multimedia Using GSM Transport

John Kelliher

2.1. INTRODUCTION

There exists, for some inexplicable reason, a popular disbelief that GSM enabled platforms are able to effectively support mobile multimedia. Even in the late 1990s there are many in the industry who claim that mixing GSM and multimedia is akin to mixing oil and water. Nothing is further from the truth.

Explaining away such a myth often requires a degree of patience, certainly some perseverance and definitely a little insight into mobile multimedia over GSM. Most certainly it requires an understanding of, and preferably a definition of, what multimedia is. Perhaps therein lies the difficulty. To date, the many attempts to define "multimedia", usually as the heterogeneous composition of voice, music, video and

Insights into Mobile Multimedia Communication
ISBN 0-12-140310-6

graphics, belies the complexity of the service creation/delivery, transport and user interaction with the environment. GSM as a mobile network standard in the context of wireless multimedia is, simply, a technically sophisticated radiocommunications network offering mobility and supporting wireless information services/highways. GSM is just an enabler of information services, which can be supported alone by the GSM standard, or in conjunction with other complementary standards such as WWW and FTP. It is the services that these systems offer and the innovation in both technological development and marketing that differentiate operators in the marketplace.

2.2. THE FUTURE IS INFORMATION

The relatively slow take up of GSM as a data service (see Chapter 10) has been attributed to a number of reasons:

- the high cost of PCMCIA modems;
- limited availability of integrated GSM PCMCIA cards combining the phone and modem in one unit;
- relative costs of service and QoS as a function of bandwidth with respect to the PSTN;
- inconsistency in the application of compression technologies;
- slow call set-up on account of analogue modems and limited availability of integrated ISDN.

While GSM has become a phenomenal success world wide as a mobile replacement for the telephone network, the explosion of the Internet and the WWW has occurred. GSM [1,2] is a wireless transport system designed in the late 1980s and early 1990s that found its origins in the CCITT blue books. The GSM service as a whole has added a level of mobility and service design that in its time was at the forefront of the technological world. Indeed its success world wide, both as an International Standard developed by ETSI's Special Mobile Group (SMG) and commercially through the GSM MoU, owes much to a vision to create services for the future. It has fulfilled many of its goals, mainly through integration with the fixed network (Figure 2.1).

2.3. THE FUTURE IS WIRELESS

Operational GSM systems already support hierarchical multi-tier networks comprising microcellular and picocellular base stations, in addition to the national coverage afforded by the macrocellular layer. Coverage in urban/metropolitan centres are enriched with microcells which, while susceptible to significant shielding by localised buildings, offer capacity enhancement and coverage extension.

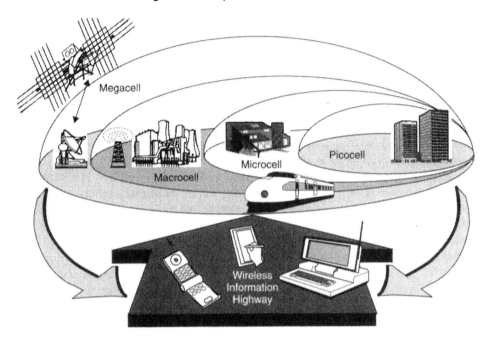

Figure 2.1 Integration with the fixed network.

Spectral efficiency remains a primary issue for GSM network operators who rapidly innovate and adopt new solutions to extending coverage and reducing interference [3,4]. Uplink space diversity is being augmented with polarisation diversity in order to reduce the number of base-station antennas on rooftops and radio masts. Receive diversity, complemented with transmit diversity to extend cell coverage in the downlink, with frequency hopping and VAD/DTX incorporated to minimise co-channel interference, are also being employed as mobile network operators strive to create spectrally efficient networks.

Intelligent underlay overlay (IUO) with controllable load sharing as a function of mobile traffic, coverage, mobile speed, channel dispersion and channel BER estimates within the hierarchical structure will ensure that GSM successfully supports the services of the future.

Network rollout programmes, including cell splitting and infill, are delivering networks that aim to exceed customer quality expectations, allowing the market in the UK to continue to grow. Coverage is being designed not just as a mobile network, but as a personal communications infrastructure for users in-building across all floors. We are also witnessing a revolution in the fundamental radio propagation models that have historically been used for designing and dimensioning the radio base-stations infrastructure of GSM networks [5]. These have been largely based on statistical methods (i.e. those of Hata and Okamura). Much work is now focused on extending the applicability of new deterministic techniques to allow accurate radio design.

2.4. THE COLLISION OF MOBILE COMMUNICATIONS AND COMPUTING

It has for some time been understood that coverage and capacity are two of the most important issues in any operator's rollout strategy. Recently, the enormous investments in the core infrastructure of GSM world wide have enabled high quality digital access to services (see Figures 2.2 and 2.3). It is the very significant investment in these capital rollout programmes coupled with the roaming capabilities of mobile networks that are creating a ubiquitous world wide information highway.

The association of voice as the primary service of GSM is a valid one, as almost all traffic on these networks is voice telephony. However, numerous initiatives are under way outside as well as within the mobile communications industry to inflect a new dimension to the perception of GSM, that of the natural wireless mobile computer highway. GSM will be an effective mobile computer data transport service by the millennium.

The addition of future bearer services within GSM, HSCSD and GPRS offer a number of opportunities if the triumverate of operator tarrifing, application interfacing and application development are to work together successfully. The evolution of the GSM system will be determined, not so much by its creators, ETSI SMG, as by the creators of the Internet, the emerging mobile computer industry and application software developers. Suffice to say that the convergence of mobile communications and computing will afford the greatest revolution to the underlying communications, because it will allow the user to create and operate within his or her own virtual home environment, probably based on Internet technologies (see Chapter 8 of this book for further details).

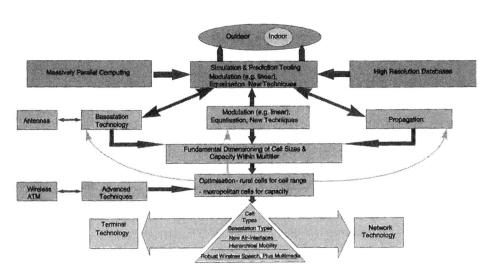

Figure 2.2 Radiocommunications environment.

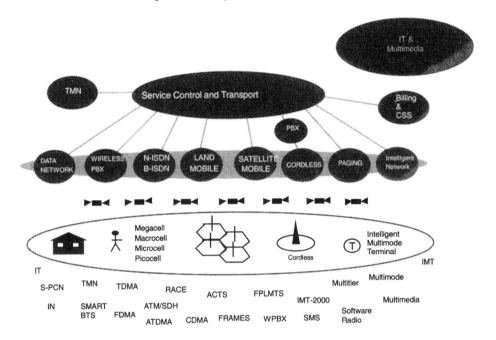

Figure 2.3 Target architecture.

2.5. GSM FOR WINDOWS

The concept of a software radio has inspired researchers for many years (software radios are discussed in detail in Chapter 35) and is now recognised as being a vital part of any future communication network. For several years now, the dominant players in the computing industry have acclaimed network computing as the ultimate alternative computing solution. They have postulated that "the network is the computer" and embraced Java as the language of the future. It is clear that the GSM-based mobile computer must embrace mobility, Java, browser, email, spreadsheet, diary and multimedia functionality.

The functionality described above is very different to that ordinarily afforded to network computers. It specifically does not relate to a diskless computer running a Java VM directly on top of the hardware. Neither does it refer to an OS taken off the client and moved to the network. GSM is the access for information, GSM is the primary transport network to the public and private data networks, using whichever computer serves the needs of the user, a 233 MHz Intel MMX Pentium notebook running Win 95, or a Nokia 9000 Communicator. The needs of the user might include:

- receiving an important fascimile while away from his office;
- surfing the corporate Intranet (e.g. to register for a training course) or the WWW to check out the latest product information from his vendors;

- publishing on the corporate Intranet or subscribing to and receiving a magazine over the Internet;
- retrieving email from a wireless ISP;
- screen sharing with a remote colleague on a field trip;
- updating the corporate programme management project plan using ftp.

GSM-based network computing is fundamentally all about "enterprise computing". It includes embracing the US$200m on-line intra- and inter-business Internet/Intranet/Extranet revolution, as well as meeting the user's normal communications needs. Importantly, the accelerating growth in on-line information services, driven by the Internet, the processing power of mobile computers and the migration of content provision to the Internet is enabling a more mobile, better informed population. There is little doubt that the growth in information services will continue.

2.6. EXISTING INFORMATION SERVICES ON GSM

GSM was not designed as a speech-based network, although speech telephony remains undisputedly its primary service in the late 1990s. It was designed for the future. Although speech is the primary service, GSM already supports a vast array of information services. These include:

- 3.4 kHz voice GSM FR (RPE-LTP);
- call waiting, call holding, call divert and silent service;
- information on demand (e.g. sports news, lottery);
- basic fax/data (including V.42bis data compression) and answerfax;
- roaming – national and international/switchable dual mode handsets;
- short message services (SMS);
- Internet to SMS and fast access Internet/Intranet via integrated ISDN;
- CLI/CLIP/CLIR;
- voice mail/answerphone;
- call centre;
- image transmisson;
- Internet email, WWW, FTP, IRC, newsgroups, electronic commerce;
- interactive screenshare;
- pre-payment;
- wireless PBX.

2.7. FUTURE INFORMATION SERVICES ON GSM

GSM is essentially following a well-defined technology timeline, and future information services will extend from the baseline that already exists [6,7]. It is important to differentiate between the technical evolution of GSM and the user market evolution.

The coupling between the two is a complex web of market pull and value driven investment pyramids. The secret of GSM's future success rests on its ability to metamorphose, to grow and embrace new technologies along the way.

Some of these new services and technologies may, for example, include:

- 3.4 kHz speech GSM HR (half-rate), GSM EFR (enhanced full-rate);
- 3.4 kHz speech GSM AMR (adaptive multi-rate), 7 kHz speech GSM W-AMR (wideband AMR);
- 14.4 kbit/s data;
- General Packet Radio Service (GPRS) and high speed circuit switched data (HSCSD);
- low bit-rate transmission video, access to information databases and intelligent camera-based telesurveillance;
- pay per view;
- virtual reality/telepresence;
- telebanking, E-Cash/SIM cash/SET;
- voice recognition/authenticated voice dialling/information retrieval;
- speech to text and text to speech synthesis;
- voice over Internet Protocol (VOIP) and wirefree webtop;
- Intranet/Extranet email, WWW, FTP, IRC, newsgroups, electronic commerce, publishing subscription service electronic newspapers;
- communications for people with special needs;
- distributed mobile computing;
- online gaming;
- Java/autonomous intelligent agents;
- integrated GPS positioning/tracking/transport information;
- training and education.

2.8. BANDWIDTH AND COMPRESSION TECHNOLOGIES

Another popular misconception is that the 9600 bps data bearer service is too slow. The dominant bottleneck on Internet connections is not the 9600 bps GSM bearer (which yields a data throughput of 8 kbps) or the 28 800 bps PSTN bearer (which yields a data throughput of 20 kbps) but the Internet itself. GSM already has a single-channel videophone running five frames per second, with an image size of 256×198 pixels and 24-bit colour (see Figure 2.4).

The challenge of real-time GSM video is not the 9600 bps bearer which allows approximately 800 bits per frame, or the stringent requirement to support 24-bit colour. Neither is it only the need to provide real-time compression ratios in excess of 500:1. The challenge is to realise the codec in software on a general-purpose platform rather than in expensive hardware. Software codecs embedded within applications on mobile computer platforms will require state-of-the-art processors. Real-time full-colour software video communications over GSM is a reality today, running on 200 MHz StrongARM^{TM} desktop computers and 233 MHz Intel Pentium^{TM} MMX notebooks.

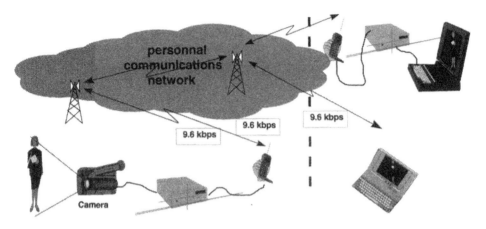

Figure 2.4 Wireless video over GSM1800.

2.9. ACHIEVING THE REQUIRED CAPACITY

The challenge of a network operator in the future marketplace will be to meet the traffic demands of the subscriber base, whether that be voice or data. Naturally, this will be compounded by any future success in any wideband data service offering HSCSD. However, the technical solutions to delivering capacity within GSM mobile communications encompass will not change but will be extended:

- upgrade the network and mobiles to support half-rate codecs;
- locate the base stations closer together (infill);
- incorporate a hierarchical network using a microcellular layer of base stations in metropolitan/urban centres;
- employ frequency hopping and VAD/DTX;
- develop adaptive antennas capable of handover between cells, while supporting spatial domain multiple access (SDMA);
- Employ H*ML, a non-volatile cache based mark-up language for Internet/ Intranet/Extranet usage, ideally suited for mobile multimedia.

Figure 2.5 demonstrates pictorially the capability of a microcellular layer to support submultiplexed voice-based traffic in Erlangs within a hierarchical microcellular/macrocellular GSM network configured for:

- 1, 2, and 4 transceivers (TRX) per omnidirectional microcell;
- spatial domain multiple access (SDMA) where multiple users operate on the same frequency on the same timeslot, shown here with an SDMA processing gain facilitating 4 users on the same 1800 MHz frequency on the same timeslot;
- full rate (FR) and half-rate (HR) (but not AMR) speech codecs submultiplexed into 16 kbps TRAU and 8 kbps TRAU respectively.

The graph clearly indicates the trends afforded by the microcellular layer alone to support high traffic call densities in urban centres. It is the relative advantage

DCS Microcell "Coverage Oriented" Traffic Densities (Erlang B - 1% Blocking)
Traffic Density in Erlangs/km²

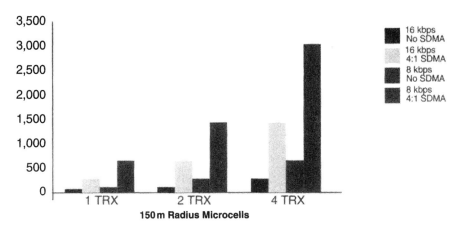

Figure 2.5 Capacity within a hierarchical cell structure employing adaptive antennas.

achieved by the application of these techniques that affords them so much interest at present.

2.10. THE FUTURE INFORMATION SOCIETY

The existing content creation, service delivery and user interaction paradigms that have been built up over several decades are all based on the push model of information transfer, whether the information be in the form of television programming, newspapers or mail drop. Services such as CLI (caller line ID) on GSM allow the user to control the information that he or she receives (e.g. by enabling the user to choose to take a call or to divert it to voicemail).

The Internet model of information encompasses both push and pull technologies. E-mail, by its very nature, is a push technology, whereas WWW is essentially a pull technology. The pull model appears to be winning through. Mobile Internet/ Intranet/Extranet (see Figure 2.5) will, in the not too distant future, allow the user to control the information he accesses, reads, transmits and publishes. Companies which organise their information best will be the winners in the future. If anyone finds that they are suffering from information overload, it will most likely be because their information systems have not been designed effectively.

It is the convergence of communications, computing, Internet and the content provider industries that are creating a new information society. This society will encompass the information-rich, controlling and having access to information, and the information-poor, who may perhaps not even own a computer. Access to information will be a commodity. GSM will bind these worlds together for people in a more mobile society.

Throughout the 1980s, and the early to mid-1990s the working practices in the UK undertook a substantial change. In the late 1990s people are travelling more, working

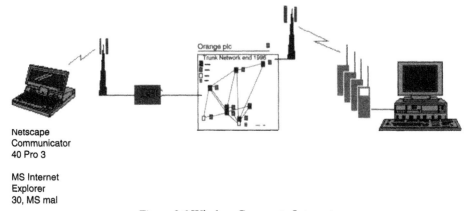

Netscape
Communicator
40 Pro 3

MS Internet
Explorer
30, MS mal

Figure 2.6 Wireless Corporate Intranet.

longer hours giving them little time for their family and for leisure. Many traditional organisations are now beginning to encourage teleworking. Such nomadic teleworkers in the late 1990s, will benefit from enhanced mobility while still maintaining considerable access to information through the Internet and Corporate Intranets or Extranets. In the future, people will restructure the way they work to enable equilibrium to return to their lives. Mobile communications and mobile computers as part of the virtual organisation will allow them to do just that. The historic "8 hour day" that is part of our national heritage will disappear and be replaced by people multitasking throughout the entire day, collecting the kids from school at 3:30 p.m., working till 8 p.m. Mobile communications and computing affords such flexibility. that teleworking and teleliving will become an intrinsic part of our future.

ACKNOWLEDGEMENTS

I have many people to thank for allowing this contribution and this vision of GSM's wireless multimedia future. Special thanks go to the Future Technology team for being the source of much inspiration, with thanks in particular to Ed Candy, John Hammac, Dr David McFarlane and Dr David Bozward for helping to make these concepts a reality. Thanks also go to the Directors of Orange plc for permission to publish this work.

WEB SITES OF INTEREST

[1] http://www.gsmworld.com
[2] http://www.etsi.fr
[3] http://www.gsmdata.com
[4] http://www.orange.co.uk
[5] http://www.mobilevce.co.uk

REFERENCES

[1] Moully, M. and Pautet, M.-B., The GSM system for Mobile Communications, 1993.
[2] Yacoub, M. D., *Foundations of Mobile Radio Engineering*. CRC Press, Boca Raton, FL, 1995.
[3] *The State of the Art 1995*. European Telecommunications and the Information Society, 1995.
[4] ACTS 96, Advanced Communications Technologies and Services, Project Summaries. European Commission, 1996.
[5] Parsons, D., *The Mobile Radio Propagation Channel*. Pentech Press, London, 1993.
[6] *The Road to UMTS – in contact anytime, anywhere with anyone*. UMTS Task Force Report, Brussels, 1st March 1996.
[7] Implementing the UK Wireless Information Society for Mobile and Personal Communications in the 21st century – an industry advisory document on the development of third generation personal communications systems. 3rd Generation Mobile Group, July 1996.

3

Cellular Mobile Radio Telephones in Japan

Kenkichi Hirade

3.1. INTRODUCTION

The inauguration of cellular mobile radio telephone services in December 1979 was the most important event in the history of mobile communications in Japan. Since then, high-quality mobile radio communications services have been provided to many users through nation-wide coverage. With recent progress in handset miniaturisation, not only mobile but also portable radio telephone services can be provided over the same cellular mobile radio network infrastructure. Customer demand is rapidly increasing for pocket-size handsets as well as a variety of mobile radio services such as data and facsimile communications.

This chapter overviews the situation of the cellular mobile radio telephone in Japan. Progress during the last 25 years is first reviewed. Current status of the network operators, frequency bands, and subscriber increase are then described. The technical features of the Japanese digital cellular mobile radio communications system known as the Personal Digital Cellular (PDC) system, are presented. This chapter is concluded with the prospects of the next-generation cellular mobile radio communication system.

3.2. HISTORY

As shown in Figure 3.1, the history of cellular mobile communication systems in Japan can be viewed as divided into four phases. The first phase is the "R&D" phase, the second the "introductory" phase, the third the "extension" phase, and the last, since 1990, is the "prosperous" phase.

3.2.1. R&D Phase

Development-directed research for the first cellular mobile radio telephone system started at the beginning of the 1970s in the Electrical Communication Laboratories (ECL) of the then government-owned network operator, NTT. The major objective of the research was to provide high-quality cellular mobile radio telephone services with nation-wide coverage by using the 800 MHz band. This frequency band was assigned primarily for the use of cellular mobile radio communications with high spectrum utilisation efficiency. The R&D covered various technologies such as high-quality voice and high-reliability data transmission, high-grade common channel radio signalling, and sophisticated cellular mobile radio network control supported by electronic switching and No.7 common channel inter-office signalling systems. The R&D also covered research of mobile radio transceiver hardware techniques including high-efficiency RF power amplifiers, frequency-agile PLL synthesisers and small low-loss antenna duplexers.

The first experimental bench test took place in 1972. The mobile radio unit developed for the test was about $26\,000\ \mathrm{cm}^3$ in volume, obviously somewhat impractical. The unit looked like a mobile radio transceiver rather than a handset, and was composed of many discrete semiconductor circuits. Power-dissipative TTL-SSI wired-logic circuits were used for the logic controller. Advancements in the mobile radio transceiver hardware techniques and applications of low-power CMOS-LSIs and small chip microprocessors with a stored program control scheme drastically reduced the size and weight of the mobile radio units. The size and weight of the mobile radio units used for the market trial test conducted in Tokyo in 1974 were

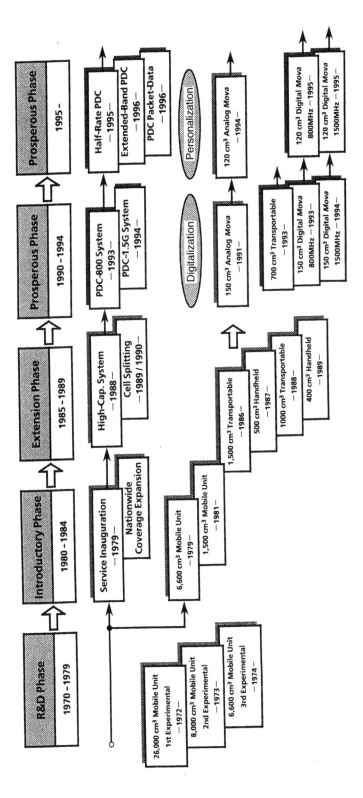

Figure 3.1 Progress in cellular mobile radio telephone in Japan.

reduced to 6600 cm^3 and 7 kg, respectively [1]. Aiming at the economisation and miniaturisation of the mobile radio units, R&D on these points started in 1977, two years before service inauguration. By advancing the mobile radio unit, the first commercial vehicular-mounted mobile radio unit with the same size and weight as the market trial version was then put into mass production in 1979. After completing administrative and legislative processes, the first commercial cellular mobile radio telephone service was inaugurated in Tokyo in December 1979.

3.2.2. Introductory Phase

The service area was gradually expanded to all major cities in Japan. Further R&D to meet the system scale requirements for cost-effective mid-to-small size city operation was conducted from 1980 to 1983 in continuation of the 1979 system inauguration. In 1983, the introduction of a cost-reduced version of the system started in mid-to-small size cities with the aim of advancing service provision to nation-wide coverage.

A simple 800 MHz PLL synthesiser for RF carrier generation was achieved by applying a prescaler counter IC and a high C/N VCO, both operating in the 800 MHz band. The use of a high-ceramic resonator was effective in miniaturising the antenna duplexer and RF bandpass filters. Application of the advanced CMOS-LSIs contributed to reducing the size and power consumption of the mobile radio unit. As a consequence, the size and weight of the mobile radio units were reduced to 1500 cm^3 and 2.4 kg, respectively, in 1981 [2].

At this time, it was dimly seen that not only vehicular-mounted mobile but also handheld portable radio telephone services could be provided within the same cellular mobile radio network infrastructure. Progress in LSI implementation technology and the application of compact packaging techniques gave birth to 1500 cm^3 transportable and 500 cm^3 handheld portable radio telephone services in 1986 and 1987, respectively [3].

Moreover, aeronautical and coastal maritime cellular mobile radio telephone systems were developed from 1983 to 1985 as extensions of the cellular mobile radio telephone system. The radio frequency bands used for the aeronautical and coastal maritime cellular mobile radio telephone systems were the 800 MHz and 250 MHz bands, respectively.

3.2.3. Extension Phase

Along with the development of the improved and economised system for use in mid-to-small size cities, other R&D activities towards the development of NTTs advanced high-capacity system started in 1980. Major objectives of the advanced high-capacity system were threefold: *capacity increase, service enhancement,* and *cost reduction.* Various kinds of advanced techniques and technologies were applied, and the air-interface protocol was completely changed from the conventional system. Since the advanced high-capacity system was primarily introduced to high-traffic-density areas, such as the cities of Tokyo and Osaka, system compatibility was the difficult problem that had to be solved to realise nation-wide services. This problem was overcome by developing sophisticated mobile and portable radio units having dual-system accessibility. The advanced high-capacity system, which overlays the conventional system, has been in operation in the Tokyo metropolitan area since April 1988 [4].

Both mobile and transportable radio telephone services were introduced using a 1000 cm³ mobile radio unit in April 1988. Progress in mobile radio units from 1979 to 1988 can be summarised as shown in Figure 3.2. Moreover, an advanced handheld portable radio telephone service was also introduced in February 1989 using a 400 cm³ handheld portable radio telephone unit. An R&D project started in September 1989 to develop pocket-size radio telephone sets. Four types of 150 cm³ 230 g pocket-size radio telephone sets were put into commercial services in April 1991 [5] under the name of *Mova*. Progress in portable radio telephone sets from 1987 to 1991 can be summarised as shown in Figure 3.3.

From 1979 to 1988, only the then government-owned operator, NTT, provided the cellular mobile radio telephone service in Japan. After the privatisation of NTT in April 1985, the Japanese cellular radio telephone market was opened up and new competitive cellular network operators, IDO and DDI Cellular group, were authorised to commence 800 MHz band analogue cellular network operations.

3.2.4. Prosperous Phase

The prosperous phase, which is considered to have started in the early 1990s, is divided into two parts. Digitalisation (first part) and personalisation (second part) are the keywords in describing this phase.

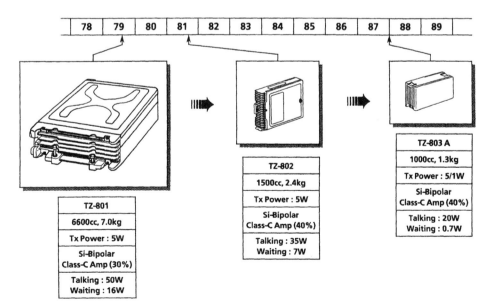

Figure 3.2 Progress in mobile units (from 1979 to 1988).

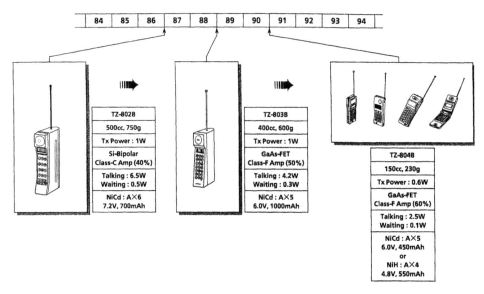

Figure 3.3 Progress in portable telephone sets (from 1987 to 1991).

Like Europe and the United States, new R&D activities for digital cellular mobile radio communication known as Personal Digital Cellular (PDC) started in 1987. In addition to the 800 MHz band, the 1500 MHz band was newly assigned for PDC system operations. Major objectives of R&D activities towards the PDC development can be summarised as follows [6]:

1. to achieve a spectrum utilisation efficiency as high as that of the advanced high-capacity system;
2. to provide not only high-quality speech but also high-grade mobile services such as data and facsimile communications; and
3. to further extend the system control capabilities with more flexible and cost-effective functional definitions.

Figure 3.4 shows the time frame of NTT's R&D activity for the PDC system, and the development activities of NTT's sister company, NTT Mobile Communications Network Inc., or NTT DoCoMo, in the context of the Japanese domestic standardisation process organised by Research and Development Centre of Radio Systems (RCR), a standardisation body for radio systems in Japan. In order to extend and enhance RCR activities, the reformation and reorganisation of RCR took place in 1992, under the name of Association of Radio Industries and Businesses (ARIB). The Telecommunications Technical Committee (TTC) tackled the standardisation of the PDC inter-node signalling protocol. The first version of the PDC air-interface standard, RCR STD-27, was established in April 1991. Its succeeding versions, RCR STD-27A, B, C, D and E contain more detailed requirements and specifications. STD-27A describes authentication and encryption processes. STD-27B details the data and facsimile service specifications. STD-27C describes a half-rate codec algorithm based upon the 5.6 kb/s Pitch-Synchronous Innovation Code-Excited-

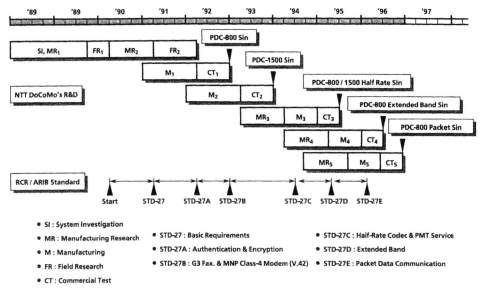

Figure 3.4 R&D and standardization schedule for PDC system.

Linear-Predictive (PSI-CELP) speech coding scheme and Personal Mobile Telecommunications (PMT) services. STD 27-D describes the technical requirements of extending the frequency band for the PDC system operation from the newly assigned 800 MHz band with 130 MHz duplex spacing to the conventional 800 MHz band with 55 MHz duplex spacing. STD 27-E describes detailed specifications for 28.8 kb/s mobile packet data communications over the PDC network infrastructure.

Along with the above standardisation activities, NTT DoCoMo developed mobile and/or portable units, base station equipment, and network switching nodes. Commercial services supported by the PDC-800 and PDC-1500 systems, which are, respectively, the 800 MHz and 1500 MHz versions of the PDC, were then inaugurated in the Tokyo metropolitan area in March 1993 and April 1994 [7,8]. The service area has since been extended to the whole of Japan. In March 1995, an enhanced 9.6 kb/s data service was introduced, and advanced mobile services have been provided within the same PDC network infrastructure [9]. Development of a half-rate speech coding scheme was finalised, and the introduction of the half-rate PDC system started in December 1995 to meet the increasing capacity demand [10]. Moreover, a new segment of the 800 MHz band was added for PDC operation, and the segment has been in use in the Tokyo metropolitan area since October 1996 [11]. At the same time, ultra small-size and light-weight digital portable telephone sets of less than 100 cm^3 and 100 g, respectively, as shown in Figure 3.5, were introduced. Since then, the market growth towards communication personalisation has been accelerated. Development of 28.8 kb/s mobile packet data communications over the PDC network infrastructure was finalised, and its commercial operation started in March 1997.

Figure 3.5 Latest fashion of *Digital Mova*

3.3. CELLULAR OPERATORS

As mentioned before, the first commercial 800 MHz band analogue cellular mobile radio telephone system was launched in Tokyo by the then government-owned network operator, NTT, in December 1979. In July 1991, NTT's Mobile Communications Sector was separated and became a new sister company, NTT DoCoMo. NTT DoCoMo was further divided into nine regional group companies in July 1992, to form the NTT DoCoMo group.

Figure 3.6 shows the Japanese operational environment of both analogue and digital cellular networks as of March 1997. The NTT DoCoMo group's coverage is now fully nation-wide. The IDO and DDI Cellular group's coverages are distinctively different. IDO's service area covers Kanto and Tokai districts, while the DDI Cellular group's covers the remaining areas, including Hokkaido, Tohoku, Hokuriku, Kansai, Chugoku, Shikoku, Kyushu, and Okinawa. There are two analogue cellular standards in Japan. The NTT DoCoMo group and IDO are using NTT's Advanced High-Capacity system; the DDI Cellular group uses the J/N-TACS systems. The J-TACS system is the TACS system modified to suit the Japanese cellular mobile radio environment. The N-TACS system is a narrowband version of the J-TACS system: J-TACS has a channel spacing of 25 kHz and N-TACS 12.5 kHz. The N-TACS system has been established in heavy-traffic areas by over-laying J-TACS operation. Moreover, the N-TACS system has been operated by IDO in Tokyo since September 1991, in parallel with NTT's advanced High-Capacity system operation.

As explained previously, both 800 MHz and 1500 MHz bands have been assigned to the PDC network operators. The NTT DoCoMo group, IDO, and the DDI Cellular group have been authorised as the network operators of the PDC-800

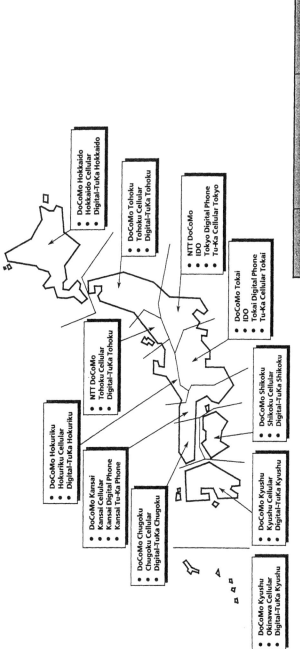

Operators	Analog System		Digital System	
NTT DoCoMo Group	NTT High-Cap.	—	PDC-800	PDC-1500
Nippon Idotsushin (IDO)	NTT High-Cap.	J/N-TACS	PDC-800	—
DDI Cellular Group	—	J/N-TACS	PDC-800	—
Digital Phone Group	—	—	—	PDC-1500
TuKa Cellular Group	—	—	—	PDC-1500
Digital-TuKa Group	—	—	—	PDC-1500

Figure 3.6 Cellular network operators in Japan.

system. The first PDC-800 system was introduced by NTT DoCoMo in Tokyo in March 1993. The IDO and DDI Cellular groups started PDC-800 services in 1994. In addition to the NTT DoCoMo group, two more digital cellular network operator groups, the Digital Phone group and the Tu-Ka Cellular group, were authorised for PDC-1500 system operations. In 1994, the PDC-1500 system was first introduced by NTT DoCoMo in Tokyo. The Digital Phone and the Tu-Ka Cellular group's PDC-1500 operations in Tokyo, Osaka and Nagoya areas also started in 1994. The authorised PDC-1500 network operator in Hokkaido, Tohoku, Hokuriku, Chugoku, Shikoku and Kyushu areas is the Digital Tu-Ka group, a joint cellular network operator formed by the Digital Phone and Tu-Ka Cellular groups.

3.4. FREQUENCY BANDS

Figure 3.7 shows the frequency bands assigned to cellular mobile radio communication services in Japan. Analogue cellular systems such as NTT's Advanced High-Capacity system and J/N-TACS use the 800 MHz band with a 28 MHz bandwidth pair with 55 MHz duplex spacing. The 800 MHz band is divided into two parts. One having a 15 MHz bandwidth pair is assigned to the NTT DoCoMo group and the other having a 13 MHz bandwidth pair is assigned to IDO and the DDI Cellular group.

The newly assigned 16 MHz bandwidth pair in the 800 MHz band, with 130 MHz duplex spacing is divided into two segments, each having a 8 MHz bandwidth pair. Similar to the analogue cellular systems, one is assigned to the NTT DoCoMo group and the other is assigned to IDO and the DDI Cellular group. On the other hand, the 1500 MHz band assignment with a 24 MHz bandwidth pair with 48 MHz duplex spacing is divided into three segments. One of the three segments, having a 4 MHz bandwidth pair, is assigned to the NTT DoCoMo group, another one having a 10 MHz bandwidth pair is assigned to the Digital Phone group, and the last one having a 10 MHz bandwidth pair is assigned to the Tu-Ka Cellular group.

3.5. SUBSCRIBER INCREASE

Figures 3.8(a) and (b) show the increase in cellular mobile radio telephone subscribers in Japan, as of March 1997. As shown in Figure 3.8(a), the subscriber increase rate was extremely low compared to the US, UK, and Nordic countries, for several years after service inauguration in December 1979. One of the major reasons for this was the high charge-of-use compared to the ordinary fixed telephone. Another reason was that services were limited to vehicular-mounted and transportable radio telephone services.

May 1987 saw the number of subscribers exceed ten thousand for the first time, seven years after service inauguration. April 1987 saw a drastic change in the subscriber increase rate when a 500 cm^3 handheld portable radio telephone was put into commercial service. The founding of the new competitive analogue cellular network operators, IDO and the DDI Cellular group, in 1988, following NTT's privatisation,

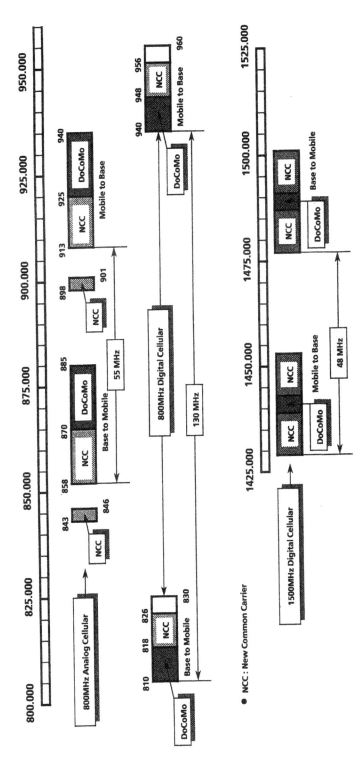

Figure 3.7 Frequency bands for cellular mobile radio telephones in Japan.

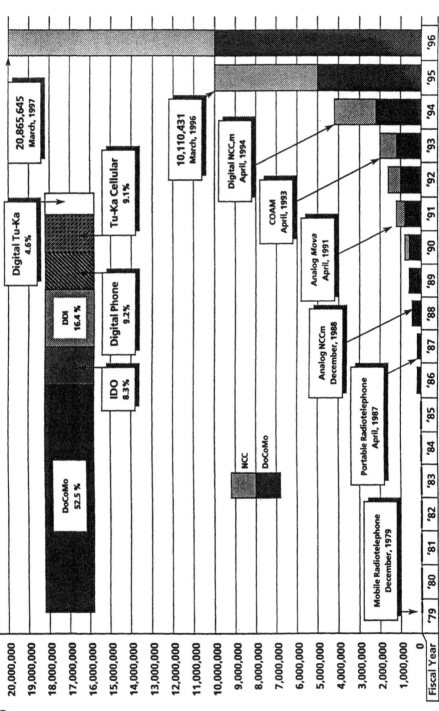

(a)

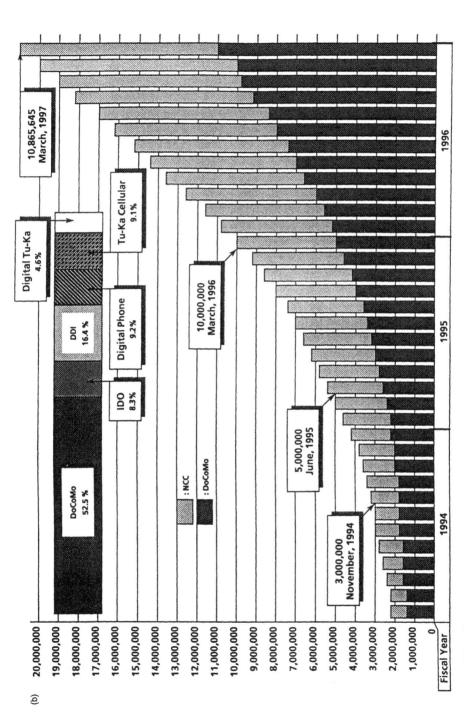

(b)

Figure 3.8 (a) Subscriber increase of cellular mobile radio telephone in Japan (from 1979 to 1996); (b) subscriber increase of cellular mobile radio telephone in Japan (from 1994 to 1996 (detailed)).

has accelerated the subscriber increase rate. The introduction of NTT's Advanced High-Capacity system, which was designed to meet the increasing demand in high-traffic density areas with a 400 cm^3 portable handset, also accelerated the subscriber increase rate. It was in December 1990 that the number of subscribers exceeded 50 000. The subscriber increase rate has been extremely high since April 1991 when the 150 cm^3 pocket-size handset called *Mova* was made available. The number of subscribers in the DoCoMo group increased beyond the one million mark in February 1993.

The PDC-800 and PDC-1500 systems have been in operation since March 1993 and April 1994, respectively. Moreover, it was in April 1994 that the customer's owned and maintenance (COAM) system replaced the rental system that had been adopted in Japan for about fifteen years. Furthermore, the founding of the two new digital cellular network operators in 1994, the Digital Phone and Tu-Ka Cellular groups, further accelerated the subscriber increase rate. As shown in Figure 3.8(b), the total number of subscribers continues to increase, and exceeded three, five, and ten million in November 1994, June 1995, and March 1996, respectively. It reached twenty million in March 1997. NTT DoCoMo group's market share is 52.5%, while those of the IDO and DDI Cellular groups are 8.3% and 9.2%, respectively. The Digital Phone, Tu-Ka Cellular, and Digital Tu-Ka groups have market shares of 9.2%, 9.1%, and 4.6%, respectively.

3.6. TECHNOLOGICAL PROGRESS

3.6.1. Radio Access Technologies

Figures 3.9 and 3.10 show the progress of radio access technologies for the cellular mobile radio telephone systems in Japan. Five steps in technological evolution can be observed over the last twenty five years. The first step was service inauguration in December 1979. Major radio access technologies employed in NTT's first analogue cellular system can be summarised as follows:

1. 25 kHz spaced analogue FM voice transmission with syllabic compander;
2. 300 b/s Manchester-coded FSK data transmission for common channel radio signalling;
3. multi-transmitter simulcasting for wide-area paging and access control.

The second step in evolution took place when NTT's Advanced High-Capacity system was introduced in 1988. Major distinctive radio access technologies adopted in the system can be summarised as follows:

1. narrow band (12.5 kHz) analogue FM voice transmission with interleaved 6.25 kHz carrier spacing;
2. two-branch space diversity reception with post-detection selection combining;
3. analogue voice-signal scrambling for speech encryption;
4. multi-transmitter simul-and-sequential casting for advanced 2.4 kb/s common channel radio signalling;

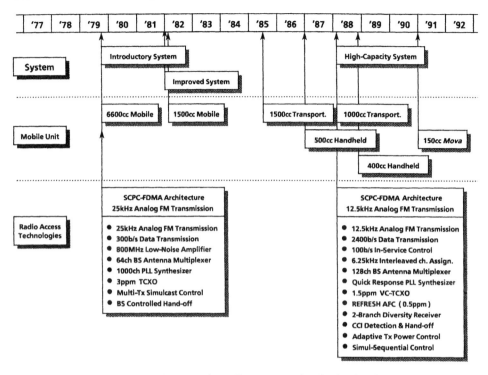

Figure 3.9 Progress in radio access technologies (pt.1).

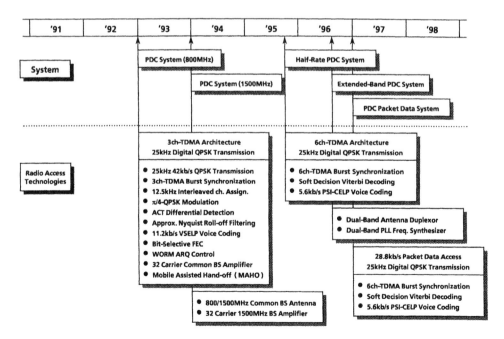

Figure 3.10 Progress in radio access technologies (pt.2).

5. 100 b/s data-under-voice transmission for channel-associated radio signalling without causing speech interruption;
6. co-channel and adjacent channel interference detection for advanced inter- and intra-cell hand-off;
7. adaptive transmitter power control at mobile station having a 28 dB dynamic range with seven 4 dB steps.

The third step saw full-rate PDC introduction in March 1993. Major radio access technologies adopted in the full-rate PDC system can be summarised as follows:

1. low-bit-rate high-quality speech codec with 11.2 kb/s Vector-Sum Code-Excited-Linear-Predictive (VSELP) coding algorithm;
2. $\pi/4$-shifted QPSK modulation and differential detection with adaptive carrier tracking (ACT) scheme;
3. two branch space diversity reception with antenna switching at handset and post-detection selection combining at base stations;
4. high-grade mobile data and facsimile communication services supported by a pooled modem scheme with forward-error-correction (FEC) coding and ARQ retransmission;
5. mobile assisted hand-off (MAHO) using idle time slots in the 3-ch TDMA frame;
6. flexible air-interface signalling protocol based upon the OSI layered model;
7. inter-node signalling protocol supporting a nation-wide roaming capability between cellular network operators.

The fourth step was the introduction of the half-rate PDC in December 1995. Major radio access technologies adopted in the half-rate PDC system can be summarised as follows:

1. ultra low-bit-rate, higher-quality speech codec with 5.6 kb/s pitch-synchronous-innovation code-excited-linear-predictive (PSI-CELP) coding algorithm;
2. soft decision Viterbi decoding for advanced differential detection in the severe land mobile radio propagation environment;
3. sophisticated air-interface signalling protocol for accessing the existing full-rate PDC system (nation-wide coverage) and the newly introduced half-rate PDC system which overlays the full-rate PDC system in urban areas with heavy traffic.

The fifth step was at the introduction of the extended-band PDC system in October 1996. Major radio access technologies adopted in the extended-band PDC system can be summarised as follows:

1. sophisticated air-interface signalling protocol for the access to the dual operation band, the newly assigned 800 MHz band with 130 MHz duplex spacing and the conventional 800 MHz band with 55 MHz duplex spacing;
2. advanced mobile radio transceiver hardware techniques including miniaturised, low-loss antenna duplexer and frequency-agile PLL synthesiser, both of which support the dual band.

3.6.2. Network Technologies

Figure 3.11 shows the progress of network technologies for cellular mobile radio telephone systems in Japan. During the first ten years, telephone service was provided by using NTT's D-10 switch. The D-10 switch is a space-division electronic switch with stored program control. Since the first commercial service was area-limited, it employed an area-designated dialling scheme. In order to provide a fully nation-wide service, No.7 common channel inter-office signalling (CCIS) supported by an advanced D-10 switch using a high-speed CPU was introduced in 1984. Enhanced services including call transfer, call waiting, ciphering, etc., have been provided since the introduction of NTT's time-division D-60 switch in 1987. At that time, the numbering capacity of the area-designated dialling scheme was also extended by employing a semi area-designated dialling scheme.

The current situation of cellular mobile radio communications in Japan is that it has already entered the mobile multi-media communications era with the introduction of the PDC system in 1993. A pooled modem control scheme supported by some sophisticated error control techniques is employed in cellular mobile switching nodes. Moreover, application of the intelligent network (IN) technology has already started with the introduction of the mobile service-control-point (M-SCP) for advanced flexible service control capabilities [12].

3.7. PDC SYSTEM

3.7.1. Network Architecture

Figure 3.12 shows NTT DoCoMo's PDC network architecture. It has a two-stage hierarchical architecture: local and transit stages. The local stage is composed of mobile control centres (MCCs) and base stations (BSs). The transit stage is composed of mobile gateway centres (MGCs). The No.7 common channel signalling network (CCSN) allows all MCCs and MGCs to access home location registers (HLRs) and gateway location registers (GLRs) as well as exchanging network control information between MCCs and MGCs. The ISDN user part (ISUP) in the No.7 common channel inter-office signalling (CCIS) system is used for the channel-associated signalling protocol. Signal connection control part (SCCP) and transaction capabilities application part (TCAP) are used for the common channel signalling protocol. Moreover, mobile application part (MAP) is used as one of the application service elements (ASEs) on TCAP.

3.7.2. Network Facilities

Figure 3.13 shows PDC's network facilities in the BSs and MCCs. Each BS is composed of transmit-and-receive antennas, low-noise tower-top amplifiers, high-power common transmit amplifier (AMP), and modulation-and-demodulation equipment (MDE). MCC consists of mobile local switch (MLS), BS control

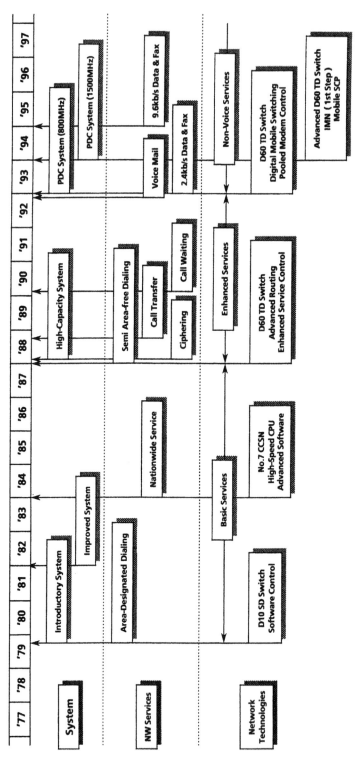

Figure 3.11 Progress in network technologies.

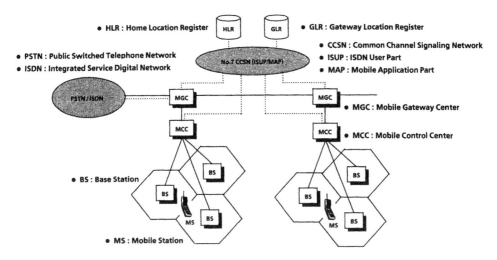

Figure 3.12 PDC network structure.

equipment (BCE), and speech processing equipment (SPE). MLS is based upon NTT's time-division D-60 digital switch, but, in addition, includes the additional service trunks (newly designed) needed to provide mobile services.

As for the full-rate PDC system, SPE consists of 648 speech codecs based on the 11.2 kb/s VSELP algorithm. A 32-bit microprocessor is used for the BCE's controller. The signalling interface between BCE and MDE is based upon LAP-B. Since SPE is located in MCC, each of the 64 kb/s digital transmission links between the BS and MCC can carry three teletraffic channels. Compared with analogue cellular systems, where each 64 kb/s digital transmission link carries only one speech channel, the transmission link lease cost of the PDC network can be considerably reduced. Moreover, since the half-rate PDC system was introduced where each 64 kb/s digital transmission link can carry twelve teletraffic channels, a drastic reduction in lease cost reduction has been achieved.

3.7.3. BS Equipment

As shown in Figure 3.14, each BS for the full-rate three-sectored PDC system with two-branch space diversity reception is composed of three transmit-and-receive antennas, three receive-only antennas, six low-noise tower-top receive amplifiers, three high-power common transmit amplifiers and three sets of modulation-and-demodulation equipment (MDE). High-gain BS antennas are commonly used for 800 MHz and 1500 MHz band system operations. In order to achieve tight spatial reuse of co-channel frequencies, the antenna beams tilt angle can be adjusted from 0 to 11°. The low-noise tower-top amplifiers must be located close enough to the six receive antennas in order to reduce the transmission loss due to coaxial cables.

AMP is made of three groups of four parallel linear power amplifiers which adopt the self-adjusting feed forward (SAFF) control scheme for ultra-linear power-

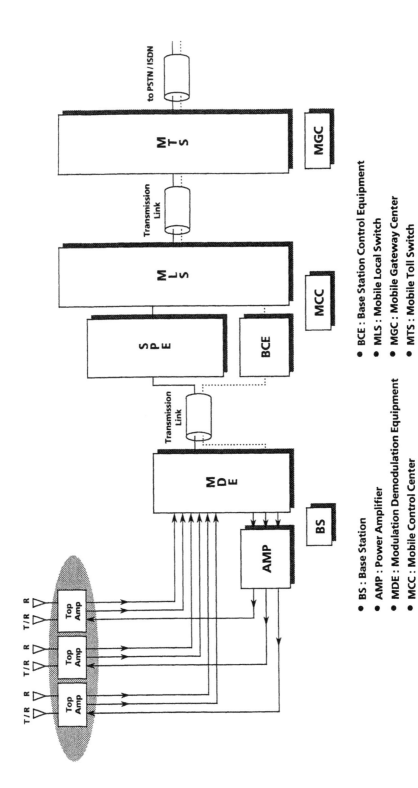

Figure 3.13 PDC network facilities.

- BS : Base Station
- AMP : Power Amplifier
- MDE : Modulation Demodulation Equipment
- MCC : Mobile Control Center
- SPE : Speech Processing Equipment

- BCE : Base Station Control Equipment
- MLS : Mobile Local Switch
- MGC : Mobile Gateway Center
- MTS : Mobile Toll Switch

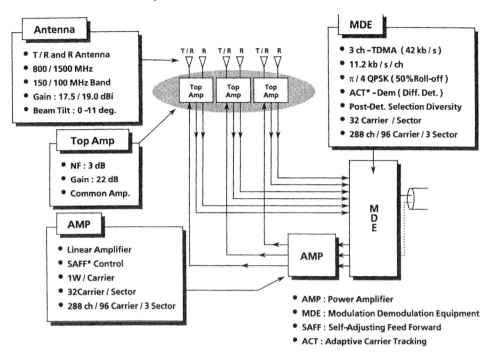

Figure 3.14 PDC base station equipment.

efficient operation. Each group has four parallel amplifiers (1-W output) handling 32 carriers. Since each sector is able to handle 32 carriers each carrying three channel TDMA signals, the total number of radio channels assigned to each sector is 96, corresponding to 288 channels in each cell. For the 11.2 kb/s full-rate system, MDE is composed of 96 transceiver cards, where signal processing for $\pi/4$-shifted QPSK modulation, ACT-based demodulation, two-branch space diversity reception, and TDMA framing/deframing are carried out. With the 5.6 kb/s half-rate speech codec system, the total number of the radio channels handled by one MDE is 576.

3.7.4. Pocket-size Telephone Set

Figures 3.15(a) and (b) show the key technologies adopted in NTT DoCoMo's digital pocket-size handset named *Digital Mova*: Figure 3.15(a) indicates a block diagram of *Digital Mova* configuration and Figure 3.15(b) indicates major technologies used in *Digital Mova*, contributing to its size reduction. They can be summarised as follows:

1. Linearised-saturation amplifier with bidirectional control (LSA-BC): to achieve high power efficiency for the $\pi/4$-shifted QPSK RF signal, a linearised GaAs-FET saturation power amplifier incorporating a high-efficiency DC–DC converter with a bi-directional control scheme is used. The RF output power is 0.8 W.

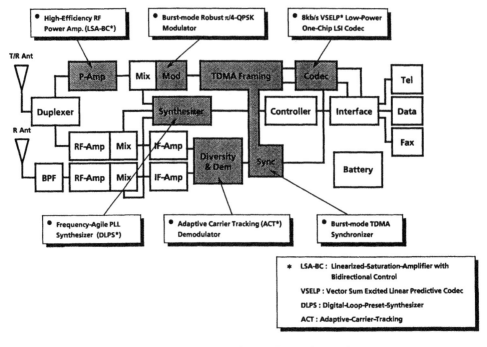

Figure 3.15(a) Digital Mova key technologies.

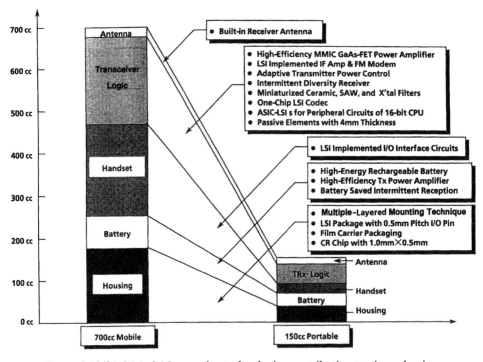

Figure 3.15(b) Digital Mova major technologies contributing to size reduction.

2. Bandwidth-efficient $\pi/4$-shifted QPSK TDMA burst-activated modem: $\pi/4$-shifted QPSK modulation takes place directly on the 800/1500 MHz burst-mode carrier by using an MMIC RF quadrature modulator. One-chip ACT demodulator CMOS-LSI is used for signal detection.

3. Low power one-chip 11.2 kb/s VSELP codec: Adoption of the latest CMOS-LSI technology and a newly devised processing-word reduction DSP architecture, which does not sacrifice the speech quality, make it possible to reduce the power consumption needed for signal processing in the low-bit-rate high-quality speech codec algorithm. A one-chip LSI codec with the processing performance of more than 10 MIPs was developed for PDC codec application.

4. Fast frequency-agile PLL synthesiser: Circuit scale is reduced by adopting a novel digital-loop-preset-synthesiser (DLPS) as a reference signal generator. High-speed channel switching can be achieved with a single VCO with 2 ms switching time.

5. High-reliability TDMA synchroniser: A high-reliability 3 channel TDMA burst-mode synchroniser CMOS-LSI was developed by using advanced application-specific IC (ASIC) technologies.

Four types of *Digital Mova*, the size and weight of which are about 140 cm^3 and 200 g, respectively, were introduced in March 1993. A long battery life is realised by using a high-efficiency power amplifier and sophisticated power saving control techniques such as adaptive transmitter power control, intermittent signal reception and voice-activated operation. The talk and idle times with the regular battery are about 80 minutes and 60 hours, respectively. With a large battery, the life times can be extended to 230 minutes and 150 hours, respectively, with the full-rate three-channel TDMA system. Therefore, the talk time of the half-rate six-channel TDMA system is expected to be about 160 minutes even with the small regular battery.

3.7.5. Data and Facsimile Transmission

Figure 3.16 shows a block diagram of the mobile data and facsimile transmission system in the PDC network. To provide high-quality and high-reliability mobile data and facsimile transmission, we must apply some type of error control. A pooled modem scheme is employed in MCC for the mutual conversion between the PDC network and the fixed PSTN/ISDN. A newly developed ARQ scheme, WORM-ARQ, is adopted for the high-quality G3 facsimile transmission that can support almost error-free fax communication even in the severe multi-path fading mobile communication environment. WORM-ARQ is an ARQ scheme using window-control operation based on reception memory, which migrates between selective repeat (SR) and go-back-to N (GBN) modes. It usually operates in the SR mode with time-out alert. When a time-out happens due to a long burst error, operation is switched to the GBN mode. Error-free facsimile transmission can be provided even during the hand-off process.

For mobile data communications, forward-error-correction (FEC) coding is employed to mitigate multi-path fading effects. The initial system operation

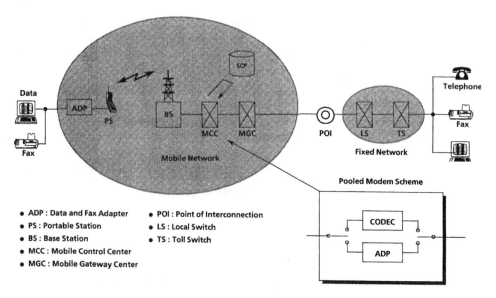

Figure 3.16 Data and fax transmission in PDC network.

supported a 2.4 kb/s MNP class-4 modem service. An enhanced 9.6 kb/s MNP class-10 modem service has been active since March 1995. To provide more effective mobile data communications, a 28.8 kb/s packet mobile data communications network, which overlays the PDC network, has been in operation since March 1997. The PDC packet data network architecture and its protocol are shown in Figures 3.17 and 3.18, respectively.

3.7.6. Service Enhancement

Figure 3.19 shows a scenario of the future service enhancement in the PDC network. Not only speech but also various kinds of mobile services have been developed. In addition to mobile facsimile and 2.4 kb/s MNP class-4 modem data services, supplementary services defined by ITU-T recommendations have been supported since the initial stage of the PDC operation. Caller identification presentation, charging advice, call waiting, call transfer, voice messaging, and three party service, which are basically included in such categories, have also been supported since the initial stage. The user authentication and data encryption mechanisms defined in the PDC air-interface protocol are very sophisticated, and it is predicted that an attacker would require a number of years to extract the original information. Pay telephone, advanced three-party service, inter-network roaming, and the personal mobile telecommunication (PMT) service have been provided since the beginning of the second stage operation in 1994. The PMT service is a kind of universal personal telecommunication (UPT) service. Since the third stage, 5.6 kb/s half-rate speech and 9.6 kb/s MNP class-10 modem services have been provided. The dual or multi-number

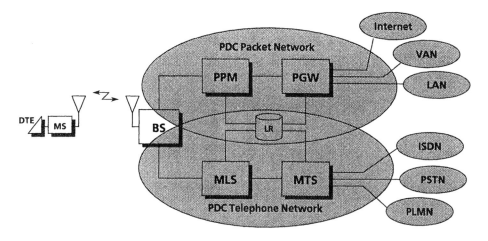

- PPM : Packet Processing Module
- PGW : Packet Gateway Server
- MTS : Mobile Toll Switch
- MLS : Mobile Local Switch

- LR : Location Register
- BS : Base Station
- MS : Mobile Station
- DTE : Data Terminal Equipment

- VAN : Value Added Network
- LAN : Local Area Network
- ISDN : Integrated Services Digital Network
- PSTN : Public Switched Telephone Network
- PLMN : Public Land Mobile Network

Figure 3.17 PDC packet network architecture.

service, virtual private network service, simplified dialling and enhanced voice messaging services were also introduced in 1995.

3.8. FUTURE PROSPECTS

Figure 3.20 shows a view of the evolution scenario of mobile radio communications. The 1980s, at the beginning of which the first-generation analogue cellular and analogue cordless telephone systems were put into commercial service and extended gradually to the world, are considered the initial stage of public mobile radio communications. The early 1990s are the growing stage, where digitalization has been an expanding trend in cellular and cordless telephone systems. These systems are considered to be the second-generation systems and will be improved and advanced toward the so-called 2.5th-generation systems with service enhancements in the mid 1990s.

To provide advanced diversified cellular mobile telecommunications services cost effectively, an intelligent mobile network (IMN) is being developed. Figure 3.21 shows an evolution scenario towards IMN. The basic IMN functions that can separate the service control from call control or bearer service control, automatic terminal roaming to/from other cellular networks, and PMT services have already been implemented within the first-stage PDC network. The 2.5th-generation system will

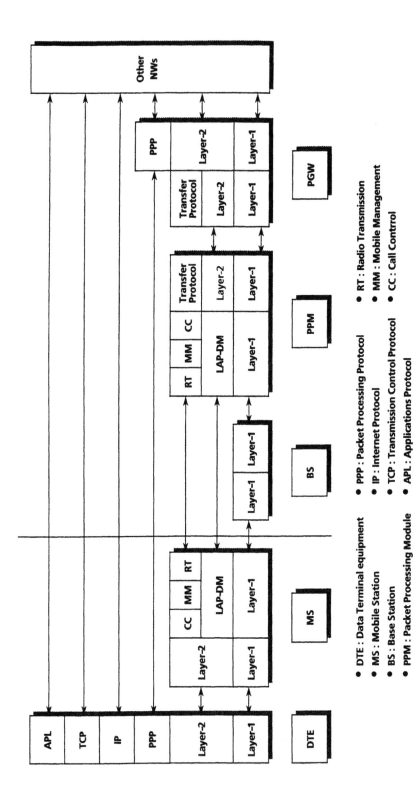

Figure 3.18 PDC packet network protocol architecture.

- DTE : Data Terminal equipment
- MS : Mobile Station
- BS : Base Station
- PPM : Packet Processing Module
- PGW : Packet Gateway Server
- PPP : Packet Processing Protocol
- IP : Internet Protocol
- TCP : Transmission Control Protocol
- APL : Applications Protocol
- LAP-DM : Link Access Procedure for Digital Mobile Communications
- RT : Radio Transmission
- MM : Mobile Management
- CC : Call Control

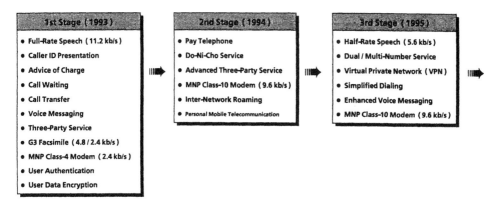

Figure 3.19 Service enhancement in PDC network.

support various kinds of advanced cellular mobile and/or portable telecommunications services including freephone and enhanced data and facsimile services. Advanced IMN services supported by the third-generation system are universal mobility, mobile multimedia and user-defined network services. The advanced telecommunications management network (TMN) will also be supported by the third-generation system.

In the 21st century, both cellular and cordless systems will be integrated into the third-generation system called various names such as FPLMTS, IMT-2000, UMTS and/or PCS. Key words in describing the major objectives of the third-generation system are "Universal", "Personal", and "Multimedia". "Digital" and "Intelligent" are also keywords describing the technologies for the third-generation system, as shown in Figure 3.22. Service area expansion to provide universal services is one of the most important and difficult objectives. Current macro-cellular mobile radio systems are mainly limited to public outdoor environments. Considering the expansion of social, economical and business activities in human lives, it is important for the third-generation systems to encompass the public, semi-public and private indoor environments as well. Roadway tunnels, subways and underground shopping areas should be covered. A universal procedure, which can allow the wireless access of portable terminals to the PBX or LAN in indoor offices, will significantly enhance our convenience. Micro-cell and pico-cell approaches are effective in achieving such purposes. Cost-effective service area expansion should be achieved by using mobile satellite communication systems with a mega-cell structure. Personal multimedia mobile telecommunications services can be provided by not only ultra miniaturised PDA-based mobile terminals but also an advanced IMN supporting universal personal mobility. A Wideband Direct-Sequence Code Division Multiple Access (W-CDMA) system and the Asynchronous Transfer Mode (ATM)-based mobile radio network are most promising as the radio access scheme and the network architecture for the third-generation system, respectively. Details of NTT DoCoMo's proposed W-CDMA system [13] are described in Chapter 33 of this book. Two-layer code assignment, adaptive transmitter power

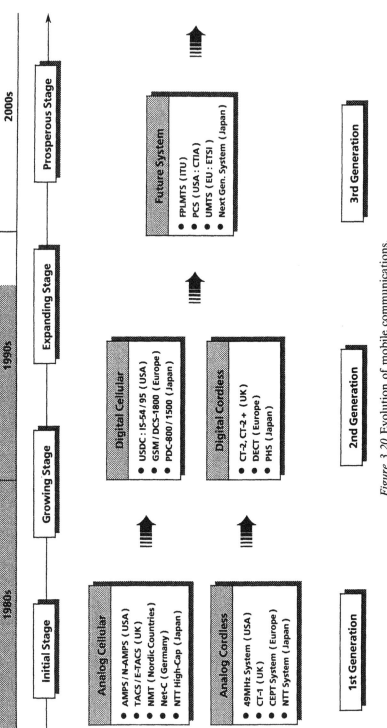

Figure 3.20 Evolution of mobile communications.

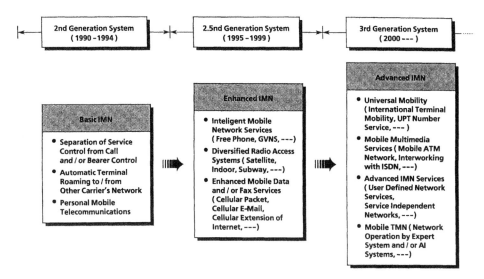

Figure 3.21 Evolution of intelligent mobile network.

control based on signal-to-interference (SIR) measurement, pilot-symbol aided coherent detection with coherent chip synchronisation and coherent RAKE combining, hybrid multiplexing in time and code, and two-stage use of concatenated channel coding are the key technologies of the system.

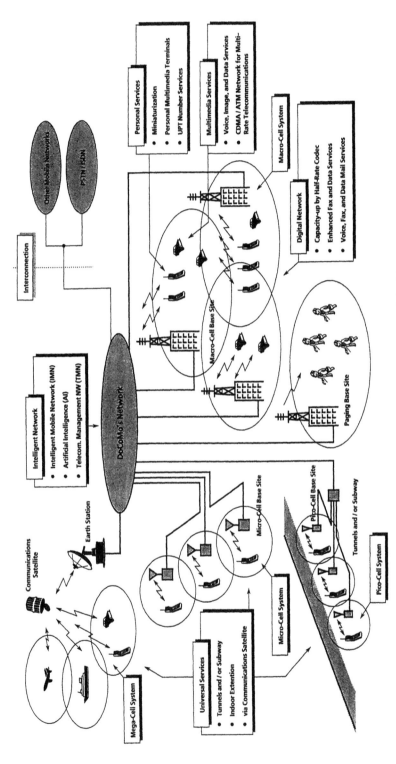

Figure 3.22 Prospect of DoCoMo's network in 21st century.

REFERENCES

[1] Ito, S. and Matsuzaka, Y., 800-MHz band land mobile telephone system. *Review of the ECL*, **25** (6), 1147, 1977.

[2] Seki, S., Kanmuri, N. and Yuki, S., New mobile radio unit. *Review of the ECL*, **30** (2): 318, 1982.

[3] Sasaki, A., Urabe, S. and Nishiki, S., Detachable mobile radio units for 800-MHz land mobile telephone system. *Review of the ECL*, **35** (1), 45, 1987.

[4] Sakamoto, M. *et al.*, High-capacity land mobile communications system. *Review of the ECL*, **35** (2), 89, 1987.

[5] Hirade, K. *et al.*, Super-compact portable telephone -*Mova*-. *NTT R&D*, **40** (7), 997, 1991 (in Japanese).

[6] Kuramoto, M. *et al.*, Development of digital cellular mobile communications system. *NTT R&D*, **40** (10), 1269, 1991 (in Japanese).

[7] Sasaki, A. *et al.*, Digital cellular mobile communications system. *NTT DoCoMo Tech. J.*, **1** (1), 1993 (in Japanese).

[8] Utano, T. *et al.*, 1.5-GHz band PDC system. *NTT DoCoMo Tech. J.*, **2** (2), 1994 (in Japanese).

[9] Akiyama, D., Shimizu, H. and Ishino, F., Advanced data and facsimile transmission scheme on PDC system. *Digest of MoMuC'95,* Bristol University, 1995.

[10] Onoe, S. *et al.*, Half-rate PDC system. *NTT DoCoMo Tech. J.*, **3** (3), 1995 (in Japanese).

[11] Onoe, S. *et al.*, Extended band PDC system. *NTT DoCoMo Tech. J.*, **4** (4), 1997 (in Japanese).

[12] Nakajima, A. *et al.*, Service enhanced mobile communications network. *NTT DoCoMo Tech. J.*, **2** (1), 1994 (in Japanese).

[13] Adachi, F. *et al.*, Wideband coherent DS-CDMA mobile radio access. *NTT DoCoMo Tech. J.*, **4** (3), 1996 (in Japanese).

4

Mobile Multi-service Data in TETRA PMR

Alistair Munro

4.1. INTRODUCTION

Mobile multimedia computing is a concept that requires integration of telecommunications and computer networking infrastructures if it is to become a reality. It offers access to services and applications using various combinations of data, text, voice, video (among others) to *anybody, anytime, anywhere*. The operator (telecommunications providers) and user (computer networking) communities have conflicting interpretations of this vision, focusing on several issues that are evolving from institutional attributes, handed out by a network operator, to subjects for debate and negotiation between a user and an operator:

Insights into Mobile Multimedia Communication
ISBN 0-12-140310-6

- *Quality of service* (*QoS* – expressing requirements for it and maintaining it) has emerged as the unifying pivotal issue. It is a tool for acquiring the resources needed to bring users of distributed multimedia applications together with network resources consistent with the communications requirements of the components of those applications.
- *Seamless integration* of all types of infrastructure. This will arrive because there will be fewer incompatible options and because the level will be driven down at which interworking of the different infrastructures occurs. The probable direction can be inferred from the interest in cell relay (ATM) to support access for the user and thus blur the distinction between multi-service support within the transmission network and at the user/network interface.
- *Intermittency*, manifesting itself at several layers. Some degree of discontinuity or transience is characteristic of telecommunications services where connections are built up and torn down rapidly and last only a few seconds on average from the users' perspective; however the infrastructure components are continuously connected with a high degree of availability. Connections between users of a computer network tend to be permanent in a virtual sense with lower expectations of availability of the interconnect – most applications will survive a significant degree of disruption. All wireless mobile systems experience some degree of discontinuity due to handover or the behaviour of the radio link. At the network layer, connection-oriented or connectionless operation will affect the physical or virtual continuity of the user/network relationship.
- *Mobility*, of terminals, users, services, and resources. For the purposes of this chapter, we are concerned with the first two, thus primarily ownership of network addresses and where they are located. This ability to roam is part of the future for fixed wireline networks and is being studied for computer networking; but it is fundamental to wireless mobile networks.

The remainder of this chapter is a discussion of the theme "mobile multimedia" and an examination of whether it exists now in *wireless* mobile communications networks. The categorisation of current standards is considered followed by an overview of requirements and mobile data architectures, focusing on the ETSI Trans-European Trunked Radio Architecture as an example. Next, we discuss mobility with a case study of Mobile IP and CDPD as illustration. A comparison of this with other evolving technologies and a view of the future concludes the chapter.

4.2. A CHARACTERISATION OF MOBILE MULTIMEDIA APPLICATIONS

Several dimensions need to be considered:

- Media and information datatypes and their encodings, covering audio (from telephony-quality speech to hi-fidelity sound), video (again with a large range of quality and rate), as well the possible combinations of these within the scope of standards such as H.320/H.261. Other information types are associated with IT applications such as e-mail or file transfer.

- Applications, or teleservices, distinguishing local services that acquire data from a remote source from those that require continuous connection to a communications bearer to achieve their purpose.
- Control, coordination and distribution mechanisms, including services for mapping names to addresses (a directory), identifying the provisioning of teleservices at specific locations, and for selecting point-to-point, point-to-multipoint, or even broadcast distribution modes.
- Supporting bearer services, considered from several perspectives, such as an OSI presentation stream, a connectionless datagram service or a synchronous 8 kHz voice bearer, among many others.
- Mobility and its implications, taking account of physical motion (of terminals), roaming between networks, and the ability of the underlying network service to provide the capabilities required by any individual user.

4.2.1. Media Types

We consider the different media, following roughly the categorisations of ITU-T I.212:

- audio, using encodings that permit reproduction at qualities from telephony through FM radio and consumer CD through to studio;
- video, with standard TV picture quality, encoded at frame rates between 1 and 30 frames per second with frame pixel dimensions of (up to) 1000 by 1000;
- basic teleservice with bulk structured data with no bounds on lifetime – facsimile, videotext, still images;
- teleservice control and teleservice-specific data – e-mail, file transfer, information services such as the World Wide Web.

There is a wide range of audio and video codecs and transmission formats. The resources needed for capture and encoding vary significantly with trade-offs of codec speed and complexity, sampling rate, or resolution to be taken into account. Control and distribution protocol will be examined below in more detail but it should be noted that several of the media encoding formats integrate important aspects of application control and management into the user data streams. MPEG is a good example of this.

4.2.2. Applications vs. Teleservices

We distinguish applications that support multimedia content viewing, editing and creation from teleservices that combine these activities with interactions over some communications service with one or more remote parties. The former can survive with non-interactive bulk delivery; the latter requires bidirectional real-time interaction. ITU-T T-series recommendations address many common teleservices. T.70 for networking support, and T.120 for conferencing are well-known examples.

4.2.3. Control, Coordination and Distribution Mechanisms

The name and address of the corresponding parties are the first items that are needed when launching an application. A connection subscription may be required in addition and this will require the address of the access point in the network. For many teleservices, particularly those operating entirely with the PSTN or ISDN, these phases can be combined – e.g. a fax call. For others, such as Internet access from a connectionless IP-capable terminal, subscription is not needed. Still others, such as Internet applications operating over a dial-up connection, require both types.

Many applications are self-contained shells that have their own control protocols. These are generally based on a well-defined (possibly private) use of some more basic service, such as RPC or X.25.

Requirements for identifying the corresponding parties vary:

- knowing the precise coordinates of the correspondents;
- gaining the knowledge through access to a well-known address or broker process, e.g. an X.500 directory or a trader;
- none, connecting directly to continuous streams where the identity of the other parties is acquired subsequently;
- unspecific well-known names, such as group addresses or distribution-list names.

Some services involve multiple applications, more or less coordinated in real time. The need for synchronisation may be critical – e.g. lip synchronisation of audio and video streams. Others, such as a whiteboard run in parallel with a videoconference, are less closely coupled. The requirements may vary dynamically as components are added or removed.

Distribution modes include:

- unicast, or point to point;
- broadcast – to all;
- multicast – point to multipoint where the address serves as a handle used by any number of subscribing terminals;
- anycast – initially point to multipoint but subsequently point to point, a type of location mechanism.

These modes impose different requirements for routing and coordination, especially if the protocols have some time bounds.

4.2.4. Basic Bearers

As indicated above, the control plane can manifest itself in various ways. Once the association between parties has been established, a bearer is required for transfer of user data. This may be an extension of the connection control protocol, be completely defined by that protocol and carried in the same stream, or it may be carried in a separate connection established during the set-up phase. Bearers may also include facilities for multiple in band streams, e.g. OSI Presentation Service, expedited data in X.25, or the "urgent" stream in TCP/IP.

The constraints on use of many bearers by terminals and switches (or routers) are specified completely by the nature of the service transported. This may be assisted by the nature of the connection supporting the bearer: 8 kHz voice streams are easy to transfer via a synchronous connection-oriented bearer but more demanding in the packet-mode connectionless case. The parameters that characterise the service – data transfer rate, delay (and its variability), residual errors (at bit, block, or message level), and so on – form an initial static definition of quality of service (QoS) that is independent of the underlying transmission mechanisms. Emerging multimedia applications expect to be able to vary these parameters dynamically during the life-time of a session and act upon individual bearers.

4.2.5. Mobility and its Implications

Users, terminals, services and supporting resources are all eligible to be mobile. We are concerned here mainly with problems presented by the motion and roaming of a terminal in a wireless mobile communications service. However, there are many related problems in the general case where any user with a predetermined package of multimedia requirements presents himself, his identification and his terminal at a network access point and demands a set of resources that may have to be moved in on demand.

If these dimensions of mobility are enabled then QoS and intermittency become factors of major importance. The possibility was discussed above that a user or application may expect to vary its demands on the communications services and it is implicit that one or other may be active or activated intermittently. However, in the mobile situation the focus comes from the infrastructure itself: the quality of the service, its availability and continuity may vary under the influence of physical and operational conditions.

4.3. WIRELESS MOBILE COMMUNICATIONS SYSTEMS

It is likely that there will continue to be four major divisions of wireless networking technology:

- *Cellular*, GSM is the major first generation digital standard [1]. These are mass public services operating in the wide area and are characterised by being very highly standardised and regulated. GSM is operated on a pan-European scale and is being taken up in other countries.
- *Cordless telephone*, e.g. DECT (the ETSI standard, also digital). These support services with local coverage and equipment is available for domestic use (a cord-less home telephone), business (small PABXs), LAN access, and public services in dense urban areas.
- *Wireless LAN*, including mainly proprietary products (NCR Wavelan is well-known) for RF and IR parts of the spectrum. Standards are still under develop-ment, including the ETSI Hiperlan and IEEE 802.11.

- *Satellite*, including Inmarsat-C, the main trunk services that support links between the major operators, and, of recent close interest, the low earth orbit (LEO) systems proposed by, among others, Teledesic to realise another vision of anywhere, any time, anyplace.

As the next generation develops, it can be expected that the approach outlined in the introduction for fixed broadband networks will be carried over to a certain extent in wireless radio mobile communications systems. There may be a point at which cells are reconstituted into packets or 64 kbit/s streams, or they may propagate as far as the terminal equipment.

In amongst this we find *private mobile radio* (*PMR*). This is very common (there are several hundred licensees in the UK), and is characterised by a simple air-interface with minimal layer separation. PMR systems are like very small low-capacity cellular systems – the distinction is mainly economic and regulatory. While PMR products and operations are subject to regulation, the scope of the standards is much narrower and is limited to the air-interface and to interworking requirements with external networks.

4.3.1. Analysis of Mobile Data Opportunities[1]

Existing PMR Services

PMR is used by civil emergency services and mobile commercial services such as taxis, fleet managers, the utilities, among others and much of its usage can be characterised as despatching. This includes direction of mobile users and employees from one job to another, and communication of status information.

Group and multicast transmissions are the norm and there is a focus on a multiplicity of small services, sometimes like electronic mail, often more interactive, with a mixture of voice and data, calls to and from external networks. Transmission of data may take place in the call control channels in limited volume; there are often additional protocols defined in traffic channels that are dedicated to data services. Terminals are usually purpose-built.

One example is RAM Mobile Data, based on the Mobitex standard from Ericsson. This is used at Heathrow by British Airways to attempt to keep passengers, their baggage and their flights in some sort of cohesion. Hutchison Mobile Data offered a service based on the Motorola RD-LAP specification. The UK DTI specification MPT1327 is used widely as well. PMR operations are regulated but not to the extent that interoperability is required, although access to the PSTN is commonplace. Thus the fixed infrastructure may well be dominated by proprietary products.

---

[1]The discussion of this section is restricted to terrestrial services but many equivalent issues can be identified in the case of satellite data communications, especially taking the emphasis on personal satellite communications in current research into account.

Cellular

Analogous requirements exist in cellular networks and serve the business professional community. They include electronic mail (local reception and central access), fax, terminal connection, and, increasingly, a general-purpose network layer service such as TCP/IP. Terminals are general-purpose (portable PCs, personal digital assistants).

GSM offers messaging using its own standard, the Short Message Service (SMS) with point-to-point and broadcast transmission modes. The specifications define a range of V, X and I series interfaces, of which the V series are being offered to support data and fax. In fact the terminal adapters tend to be proprietary and to integrate the network termination without exposing the control-plane or user data paths.

The present situation with GSM illustrates the collision of the telecommunications community with the computer networkers again. All the divisions are perpetuated in a form that is becoming increasingly less relevant to the computer networking requirements; the services will be expensive to use; seamless integration with fixed networks is difficult. Developments in GSM Phase 2 will improve data services (e.g. aggregation of traffic channels to achieve higher bandwidths). However, it is likely that this dreadful mess will be tidied up only with the completion of the Universal Mobile Telecommunications Service programme (UMTS) complemented with European Commission funding from the ACTS programme through projects concerned with integration with ATM broadband networks and development of the GSM Packet Radio Service.

Cordless

The data requirements outlined above apply equally to cordless telephones. DECT is better prepared to meet them because packet mode services are defined as part of the base standard [2]. The first releases of the ETSI specifications offered many opportunities for "evolution and innovation" thus leaving some doubt about eventual interoperability. However, the completion of work on profiles for generic use (GAP) and public access (PAP) has resolved many of the issues both in use of the air interface as well as access to the PSTN and ISDN, including packet mode access. As far as actual realisations are concerned, there are some LAN-access products that are usable in the office automation environment. There is no evidence of any offerings comparable to those of GSM at this stage. Indeed, most DECT products are conventional voice terminals.

In spite of intense industrial interest, DECT services are still in their infancy and mobile data has not been a high priority. The interest is dominated by the opportunity for public urban use in pico-cell/microcell configurations. This should give an economic alternative to GSM and, possibly, reduce the load on it. However, the concept cannot work without seamless interworking (including flexibility in the radio part of the terminal), roaming, number and service portability, and a considerable effort in defining the regulatory constraints on operators.

Deploying DECT in the private situation is also problematic. Apart from the expected difficulties of network planning, the service is useless unless calls can be made into and out of the cordless environment. A secondary level of switching linked to existing PBXs will achieve this for voice and fax but not, in general, for data or messaging.

Wireless LAN

By contrast with the services discussed above, the WLAN is, firstly, a provider of packet-mode data services and, secondly, a standard wireless extension to an existing LAN supported by existing routing and management infrastructure. These features, plus the operation in unlicensed spectrum, make many issues of service provision and interoperation hitherto discussed irrelevant.

WLANs are often operated as bridges between existing wired networks but much of the marketing of WLAN products is based on the assumed ease of installation without wires to provide a service to a desktop, or easily movable, workstation. Transparent support for terminal mobility for portable workstations within a single LAN subnetwork (at the very least) is a necessity in this situation. User and service mobility within the same technology but outside the home LAN subnetwork are more demanding – Mobile IP (see below) is a possible solution but it is not well enough established at this stage.

4.3.2. Wireless Mobile Data – Architecture and Performance Issues

Researchers have proposed a large number of schemes for packet mode transmission of data in wireless channels. These range through all the dimensions of multiple access architectures: FDMA, TDMA, CSMA (DSMA), CDMA, FHMA, PRMA, and various hybrids. The problem is essentially one of achieving acceptable capacity with very limited channel resource, and many schemes are based on shared use. There are circumstances that make this more difficult:

- collisions are perceived by the receiver;
- coverage may be poor: a transmitter may not be heard, a receiver may not hear; and it may be difficult to locate a station;
- propagation delay may be sufficient to require additional synchronisation measures;
- the energy of received transmissions varies widely, leading to power capture;
- physical effects (multipath fading, Doppler shift) affect error rates and distribution.

The conventional model for CSMA-CD provides an initial set of upper and lower bounds for throughput against offered load under most of the multiple access proposals. The worst case combination of the effects listed above with a lazy or busy base station with high latency will cause throughput to reduce to that achievable with slotted ALOHA. The upper bounds can be refined by taking power capture effects into account. This reduces the tendency to unstable behaviour under high offered load but results are disputed. Where capture works to favour throughput, this is obtained at the expense of very long transit delays indeed for mobiles that are drowned out by their more powerful peers.

Residual error rates in the channel vary widely and depend on many aspects of the physical environment, such as received signal levels, noise, interference, velocity, physical composition and layout of objects in the locality. Thus poor averages will coexist with periods of good transient throughput. FEC is essential but must be complemented by a higher layer detection scheme to combat the probability of undetected errors. This leads to a significant overhead over and above user data and eventual reduction of throughput accordingly.

All wireless mobile systems have a random access element at some point in the relationship between a mobile station and a fixed base station. The common channel signalling protocols are specified in such a way that the contention phase is very short and generates very low traffic levels even for large populations. Simple use of these random access schemes for bulk data transfer tends not to scale, with resulting instability. Various avoiding measures can be taken by reserving in advance the common channel "time" for known volumes of data, either statically or on demand, or by redirecting data streams to different channels to reduce peaks of offered load. The objective is to reduce transit delay and increase utilisation. However, hybrid schemes may lead to increased protocol overhead and wastage (how can a reservation be cancelled?) and defeat the object.

4.4. PROPOSAL: OPPORTUNITIES FOR MOBILE MULTIMEDIA IN PMR

The main proposal in this chapter is that private mobile radio is the best environment to establish wireless mobile data and multimedia services. To support this the analysis presented above has considered:

- global influences such as QoS management, seamless integration, intermittency and mobility;
- characteristics of multimedia applications;
- existing PMR systems alongside second generation cellular and cordless standards together with wireless LANs;
- issues in multiple access protocols for mobile data.

To summarise the main features and conclusions:

- Multimedia teleservices are very diverse in requirements, structure, protocol and behaviour.
- Mobility affects the potential quality of the delivered services, which may range from a higher variability of key parameter values to a complete inability to support specific requirements.
- Established public wireless mobile services are dominated by the concerns of the mass market.

- Private mobile wireless services are characterised by integration of telephony and data services, and flexibility in, with the possibility of faster change of, operational characteristics and configuration.

Three other observations can be made:

- GSM, as the first international digital cellular system, had to be specified completely in terms of technology and services because it could not be allowed to fail. Having succeeded (at enormous cost, which has to be paid for somehow), with the expectation of 400 million users by the year 2000, it cannot be changed easily. DECT has to be treated the same way, with a great deal of pain on the way, and, if it succeeds, then it will be bound by the market in a similar way.
- PHS from Japan and other standards from the US represent severe competition in both public and private domains. PHS in particular is structured to allow significant flexibility in the radio part with the complexity distributed in the infrastructure. Furthermore there is a tradition in the US of specialised mobile radio services that are reminiscent at a technology level of PMR as it is described in this chapter. In a suitable regulatory environment, these could, of course, include the GSM or DECT air-interfaces with a more flexible infrastructure.
- Mobile multimedia teleservices are in their infancy and it is likely that there will be dead ends in the future development paths. As knowledge improves, it will be essential that the communications infrastructure be adaptable to reflect the maturing requirements.

Taking these considerations into account, the choice of PMR as the support for mobile multimedia is a straightforward recommendation.

4.5. TETRA – A CASE STUDY OF MOBILE DATA FOR MULTIMEDIA

Having selected PMR, the question arises of the choice of technology. It is assumed that it should be a public standard and digital. The combination of telephony with data is important, so current wireless LAN services should be excluded at present. PHS could be a candidate as well but its introduction would cause significant disruption due to spectrum allocation. Thus we should look at technology that follows the traditional PMR route, of which TETRA is the present state of the art.

The ETSI Trans-European Trunked Radio Architecture for PMR (TETRA) is one of the first general purpose digital wireless mobile communications standards to integrate data (packet and circuit mode) and voice services as part of the initial base standard [3,4]. TETRA comes in two parts: the combined voice and data part (V + D); and the so-called "packet data optimised" part (PDO). It is expected to be operated in the conventional trunked environment but, acknowledging the requirements of the emergency services, there is also a "direct mode" operation.

V + D is a narrowband TDMA system with a control architecture based on slotted ALOHA with reservation. It is typical of many PMR systems although it has been layered in the ETSI way. PDO has inherited many of the attributes of mobile data specifications (such as Motorola RD-LAP or Ericsson Mobitex). PDO and V + D provide connection-mode and connectionless data services and are identical at the network layer interface.

V + D contains few surprises and it is expected that initial products will focus on it. PDO is a major innovation in packet mode mobile data: it arrives at a time when mobile data has had some successes on a small scale but larger failures in the wider market so the overall requirements and potential had to be assessed thoroughly. Serious efforts were made to create an environment that would encourage multiple services and flexibility in using the air interface, so we give it more detailed consideration in section 4.5.1. The other issue is mobility support in the infrastructure and how to achieve it for data services. This is discussed with reference to existing models in section 4.5.2.

4.5.1. Optimisation for Packet Mode – TETRA PDO Protocol Stack

The PDO MAC is a random access system that can be parameterised in several different dimensions:

- random access – using digital sensing (via downlink busy status) for contention resolution (DSMA);
- addressing – by individual, group or broadcast;
- reservation – uplink time can be reserved in advance optionally.

The uplink resource can be allocated flexibly by choice of specific parameter values. At one extreme it can be allocated sequentially to individual mobiles by time division (possibly dedicated to just one mobile). At the other it can be operated as a CSMA system, taking advantage of statistical multiplexing gain from the (presumed) bursty nature of certain types of data traffic. The configuration can be tailored to the mix of traffic requested by the applications.

The focus of the PDO engine is the *Mobile Link Entity* (MLE), represented by the large box in Figure 4.1. All information flows through the MLE. It is responsible for registration and cell changes (prepare, update, restore). It serves three processes: mobility management (MM), connection-oriented data transfer (CONP), and connectionless data transfer (S-CLNP) as shown in Figure 4.2. It receives system information (TL-Broadcast) and communicates with the monitor of the radio subsystem (RSSI).

The MM process will signal disruption of the link layer (MLE-Break, MLE-Resume) to the data clients (CONP, S-CLNP) so that internal processes can be suspended until the DLC is restored. It will also control some of the state of the MLE based on management messages received from the infrastructure as well as through decisions made internally.

The PDO data-link control (DLC) is based on a modified unbalanced form of the ITU-T LAP protocol. Uplink transfers are expected to suffer higher error rates and the uplink direction incorporates the full sliding window protocol accordingly. The downlink direction is a stop-and-wait protocol, reflecting the anticipation of good downlink quality as well as minimal buffering in the mobile.

The design guideline for the DLC and the MLE was simplicity. The residual error rate, including the probability of undetected errors, delivered by the DLC is

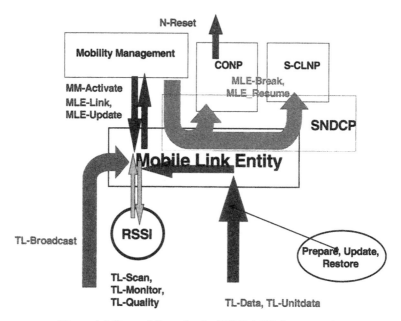

Figure 4.1 Control flows in the TETRA PDO protocol stack.

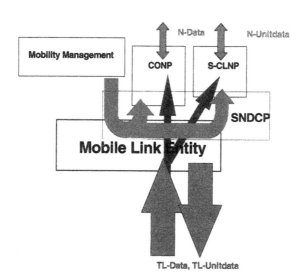

Figure 4.2 Data flows in the PDO protocol stack.

negligible, as is the probability of mis-sequenced or duplicated packets. This meets the conditions necessary for supporting OSI CONS; and it is far in excess of what is needed for CLNP or IP.

CONP provides support for the OSI CONS. S-CLNP means specific (to TETRA) connectionless network protocol. It is modelled on OSI CLNP and provides a unit data service with short TETRA addressing, sub-addressing to distinguish protocols and a certain amount of status information. Client connectionless protocols map into the S-CLNP through a convergence sublayer.

In aggregate, these elements can be combined to provide a wide range of services of different qualities of service. Having said this, TETRA is a new standard and the precise details of how to achieve this are still a matter for further study.

4.5.2. Mobility Support for Mobile Data

Mobility and the processes that manage it are fundamental to wireless mobile communications systems. The terminal mobility associated with access to radio channels is complemented by roaming, which is related to the subscriber's relationship with network administrations.

The ability to roam is a general requirement. For a telecommunications operator it means the surrender of a number to the subscriber. This is a desirable convergence with computer networks where systems own their own network addresses already. However, most computer network architectures do not cope with mobile systems. The Internet is especially poor at this although certain efforts have been made to adapt its routing architecture.

Mobile IP

There has been an IETF working group concerned with IP and mobility for some years [5]. The results have been published in various revisions as an IETF draft. The current version has the following features:

- A mobile host may attach itself to any IP network.
- It receives packets via a "foreign agent" that maintains routing information for mobile hosts on the current network.
- A "home agent" maintains routing information for foreign agents that are supporting mobile hosts belonging to the home agent's network.
- Packets received by the home agent on the home network for the mobile host are forwarded by encapsulation in a tunnel to the foreign agent.
- Packets transmitted by the mobile host are forwarded using normal IP routing.
- Extensions to ICMP protocols and additional protocol exist to support registry of mobile hosts with the agents.
- Mechanisms are incorporated to authenticate mobile hosts and to provide a basis for billing.

The extensions will cater for mobility in almost all circumstances with special consideration of overlays over point-to-point networks such as mobile cellular communications systems. It is envisaged that it will be used in wireless LAN applications as well.

Mobile IP refers here to IP version 4. The new generation of IP (IP version 6) has adopted this approach with some detailed changes to take advantage of the new addressing architecture. In respect of the evolution of IP and its relationship to the telecommunications world, the final solution for handling mobility is still some way off, especially if one takes into account the evolution of the fixed broadband network with ATM, and the deployment of UPT intelligent networks that will handle mobility transparently. The requirements may change radically or disappear.

Cellular Digital Packet Data

The mobile data service, CDPD, that coexists transparently with voice traffic on AMPS networks defined the early versions of Mobile-IP [6]. Detailed characteristics have diverged but the CDPD mobility management is still a good case study of the practice. CDPD retains the concepts of home agents and foreign agents and much of the architecture details are the same, including the routing of inbound packets to the home network – paging mechanisms used for voice are not used. There are several important differences:

- It is impractical to use the ICMP extensions to handle the lower layer aspects of mobility. It is unnecessary as AMPS does this anyway, as well as doing authentication.
- Routing information in the infrastructure is maintained using OSI routing concepts and protocols.
- Inbound IP packets are encapsulated in OSI CLNP protocol.

CDPD is a true IP connectionless network service from the users' point of view and provides an excellent starting point for other mobile networks. We will examine it in some detail here.

The Cellular Digital Packet Data system specification (CDPD) defines vendor-independent, multi-protocol, multi-application support for wireless packet data connectivity to mobile data communications users. It implements this goal by using capacity not used by voice in the AMPS system.

CDPD is defined as a peer multi-protocol extension of existing connectionless data-networks and ensures compatibility with existing technology and applications with minimal change to users' existing configurations.

The main points to note are:

- CDPD has to be completely transparent to any other AMPS services. Mobility management for the data service in particular is entirely separate from the corresponding service for voice calls.
- CDPD provides IP and has made certain choices about how it does it. This may be sub-optimal when the new generation of IP is deployed.
- Whether the service is IP or CLNP, all forward (downlink) traffic into the CDPD network is sent to a home system, then encapsulated and forwarded in a common protocol to the serving base-station (or equivalent). But, reverse (uplink) traffic is carried as native IP or CLNP in the infrastructure from the base station outwards.
- The CDPD architecture provides a network service. Thus IP "management" protocols such as ARP and ICMP are not needed, nor supported, in CDPD. This could be unfortunate – e.g. the "ping" facility depends on ICMP ECHO, or the router discovery extension in Mobile IP.

- CDPD is strictly connectionless. Thus a connection-mode network service is absent and support is provided at the transport layer by COTS + TP4. X.25 and its relations can be supported over IP, but there are no known standards.

It introduces two primary additional components not (normally) present in fixed networks: an airlink protocol and mobility management. The airlink protocol ensures efficient use of cellular channels. It is designed to support a connectionless network layer service that has no impact on end-to-end transport and higher layer protocols. In particular, it is a requirement that there be no changes to existing network software above the network layer and that commercially available connectionless routers be usable without change.

Mobility management provides continuous reliable communications to mobile subscribers, in particular simple and reliable seamless cell transfer mechanisms. It permits roaming between CDPD networks operated by different providers and appropriate interworking procedures are specified. CDPD network services include transfer of user data, associated applications and service provider support functions (authentication, access control, accounting, management, store-and-forward messaging, directory services (white pages)).

The data services are connectionless, supporting ISO CLNP (ISO 8473) and IP. A connection mode service is not supported at the network layer; it is expected that the ISO Class 4 Transport Protocol or TCP will operate at the end-to-end layer to provide this service in which the CDPD service provider is not involved.

Broadcast and multicast services are provided as value-added network applications (i.e. not fundamental in the basic connectionless service). A name/address resolution service may be provided also to allow human-readable names to be mapped to network addresses. A subscriber location service is also envisaged.

CDPD network layer reference model
Figure 4.3 shows the building blocks in the CDPD network. Without describing the functions in detail, important points include:

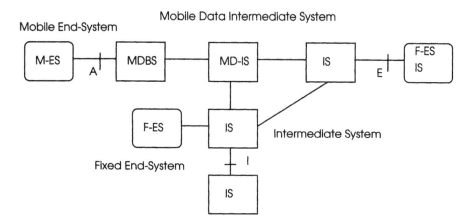

Figure 4.3 Functional components of the CDPD system.

- Mobile end-systems (M-ESs) change their subnetwork point of attachment (SNPA) at will and traditional routing functions cannot be used. The M-ES contains mobility management functions that track the current SNPA and route data accordingly. It contains a radio resource management function that discovers and maintains connectivity to a suitable SNPA.
- The intermediate systems (MD-Is) are the only entities that are aware of mobility and exchange location information by means of a *mobile network location protocol*. A MD-IS communicates with a relay function in the mobile data base station (MDBS).
- The MD-IS contains the mobile home function (MHF) that anchors an M-ES in a home area. The MHF provides a packet forwarding service (in the forward direction only: it is not involved in the return transmission), encapsulating user protocol and sending it to the MD-IS that is currently serving the addressed M-ES.
- The MD-IS contains the mobile serving function (MSF) that routes packets to addressed visiting M-ES served by the MD-IS and performs management functions such as authentication, authorisation and accounting.
- The MDBS operates the airlink protocol to the M-ES attached to the cell it is serving. It manages the radio interface in the cell for CDPD users.

Mobility management in CDPD

Mobility management is concerned with maintaining M-ES location information and routing network packets according to this information. It is closely related to the RRM functions described above.

A *cell* is defined by its relationship to a single MDBS. There may be multiple cells per MDBS. Each MDBS is associated with an MD-IS. The collection of MDBSs managed by an MD-IS is termed a *routing area subdomain*. A CDPD *domain* is identified by a collection of MD-ISs.

The routing area subdomain corresponds to the ISO concept of a routing domain (as defined in ISO TR9575, *OSI Routing Framework* and ISO10589, the *IS-IS Intra-domain Routing Protocol*). In this architecture, the MD-IS corresponds to the level 1 IS, routing to ESs and outwards to level 2 ISs.

Each M-ES is a temporary member of a routing area subdomain associated with the MD-IS that is managing, through the MDBS, the cell in which the M-ES currently has a channel stream. It is also a member (logically) of one or more home areas, associated with the MHF and MSF functions in a home MD-IS. The NEIs supported at an M-ES identify the respective home MD-ISs. Mobility support is provided at the data-link layer, for transparency within the cells managed by the MDBSs associated with a single MD-IS, and at the network layer, for transparency over all CDPD domains. As well as cell selection (see RRM above), mobility management comprises location updating and redirection and forwarding. As noted in the reference model, the latter two functions are part of the MHF.

The support for multicast is defined as an extension to the MHF. Two other mobility management protocols are especially important: the process by which a new mobile registers (including the exchange of authentication data, establishing that it is eligible to access services and able to provide them), and the location of mobiles in the CDPD network.

(Mobile) network registration protocol (MNRP)

The purpose of the registration protocol is to inform the MD-IS about the configuration of M-ESs it is serving. This will include the SNPA and the network protocol entities. The protocol functions are logically the same as the ES-IS routing exchange protocol (ISO 9542) that supports CLNS, violated and adapted to suit the mobile wireless environment.

It provides the following services:

- *notification* of reachability and registration of Network Entity Identifiers (NEIs) associated with an M-ES to a serving MD-IS;
- *authentication* of NEIs at an M-ES to the serving MD-IS;
- *confirmation* by an MD-IS of its willingness to provide services to an M-ES;
- *deregistration* by an M-ES of NEI services with the serving MD-IS.

The protocol operates over the SNDCP, using SNPA addressing.

(Mobile) network location protocol

The MNLP is the complement of the MNRP and operates in conjunction with it and the RRMP to complete the functions of mobility management. It provides for the exchange of location and redirection information between MD-ISs as well as for the forwarding and routing of messages to roaming M-ESs.

The MNLP is the single inter-MD-IS protocol used in the CDPD Network. It operates over CLNP using NSAP addressing together with protocol identification (following the principles of OSI TR9577, *Protocol Identification in the Network Layer*). It supports the following services:

- *notification* to MD-ISs of an M-ESs routing area subdomain and registered NEIs together with associated authentication information;
- *confirmation* between MD-ISs of their willingness to forward packets to an M-ES;
- *forwarding* of packets from the home MD-ISs to the current MD-IS serving an M-ES.

The services are implemented by the MHF and MSF.

4.6. CONCLUSIONS

The important points of the discussion in this chapter are the following:

- Convergence of telecommunications and computer networking includes wireless mobile communications networks. It will be necessary to take account of QoS, seamless integration, mobility and intermittency in a uniform way.
- Data services are not provided uniformly in mobile communications networks. Only the most recent PMR standard makes more than a token effort towards integrated multi-service provision.
- Performance and robustness of mobile data systems is a major concern.
- Data communications protocols and associated routing systems do not accommodate roaming.

These represent greater or lesser barriers to achieving mobile multimedia communications seamlessly in wireless mobile networks. However, they illustrate that there are still problems on the fixed network side. The relatively low quality of the radio environment is a concern for any service. Services that require timely, 100% accurate, delivery could suffer badly in this situation.

It is significant that it has not been necessary to examine multimedia services in detail. In the case of TETRA PDO, attention to the parameters and configuration of the service at the air interface, and a focus on common service boundaries were seen as the main enabling factors in allowing QoS negotiation to determine the resources needed for a possible multimedia application.

The success of TETRA as a standard is still unassured. At an industrial level there is a Memorandum of Understanding concerning introduction of services and products – as suggested above, these focus on V + D. The introduction of PDO is still under discussion at a regulatory level. In spite of these doubts, TETRA still represents the most flexible environment for the introduction of mobile multimedia at low risk, and, possibly, a longer term competitor on the European stage.

REFERENCES

[1] Mouly, M. and Pautet, B., The GSM system for mobile communications. Published privately, 1992.
[2] Digital European Cordless Telecommunications (DECT) common interface. ETS 300 175, parts 1–9, ETSI, May 1992.
[3] TETRA packet data optimised, Part 2: Air interface. prETS 300 393-2, ETSI, 1996.
[4] What is TETRA? Information published by TETRA MOU, 1997, referenced via URL http://www.tetramou.com/ppprenta.htm.
[5] IP mobility support. IETF draft, 1996. Via http://sunsite.doc.ic.ac.uk/computing/internet/internet-drafts/
[6] Cellular digital packet data, system specification. V1.0, CDPD Industry Co-ordinator, July 19 1993, Costa Mesa, CA.

5

Multimedia Over Mobile Networks Using the H.324 Family

John Bassil and Mike Whybray

5.1. INTRODUCTION

As mobile terminals become more common in society, the expectations placed on them are becoming more demanding. All of the standard applications of a desktop computer and also on-line applications such as Internet access are now available to the mobile worker. The migration of applications such as video conferencing and interactive multimedia services from fixed networks (such as ISDN and General Switched Telephone Network, GSTN) to mobile networks (such as GSM and DECT) is an important next step. This chapter looks at some of the problems unique to mobile channels and describes studies of the effectiveness of the GSTN multimedia conferencing standard H.324 running over a single GSM channel.

GSM networks currently offer two basic modes of data bearer services: non-transparent mode and transparent mode. In non-transparent mode, automatic retransmission requests (ARQ) to combat transmission errors are handled by the GSM network itself (with a layer 2 retransmission protocol known as radio link protocol, RLP), meaning that the bit error rate of the received data after transmission is similar to that of the GSTN. This incurs variable delay and data rate penalties

Insights into Mobile Multimedia Communication
ISBN 0-12-140310-6

which reduce the usefulness of non-transparent mode for extremely delay-sensitive services such as real time communication. The transparent mode does not offer error protection, so the network delay between packet transmission and reception is fixed (and restricted to the propagation and interleaving delays). In this case the onus is purely on the network user to provide error protection. This may again incur a variable delay penalty but this is controllable by the receiving end which can choose how to react to the error.

5.2. H.324 STANDARD

The ITU H.324 family of standards ([1] and Figure 5.1) define a terminal which may be used for multimedia applications over the GSTN. This family of standards consists of highly efficient video and audio compression algorithms (H.263 and G.723.1 respectively), a multiplex protocol (H.223), a control protocol (H.245) and a definition of the modem used for transmission (V.34). Also included is the facility for multiple data channels which may be used for multipoint multimedia data transfer (T.120). Given the suitability of this family to low bit rates, it seems reasonable that it may also be used in conjunction with mobile networks, such as GSM. In practice, due to the differing error performance of fixed and mobile networks, some modifications may be necessary to improve the robustness of the bitstream: these are currently being studied by the ITU for inclusion in the H.324 Annex C recommendation for multimedia telephone terminals over mobile radio. The main proposal in this is a new mobile multiplex structure known as H.223 Annex A [2].

At the time of writing (mid-1997), the technical details of this H.223 Annex A were not fixed. A draft had been developed in 1996 based on a fixed length frame structure to give robust synchronisation, and a flexible framework for use of ARQ

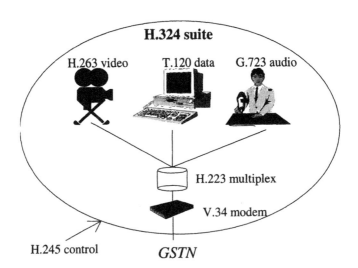

Figure 5.1 Protocols used in the H.324 Multimedia Terminal recommendation.

and forward error correction using rate compatible punctured convolutional (RCPC) codes [2]. However, the proposed system had grown in complexity as the problems of mapping data from the variable length packets of H.223 to and from the fixed length packets of H.223 Annex A became apparent. Although H.223 Annex A would work appropriately on a mobile link, it was recognised that in many cases this would be connected to a fixed (GSTN) link system and the complexity and delay introduced because of the mapping problem were unacceptably high.

Thus at the March 1997 meeting of ITU-T Study Group 16, H.223 Annex A was not accepted, and a process of redesign began. H.324 Annex C always intended to allow an alternative to the use of H.223 Annex A, which was to have the mobile link provide an adequately low error rate by whatever means to allow use of the existing H.223 recommendation. This may not provide as high a throughput in error conditions as an optimised combined multiplex and error control strategy, but has the great advantage of requiring no mapping of data from one multiplex to another at the fixed/mobile network interface (hence no extra delay and cost at the interface), and also allows essentially the same terminal to be operated on both networks, reducing manufacturing costs. The new approach to developing H.223 Annex A was to be to start from the basic H.223 multiplex, and progressively add optional error hardening features such as longer synchronisation flags, a better error protected header, and so on. Also, to try and simplify the draft Annex A to become the highest level of this hierarchy of levels of robustness.

At the time of writing, the above work was still in progress, but the rest of this chapter considers the effectiveness of operating the existing GSTN standard using H.223, equivalent over mobile networks, equivalent to the lowest level of the proposed hierarchy of robustness. The operation of H.223 and H.263 is considered in more detail, and some experimental results presented.

5.3. ERROR RESILIENCE OF H.263

The H.263 video coding algorithm was developed by ITU for low bit rate audio-visual communications at px64 kb/s. The algorithm relies on knowledge of the statistical properties of image sequences to compress the data used to recreate that sequence, i.e. the sequence is *not* treated purely as random data. Compression is achieved using a process of motion compensated inter-frame prediction to exploit the temporal correlation, transformation using the discrete cosine transform (DCT) to exploit the spatial correlation, quantisation and variable length coding. For more details on H.263 refer to Chapter 19 of this book. This H.263 bitstream is then passed down into the H.223 multiplexer where it is packetised ready for transmission over a V.34 modem. The top layer of the multiplex, known as adaptation layer 3, AL3 (for video information), precedes the data by a header, which may optionally contain a 7- or 15-bit sequence number, and succeeds it with a 16-bit cyclic redundancy code for error detection (see Figure 5.2). The second stage of the multiplex encloses the adaptation layer packet in a multiplex layer packet, the header of which is protected by a 3-bit checksum and indicates the format of the data in the packet (e.g. whether it is purely video or video multiplexed with audio and/or data).

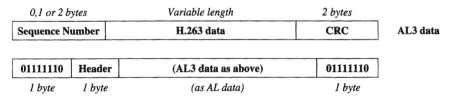

Figure 5.2 Multiplex and adaptation layer packets.

Opening and closing flags are appended to the multiplex packet for delimitation purposes and bit-stuffing is used in the body of the packet to prevent flag emulation.

In the event of a burst of errors occurring in the bitstream one or more packets may be corrupted and may affect either:

- the multiplex header information, in which case the error may be picked up by the 3-bit CRC;
- the multiplex packet payload information, which will result in the adaptation layer rejecting the information and possibly requesting a retransmission;
- the opening or closing flag, in which case the packet delimitation will not be correctly detected.

Use of a retransmission policy for video data prevents entire frames of inter-coded video data being lost. This is important since otherwise picture artefacts may occur due to loss of synchronisation between the encoder reconstruction framestore (used to calculate the inter-information) and the decoder reconstruction framestore. A trade-off must be made between delay caused by retransmission and the perceived effects of errors, depending upon which would make the resultant conversation feel more natural. Various techniques are possible to conceal errors if retransmission is undesirable, such as replacing a known errored block with the equivalent block from a previous frame known to be free of errors or sending of an H.245 fast update request (in which a complete intra picture is transmitted). It is also possible to increase the number of intra-blocks per frame in order to limit the effect of errors at the decoder, at the expense of compression efficiency.

Control of the retransmission window at the encoder and the timer at the decoder may be exercised in order to ensure that the delay does not become too large in high error conditions. The decoder may need to have some local knowledge of the error properties of the mobile link; after loss of several consecutive frames, it can then send a fast update request message to receive an intra frame. A fast update request can also be used where the decoder detects loss of synchronisation, e.g. if a DRTX (decline request for retransmission) message is received. Whilst the decoder waits for the fast update intra frame to arrive, it is put into a freeze state in which the picture should be free of errors.

Several extensions to H.263 were "determined" by ITU Study Group 16, and due for "decision" in January 1998. Two in particular are means of increasing resilience of the overall system to errors – Annex K (Slice Structured Mode) and Annex N (Error Resilient Mode by Backward Channel Operation). The first helps prevent the propagation of errors in one part of the picture into other parts by dividing the picture up into independent "slices", which are coded completely separately to other slices. This contrasts to the normal H.263 group of blocks mode, where the data for different areas is separate but the process of motion compensation can still allow errors in one area to propagate into an adjacent area. The second Annex (N) allows the encoder and decoder to use more than the usual single prediction store, effectively keeping a longer history of past reconstructed pictures. A decoder detecting loss of a group of blocks or slice can request an update based on a specific numbered prediction frame which it knows it has decoded correctly. Although not providing as efficient compression as the most recent (lost) prediction frame could have provided, the compression achieved is still rather higher than that of the alternative, which is an Intraframe coded update.

Lower down in the protocol stack, at the multiplex level, various parameters may be adjusted to optimise performance in the presence of errors. These may include packet length, retransmission buffer size, static or dynamic packet sizing, etc. When considering changing packet size to improve error performance it is important to take account of the following two factors:

- The longer the packet, the more likely there is to be at least one bit error within the packet.
- The shorter the packet, the greater the overhead imposed by the error correction and sequence number fields.

These multiplex level aspects were investigated using the existing H.223 multiplex.

5.4. RESULTS

We have developed a software-based implementation of the video coding and multiplex parts of H.324 written in C and running under Windows NT and 95 on Intel-based platforms such as the Pentium and the Pentium Pro and also on the RISC-based Power PC platform. This has been used for transmission over the GSM network to investigate the effects and the interaction between the network performance and the algorithms. Two types of test were performed using firstly a GSM network error simulator to simulate fading effects etc., and secondly using the real GSM network to test "live" transmission.

The error simulator was used to transmit live video using H.263 encapsulated within the H.223 multiplex over a 9.6 kb/s GSM data channel. This simulates common error profiles encountered on GSM channels using a number of DSP cards connected to a host PC to simulate both transparent and non-transparent modes. For this work, the following error profiles were used over simulated transparent mode GSM:

- typical urban, 3 km/h, interference level 12 dB;
- typical urban, 50 km/h, interference level 9 dB;
- typical urban, 50 km/h, interference level 12 dB.

Detection of an error by the adaptation layer, which may occur either due to a corrupt multiplex header or due to a bad CRC at the adaptation layer data, means that the adaptation layer must either request a retransmission or "work around" the problem, e.g. by using video error concealment. (If the multiplex header of the received packet is corrupted, the multiplex does not know for which service(s) the packet is destined and the packet must be discarded). If no form of error conceal-ment is used, a sequence of selective reject (SREJ) messages is sent to the transmit-ting end to request retransmission of the corrupt packet(s). In this instance, one of three possible scenarios presents itself:

- the retransmitted packets are received correctly;
- nothing is received (in the case of a deep fade);
- the packet is received corrupted once again.

In the cases where either a freshly corrupted packet or nothing at all is received, the picture is seen to "freeze" momentarily. At this point, the decision is left to the adaptation layer as to whether to pass an empty AL-SDU or an errored AL-SDU plus appropriate error indication. In this latter case, an error concealment method may be adopted to prevent semi-random blocks appearing on the screen or the intra refresh can be relied upon to eventually re-synchronise the decoder with the encoder. If further corruption is detected a fast update may be requested to transmit a full intra frame: this does however incur a performance penalty and should be used sparingly.

As shown by Figure 5.3, investigations of error rates and overall throughput at different packet sizes revealed the expected characteristics of packet size against error performance with a trade-off being made between the high percentage overhead for small packets and the increased probability of errored packets for larger packet sizes. The effect on overall (error corrected) data throughput versus packet size was a broad peak at around a packet size of 32 bytes for all three profiles. Note however that throughput was also affected in our experiments by finite processing power in

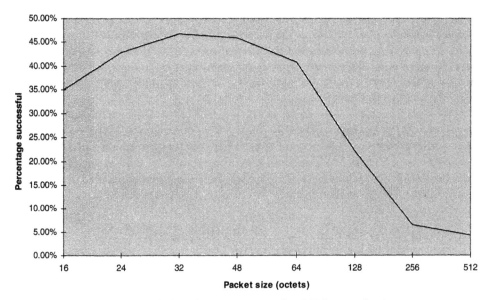

Figure 5.3 Typical performance curves for GSM error simulator.

the PCs during the encoding, multiplexing and decoding, which had the effect of shifting the peak to a lower packet size than would be the case with unlimited processing power. Hence this peak value should not be taken as a definitive result – further work is needed.

The system was also tested over the "live" GSM network, both using one timeslot and, in an attempt to understand what the future will offer, using multiple timeslots concatenated using the channel concatenation properties of Microsoft's Windows NT v4.0 operating system. Running over one timeslot in non-transparent mode, the software achieves a frame rate of about 3 frames/s: note however that this is dependent upon the error conditions of the channel.

5.5. CONCLUSIONS

It is now possible to transmit compressed video data at several frames per second over a single GSM channel. To extend this concept to multimedia conferencing one needs to protect against the unique errors prevalent in wireless services. We have shown a solution that will interwork with current GSTN-based terminals by slight modification of certain parameters of the H.223 multiplex in the H.324 standard. This approach retains compatibility with existing H.324 GSTN terminals and introduces no extra delay by avoiding any requirement to map data between a mobile and fixed network multiplex.

The current restriction to one channel for GSM communication means that opportunities for video conferencing are limited. However, other applications such as remote surveillance and access to multimedia databases are possible, even at these extremely low bit rates. As the video compression standard H.263 evolves to H.263L, the compression ratio will improve, and there will be additional optional modes to improve error resilience and recovery. There will also be new options available such as the option to dynamically alter the video resolution mid-call, which would lend itself well to a security type application for example. In parallel with this work, the GSM data capability is evolving, both in terms of increased bandwidth over a single channel and towards the high speed circuit switched data service (HSCSD), offering the possibility of up to four concatenated channels in non-transparent mode and up to eight in transparent mode. A combination of these improvements will enhance the potential for both video conferencing and other types of multimedia video transmission such as interactive multimedia services and remote assessment.

REFERENCES

[1] Recommendation H.324 – Terminal for low bit rate multimedia communication. ITU, November 1995.
[2] Recommendation H.223 Annex A – Multiplexing protocol for low bit rate communication over error-prone channels. ITU, December 1996 and revisions.

6

Enabling Future Computer Applications Using GSM Phones

G. J. Proudler

6.1. INTRODUCTION

Digital cellular radio systems can provide communication links for computers, but until recently there was no comprehensive way to control a mobile phone from a computer. In 1994 a working group from Ericsson, Nokia and Hewlett Packard defined a set of commands for interaction between computers and digital cellular phones, specifically GSM phones. The commands were adapted in 1995 and adopted in 1996 by the European Telecommunications Standards Institute [1].

Insights into Mobile Multimedia Communication
ISBN 0-12-140310-6

The commands are written in the "AT" style, and allow the computer to find out what is happening at the phone, to control the phone functions normally controlled by a human user, and to control the sending and receiving of messages. This chapter introduces the types of applications that are enabled by some commands.

Precise details and examples of all the commands are in the ETSI standards: GSM 07.05 [2] deals with messaging while GSM07.07 [3] deals with handset identification, call control, network services, handset status, and control of a handset. Those standards are the definitive reference for any and all of the commands mentioned in this chapter, and always take precedence over any statements in this chapter. They should always be consulted to determine the precise definition of commands.

6.2. APPLICATIONS USING MOBILE PHONES

Applications can make it easier for the user to operate the radio or can themselves use the radio for mobile communications. An application needs to know what is happening at the radio, control the functions normally controlled by a human user, send and receive data, send and receive fax, and send and receive short messages.

6.2.1. Phonebooks

Computer manufacturers want to encourage the use of laptop or palmtop computers, or personal digital assistants, for storage of phonebook information. These devices generally have a user-friendly interface for storing phone numbers, and can store other information associated with a phone number (such as address and the context of the relationship to that individual). In the past a user had to read the desired phone number from the computer screen and type it into the phone. The usefulness of the computer would obviously be enhanced if the user could select a destination on the computer and an application automatically dialled the number on the phone.

Only a few commands are required to allow the phonebook in the computer to communicate with the phonebook in the phone: listing, reading, writing and deleting entries. However, the phone contains several different memories, and a user might want to store different parts of telephone numbers in different locations. Personal numbers might be stored in the SIM, since that identifies the user, while business numbers might be stored in the phone *per se* if the user has a particular phone (and phone number) for work. All of these need to be taken into account.

6.2.2. Expanded User Interface

Ergonomic details and cost usually restrict the size of screen and keypad on the phone. There are often too many functions to allow single-key-press access, so multiple key presses are used for less common functions, and the information on the display is often terse.

A computer, even a small one, usually has a bigger display and larger number of keys, and offers the potential of an easier phone interface if the phone controls and display can be emulated on the computer. This requires access to phone controls, to the phone display, and to other phone indicators not on the display. Then the computer can track events at the phone, learn keystrokes for less common functions, and even provide full control of the phone from the computer. In particular, an expanded user interface would make it easier to send and receive short textual messages over the phone.

6.2.3. Voice Messaging

For most users, a phone is primarily a mechanism for voice communication. This functionality can be improved by the use of a computer. The radio transmitter, radio receiver, phone microphone, phone loudspeaker, and computer memory can be interconnected in several ways. The computer can be used to record a conversation and to act as an answer phone, using the phone as both the communication means and the audio input and output.

6.2.4. Fax and Data

If the phone supports fax or data communications, obviously the computer can use the phone as a modem.

6.3. SUPPORTING COMMANDS

6.3.1. Information about the Phone

When an application starts, it must decide whether the phone can support the application. It must get the details of the phone, and possibly match those details to a database to determine the capabilities of the phone.

The first question is whether the phone supports a suitable command set. This can be done using the command *+GCAP complete capabilities* which is part of V.25ter. If the response includes *+CGSM* then the phone supports the ETSI command set. Of course, the phone may not support V.25ter, in which case +GCAP will not be recognised. Then the only option is to try a mandatory command from some command set and check for the correct response. Since control of phonebooks is likely to be a standard feature, a good choice would be a phonebook command. In the ETSI set, *+CPBS phone book memory storage* is mandatory if phonebook read or write is supported, and would be a suitable choice.

The most important attribute of a phone is its phone number. This might be different according to the information class currently in use at the phone since a phone might have a different number for voice calls, data calls, or fax calls. Also, the number might be in a local format (country specific, for example) or in full international format (ITU-T E.163). The ETSI command set uses +*CNUM subscriber number* to make the phone return its current phone number. It may be important to know the phone manufacturer, or the model type, the revision of that model and its serial number. The ETSI set does this with a series of commands: +*CGMI manufacturer identification* returns a text string that a manufacturer can use to represent its name in whatever way it wishes. Similarly +*CGMM model identification* returns a text string that a manufacturer can use to represent the model type, +*CGMR revision identification* returns a text string for representing the revision of the model, and +*CGSN serial number* returns a text string for representing the serial number. All of these can be chosen without reference to an external authority.

6.3.2. Network Interaction

An application probably needs information about the cellular operators available to the phone, and may want to select certain options available in the network. It may choose to use a certain operator based on a knowledge of tariffs, or based on the degree of service or extent of service associated with particular operators. It may attempt to provide the lowest cost communications for the application, or select the best available data service, whatever. All of these depend on the exact nature of the application and the database incorporated into the application.

The ETSI command set uses +*COPS? operator selection* to provide information about network operators and their availabilities. The operator name can be written in long format alphanumeric, short format alphanumeric, or as a number. The status of available operators is returned by COPS = ?, listing available operators and showing whether the operator is unknown, available, current or forbidden. The current operator is returned by +COPS?.

+*CREG network registration* shows whether the phone is "connected" to the current operator. The state of the connection could be: not-registered and not attempting to register, registered at the home network, not registered but attempting to register, registration-denied, unknown, registered outside the home network. Obviously communications are possible only if the phone is registered with the operator. If this is not the case, the application must select another operator or present an error message to a user, and ask for the problem to be resolved. In that case the application might prompt with the telephone numbers of the service agents for a particular operator.

6.3.3. Configuring the Channel

The network provided by the chosen operator needs to be configured. First of all, the application may wish to select the *QoS* characteristics of the protocol used over the radio channel. A reliable service provides a low error rate but indeterminate throughput and latency. An unreliable service provides an unknown error rate but fixed throughput and latency. The choice depends on the particular application. A real-time application that is tolerant of errors would use the unreliable service. A conventional data application would probably be intolerant of errors but latency would be of little concern, and hence the reliable service could be used. In the ETSI set, *+CRLP radio link protocol* allows the application to select either a reliable service or a non-reliable service.

If the phone supports data communications, the network must be notified of the required land-based service. This involves selecting a particular land-based service and the characteristics of that land-based service. In the ETSI set, *+CBST bearer service type* selects between asynchronous modem, synchronous modem, PAD, packet access and the speed of the chosen service. GSM in particular allows identification information to be presented to a user, and the ETSI set provides four commands to use that information. They allow each end of a communication link to enable or disable presentation of information about either end of the link. *+CLIP* controls whether "calling number" information is presented to the phone when the phone accepts an incoming call, provided the caller has used *+CLIR* to allow such information to be presented. Similarly, *+COLP* controls whether "called identity" information is presented to the phone when the phone makes an outgoing call. Other commands can tell the network how to treat incoming calls. Arguably the most powerful are control of forwarding, depending on the current mode (data, voice, fax, etc.) of the phone. ETSI uses *+CCFC call forwarding number and conditions* to allow a distinct forwarding number for each combination of mode and situation. To use this command, the phone is put into the relevant information mode with +CSIC, and then +CCFC is repeatedly used with different values of reason and individual forwarding number. Calls can be forwarded because of various reasons, including mobile busy, no reply, not reachable. Alternatively *+CTFR call transfer* tells the network to transfer an existing call to a phone number included in the command syntax.

6.3.4. Configuring the Phone

Probably the most important configuration decision is to dictate how the phone deals with unsolicited reporting of events. Should the phone pass such indications immediately or buffer them? Which events should be reported to the application? An application may decide to request immediate indications when the computer is switched on, but need to request buffering if the computer is about to be switched off (and is unable to wake up in response to external signals).

The ETSI command is *+CMER mobile event report*, and is complex. The command determines what happens to result codes caused by events at the phone keypad, the phone display, or the other phone indicators.[1] All result codes are treated in the same way, but reporting of individual classes of events can be enabled or not.

- The parameter "buffer" determines what happens when +CMER is executed. Should any existing results (stored in a buffer) be cleared or should they be flushed to the application? If the command allows result codes to be passed to the

---

[1] See section 6.3.6, "Accessing phone resources", for more information about the keypad, display and indicators, and the commands to use them.

application, then the choice is either to clear existing result codes from the buffer or flush existing result codes to the application.

- The "mode" parameter determines what happens to the result codes. The choice is:
  - buffer the result codes, so that they are stored for later access by the application. These buffered codes can be flushed to the application by a subsequent use of +CMER that sets the "buffer" parameter to flush the buffer *and* does not select this "mode" option.
  - pass result codes directly to the application, but throw away result codes that occur when the link between computer and phone is busy.
  - pass result codes directly to the application, but buffer result codes that occur when the link between computer and phone is busy and flush them when the link becomes free.
  - pass result codes directly to the application, but use the standard in-band embedded technique to pass result codes that occur when the link between computer and phone is busy.
- The "keypad" parameter determines whether the pressing of buttons on the phone keypad should be reported. The names of the buttons are the same as used in the command +CKPD. The choice is:
  - don't report events at the keypad.
  - report only those keypad events that were not caused by the application itself. If any buttons are being pressed at the instant the +CMER command is activated, report those events.
  - report all keypad events, irrespective of their cause. If any buttons are being pressed at the instant the +CMER command is activated, report those events.
- The "display" parameter determines whether changes on the phone display should be reported. The position of the change on the display is shown by a number which is that of the appropriate display element in a list starting from top left of the display through to the bottom right. This is the convention as used by the command +CDIS. The choice is:
  - don't report changes at the display.
  - report only those display changes that were not caused by the application itself.
  - report all display changes, irrespective of their cause.
- The "indicator" parameter determines whether changes on the phone indicators should be reported. The particular indicator is identified by a number which is the order of the indicator in the list produced by the command +CIND=?, which returns a list of all indicators and their possible states. The choice is:
  - don't report indicator changes.
  - report only those indicator changes that were not caused by the application itself.
  - report all indicator changes, irrespective of their cause.

A similar requirement is to control whether the phone returns intermediate result codes to the application during connection negotiation for data calls. These confirm the type of call which has been set up and indicate which type of data call has been negotiated. ETSI provides *+CR service report* which indicates async transparent, sync transparent, async non-transparent or sync non-transparent.

6.3.5. Making a Call

This is more complex than might be expected because of differences between traditional networks and modern networks. In the traditional PSTN, the call was established and then the information type (voice, data, fax) was declared. In newer networks such as GSM, the network needs to know the information type before the call can be established since different network resources are required for different information types. The peculiar method of call control used in the ETSI command set is the result of reconciling these contradictory requirements, so that traditional applications will still operate.

In the call control procedure, the nature of the connection must be declared first before starting or accepting a connection. This is necessary so that appropriate network resources can be allocated. The declaration is whether the connection will use a single type of information, or will be voice followed by data, or will alternate between voice and fax or between voice and data. The originator of a call will simply declare the type of call that is required. The terminator of the call will know the type of the proposed call from the unsolicited result code +CRING. Both originator and terminator must then ensure that current settings are correct (no action is needed if current settings are already as required). Before originating or accepting a call, the originator/terminator must have declared any non-voice type using a conventional *+FCLASS* command and declared the call mode using *+CMOD call mode*. Note that +CMOD must be changed back to *single mode* at the end of any alternating call. After ensuring that its current settings are correct, the originator dials the destination number. If the dial string finishes with a semicolon (;) the call is initiated as a voice call. Otherwise it is initiated as a data or fax call, depending on the previous settings.

When the first "call" is finished, the connection can be terminated or (if relevant) the other mode of call can be started using the same connection. When that second call is finished the call must be terminated, unless the call mode is alternating voice/ data, when the mode can continue to switch between voice and data.

A new "hangup" command *+CHUP hangup call* is defined by ETSI to assure connection hangup. Other methods of hangup are not guaranteed to take down the connection, because of the historical context of those commands and the complications of alternating call modes.

Dialling is done using the standard *D dial* command. If the connection is a single voice call, or is a dual type call and the first call over the connection is voice, the dial command must end in a semicolon (ATD...;). If the connection is a single non-voice call, or is a dual type and the first call is not voice, the dial command must not end in a semicolon (ATD...).

Dialling from a Phonebook

An application can request dialling of a phone number in a particular location in phone memory or the phone number corresponding to a name. The commands

D > ... cause the phone to dial a number stored in phone memory. Since a phone is logically composed of three separate parts, it is possible to specify the memory in either the *M*obile *E*quipment (the phone *per se*), the *S*ubscriber *I*dentification *M*odule (the SIM card), or the *T*erminal *A*daptor (an adaptor connected to the phone). It is also possible to dial a number matched to a particular alphanumeric string.

The command must end with a semicolon (;) if the connection is a single voice call, or is a dual type call and the first call over the connection is voice. Otherwise it should not end with a semicolon.

Extended Error Reports

Some phones may support extended error reporting.

If the phone has extra information about the last unsuccessful attempt to initiate or receive a call, the command +*CEER extended error report* causes it to pass a textual string with that information to the application.

6.3.6. Accessing Phone Resources

The most basic control function is to switch the phone on or off, and ETSI does this with +*CFUN functionality*. The phone is switched on by selecting "full functionality" and switched off by selecting "minimum functionality", instead of just "on" or "off", because of difficulties of definition – a phone may have several degrees of "on" and "off". Other degrees of functionality are also provided where various RF circuits may be disabled.

Before attempting to take control of a phone, the application may wish to check that the phone is not in use. This can be done in ETSI with +*CPAS phone activity status*. This replies with

- *ready* (commands will be accepted by the phone);
- *unavailable* (commands will not be accepted by the phone);
- *unknown* (commands may or may not be accepted by the phone);
- *ringing* (the phone will accept commands, but there is an incoming call);
- *call in progress* (the phone will accept commands, but a call is in progress);
- *asleep* (the phone will not accept commands because it is "switched off").

So if a call is in progress or the phone is ringing, the application may choose to wait and try again in a few moments.

Normally (of course) a phone is operated by a user with its normal switches and keypad. However, an application providing an expanded user interface to the phone may wish to deactivate certain controls actually on the phone, to prevent accidental interference. ETSI command +*CMEC mobile equipment control mode*, in conjunction with others, does this by allowing the application to take exclusive control of the phone keypad, the phone display, or other phone indicators (such as independent lights or audio buzzers). However, dual control (both the application and the phone) is an option. This may be useful for control of a display or indicators.

- The ETSI command *+CKPD keypad control* allows the application to simulate key presses on the phone itself. This important command is useful when an application is simulating the phone keypad, or when the application needs to use some special sequence of key strokes to activate a special phone feature not covered by an individual "AT" command. + CKPD allows the application to specify the key which is pressed and the time for which it is pressed. This is because some keys change function depending on whether they are merely touched or held down. A common set of keys is defined within the command. Colon (:) is used as an escape character so that other keys may be defined – a colon followed by a byte is the IA5 58 character defined by that byte. A semicolon (;) is the escape character for entering a string.
- The phone textual display may be read and written using *+CDIS display.* + CDIS passes characters which are part of a character set (defined by *+CSCS select character set*) and addresses characters from the top left corner of the display through to the bottom right hand corner of the display. + CDIS = ? returns the number of elements in the display. + CDIS? returns the current text on display and + CDIS = writes text to the display. The character sets (obviously) define the visual symbol appropriate to a particular character. The sets supported by the command set are the GSM character set, international alphabet number 5 (IA5), PC character sets, some ISO 8859 character sets and hexadecimal representation of characters.
- The other indicators on a phone may be read and written with *+CIND indicator.* This is a "catch all" command intended to address all indicators, whether visual or audio, other than the main display. Some common indicators are defined, but it is certain that not all possibilities are covered! Naturally, the values associated with any particular indicator depend on the precise nature of that indicator. Generally, however, the value 0 is taken to mean off, and higher values indicate progressively more and more on. + CIND = ? returns a list of the indicators and the values supported by each indicator. + CIND? returns a list of the current values in the same indicator order as "+ CIND = ?". + CIND = allows a list of values to be written to the indicators. The list must be in the same indicator order as "+ CIND = ?". A null value in the "+ CIND = " list implies that the indicator is to retain its present value.

It is also necessary, at times, to disable certain features of the phone. In the ETSI command set, *+CLCK lock* allows certain types of call to be barred or enabled, can disable the phone keypad, disable the SIM, and can lock the phone to the SIM. + CLCK uses a password(s), to ensure that the locking or unlocking is authorised. Each lockable facility can have its own independent password, which is stored somewhere in the phone.

- If the phone keypad is locked, many of the normal controls will not operate. If the phone is locked to the SIM, it will not operate with any other SIM card.
- If a particular type of call is barred, the phone will refuse to either initiate or receive that type of call. All outgoing calls, outgoing international calls, outgoing

international calls except to home country, all incoming calls and incoming calls when roaming are a few types of barred calls.

Phone password(s) can be changed through the normal phone menu or in ETSI by the use of *+CPWD password.* +PWD requires use of both the old password and the new password. It is the responsibility of the application to ensure that a new password is the actual password intended by a user. +CPWD=? returns a list of the facilities that may be locked, and the maximum length of each password.

Using the Phonebook

Accessing the phonebook is likely to be a popular application, requiring the means to read, write and delete entries which may be stored in memory in various places.

The ETSI command *+CPBS phonebook storage* allows the application to deal with phonebooks in either the phone *per se* (the *M*obile *E*quipment), the SIM or the *T*erminal *A*daptor.

Once the memory has been selected, *+CPBR phonebook read* allows reading of all entries from one selected index through to a second selected index. Each entry consists of some text and a number in the format previously selected. +CPBR=? returns the maximum lengths of numbers and text fields that are supported by the phone.

+CPBW phonebook write allows a number (in the selected format) and associated text to be stored in a stated index in the selected memory. This command is also used to delete entries from the phonebook: if the command parameters are empty, all data at the index are deleted.

6.3.7. Dealing with Mobile Messages

Actual message packets (protocol data units) contain protocol information such as addresses as well as a message. Phones normally allow a user to enter these elements (including, of course, the actual characters of the message) using the phone keypad and then assemble the packet within the phone (invisibly to the user). An application could also work this way and cause the phone to assemble the PDUs, but alternatively might communicate with the phone using entire PDUs, and do assembly and disassembly in the application. The main differences between the methods are:

- mimicking a human user dictates that messages are comprised of characters. It also uses extra commands to specify other PDU fields to the phone (these are required in any case to allow applications to configure a phone so that a user can do messaging with the phone keypad).
- using entire PDUs allows messages to contain arbitrary data and uses generic commands only.

In both cases the application must "understand" the contents of a PDU and provide the necessary information.

Most other messaging commands are also generic: sending messages immediately, sending messages previously stored in the phone, receiving messages, deleting messages, specifying where messages are to be stored, and so on. So while the following ETSI commands were developed for GSM, many of them are not GSM specific.

First of all, the actual messaging service must be selected. In the ETSI set, +*CSMS select message service* selects a particular messaging service (GSM, CDMA, etc.) and also determines whether messages may originate at a mobile (*mo*), terminate at a mobile (*mt*), or support broadcast messages (*bm*). +*CPMS preferred message storage* specifies which of the phone memories are to be used for received messages.

+*CMGF message format* specifies either PDU mode or text mode. In PDU mode, messaging commands operate on a data structure that includes all the fields associated with a message, and a phone does not need to parse the message. In text mode, messaging commands operate only on the message part of the PDU (the field that is the purpose for the message). If text mode has been selected, the character set that defines the symbols represented by individual bytes is dictated by +*CSCS select character set* (described previously).

+*CNMI new message indications* is a powerful command that determines how newly received messages should be indicated. Should an indication be passed to the computer, or should the message itself be passed? If indications are requested, should they be passed immediately or buffered? An application may decide to request immediate indication of a message (or the message itself) when the computer is switched on, but need to request buffering if the computer is about to be switched off (and is unable to wake up in response to external signals).

Separate fields determine what happens to unsolicited result codes, point-to-point messages, broadcast messages, delivery reports. Another field (*buffer*) determines whether existing unsolicited result codes are flushed or discarded when the command executes.

- *bfr* determines what happens to existing unsolicited result codes when the command executes *and* the *mode* parameter has selected an option where data is passed to the application. The choice is:
  - flush to the application;
  - delete.
- *Mode* determines what happens to unsolicited result codes. The choice is:
  - buffer in the TA. If the TA is full, buffer elsewhere or discard. These buffered codes can be flushed to the application by a later use of +CNMI that sets the "buffer" parameter to flush the buffer and does *not* select this mode option.
  - discard when the computer-phone link is in use, otherwise pass them directly to the application.
  - buffer when the computer-phone link is in use, and flush them to the application when the link becomes free. Otherwise pass them directly to the application.
  - pass directly to the application using specified inband techniques.
- *mt* determines what happens to new point-to-point messages. *bm* determines what happens to new broadcast messages. In both cases, the choice is:
  - do nothing;
  - indicate new messages to the application using unsolicited result codes;
  - pass new messages to the application.
- *ds* determines what happens to delivery status reports. The choice is:

- do nothing;
- indicate new status reports to the application using unsolicited result codes.

There are two commands for sending messages. One (+CMGS *send message*) passes a message to the phone and requests immediate transmission. The other (+CMSS *message store send*) requests the transmission of a message that has previously been stored at some *index* in a memory in the phone. These messages are written to memory using the command +CMGW *message write*, and the return value is the index (position) of the message in that memory. The type of memory is that previously selected with +CPMS. The messages currently stored in a phone are listed by +CMGL *message list*. This allows the user to select the type of message to be returned. Individual stored messages may be read using +CMGR *message read* from location *index* in the memory type selected previously with +CPMS. The command returns the status (read, unread, sent, unsent) of the message as well as the message itself. Messages are deleted using the command +CMGD *message delete* which specifies a particular index.

Most messaging services are likely to include a service centre, so the command +CSCA *service centre address* allows its telephone number to be written. Finally, the commands +CSAS *save settings* and +CRES allow different profiles of settings related to short messages to be stored and recalled. This is intended for use in phones (or pagers) which deal with different message delivery services, each of which may require different options depending on the service. For example, one messaging service may relate to financial information, another to personal information, etc. A user may want personal information to be presented immediately, but want only indications of financial messages. The personal messages may be stored in the ME, and the financial messages stored in the TA.

6.3.8. Commands from Existing Standards

Some other command sets have individual commands which are useful in this context: in particular +CBC *battery charge* and +CSQ *signal quality*, both of which are self explanatory. Another set of commands from TIA is useful for directing the flow of audio in the computer and the phone. This particular command set allows various combinations of audio paths, so that transmitted audio may come from the phone microphone or the computer, received audio may go to the phone loudspeaker or the computer, and various combinations thereof.

6.4. CONCLUSIONS

This chapter provided an overview of the interaction between computers and GSM cellular phones. The procedure for setting up and making a connection was illustrated. The main features, some of the important control commands and their usefulness was mentioned. With the launch of GSM PhaseII additional multimedia computer applications can be provided using the GSM cellular phone.

REFERENCES

[1] ETSI: (postal address) F-06921 Sophia Antipolis CEDEX – FRANCE, (office address) 650 Route des Lucioles – Sophia Antipolis – Valbonne – FRANCE, (email) secretariat@etsi.fr, (telephone) +33 92 94 42 00, (fax) +33 93 65 47 16.
[2] GSM 07.05: Short Message Service (SMS) and Cell Broadcast Service (CBS).
[3] GSM 07.07: AT command set for GSM Mobile Equipment (ME).

Part 3

Third Generation Mobile Multimedia Systems

Part 3 discusses the capabilities and services to be offered by second and third generation mobile multimedia systems. The problems and difficulties with current solutions are reviewed and the requirements for future systems are suggested.

In particular, Chapter 7 discusses the design of the InfoPad, a new portable multimedia terminal test-bed. In Chapter 8, the authors describe the concept of real-time multimedia access to the Internet, which is expected to be one of the major applications for wireless multimedia.

Chapter 9 discusses the subject of wireless ATM, which is seen by many as a key enabling technology for multimedia personal communications. The chapter provides an overview of the subject, followed by a list of service requirements and an example of wireless ATM architecture. Specific issues relating to the PHY, MAC and DATA LINK layers are also described in this chapter.

The current status of wireless voice and data systems around the world is presented in Chapter 10. The wireless data capabilities of systems such as GSM, GSM 2+, GSM EDGE, Hiperlan, and UMTS are discussed and shown to meet most of the expected requirements for mobile multimedia. Global estimates of the number of wireless voice and data users are given together with future predictions for GSM and UMTS. The chapter ends with an overview of the wireless cellular and LAN activities currently underway within the EU ACTS programme.

Finally, Part 3 ends with a chapter dedicated to the problems of security in a mobile multimedia network.

7

InfoPad: A Portable Multimedia Terminal

Brian C. Richards and Robert W. Brodersen

7.1. INTRODUCTION

Making multimedia access available on portable terminals, removed from the desk-top, does not mean that the capabilities of a desktop machine have to be available. By taking advantage of modern high speed networks and wireless technology, a wealth of computing and information services can be provided by a portable term-inal with flexible I/O capabilities and minimal compute power. This philosophy has

Insights into Mobile Multimedia Communication
ISBN 0-12-140310-6

been embodied in a new portable multimedia terminal, InfoPad. InfoPad has been designed to take advantage of high-speed wireless networking capabilities in order to reduce the amount of local computation required in the terminal.

The InfoPad system, as illustrated in Figure 7.1, consists of network-based servers that communicate through a wireless link to a variety of I/O devices on portable InfoPad terminals. Computing resources on the network provide access to global data and specialised compute servers, such as speech and handwriting recognizers. The network also provides the applications that access these services, and the underlying servers that provide terminal functions such as the windowing system [1,2].

7.2. THE INFOPAD SYSTEM MODEL

The InfoPad terminal is fundamentally an I/O device with several features to support multimedia. The terminal does not contain a user-programmable CPU or any mass storage, instead relying on network servers to provide all of the pad's functionality. This includes multimedia I/O servers for graphics, video, pen and audio, and applications which build upon these servers to provide a friendly user interface [3,4].

7.2.1. From Wireless Data Transfer to Wireless I/O

By moving all computation to the network, the wireless link carries only fault-tolerant I/O information, rather than critical, error-sensitive data. If the wireless network were required to carry error-free data using a retransmission protocol such as TCP/IP, bit error rates on the order of 10^{-5} to 10^{-2} would lead to frequent retransmission of packets, introducing undesirable latency into the system. For the playback of audio and video information, this leads to significant jitter and hence the requirement for large memory requirements to counteract these effects. For pen-

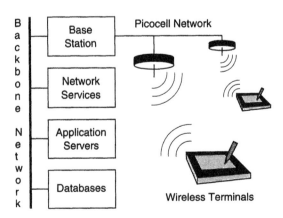

Figure 7.1 The InfoPad system: wired computation and wireless I/O.

based applications that draw "ink" on the screen, latency can be on the order of seconds.

The InfoPad terminal, shown in Figure 7.2, avoids many of the latency and jitter problems by selecting fault tolerant input and output data stream formats, and minimising the amount of data that requires error-free reception. This has a beneficial side effect of minimising the memory buffering requirements for video, graphics and audio applications. Also, the operation of the terminal will degrade gracefully as bit error rates increase.

7.3. SYSTEM DESIGN FOR LOW POWER

Design goals for the InfoPad terminal included making the physical unit compact, lightweight and portable, with a long battery life – extending to several hours. This resulted in the use of several low power design strategies to minimise the total power consumption in the terminal.

7.3.1. Terminal and Base Station Architecture

InfoPad's predecessor, the IPgraphics terminal, contained most of its data manipulation functions on a single controller chip. In the InfoPad pad, one of the design

Figure 7.2 The InfoPad portable wireless terminal.

goals was to make the unit flexible so that new features can be added without a complete redesign. This led to the definition of a low-voltage bus architecture.

The InfoPad terminal uses a conventional 8-bit data and 8-bit address bus architecture running at 1 MHz. Given the target radio bit rates of 1–2 Mb/s, this slow speed is more than adequate to handle the uplink and downlink data. Also, this bus operates at 1.5 V to minimise the power consumption [5,6]. Custom ASICs provide voltage level conversion where required to commercial 3.3 V or 5 V systems.

Most of the data transferred on the low power bus is organised into packet datagrams of variable length which are sent between "channels", defined by one pair of 8-bit data and 8-bit control registers per channel. The datagram begins with a start of packet indicator sent to the desired channel control register, and also provides a 6-bit data type field. This is followed by one or more bytes of data written to the data register, after which an end of packet command is written to the control register, along with possible error status information. The 8-bit bus supports several bus master devices, with a request/grant priority scheme. Datagrams sent over the bus can be interleaved, and in general, a given bus master is restricted to accessing the bus no more than once every other cycle. This prevents one device from blocking data transfers between other devices.

A given device on the data bus has one or more channel register pairs, to support multiple data streams and packet types. At start up, input devices which send data packets are programmed to define which channel of a device on the low power bus is to receive a given type of data. As an example, pen input data could be sent to the controller CPU for processing or directly to the radio transmitter interface to send to the base station.

Both the InfoPad terminal and the current implementation of the base station unit, which connects to the wired network, share the same bus architecture. This greatly simplifies the system design. Figure 7.3 shows both the base station and terminal block diagrams, each with several functions built around the low power

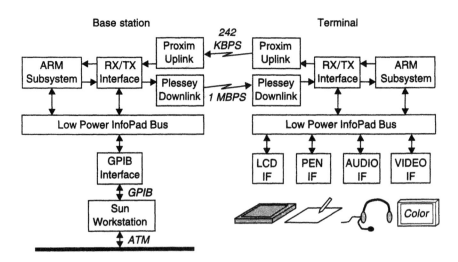

Figure 7.3 The architecture of the InfoPad base station and portable terminal.

bus. Both the terminal and the base station contain a 20 MHz ARM 60 CPU to implement system control functions. A custom interface IC is used to interface the CPU with the low speed bus, providing an asynchronous interface to separate the high memory bus speeds from the slower I/O circuitry.

In addition to the ARM CPU, both the base station and the InfoPad terminal have two radio data interface ASICs, one for packetising data for transmission and one for forwarding received data to the low power bus. Currently, to support experimental protocol research, a Xilinx PLD is used to interface 8-bit transmit and receive data streams with the uplink and downlink radios from Proxim and Plessey, respectively.

In the base station, a PLD-based interface is used to communicate with a SUN workstation through a GPIB interface. The workstation serves as a gateway to the wired network, and can provide some of the remote computing to implement underlying terminal multimedia functions. In the future, the workstation could be replaced by a custom ATM or Ethernet interface with built-in radios. This would reduce the functionality needed in the base station and help to minimise the cost of the wired infrastructure required to support the InfoPad terminals.

7.3.2. InfoPad Custom VLSI ASICs

Custom ASICs provide the basic multimedia I/O functions for the InfoPad terminal, including black and white graphics, colour video, audio, and pen input. Each of the functions shares a common bus interface circuit to connect to the low power bus.

Most of the custom ICs have mixed power supplies, typically operating the core circuitry at 1.5 V, and driving the pads with either a 1.5 V or a 5 V logic levels. In addition, some I/O pads have separate supplies so that pads can either be completely turned off to save power, or operated at other supply voltages.

7.3.3. ARM Subsystem

The ARM 60 CPU and support circuitry is an exact copy of the ARM PIE development system from Plessey. The system has 512K of SRAM, which is sufficient for the limited control functions that are implemented in the CPU. This system consumes roughly 800 mW at 5 V under typical operating conditions, and can benefit in the future from a drop to lower voltage operation with a more highly integrated version of the ARM CPU [7].

A simple event-based operating system has been developed to support the real-time and control tasks for the InfoPad terminal. Functions include controlling data transmission sequences, selecting which type of packet should be sent to the radios at a given time. The CPU can also handle management of buffers for tasks such as synchronising video and audio data streams.

For the case when critical control information must have guaranteed transmission between the InfoPad terminal and the base station, the ARM CPU implements a simple retransmission protocol. This can be used to exchange statistics such as power levels and bit error rates to the network support services which control pad power-up, authentication, and mobility issues. Also, if pen or speech data must be error-free, this data can be forwarded to the ARM CPU, from which it would be transmitted to the base station.

One more application of the CPU is to initialise and test the custom circuitry on both the terminal and the base station. Many of the ASICs have diagnostic registers to verify basic functionality. Also, the audio, pen and radio interfaces must be programmed to specify data formats and to specify the destination on the low power bus for their respective data packets.

7.4. CUSTOM I/O SUBSYSTEMS AND ASICs

Most of the input and output functions are implemented using ASICs, to take advantage of low power component libraries, and to increase the parallelism of the system to allow for slower clock speeds. The InfoPad terminal contains 14 1.5 V ASICs, each with a typical power consumption in the 0.1–10 mW range. Custom circuits have been designed for all devices attached to the low power bus, ranging from the ARM interface to the individual pen, audio, video and text/graphics interfaces.

7.4.1. ARM Interface

The ARM CPU interface ASIC serves primarily to isolate the 20 MHz CPU circuitry from the slow 1 MHz data bus. When writing a datagram, the ARM can send a sequence of 8, 16, 24 or 32 bit writes, which will be disassembled and queued in a FIFO to synchronise the data with the 8-bit bus. The ARM reads a datagram with 32-bit reads, which are padded if the number of bytes received is not a multiple of four. The ARM can also perform direct reads and writes to the low power bus, to test for device status or to initialise registers.

The ARM interface also serves as an interrupt controller, provides address decoding for eight I/O devices, and generates several system clocks from the 20 MHz ARM interface. These signals are delivered at 1.5 V to accommodate the low power bus, and reduce the amount of external logic required in the pad.

The ARM interface ASIC also doubles as a general-purpose low power bus interface. Data can be pushed into a FIFO for transmission to the low power bus, and data written to the chip can be popped from a second FIFO. This chip is used in the current base station in conjunction with an XILINX XC4008 to implement an GPIB interface to a SUN workstation.

7.4.2. RF Interface

The radio interface consists of a pair of ASICs which manage data packetisation for transmission and process the incoming radio packets, formatting them for the 8-bit bus. The transmitter circuit collects datagrams from the low power bus, queuing the data in a 32K byte static RAM. Since datagrams can be interleaved from several sources, the transmitter must buffer any incoming packets to assemble complete datagrams before transmission. Once a complete packet is stored in memory, an interrupt is sent to the ARM processor to inform the control software that one or more packets is ready to send. The ARM will then request that the transmitter send a specific packet. This allows the controlling software on both the pad and the base station to determine which type of data is to be sent, and when to send it.

When a packet of data is ready to be sent, the transmitter ASIC presends a programmable header to the radio packet data. A unique terminal identifier byte can be sent, followed by an optional sequence number. The data packet type follows, containing the 6-bit datagram type identifier which is included in the start of packet command on the low power bus. Then, depending on the data type, a two byte data length field follows, and a CRC for the header. The payload data is then sent, followed by a second CRC for optional payload protection.

The receiver circuit is more straightforward, stripping off this header, and forwarding the data to the low power bus in the order received. The packet *type* is forwarded as part of the start of packet status word, and the CRC error status is calculated and included with the end of packet status word. To determine where to send a given packet, the *type* is looked up in memory to select a particular channel address.

The radio interface circuitry is designed to accommodate a variety of radios, including those listed in Table 7.1. The InfoPad terminal uses a 625 kb/s frequency hop radio from Plessey for the downlink, and a slower 242 kb/s radio from Proxim for the lower bandwidth uplink. The terminal will operate the Plessey system primarily as a receiver, consuming about 400 mW when active. Similarly, the Proxim radio will generally be used only in transmitting mode, consuming 825 mW when active. Due to radio operation in different bands, full duplex operation can be attained. Several other radio choices are available. A DECT chip set is being evaluated, using commercially available parts. Infra-red solutions are also being evaluated, which may prove suitable for some locations.

A major research focus closely tied to the InfoPad project involves developing a monolithic CMDA radio that provides up to 2 MHz bandwidth to each user, and is capable of providing 100 MHz of aggregate bandwidth [8]. Most of the circuitry runs at 1.5 V, and is currently capable of operating above the 1.1 GHz RF band using 1.2 μm CMOS technology, with target power consumption in the range of 100–150 mW. In the future, advanced technologies will allow these radios to be implemented for higher RF bands.

Table 7.1 Radio alternatives for the wireless network.

Radio	Scheme	Data bandwidth	RF band
Plessey	Freq. hop	1 MHz aggregate	2.4–2.5 GHz
Proxim	Spread/TD MA	242 or 121 kHz aggregate	902–928 MHz
DECT	FM/TDMA	1 MHz aggregate	1.88–1.90 GHz
Custom "SAM"	CDMA	100 MHz aggregate	>1.1 GHz
IR	TDMA	100 MHz aggregate	IR

7.4.3. Video Interface

A full motion video interface is provided which comprises five 1.5 V ASICs [9]. Video compression is implemented using three 8-bit vector quantisation (VQ) frame buffers, one each for the Y, I and Q image components. A video timing controller and bus interface chip processes the incoming data, forwarding coded image data to the three frame buffer chips, each of which contains a video rate VQ look-up table to decode in real time. The frame buffers output Y, I and Q components which are fed to a custom colour space translator, which converts the pixel data to RGB format, and includes three low power video DACs.

The power consumption for the five-chip video circuit is in the tens of milliwatts, which is dwarfed by the power drawn for the LCD panel. Without the backlight, the display consumes 900 mW, and the backlight consumes an additional 1.7 W. Options for reducing this power include using a head-mounted display or reflective colour LCD technologies, which are still under development.

7.4.4. Text/Graphics Interface

A 640- by 480-pixel black and white LCD is used to display text and graphics information, using a customised X11R6 server running on a remote platform. The graphics primitives draw data to a 1-bit deep frame buffer, writing 32 pixels of data at a time, organised as a row of consecutive pixels. A single LCD controller chip provides four raw data formats for drawing data on the screen. The basic drawing entity is the same 32 pixel wide by 1 pixel high block, corresponding to a long word of data. The first three formats, shown in Figure 7.4, draw rows, columns and rectangular regions of these blocks on the LCD screen, regardless of potential data errors. The fourth format, called a protected block, only draws data on the screen if the data passes a CRC test. This latter mode allows the screen to be refreshed slowly, to clean up possible errors drawn with the "error tolerant" modes.

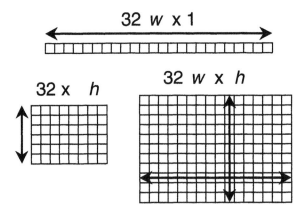

Figure 7.4 Horizontal, vertical and block-shaped regions of pixels, based on a 32 × 1 pixel long word.

The frame buffer is implemented using four custom 1.5 V SRAM ASICs which provide 32-bit memory access, and include RAM bank decoding circuitry. By using a wider data word, the SRAM can be run slower to reduce power consumption. Also, the regularity of common text and graphics operations can result in reduced transitions in the data read from the frame buffer.

The power consumed by the text/graphics ASICs is less than 10 W at 1.5 V, again insignificant compared to the display power requirements. The SHARP LCD panel used in InfoPad consumes 410 mW, with an optional 1.7 W required for the backlight. One option for reducing this power is to use LCD panels with memory, which eliminate the constant refresh power consumption.

7.4.5. Audio and Pen Interface

Audio I/O and pen input functions are combined on a single chip, due to their simplicity and shared signals. The audio data is delivered through an 8-bit, 8 kHz mu-law codec, for a monaural input and output channel. The pen interface connects to an off-the-shelf tablet from Logitec, to provide wireless pen input capabilities. The tablet draws 50 mW, and uses a powered pen. The pen and audio devices can be shut down under CPU control to save additional power.

7.4.6. Other Circuitry

Aside from the custom ICs and the ARM 60 subsystem, over 2.5 W are consumed by the remainder of the circuitry on the InfoPad terminal. Part of this is attributable to the Xilinx PLD circuit, along with related EPROMs. These can be replaced in the future with custom circuits, but are used at present to support protocol development research.

Much of the remaining circuitry on the InfoPad terminal is dedicated to 7 switching power supplies in the terminal. Although each of these supplies can approach 90% efficiency, the off-the-shelf components were not optimised for the power requirements in the InfoPad, resulting in 25% of the measured power being consumed by the supplies. The efficiency of these supplies can be increased substantially by customising the supplies for the targeted power consumption [10].

7.5. PACKAGING THE INFOPAD TERMINAL

In parallel with the low-power requirements, one of the design goals for the InfoPad terminal was to produce a compact, portable unit that can be used to evaluate different air interfaces and to experiment with real world applications. To encourage the development of portable multimedia applications, the package should have good ergonomics and not be awkward to use or clumsy.

To support compact electronics and package design, a research effort is in progress to build a framework which integrates electrical and mechanical CAD tools

[11]. Information from the PCB design tools and part libraries can be carried over to the mechanical modelling core to accelerate the design process. Design information has been transferred between mechanical and electrical CAD tools to evaluate component placement, and to help optimise the design of the circuitry and the packaging.

Once packages have been described, a parallel research effort involves using feature-based package descriptions to automatically drive a programmable HAAS mill to quickly generate package prototypes. By using a rich set of manufacturable features including curves, rapid prototyping of custom packages can be achieved. Given initial circuitry size estimates, custom packaging turn-around times of two days have been demonstrated, for the case shown in Figure 7.5(a). This gives both the mechanical engineer and the electrical engineer increased opportunities to experiment with design options.

As an alternative to milling plastic, the same mill equipment has been used to create moulds from aluminium, which can be used to create many injection moulded plastic cases. The aluminium mould can be used to rapidly create more than 100 cases, with little more engineering effort and labour than is needed to create a single plastic case. A commercial foundry was selected to produce 50 cases, shown in Figure 7.5(b), given the milled and polished mould.

7.6. CONCLUSIONS

The InfoPad terminal, as the most visible portion of the InfoPad system, is being designed to account for multimedia capabilities, ergonomics, size, weight, low price, and long battery life. The terminal must be able to tolerate a noisy wireless environ-

(a)   (b)

Figure 7.5 (a) The first version of the InfoPad terminal in 1995 used a milled plastic case. (b) This was replaced by an injection-moulded case created using an automatically manufactured mould.

ment in order to offer high-speed access to remote databases and computation servers. Ultimately, the InfoPad terminal and related support services must be carefully combined to make a multimedia device that the end user will want to make a part of everyday life.

ACKNOWLEDGEMENTS

The InfoPad terminal design is but a small part of a large system design effort on the part of many faculty, student and staff members, led by Professor Robert W. Brodersen. The terminal is being designed by the PadGroup team including Brian Richards, Trevor Pering, Syd Reader, Sue Mellers, Ken Lutz, Roger Doering, Ian O'Donnell, Shankar Narayanaswamy and Tom Truman, building upon the efforts of Tom Burd and Anantha Chandrakasan from the IPgraphics terminal. The InfoPad Project is supported by ARPA and several industrial corporations.

REFERENCES

[1] Le, M. T., Burghardt, F., Seshan, S. and Rabaey, J., InfoNet: the networking infrastructure of InfoPad. In *Proceedings of COMPCON* 1995.

[2] Burstein, A., Long, A. C., Narayanaswamy, S., Han, R. and Brodersen, R. W., The InfoPad user interface. In *Proceedings of COMPCON* 1995.

[3] Narayanaswamy, S. *et al.*, Application and network support for InfoPad. *IEEE Personal Communications Magazine*, March 1996.

[4] Long, A. C., A prototype user interface for a mobile multimedia terminal. *Proceedings of the 1995 Computer Human Interface Conference*, May 1995.

[5] Chandrakasan, A. P., Sheng, S. and Brodersen, R. W., Low-power CMOS digital design. *IEEE Journal of Solid-State Circuits*, 27 (4), 473–484, 1992.

[6] Burd, T., Low-power CMOS library design methodology. Masters thesis, ERL U.C., Berkeley, June 1994.

[7] Burd, T. D. and Brodersen, R. W., Energy efficient CMOS microprocessor design. In *Proceedings of the 28th Annual HICSS Conference*, January, 288–297, 1995.

[8] Sheng, S., Brodersen, R. W. *et al.*, A monolithic CMOS radio system for wideband CDMA communications. In *Proceedings of the Wireless '94*, Calgary, Canada, July, 1994.

[9] Chandrakasan, A. P., Burstein, A. and Brodersen, R. W., A low-power chipset for portable multimedia applications. In *Proceedings of the IEEE International Solid State Circuits Conference*, February, 1994.

[10] Stratakos, A., Sanders, S. and Brodersen, R., A low-voltage CMOS DC-DC converter for a portable low-powered battery-operated system. In *Proceedings of the IEEE Power Electronics Specialists Conference*, 1994.

[11] Wang, F., Wright, P. and Richards, B., A multidisciplinary concurrent design environment for consumer electronic product design. *ISPE Concurrent Engineering: Research and Applications*, 4 (4), 1996.

8

Mobile Multimedia Access for the Internet

V. J. Hardman and S. Hailes

8.1. INTRODUCTION

The use of desktop multimedia conferencing is becoming increasingly important to business and academia, and is expected to be attractive to home users. There are two

Insights into Mobile Multimedia Communication
ISBN 0-12-140310-6

major reasons behind this trend: the perceived need already exists, since multimedia conferencing was being used before the Internet arrived, and the emergence of low-cost multimedia-capable PCs and a suitable low-cost global network has enabled the perceived need to be realised for many users.

The Mbone is a multicast overlay on some of the high-speed parts of the Internet, and it provides multi-way communication in a way that scales to many participants. New Mbone/Internet multimedia applications are evolving from multimedia conferencing and the World Wide Web: distance learning, telepresence and video-on-demand. These new applications have a set of more demanding requirements (multi-way higher-quality audio, multi-way higher-quality video, interface integration and user mobility) which are driving advances in the underlying systems: audio, video and text. Many multimedia piloting applications use freely available Mbone tools for communication: multi-way audio at a range of different qualities (RAT), a video tool (vic), and shared text (nte) and whiteboard facilities (wb).

Since desktop multimedia conferencing is becoming a required business tool, there will be a demand for the service when users cannot be at their desks. It is probable that the multimedia tools in both wired and wireless networks will need to be very similar, and that a single system must be capable of interacting with both fixed and mobile terminals. Unfortunately, there is a major problem in achieving this sort of transparency, since mobile hosts lack both computational power and, more significantly, communication bandwidth. Various approaches have been suggested to reduce the mismatch of multimedia traffic requirements and mobile network capabilities, such as degrading the information sent to mobile hosts, or employing high levels of source compression. Multimedia traffic is often sensitive to delay and jitter, and this means that delay penalties incurred when hosts cross cell boundaries must be kept very low. A fundamental premise in mobile systems design is that cell size must be reduced in order to reduce the amount of sharing and so to increase individual share of the bandwidth. The consequence of this is that the number of hand-offs increases.

As can be seen, there is a potential market for mobile multimedia services, but the technical problems which must be overcome to realise this vision are considerable. In this chapter, we discuss the problems and constraints which realistic implementations of mobile multimedia applications face and then outline an approach to their provision, which involves three major elements. The first are transcoding gateways at the boundary between wired and wireless networks which hide differing network characteristics from applications and provide seamless mobile access to the Internet/Mbone. The second is the use of hierarchical coding and packet-loss protection mechanisms which are specifically tailored to deal with media packet loss in the Internet, but which we believe will also have beneficial effects in the wireless environment. The final element is a mobile multicast protocol, derived from existing approaches, but fixed to allow for host mobility.

8.2. MULTIMEDIA APPLICATIONS

Networked multimedia applications can be between two or more people or between people and systems. People-to-people applications include both interactive conferencing (multimedia conferencing, distance learning and telepresence) and conference broadcasts. People-to-systems applications can be further classified as interactive (such as WWW access), or distribution (such as video-on-demand) [1].

Often a single application requires a combination of the above categories. Conference multicasts often include question and answer sessions at the end of the presentation. Distance learning and multimedia conferencing often also require access to stored material provided by a multimedia server in the form of still and moving images. The technology for mobile WWW access is already in the market place, though location-sensitive WWW access is still a research topic [2]. Other networked multimedia applications do not yet support user mobility, because of the large demands multimedia can place on bandwidth

8.2.1. People-to-People Applications

Multimedia Conferencing

Since the first audio cast of IETF in 1992, multimedia conferencing on the Internet/ Mbone [3] has been used on a regular basis by both academics and industry. Multimedia conferencing is generally taken to consist of interactive real-time audio, video, and shared text. Separate programs or "tools" are used to provide the individual media in order to retain the flexibility of using a combination of the possible media (audio, video and shared text).

The MERCI project [4,5] pilots multimedia conferencing facilities (Figure 8.1) throughout Europe (and beyond), with sites in England (UCL), Sweden, Norway, Germany, France, Belgium, and Canada. The tools commonly used are RAT (robust audio tool) [6], or vat [7], vic (video) [8], and wb (whiteboard) [9] or nte (shared text) [10]. Session information can be obtained from the session directory tool (sdr) [11], which identifies multimedia conferencing traffic on the Mbone. There is potentially a very large market for mobile access to multimedia conferences as users often find themselves located somewhere where Mbone access is not available.

Conference multicasts (such as Globecom [16]) use the same media as interactive conferences: audio, video and possibly shared text. The application for most of the time is a one-to-many application, with all the traffic coming from a single sender to a group of receivers, although the usual question and answer sessions at the end of a broadcast mean that multimedia conferencing facilities are also required. Mobile access to streamed media applications is likely to be attractive for the same reasons as put forward for interactive multimedia conferencing.

Distance Learning and Telepresence

The old model of education is changing from a one-off intensive period in a life to continuing training throughout a career, lifelong learning [12]. Many higher education establishments are rising to meet this need by offering distance learning activities.

The ReLaTe project pilots remote language teaching over SuperJANET between University College London, and Exeter University [13]. Using real language teachers and real students, small group lessons in French and Portuguese have been run

Figure 8.1 MERCI Multimedia Conferencing, using audio, video and shared whiteboard
tools.

during term time over the last 2 years [14]. Distance learning requires the same
components as multimedia conferencing: audio, video, and shared text. However,
when used in small group situations, where increased usability is required, a number
of enhancements are needed to the basic tools used: hands-free operation, lip synchronisation and an integrated interface.

Remote surgery demonstrations have also been piloted by projects with partners
at UCL; the MERCI project ran a three-site demonstration to medical students in
novel surgery techniques. Multimedia mobile access to distance learning technology
will prove extremely attractive to busy professionals, especially if they can attend a
lesson at their own convenience, rather than having to attend a regular weekly
meeting at a particular location.

The visualisation of complex data, such as WWW or library material, also
benefits from allowing users in the virtual world to communicate in real time.
Consequently, virtual world piloting applications, such as COVEN (COllaborative
Virtual ENvironments) [15], use multicast audio and video facilities; when users in
the virtual world get close together, they automatically join the same multicast group
so they can communicate. Mobile access to virtual worlds is likely to be attractive to
the entertainment industry.

8.2.2. People-to-System Applications

WWW Access and Media Distribution

WWW access involves a range of clients accessing material from a much smaller number of servers, and generally consists of still images and text. It has however now evolved to include stored audio and video. Real-time media are transferred as files to clients, and played back using an appropriate player. Mobile access for these real-time media does not require any extra special provision than a packet service on top of the mobile link.

Video-on-Demand/Streamed Media

Video-on-demand applications consist of audio and video, and possibly shared text, that is retrieved from a remote server. The application is characterised by most of the traffic being from the server to the clients, although interactivity in the form of controlling the playback etc. of the media is available. Popular players include RealAudio [17], RealVideo [18] and Netshow [19], where lots of buffering is used at the receiver to improve performance. A VoD service might be offered over the wide area from a single server, but network-related performance can be improved by employing suitably located proxy servers to locally cater for concentrations of clients.

8.3. NETWORK TECHNOLOGY

The Internet is a shared packet network that has spread to most parts of the globe. The World Wide Web has become the most popular use of the Internet, and real-time interactive and stored material access are expected to be even more attractive. Real-time interactive communication requires a different underlying model to the client/server one used in the WWW; a peer-to-peer model is needed. Real-time media transfer protocols designed for stored material access are also not suitable for inter-active communication.

The ability to communicate via groups is a natural requirement for real-time interactive communication. Circuit-switched networks find it difficult to support scaleable multi-way communication. Chained multi-point conference units are used in circuit-switched telephone networks to provide limited multi-way facilities. The Mbone [3] is a multicast backbone over some of the high-speed parts of the Internet, which provides multi-way communication that *scales*; conferences with very large numbers of participants are possible. A variety of Mbone routing protocols exist, which provide different levels of scalability, and incur different amounts of overhead. In the UK, the Mbone is moving from a pilot to a service, and bandwidth will be better provided as a result.

8.3.1. Internet

The Internet is a shared packet network, where information is sent in independent packets or datagrams, and datagrams from many users are multiplexed and individually routed at nodes (connectionless data transfer). Unlike many broadband networks specifically designed to carry multimedia traffic (where the packet size is fixed), Internet datagrams are of variable length. The minimum viable size is governed by consideration of header overhead to payload, and the maximum size by the limits set in routers; greater than half a kilobit in a datagram results in fragmentation before transmission by routers. The Internet Protocol (IP) allows traffic to be sent across many different network technologies by using the same network protocol (Figure 8.2).

Congestion occurs in the Internet when there are too many packets in the net-work, and can result in either transient loss patterns or longer term loss rates [20]. The critical factor is not the size of packets (number of bits), but rather the number of packets per second arriving at a router, since they must all be individually routed, based on the address each carries in its header. Best-effort transmission means that datagrams can be lost, and that delay is unpredictable. This means that packets losses are always present, and that jitter occurs in the received flow of datagrams. The transport protocol needs to be able to provide a reliable service on top of the "best-effort" IP transmission when required to do so, and applications can choose either a reliable service, Transmission Control Protocol (TCP) [21], or a lightweight unreliable service, User Datagram Protocol (UDP).

TCP achieves reliability by requesting retransmission when datagrams have been lost, and by reducing the throughput rate in response to packet loss within the flow. Information about packet loss at the receiver is communicated to the sender within ACK messages, and senders can often back off to almost 0 bit/s. This practice is very important for a shared network like the Internet, because it means that congested backbone links are fairly shared between all contending users.

Real-time media, such as audio and video, cannot benefit from retransmis-sions in interactive applications. The media have a maximum delay associated with them, and retransmitted traffic arrives too late to be played back. Since TCP backs off, it is not suitable for real-time traffic; audio and video have a minimum bandwidth required to transmit something meaningful, and speech in particular has a base rate (for each compression algorithm family) below which the source signal is distorted too much. Consequently, real-time media do not use TCP, but *UDP* (UDP provides a very lightweight service, and neither re-transmits lost data nor backs off in the presence of lost datagrams within a flow) (see Figure 8.2). Media that use UDP as the underlying transport protocol will suffer from packet loss.

Real-time media tools are themselves directly responsible for providing appli-cation layer framing (ALF) [22] for the transport protocol used. The Real-Time Protocol (RTP) [23] has been developed within the IETF and provides facilities such as sequence numbers and timestamps which are required by real-time media.

8.3.2. The Mbone

The Mbone is a multicast backbone over some of the high speed parts of the Internet, that provides multi-way communication that scales. Multicasting [24] is the ability of a sender to send a single data packet to a number of receivers in

Conference control	Audio	Video	Shared tools	WWW	E mail
RSVP	RTP and RTCP			HTMP	SMTP
	UDP			TCP/IP	
IP					
Integrated services forwarding					

Figure 8.2 Internet Protocol Stack.

such a way that the packet only traverses each network link once, or as closely to once as is possible (see Figure 8.3). The multicast routing algorithm constructs a distribution tree from the sender to all interested receivers; a different source distribution tree is constructed for each active source. In this way, the communication scales to many participants, since only those parts of the network that need to carry traffic do so.

In meshed networks, routers must use multicast protocols to ensure at most one traversal of any link for each multicast packet. Receivers announce interest in receiving packets from a particular multicast group, and multicast routers conspire to deliver datagrams to them. Datagrams are sent to a multicast address, which identifies the packets as belonging to that group, and is the means by which multicast capable routers know Mbone from Internet traffic. This level of indirection between sender and receivers is important, because it means that hosts can join and leave the multicast group at will (dynamic group membership).

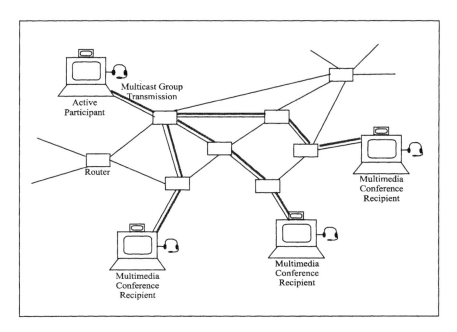

Figure 8.3 A per-source distribution tree between the sender and receivers.

Internet Group Management Protocol (IGMP) [25] operates over the local sub-network between hosts and multicast router(s). IGMP enables a multicast router to monitor group presence on the local subnet; when a router receives a multicast packet from a local host, it then knows which interfaces correspond to group members. The current Mbone (and Internet) delivery model is "best-effort", which may result in packets being lost. Resource reservation in multicast networks can be engineered using the ReSerVation Protocol (RSVP) [26]. Receivers set up guaranteed links back to the packet's source, since the assumption is that more datagrams will be traversing the same paths.

8.4. MULTIMEDIA SOURCE AND NETWORK REQUIREMENTS

Different applications need different media quality; interactive conferencing requires far lower video quality than video on demand. Different qualities consume different amounts of bandwidth; the newest compression algorithms produce variable bit rates for a desired quality [27], and audio techniques such as silence suppression result in even more bursty traffic [28]. These considerations are further complicated in a multi-way environment, where the amount of bandwidth required is difficult to determine and subject to change.

A shared packet network (and especially a "best-effort" delivery model) will always suffer from congestive packet loss, and lightweight protocols (UDP/IP) that are suitable for real-time traffic do not provide any repair facilities. A technique commonly used to provide rapid take up of new technology is that of a software-only strategy on general-purpose computing facilities. Multimedia, and particularly inter-active multimedia, place stringent requirements on the host – both in terms of computing resources, and in terms of deadline scheduling [6].

8.4.1. Multimedia Quality and Bandwidth Requirements

Audio Quality

Audio compression algorithms are commonly designed to produce a specific quality. There is a variety of different qualities possible, ranging from DAT and CD quality audio (sampling frequencies 48 and 44.1 kHz), to FM radio (32 kHz), AM radio (16 kHz) and Toll or telephone quality (8 kHz). Real-time interactive audio is usually transmitted using telephone quality speech codecs, which produce a bit rate of 64 down to 8 kbit/s, depending upon the complexity of the algorithm used. Stored audio is often of high quality (state-of-the-art is 64 kbit/s per channel), although the requirement is for a range of qualities to be available; stored CDs may form part of a browsing facility, but only if the audio is delivered in this instance at much lower quality.

Audio Protocols and Bandwidth Requirements

Audio files are transmitted using TCP/IP, which includes retransmission facilities. Interactive and streamed audio are transmitted in media independent datagrams using the lightweight protocol UDP. This means that reliability requirements can be applied on an end-to-end basis according to the individual needs of the application. Interactive applications use the Real-Time Protocol (RTP) [23], which provides facilities such as timestamps and sequence numbers, as well as session information via the Real-Time Control Protocol (RTCP) [23]. Delay tolerant applications, such as video-on-demand, may use RTP for media delivery and a new protocol (Real-Time Streaming Protocol – RTSP) for stream control [29].

Packet rate is related to congestion, (and therefore to packet loss), and this means that for a real-time flow a trade off exists between packet rate and delay. The delay equal to the packet size must be added to the end-to-end delay before jitter considerations are taken into account [30]. Audio samples are collected into packets of 20, 40, 80 and even 160 ms worth of audio; the size chosen reflects the user's current assessment of the packet loss conditions on the network. The actual size in bytes of each time length of audio changes according to the compression algorithm used, but since IP packets have a large header (IP, UDP, RTP headers) to carry globally unique source and destination addresses, there is a self-regulatory minimum size in bytes for audio packets.

The picture is further complicated by the behaviour of individual audio compression algorithms during packet loss; higher compression algorithms use higher amounts of implied state in the codec to achieve the compression advantage, state which must be replenished after loss. Consequently, there is a complex relationship between the compression algorithm used, the packet size, and the perception of packet loss.

Video Quality

Video compression algorithms are designed to produce a specific frame resolution, but sometimes the frame size and the frame rate can be altered as required. Real-time video for interactive conferencing is commonly transmitted using H.261 (and more recently H.263) at QCIF size (or possibly CIF). The frame rate can be adjusted from 1 frame a second up to 30 frames per second. For about 5 frames QCIF per second, this results in 128 kbit/s. The Mbone variant of H.261 does not transmit intra-frames, but uses inter-frames only, as this results in better performance after packet loss [8]. Higher quality video transmission may be achieved using MPEG compression, or a range of other similar algorithms. Video is transmitted over UDP using RTP [23], and a single video frame may be spread across a number of datagrams.

Multi-way Video Bandwidth Requirements

Video is normally transmitted in a multi-way environment in an *ad hoc* manner; new participants transmit video as they feel like it. For small group sessions involving up to three or four parties, this is mostly not a problem for reasonably low frame rates, but generating more traffic than this is likely to produce packet loss in the UK, as there is a bit rate limit in the routers of 512 kbit/s.

8.4.2. Loss and Delay Limits in Real-time Multimedia

The Perception of Packet Loss in Audio and Video

Jitter in the Mbone can be quite high in some portions of the network (100–200 ms) [20]. For packet speech systems a delay is inserted into the receive buffer at each end and the round trip delay is twice this value. The limit on round trip delay for interactive conversations is 400–600 ms in the absence of echoes [31], and this limit is often approached by some parts of the Mbone. Delays across a local network are however very small (10 ms) [20].

The effect of packet loss on audio is to cause a gap in the output, since the relative timing relationship between the source and sink needs to be maintained. The human perception of the loss depends upon both the length (and the position in the audio material), and frequency of the loss [30]. Packet sizes commonly used in Internet audio produce significant impairments, and for speech this results in a large reduction in intelligibility [30]. Packet loss repair techniques that operate purely at the receiver fail for many Internet packet sizes, because packet sizes are such that the basic underlying assumption of all these techniques (that speech characteristics have not changed appreciably) no longer holds true for many of the losses. Combined source and channel techniques, such as interleaving, rely on spreading the components of a packet at the transmitter across a number of packets. Consequently the packet loss results in a small number of much smaller gaps over which packet loss repair techniques can be successfully used. The disadvantage associated with this technique is that it necessarily incurs an additional extra delay, which often cannot be tolerated in an interactive conversation. The effect of packet loss on video depends on the compression and packetisation methods used. For example, errors are more pronounced when inter-coded frames are present.

The Implications of Packet Loss and Delay for Different Multimedia Applications

The most delay tolerant multimedia application is WWW access. Material is stored and retrieved using HTTP on top of TCP/IP, and packet loss is repaired by re-transmissions. Access to real-time media occurs after files have been downloaded, and so the delay between requesting the file and beginning playback can be varied substantially according to the congestion levels on the network.

Conference broadcasts (one-to-many) are also fairly delay tolerant. Recent techniques developed for multicast transmission include a variation on the scaleable reliable multicast protocol developed for shared white-board communication [9]. Based upon the concept of shared state (all members of the multicast group retain all previous transmissions), retransmissions are used to repair lost packets, and hosts local to the host requesting retransmission are arranged to be the first ones to retransmit. The major problem with this scheme is that care must be taken not to request retransmission from all other hosts in the multicast group or from a significant number at the same time. Video-on-demand applications are less delay tolerant than conference broadcasts, since material can be selected for playback in any order, and access time to material must be minimised. Interpolation techniques provide good error control whilst minimising delay [32].

Interactive real-time media used in multimedia conferences or streamed media used in video-on-demand applications suffer more from network effects like packet loss and delay. Packet loss occurring in interactive environments is usually eased by incorporating redundancy in the media [30]. This method does not incur the extra delay penalty for receivers experiencing very low loss rates that all other schemes result in since it suits the heterogenous nature of Mbone communication. Packet loss occurring in streamed media can be repaired more successfully by using some of the techniques developed for repairing bursts of bit errors in mobile communications [33].

8.5. PROVIDING MULTIMEDIA OVER THE INTERNET

Mbone tool developers produce freely available software-only systems that can run on a wide variety of general-purpose computing platforms. This strategy has enabled a critical mass of users to be established. Most workstation operating systems are supported, together with Linux, FreeBSD, Windows95 and WindowsNT for personal computers. Generally all that is required on top of a standard PC is a camera and video card, and a headset and audio card.

8.5.1. Robust Audio Tool (RAT)

The Robust Audio Tool (RAT) [6] is a second generation audio tool for the Mbone. It is intended to provide multi-way communication for interactive applications. It is not aimed at the point-to-point transfer of buffered audio provided by RealAudio [17]. RAT provides packet loss robustness in the form of redundancy [30], but also improved workstation scheduling performance [34] and improved hands-free operation for small conferences [6]. RAT has recently been enhanced to provide higher-quality audio for interactive conferences in the form of 7 kHz audio compression capabilities, and higher-quality audio codecs are planned.

Interactivity plays a large role in determining which method of packet loss repair can be used. Many schemes involve repairing errors from information in a sequence of packets, such as FEC or Interleaving. If delay can be tolerated, then retransmission schemes are also of use, but need care to implement in a multicast environment. Which method is suitable for a given application depends upon the maximum delay that can be tolerated by the application. Highly interactive conferencing requires minimum delay, and consequently, media dependent redundancy is used, since the delay increase is at most 1 packets worth. RAT uses low bandwidth redundancy in the form of LPC, or other low bit rate version of the speech, such as GSM [30]. Packet loss robustness is not, however, the panacea of packet loss in the Internet/Mbone. If congestion is persistent, then applications should try to reduce the offered load onto the network, preferably in response to packet loss information from receivers [6].

8.5.2. Mbone Video

Most Mbone applications that use video communication use a video tool called vic
[8], from Lawrence Berkeley labs in the US. Vic offers a variety of compression
algorithms, such as H.261, CellB and nv, at many different image sizes (QCIF,
CIF and SCIF). The frame rate and resolution can be varied, although a common
bit rate and frame rate is 2 frames/s at 128 kbit/s, using a QCIF. There is currently
no robustness to packet loss in any Mbone video tools. Recent research has reported
the use of a scaleable video codec, based on wavelets, which uses receiver-driven
layered multicast to accommodate a wide variety of users in the heterogenous
Mbone environment [35].

8.5.3. Shared Whiteboard

A shared whiteboard facility, such as wb from Lawrence Berkeley Labs in the US,
enables users to type text, draw screen annotations and to load in still (postscript)
images. The Mbone protocol used is not RTP, but rather scaleable reliable multicast
[9], where retransmissions for lost packets are from the nearest members of the
multicast group, and where requests for retransmission are produced in a scaleable
way (if the packet has been lost near the source, multiple receivers from the potentially infinite group may want a retransmission. This could potentially lead to an
explosion of traffic requesting the retransmission). A shared text only tool has been
developed at UCL, which uses redundancy to protect typed text from packet loss
(each change in a line results in the whole line being retransmitted) [10].

8.6. NETWORK COMPONENT CANDIDATES FOR MOBILE MULTIMEDIA

Providing Mobile access for the Mbone involves not only identifying a suitable
candidate technology for the mobile link, but also identifying the impact of host
mobility on the fixed part of the network. In particular, an impact on the Mbone
routing algorithm is expected, since distribution trees are calculated on a per source
basis. A mobile host will effectively appear as a number of separate hosts, who join
and leave the multicast group as the mobile host migrates around the edges of the
Mbone.

8.6.1. Mobile/Cordless Candidates for Mbone Multi-way Access

There is a distinction between local area wireless environments (where there is adequate bandwidth to support multimedia) and wide area systems (where, although
mobile host connection to the Internet has been initially researched, the bandwidth is
so limited that multimedia is impractical at present). Developments to cellular telephony systems are expected soon, which will allow wide-area packet-based communication at data rates suitable for multimedia. Of the commercial local area wireless
providers, three suppliers hold more than two thirds of the market: Aironet, Proxim

and Lucent. Their products operate in the 2.4 GHz band (IEEE 802.11) at rates of up to 2 Mbit/s. Research is proceeding into much higher rate (185 Mbit/s) wireless LANs [36]. In this section we outline current and future European standards, since these have the largest global following. Much of the material in this section was prepared with the aid of reference [37].

European Technologies: DECT

The Digital European Cordless Standard (DECT) evolved in the mid 1980s, under the auspices of the European Conference of Telecommunication Manufacturers (ECTEL). Their initial aim was to extend the standards used for cordless communication to support multiple simultaneous users and data rates far greater than 32 kbit/s speech (actually, up to 552 kbit/s unidirectional). DECT was designed to allow future interworking with GSM-based systems (GSM phase 2 +).

European Technologies: GSM

GSM is deployed in over 100 countries at the original 900 MHz frequency and now at 1800 MHz (European PCN networks) and at 1900 MHz (US PCS networks). There are approximately 10 million subscribers, with a projected figure of at least 100 million in 120 countries by the year 2000 [38].

GSM has had two major phases (1 and 2), with a third due into service by 1998 (phase 2 +). This third phase is important in the provision of multimedia services, because it includes DECT access to GSM, PMR/PAMR, Virtual Private Networks (VPNs), and High Speed Circuit Switched Data (HSCSD) and the General Packet Radio Service (GPRS).

The current GSM data rate of 9.6 kbit/s is inadequate for multimedia. GSM phase 2 + [39,40] proposes two different mechanisms: High Speed Circuit Switched Data (HSCSD) and General Packet Radio Service (GPRS).

HSCSD is based on the aggregation of $n \times 9.6$ kbit/s independent sub-channels (where $n = 1,...,8$). This will typically be used to provide a data rate of 64 kbit/s, with the potential to connect to ISDN networks. Although this requires changes to the link layer, the advantage of supplying bandwidth in 64 kbit/s chunks lies in the ease of testing, and the compatibility of the existing wired PSTN networks.

GPRS is specifically designed to take account of the highly bursty nature of most data applications, and multimedia applications in particular. Charging will be by transmission volume, not by connection time, which is a much more natural and acceptable mechanism when traffic is highly bursty. The main objective of GPRS is to provide access to a range of data networks, from X.25 and CLNP to TCP/IP. The key problem to be solved in GPRS is routing, which will use encapsulation and a home-agent-like node (similar to mobile-IP[1]). Furthermore, the wired GPRS backbone network will use IPv6 (a connectionless protocol), and will appear as a subnet, i.e. the GRPS gateway will appear to be a standard IP router. However, at the

---

[1]Note that GPRS will not simply be an implementation of mobile-IP, though it will be similar, because of the need to support multiple protocols.

gateway, the GRPS system will encapsulate messages in its own protocol. Overall, GPRS data rates will be up to 115 kbit/s in the wide area.

European Technologies: BRAN

The Broadband Radio Access Networks (BRAN) project [41], aims to define standards for networks and systems having peak rates of at least 25 Mbit/s (due to complete in 2005). BRAN will support both circuit-oriented and packet-oriented protocols (such as ATM and IP). Its main aim is to produce specifications for high-quality fixed radio access networks, and to exploit HIPERLAN [42]. BRAN's aims include the support of ISDN-like, Internet and video-on-demand (VoD) services.

UMTS

The Universal Mobile Telecommunications Standard (UMTS) defines worldwide personal mobile communications, which will have the quality of today's wired access networks. The aims of UMTS (to be introduced in 2005) are ambitious and include the following [38]:

- flexible bearer and bandwidth-on-demand services up to 2 Mbit/s (local access) and at least 144 kbit/s (wide area coverage);
- mixed traffic types and usage-based charging for mobile multimedia applications;
- wideband wireless local access enhancements to fixed networks.

Other Techniques

There are several other approaches, such as wireless ATM [43,44,45,46], which even has prototype versions in existence (e.g. RATM by the Olivetti Research Laboratory). Standardisation has been slow, but ETSI RES 10 sub-technical committee have it as an active work item, with the aim of standardising ATM networks with a peak bit rate of 24 Mbit/s, based most probably on a microcellular structure.

While some uncertainties remain, it is likely that GSM (or an evolution of GSM) will be a suitable candidate for Mbone mobile access.

8.6.2. Mbone Routing Algorithms

Multicasting has been used since 1992, either via distance vector multicast routing protocol (DVMRP) [47,48] or via Multicast extensions for Open Shortest Path First (MOSPF) protocol [49]. More recently, other multicast routing protocols have been designed (and partially deployed) which improve the efficiency and scalability of the routing protocol: Core Based Trees (CBT) multicast protocol [50,51]) and Protocol Independent Multicast (PIM) protocol [52,53]. Both CBT and PIM also aim to be independent of the technology underlying the IP network (and native routing protocols). Neither CBT nor PIM take any account of host mobility. Different multicast routing algorithms use different techniques for establishing a distribution tree. The following material is taken from [33]. The algorithms can be classified into source-based tree algorithms and shared tree algorithms:

- Source-based trees (SBT) are rooted at the sender, and the tree is composed of the shortest paths between the sender and each of the receivers in the multicast group. Multicast algorithms that use SBTs require that a separate tree be constructed for each active source in each multicast group.
- Centre- or Core-Based Trees (CBT) use a single shared tree for all senders in a group. The tree is a shortest path tree rooted at one of a set of nodes called cores. Algorithms utilising CBTs construct a single tree for each group, regardless of the number of senders.

Two protocols that implement shared trees are CBT [51] and the PIM Sparse Mode (with packet delivery along shared trees (PIM-ShT)). Protocols that implement source-based trees are PIM Sparse Mode (with packet delivery along source-based trees (PIM-SBT)), PIM Dense Mode (PIM Dense) [52] and DVMRP [47].

While PIM is currently the Mbone routing algorithm of choice, the prospect of rapid user migration between cells indicates that a better choice would be the more stable tree offered by CBT.

8.7. MOBILE MULTIMEDIA

Wide area mobile environments have restricted bandwidths (< 115 kbit/s) and bit error rates of as high as 10^{-2}. Internet multimedia systems are based around the use of connectionless IP. A connectionless approach is feasible because the wide area loss rates are moderate, and because there is adequate bandwidth to carry the information. If we are to have seamless mobile access for the Mbone/Internet, then the problems of high error rates and low bandwidth must be addressed.

8.7.1. Coping with Low Bandwidth

The problems facing multimedia coding algorithms in mobile environments are considerable: there are large delays arising from the basic radio access methods, but there is also low bandwidth availability, and very high error rates. This means that the same coding schemes cannot be used for both wired and wireless access. There are two possible approaches to providing seamless multicast mobile access to the Mbone:

Transcoders

Transcoders are application entities probably placed at the boundary between the wired and wireless network to translate from wired to wireless coding schemes. However, the precise placement of such transcoders is an open research question particularly when considering multicast trees [54].

Real-time multicast mobile access is easily provided by using multicast IP, and particularly RTP over UDP. The use of IP datagrams which have quite large headers means that either a relatively large minimum payload needs to be used (which lead to increased delay for interactive real-time communication), or that some form of header compression (before transmission over the multicast links) must be investigated.

Hierarchical Compression Algorithms

Hierarchical compression algorithms provide a number of output channels, where there is a base channel, and subsequent channels serve to improve the quality of the base. At each node in the multicast tree, only those channels which can be successfully transmitted without loss are forwarded. The approach planned for available bandwidth variations that occur in the Mbone is to use congestion control to manage variations in loss characteristics both across the multicast group and over time [6]. The channels are broadcast out over different multicast groups, and receivers determine whether or not they receive the full quality by joining more or less of the multicast groups. The use of congestion control is not however applicable to mobile network access, because information that the sender needs regarding the spread of loss rates across the multicast group will not be available early enough to be of any use (mobile cells suitable for multimedia may be very small, and hence the loss rates being seen over the mobile link will vary very quickly).

These considerations mean that some form of compression management should be accomplished at the mobile support station (MSS), along with header compaction [55]. Compression management could initially be transcoding, but in the longer term it will use hierarchical compression schemes, and explicit selection of a sub-set of available multicast groups. This means that all (or most) of the groups will be forwarded over wired links from the sender, but perhaps only the base channel will be transmitted down the sub-tree from the core towards the wireless links. The reduced quality will be either because there is insufficient bandwidth, or because the user is unwilling to pay for better quality.

8.7.2. Coping with a High BER

The mobile channel produces large numbers of bit errors because of interference, fading, and signalling disruptions. Various protocols have been developed to counteract this error level, but these introduce extra delay into the system and/or reduce the bandwidth available for source traffic. The following section describes the reasons for bit error rates in GSM as the expected access technology (figures are taken from [39]).

Real-time media tools have a certain amount of packet loss protection in order to cope with best-effort transmission over the Internet/Mbone. These packet loss protection mechanisms are ideal for use with lower-level BER improvements provided within GSM.

GSM

GSM operates in two different modes: transparent mode (which incorporates forward error correction), and non-transparent (which also includes a layer 2 protocol between the mobile and network protocols).

In *transparent mode*, error protection is via channel coding. There are no retransmissions and the overall delay through the coder is approximately 150 ms. However, since the BER can be as high as 10^{-2}, this service is only useful where there are higher-level error correction facilities (e.g. X.25), or where the services are asynchronous, and operate at low enough bit rates that FEC is viable.

In *non-transparent mode*, a GSM-specific radio link protocol (RLP) is used with the channel coding scheme; erroneous frames can be retransmitted. A one-way coding delay is approximately 200 ms (round-trip delay of 400 ms). When used in tandem connections, delays of up to about 1 s are possible, though unpredictable, and the BER is less than 10^{-6}.

Since GSM2 +: HSCSD aggregates GSM channels, it has the same transparent/non-transparent error controls, but slightly higher delays, because of the need for synchronisation of different streams at the receiver, and extra signalling (non-transparent mode). No accurate figures are available for GPRS, and these figures are taken from [56]. Since the BER requirement is less than 10^{-9}, interleaving delays (at a depth of 4) are approximately 20 ms, and channel access delays in the range 12–22 ms, for a successful situation. Worst case delays could be several seconds. The delay introduced by the physical layer ARQ depends on the number of retransmissions, each introducing an extra delay of 12–22 ms.

Packet Loss Protection for Real-time Mobile Access

As with packet loss protection in wired network, packet loss protection for the mobile portion of the network is dependent upon the level of interactivity required by the application. Consequently, interactive applications should use redundancy, and less interactive applications interleaving. Extra packet loss protection, by using either more layers of redundancy or interleaving, can be added by the transcoding gateway at the mobile support station.

8.7.3. Mobile Multicast Scheme and Mbone Routing Algorithms

This section identifies what is required of a mobile multicast protocol, and proposes a scheme which satisfies these requirements.[2]

Requirements of a Mobile Protocol for a Multicast Environment

The primary requirement of any mobile protocol is to support host mobility. A host is defined to be mobile if it can maintain its network connections as it migrates around the local or wide area network. This requirement can be given two interpretations: operational and performance transparency.

Operational transparency means that users do not need to perform any special actions due to host movement. It can only be achieved by detecting host movement and performing actions that ensure service continuation at the mobile host's new location. A protocol that needs hosts to be reset every time they migrate cannot be

---

[2]This work was originally done as part of an advanced MSc by Al-Sabi, Lee, Corona-Olmedo, Papadimitriou and Sanchez-Vivar [37].

considered mobile but merely portable. Operational transparency by itself does not necessarily result in a useful service, since it makes no quality of service guarantees. If a mobile host moves to a new location, the services supplied by the protocol should continue with similar performance as before.

Performance transparency is therefore another requirement. Methods of ensuring that a mobile protocol has performance transparency include optimum routing of packets to and from MHs, efficient and robust migration procedures and efficient use of network resources, such as transmission bandwidth and processing. Moreover, because the bandwidth of the wireless links is a scarce resource, protocol-originated overhead in the transmissions over these links must be limited.

A mobile protocol must also consider cost, host impact and infrastructure requirements. Practical considerations also mean that a protocol must be back-wards-compatible with existing protocols that have been developed for the Mbone. It is for these reasons that a modification to the existing IGMP local network Mbone protocol is proposed.

The Internet Group Management Protocol

The Internet Group Management Protocol [25] is used by LAN hosts to inform locally connected multicast routers of their host group membership. IGMP messages are encapsulated in IP datagrams, and sent to the router(s). A host interested in any sources in a multicast group, previously caused the multicast flows from all sources to be forwarded onto the local network. The latest version of IGMP (version 3) uses source-specific messages, so that multicast group sources can be selectively received. This means that group-source membership information (gathered via IGMP) can be used by the multicast routing protocol to discard datagrams from unwanted sources, and to prune the associated multicast tree(s); it is possible to achieve a significant reduction in network traffic if all the flows cannot be forwarded onto the local network.

Multicast routers send Host Membership Query messages to all hosts on the local network, to discover which are interested in receiving flows from existing multicast groups. An IGMP V3 host that receives a query, responds with Group-Source Report(s). IGMP V3 routers periodically send Host Membership Queries, to refresh knowledge of group-source membership on the local network. If no reports are received for a group (or group-source) after two queries, the remotely-originated flow(s) for that group (or group-source) are pruned. When a host *joins* a group (or group-source), it immediately transmits a Group-Source Report without waiting for a query, and repeats it after a short delay to ensure the report was received. When a host *leaves* a group, it sends a Group-Source Leave Message. When leave messages are received by a multicast router, it issues a Host Membership Query to the group (or group-source) that is being left. If no membership reports are received after a time-out, then the group (or group-source) is pruned.

A Mobile Multicast Protocol (MMP) Suitable for Multimedia

The solution presented here proposes modifications to the Internet Group Membership Protocol (IGMP), and a network layer handoff mechanism that transfers state information during cell switching by a mobile host. The scheme proposed also assumes that the multicast routing at the static network is performed by the CBT protocol, since other Mbone routing algorithms will suffer from excessive tree recalculation when hosts are mobile.

MMP assumes a GSM-like cellular radio coverage system, where access to the fixed network is provided by a mobile support station (MSS). Each MSS covers a single cell, and communication between the MSS and the set of mobile hosts in the cell is assumed to be via bi-directional unicast. Each MSS also includes a CBT router function. The MMP that operates between an MSS and the MHs in the cell is a simplified version of IGMP, where the Host Membership Queries have been abandoned and each MH explicitly reports to the MSS if it wishes to join or leave a multicast group, using Group Reports or Leave Messages. No information about multicast sources is included in the IGMP messages, as the CBT protocol precludes the use of this source-specific information.

A Multicast Support Data Base (MSDB) (located at each MSS) associates MHs in the cell with multicast groups. MMP updates this data-base according to the IGMP messages it receives from the local MHs. According to the contents of the MSDB, MMP communicates with the CBT router function at the MSS to request a join or a pruning of a multicast group. The MSDB is also consulted by the MMP function upon the receipt of a multicast packet by the MSS. The information retrieved from the MSDB identifies MHs in the cell that belong to the multicast group specified by an arriving datagram. Using this information, the MSS delivers the datagrams to the MHs via multiple unicasts.

The MMP changes the IGMP functionality at the MSS by omitting querying messages and introducing the MSDB. IGMP behaviour at the MH end is also slightly different to the standard IGMP V3, since the MHs do not have to respond to membership queries (since they never receive any), and cannot select or drop specific sources within a multicast group. This second restriction also applies to fixed hosts.

The modifications to the IGMP function are most evident during handoff. When an MH moves to a new cell, the MMP at the original MSS sends a message to the target MSS, informing it about the groups the moving MH has subscribed to. The target MSS uses this information to join any new groups. The handoff message enables the target MSS to have all the multicast streams available before the arriving MH is ready to receive traffic in its new cell (or at a very short interval afterwards).

When the MH moves out of its current cell boundaries, the MMP function at the original MSS is notified that the MH is migrating to a new MSS's cell by a service primitive assumed to be provided by the underlying Data-Link layer. There are a few assumed service primitives that are critical to the MMP: notifications that an MH can receive data, an MH is moving to another cell, an MH is disconnected for an unknown reason, or has already registered in another cell.

During handoff, the time interval after an MH has switched to the frequency of a new cell and before it can receive data packets is called the open interval gap. The open interval gap is mainly composed of the time needed for channel assignment, and time for synchronisation at the new cell. During the gap, the target MSS may receive multicast packets for one or more of the groups the migrating MH is interested in. The MSS may receive these packets either as a result of other MHs present in the cell, or because the MMP handoff message has triggered a CBT join. However, the migrating MH misses all packets transmitted by the target MSS during the open interval gap.

To restrict packet loss, the MSS has new buffering for each migrating MH. When an MSS receives an MMP handoff message for a migrating MH, the MMP dynamically allocates buffer space for the host and stores any multicast messages for the MH. When the MH can receive data in the new cell, the buffer contents are sent to the MH and the buffer space de-allocated.

8.8. CONCLUSIONS

Multimedia conferencing has evolved towards more demanding applications, such as distance learning and telepresence, which are placing more stringent demands upon the media: higher-quality audio and video, and user mobility. The underlying media engines are being developed to meet these new requirements, and a range of research in a number of areas is in progress, error protection mechanisms to deal with high levels of packet loss, and congestion control which uses hierarchical coding algorithms spread across a number of multicast groups.

Multicast mobile access is difficult to accomplish, because of low bandwidth availability and high error rates. It is not certain which current mobile technology will persist to provide packet-based access, but it is likely to be something based on GSM: either HSCSD or GPRS. This chapter has identified the probable error rates that such technology will be able to offer to higher layers, and subsequently delay budgets.

Current Mbone routing strategies do not take mobile access into account, and it is expected that user mobility will create problems for routing tree stability. There are three aspects to designing mobile multimedia access for the Mbone, coping with low bandwidth, coping with high error rates, and a multicast mobile access protocol. Low bandwidth will be catered for by special provisioning at the mobile support station. This is expected to initially be a transcoding function, but later MSS-driven use of hierarchical compression algorithms, so that the multicast tree does not carry excess traffic further than is required. High error rates will be catered for by a combination of error protection mechanisms at the data-link layer, and Mbone-suitable application-specific error protection mechanisms, of which we believe redundancy to be the most likely candidate for interactive applications. A full treatment of the topic would require consideration of issues such as QoS negotiation, charging mechanisms and battery power availability, but the scope of this chapter does not permit this. Work on these areas is in progress at UCL, and we believe that our solutions will integrate well with the proposals in this chapter.

REFERENCES

[1] Fluckiger, F., *Understanding Networked Multimedia*. Prentice Hall, Englewood Cliffs, NJ, 1995.
[2] Geoffrey, M., Voelker, M. and Bershad, B. N., Mobisaic: An information system for a mobile wireless computing environment. In Imielinski, T. and Korth, H. (eds), *Mobile Computing*. Kluwer Academic Pubilshers, New York, 1996.

[3] Macedonia, M. R. and Brutzman, D. P., Mbone provides audio and video over the Internet. *IEEE Computer*, April, 1994, 30–36, 1994.

[4] Handley, M., Kirstein, P. and Sasse, M. A., Multimedia integrated conferencing for European Researchers (MICE): piloting activities and the conference management and multiplexing centre. *Computer Networks and ISDN Systems*, **26** (3), 275–290.

[5] Kirstein, P. T., Sasse, M. A. and Handley M.J., Recent activities in the MICE conferencing project. In *Proceedings of INET95*, Honalulu, Hawaii, 1995. http://info.isoc.org/ HMP/PAPER/166/ps/paper.ps.

[6] Hardman, V. J., Sasse, M. A. and Kouvelas, I., Successful multi-party audio communication over the Internet. *Communications of the ACM*, **41** (5), 74–80, 1998.

[7] Jacobson, V., VAT manual pages. Lawrence Berkeley Laboratory (LBL), February, 1992. Also http://www-nrg.ee.lbl.gov/vat/.

[8] McCanne, S. and Jacobsen, V., vic: a flexible framework for packet video. In *Proceedings of ACM Multimedia '95* San Francisco, CA, pp. 511–522, November 1995.

[9] Floyd, S., Jacobson, V., McCanne, S., Liu, C. and Zhang, L., A reliable multicast framework for light-weight sessions and application level framing. *IEEE/ACM Transaction on Networking*, 1995.

[10] Mandley, M. and Crowcroft, J., The network text editor (nte). In *Proceedings of ACM SIGCOMM*, 1997.

[11] Handley, M., On scaleable Internet multimedia conferencing systems. University of London PhD thesis, November 1997.

[12] Dearing, Sir R. *et al.*, 'Report of the National Committee into Higher Education' available as http://www.leeds.ac.uk/educol/ncihe/

[13] Buckett, J., Campbell, I., Watson, T. J., Sasse, M. A., Hardman, V. J. and Watson, A., ReLaTe: Remote Language Teaching over SuperJANET. *Proceedings of UKERNA 95*, Networkshop, March 1995.

[14] Watson, A. and Sasse, M. A., Evaluating audio and video quality in multimedia conferencing systems. *Interacting with Computers*, **8**, 255–275.

[15] Collaborative virtual environments (COVEN). ACTS project AC040, 4th Framework.

[16] Perkins, C. and Crowcroft, J., Real-time audio and video transmission of IEEE Globecom 96 over the Internet. *IEEE Communications Magazine*, 30–33, April 1997.

[17] RealAudio. http://www.real.com/

[18] RealVideo. http://www.real.com/

[19] Netshow. http://www.microsoft.com/netshow/

[20] Handley, M., An examination of Mbone Performance. USC/ISI Research Report ISI/ RR-97-40.

[21] Postel, J., Transmission control protocol. STD 7, RFC 793, September 1981.

[22] Clark, D. and Tennenhouse, D., Architectural considerations for a new generation of protocols. In *Proceedings of ACM SIGCOMM*, September 1990.

[23] RTP: a transport protocol for real-time applications. Audio-Video Transport WG, rfc 1889, 2/1/98. RFCs and Internet drafts can be found at http://www.ietf.org/

[24] Deering, S., Host extensions for IP multicasting. *Request for comments rfc 1112*, Internet Engineering Task Force, August 1989.

[25] Cain, B., Deering, S. and Thyagarajan, A., Internet Group Management Protocol Version 3 (IGMP v.3), August 1995. draft-cain-igmp-00.txt.

[26] Zhang, L., Deering, S., Estrin, D., Shenker, S. and Zappala, D., RSVP: A new resource reservation protocol. *IEEE Net*, **7** (5), 8–18, 1993.

[27] Vetterli, M. and Kovacevic, J., *Wavelets and Sub-band Coding*. Prentice Hall, Englewood Cliffs, NJ, 1995.

[28] Brady, P. T., A technique for investigating the on-off patterns of speech. *Bell Systems Technical Journal*, **XLIV** (1), 1–22, 1965.

[29] Schulzrinne, H., Rao, A. and Lanphier, R., Real time streaming protocol (RTSP). IETF INternet Draft, 21 November 1997. draft-ietf-mmusic-rtsp-06.txt/ps.

[30] Hardman, V. J., Sasse, M. A., Watson, A. and Handley, M., Reliable audio for use over the Internet. *Proceedings of INET95*, Honolulu, Oahu, Hawaii, September 1995.

[31] Brady, P. T., Effects of transmission delay on conversational behaviour on echo-free telephone circuits. *Bell System Technical Journal*, January, 115–134, 1971.

[32] Jayant, N. S. and Christensen, S. W., Effects of packet losses in waveform coded speech and improvements due to odd–even sample interpolation procedure. *IEEE Transactions on Communications*, **COM-29** (2), 101–109, 1981.

[33] Al-Sabi, M., Lee, S.-C., Corona-Olmedo, R., Papadimitriou, G. and Sanchez-Vivar, E., Multicasting in networks with mobile hosts. Advanced MSc report, available as http://www.cs.ucl.ac.uk/research/mobile/archives/ucl/mobile/dcnds-multicast-96/

[34] Kouvelas, I. and Hardman, V. J., Overcoming workstation scheduling problems in a real-time audio tool. *Proceedings of the USENIX Annual Technical Conference*, Anaheim, CA, January 6–10 1997.

[35] McCanne, S., Jacobsen, V. and Vetterli, M., Receiver-driven layered multicast. In *Proceedings of ACM SIGCOMM*, 1996.

[36] Skellern, D. J. *et al.*, A high-speed wireless LAN. *IEEE Micro* January/February, 1997.

[37] Al-Sabi, M., Lee, S.-C., Corona-Olmedo, R., Papadimitriou, G. and Sanchez-Vivar, E., Multicasting in networks with mobile hosts. Technical report for an Advanced MSc in Data Communications, Department of CS, UCL, September 1996.

[38] Berg, G., Ericsson 'GSM evolution towards UMTS', available as http://www.ericsson.se/systems/gsm/umts.html, 1998.

[39] Nieweglowski, J. and Leskinen, T., Nokia 'Video in mobile networks', available as http://www.club.nokia.com/mobile_office/library/video.html, June 1996.

[40] Hämäläinen, J. and Kari, H. H., Nokia 'Packet radio service for the GSM network', available as http://www.forum.nokia.com/nf/magazine/papers/gprsbook.html

[41] Broadband Radio Access Networks (BRAN) homepage, accessible via www.etsi.fr

[42] ETSI, HIPERLAN system definition. Technical Report ETR 133.

[43] ATM Forum Technical Committee, Private network–network specification interface v1.0 (PNNI 1.0). March 1996.

[44] ATM Forum Technical Committee, ATM user–network interface (UNI) Signalling Specification v. 4.0, January 1996.

[45] Proceedings Wireless ATM Networking Workshop, Columbia University, June 1996.

[46] Mikkonen, J. Wireless ATM overview, available as http://www.forum.nokia.com/nf/magazine/papers/wireless.html

[47] Waitzman, D., Partridge, C. and Deering, S., Distance vector multicast routing protocol. *RFC 1075*, November 1988. RFCs and Internet drafts can be found at http://www.ietf.org/

[48] Pusateri, T., Distance vector multicast routing protocol, draft-ietf-idmr-dvmrp-v3-06, March 1998, Expires: September 1998. RFCs and Internet drafts can be found at http://www.ietf.org/

[49] Moy, J., Multicast extensions to OSPF, *RFC 1584*, March 1994.

[50] Ballardie, A., Core based trees (CBT version 2) multicast routing: protocol specification. *RFC 2189*, September 1997.

[51] Ballardie, A., Core based trees (CBT) multicast routing architecture. *RFC 2201*, September 1997

[52] Estrin, D., Farinacci, D., Helmy, A., Thaler, D., Deering, S., Handley, M., Jacobson, V., Liu, C., Sharma, P. and Wei, L., Protocol independent multicast-sparse mode (PIM-SM): protocol specification. *RFC 2117*, June 1997.

[53] Estrin, D., Farinacci, D., Helmy, A., Thaler, D., Deering, S., Handley, M., Jacobson, V., Liu, C., Sharma, P. and Wei, L., Protocol independent multicast-sparse mode (PIM-SM): protocol specification, draft-ietf-idmr-pim-sm-specv2-00.txt, September 1997.

[54] Hardman, V. J., Sasse, M. A. and Hailes, S., Multi-way effective distance learning at low-cost. EPSRC Project #Gl, 1996.

[55] Degermark, M., Engan, M., Nordgren, B. and Pink, S., Low-loss TCP/IP header compression for wireless networks. *Proceedings Mobicom*, 1996.

[56] Hämäläinen, J., Jokinen, H., Honkasalo, Z. and Fehlmann, R., Multi-slot packet radio air interface to TDMA systems – variable rate reservation access (VRRA). In *Proceedings of Personal, Indoor and Mobile Radio Communications (PIMRC)*, 366–371, 1995.

9

Wireless ATM: An Enabling Technology For Multimedia Personal Communication

D. Raychaudhuri

9.1. INTRODUCTION

In this chapter, we provide an architectural overview of "wireless ATM", which has recently been proposed [1–3] as a key enabling technology for personal multimedia systems. The proposed wireless ATM system is intended to provide mobile personal terminals with high-speed cell-relay services qualitatively similar to those provided by future B-ISDN/ATM, thereby facilitating seamless support of network-based multimedia applications on both fixed and portable terminals. Such a general multimedia computing and communication ("C&C") scenario is shown schematically in Figure 9.1.

Insights into Mobile Multimedia Communication
ISBN 0-12-140310-6

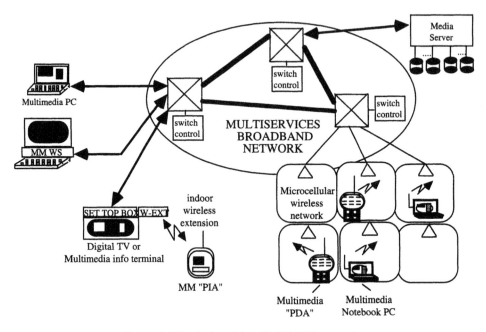

Figure 9.1 Typical multimedia "C&C" scenario.

In this environment, a variety of multimedia computing devices (both fixed and portable) communicate with each other and/or access information from remote media servers. In such systems, network-based multimedia services are central to the application, and should be equally accessible to both static and mobile users. It is recognized that in order to maximize utility and convenience, it is essential to design a seamless networking and software architecture which incorporates both wired and wireless portions of the system. Although there may be a number of viable system design approaches, a general solution should provide flexible multimedia integration with quality-of-service (QoS) control all through the network protocol and computer software stacks. ATM network technology provides a useful basis for such a system since it was designed for service integration and explicit quality-of-service support. Our conceptual view of such a multimedia system architecture [4] based on ATM is shown in Figure 9.2.

Note that the system architecture is centered around a high-speed network with unified wired + wireless services via standard ATM network and signaling/control layers. Special medium access, data link and network control requirements of the radio channel are accommodated in new wireless-specific layers denoted as WATM MAC, DLC and control, just above the wireless physical layer. The figure also shows that a typical multimedia scenario will involve three types of terminal/computer equipment connected to the network: media server, (fixed) computer/workstation and (portable) personal terminal. A new software stack consisting of a multimedia-capable OS, ATM application programming interface (API), multimedia/mobile middleware and interpretive script-based application program is proposed for the personal terminal. Detailed discussion of each hardware, protocol and software layer shown in Figure

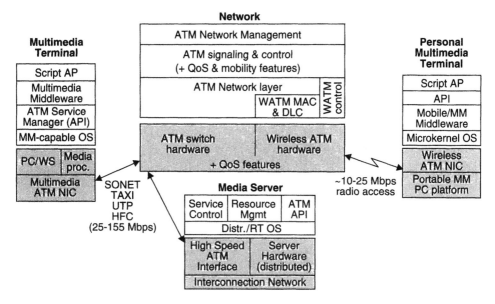

Figure 9.2 Conceptual view of proposed multimedia system architecture.

9.2 is beyond the scope of this chapter. In what follows, we focus on the wireless ATM network necessary to support multimedia communication at the personal terminal.

9.2. WIRELESS ATM OVERVIEW

An architectural view and design approach for wireless ATM can be found in [3]. As discussed in that paper, wireless ATM may be viewed as a general solution for next-generation personal communication networks capable of supporting multimedia. The technology is considered to be applicable to both microcellular personal communication services (PCS) [5] and wireless LAN scenarios. Of course, acceptance of a specific "third-generation" wireless system for the public PCS scenario involves a relatively lengthy process of technical evaluation, standardization and spectrum allocation. Nevertheless, it is important to accelerate development of such an integrated services wireless network for future PCS in order to avoid a potentially serious mismatch between wireless and wireline network capabilities as broadband services become more generally available.

In the following subsections, we review the basic concepts of wireless ATM and briefly discuss important design issues at each functional layer.

9.2.1. Service Requirements

Service classes under consideration for ATM are ABR (available bit-rate), VBR (variable bit-rate) and CBR (constant bit-rate). These three transport services are intended to support a range of voice, video and data applications with applicable

Table 9.1 Typical service requirements for wireless ATM

Traffic class	Application	Bit-rate range	QOS requirement
CBR constant bit-rate	• voice • digital TV	32 kbps–2 Mbps	• isochronous • low cell loss • low delay jitter
VBR variable bit-rate	• videoconference • multimedia comm.	32 kbps–2 Mbps (avg.) –6 Mbps	• low delay jitter • statistical mux • moderate cell loss
ABR available bit-rate	• interactive data • client–server	(peak) 1–10 Mbps	• low cell loss • higher delay OK • high burst rate

QoS controls. Wireless ATM should be designed to work within the same framework, although quantitative limits on service parameters and QoS will be different due to radio channel constraints. Table 9.1 summarizes typical targets for wireless ATM service capabilities.

9.2.2. Wireless ATM Architecture

The basic approach in wireless ATM is to use standard ATM cells for network level functions, while a wireless header/trailer is added on the radio link for wireless-channel-specific protocol layers (medium access control, data link control and wireless network control) as shown in Figure 9.3.

ATM services with QoS are provided on an end-to-end basis, and standard ATM signaling functions are terminated at the mobile unit. This approach minimizes

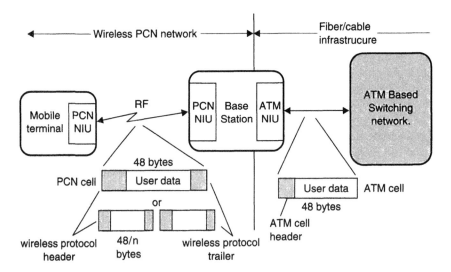

Figure 9.3 Interface between wireless and fixed ATM network.

complexity at the fixed-to-wireless network interface, since no major transcoding function is required at the base station.

The wireless ATM system under consideration is composed of the following major elements (design approaches for each of these is given in the following sub-sections):

- high-speed radio modem (~10–25 Mbps);
- hierarchical ATM interconnection network;
- ATM cell relay (ABR, VBR & CBR) services end-to-end;
- new wireless MAC and data link control layers, incorporated into ATM protocol stack;
- standard ATM network and control functions + augmentations for mobility support.

9.2.3. Radio Modem

Selection of a modulation method and set of bit-rates for macro/micro/pico cell operation is a basic design issue for next generation wireless networks. A reasonable level of ATM service could be provided on a wireless channel of say 8 Mbps, a bit-rate that can potentially be achieved in pico/micro cellular environments with conventional unequalized QAM modulation [6]. Higher bit-rates (e.g. 25 Mbps +) may be achieved using equalized QAM or multicarrier modulation methods such as OFDM [7]. Another key requirement for the modem is fast burst acquisition capability, consistent with the short ATM cell size.

Note that the use of transmission bit-rates below the nominal B-ISDN specification of 155.5 Mbps for wireless ATM access is similar in spirit to several current ATM Forum proposals [8] which aim to standardize a lower bit-rate (e.g. 25/50 Mbps) for twisted-pair copper access. It is possible that as bit-rate limitations of PCS and LAN pico/micro/macro cell scenarios become clearer, a well-chosen set of wireless channel access speeds could be incorporated into applicable ATM standards.

9.2.4. Interconnection Network

Figure 9.4 illustrates a typical backbone network structure composed of several hierarchical layers of ATM switches and multiplexors. It is noted that the ATM multiplexors at the leaves of the interconnection network are used to support several base stations in pico/microcell environments; macrocells with large traffic volume may be connected directly to an ATM switch port, as shown.

9.2.5. Protocol Layers

The multiservices wireless network follows a protocol layering harmonized with that of ATM. The approach adopted here is to incorporate new wireless-channel-specific physical, medium access control (MAC), data link control (DLC) and wireless

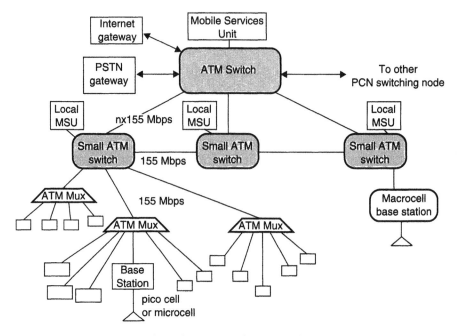

Figure 9.4 Typical ATM interconnection network for wireless ATM.

network control layers into the ATM protocol stack as shown in Figure 9.5. This means that regular ATM network layer and control services such as call set-up, addressing, VC multiplexing, cell prioritization, ECN, etc. will be used for mobile services as well. Of course, a limited number of mobility extensions to ATM network and control layers will be required to support additional functions such as location management, handoff, QoS specification/renegotiation, etc.

9.2.6. Medium Access Control (MAC)

As discussed earlier, the wireless ATM access channel is required to handle a mix of ATM services, including connectionless ABR as well as connection-oriented VBR and CBR. Thus the adopted MAC approach must provide mechanisms to deal with each of these ATM services at reasonable quantitative quality-of-service (QOS) levels. As indicated in [1,3] a reasonable solution can be obtained by extending demand assigned TDMA protocols originally proposed for satellite network scenarios (such as CPODA [9]). A fairly general MAC framework for CBR, VBR and packet data can be realized with the dynamic TDM/TDMA channel format outlined in Figure 9.6.

The channel frame contains downlink preamble, control and data, followed by uplink control, ABR data , VBR data and CBR data slots. A group of $M > 1$ frames (e.g. $M = 12$) constitutes a MAC superframe, which is used to obtain a reasonable granularity for bit-rate assignments (e.g. $n \times 32$ kbps). Boundaries between subframes may be varied gradually with time in order to accommodate changing traffic needs.

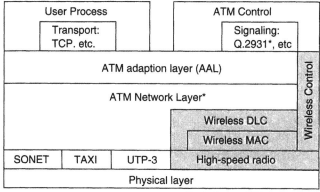

Figure 9.5 Relation of wireless network protocol layers and ATM.

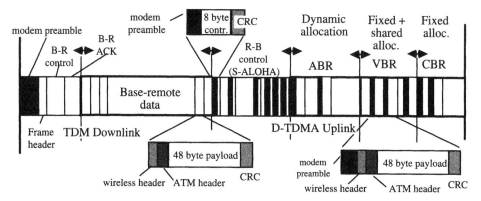

Figure 9.6 Dynamic TDMA frame format for wireless ATM.

CBR calls are assigned contiguous sequences of slots in one or more frames of a superframe; the position of CBR slots in a frame is kept relatively static in order to facilitate operation of low complexity telephone terminals. VBR calls are assigned slots with the aid of a statistical multiplexing algorithm based on usage parameter control (UPC). ABR calls are handled on a burst-by-burst basis with dynamic reservation of ABR slots or unused CBR/VBR slots in each frame. Note that CBR and VBR calls may be blocked, while ABR virtual circuits are always accepted, subject to applicable source rate flow controls. Performance results obtained via software emulation indicates that channel throughput of the order of 50–70% can be achieved, depending on traffic mix and QoS objectives.

9.2.7. Data Link Layer

The DLC protocol used in wireless ATM is designed to provide an additional layer of error protection to ATM services, via several alternative retransmission options [10]. DLC protocols are applied not only to packet-mode ABR services, but also to stream-mode CBR and VBR services. For ABR, DLC operation follows traditional SREJ ARQ procedures on a burst-by-burst basis, without time limits for completion. On the other hand, for CBR and VBR, the DLC operates within a specified timeout interval that is specified by the application at call set-up time. In this case since CBR or VBR allocation is periodic, additional ABR allocations are made at the MAC layer to support retransmitted cells. The packet sequence number required for DLC error recovery is contained in the 2-byte wireless header shown in Figure 9.7. This header also contains fields supporting other wireless network functions such as service type definition, handoff recovery and cell segmentation. Preliminary emulation results for DLC/MAC operation indicate that the proposed DLC procedures are quite effective over a range of typical channel fading models.

9.2.8. Signaling and Control

Selected extensions to ATM signaling syntax (currently Q.2931) are likely to be useful for mobility support [11]. Specific mobile network functions requiring signaling support include address registration for mobile users, wireless network QoS parameter specification and handoff. A basic issue to be addressed is that of designing wireless network control and ATM signaling to support relatively transparent handover operations. In general, this will require re-routing of the ATM connection from one base station to another, while also moving wireless network state information (e.g. DLC buffers and MAC permissions) to smoothly resume communication with a minimum of cell loss. Wireless and wired network service parameter and QoS specification/renegotiation during call setup and handoff are other important considerations in the mobile scenario.

9.3. PROTOTYPING ACTIVITIES

We are currently developing a wireless ATM prototype incorporating an 8 Mbps (2.4 GHz ISM band) radio, dynamic TDMA medium access, new data link control protocols and ATM network mobility enhancements. The prototype wireless network will subsequently be integrated into our ATM-based "multimedia C&C

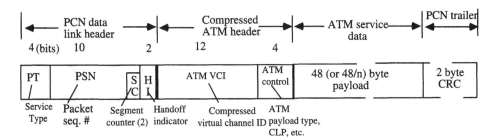

Figure 9.7 Example PCN data link packet format.

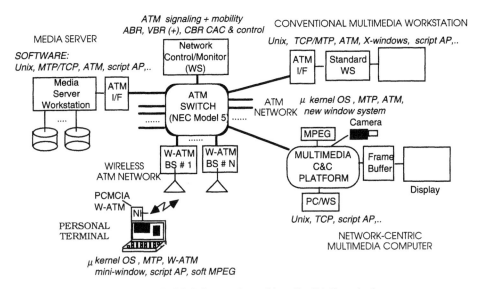

Figure 9.8 NEC Princeton's multimedia C&C testbed.

testbed", which demonstrates advanced multimedia applications on both workstations and mobile personal terminals. An outline of major hardware and software components planned for this experimental testbed is given in Figure 9.8.

9.4. CONCLUDING REMARKS

High-speed wireless networks capable of supporting integrated voice, video and data services have been identified as a key enabling technology for future multimedia systems. A specific "wireless ATM" network has been proposed to meet this need, and an architectural overview of this system has been presented. Further design and prototyping work is ongoing at NEC Princeton in order to experimentally validate the proposed wireless ATM and multimedia system architectures.

REFERENCES

[1] Raychaudhuri, D. and Wilson, N., Multimedia personal communication networks: system design issues. In *Proceedings of the 3rd Workshop on Third Generation Wireless Information Networks, Rutgers University,* April 1992, 259–288. Also in Holtzman, J. M. and Goodman, D. J. (eds), *Wireless Communications*, Kluwer, New York, 289–304, 1993.

[2] Raychaudhuri, D., ATM based transport architecture for multiservices wireless personal communication networks. In *Proceeding of ICC*, 317.1.1–7, 1994.

[3] Raychaudhuri, D. and Wilson, N., ATM based transport architecture for multiservices wireless personal communication network. *IEEE J. Selected Areas in Comm.*, October 1401–1414, 1994.

[4] Raychaudhuri, D., Multimedia networking technologies: a systems perspective. In *Proceedings of the CTR Multimedia Networking Symposium*. Columbia University, NY, 28 October 1994.

[5] Cox, D. C., Wireless network access for personal communications. *IEEE Comm. Mag.*, December, 96–115, 1992.

[6] Valenzuela, R., Performance of quadrature amplitude modulation for indoor radio communications. *IEEE Trans. Commun.*, **COM-35** (11), 1236–1238, 1987.

[7] Alard, M. and Lassalle, R., Principles of modulation and channel coding for digital broadcasting for mobile receivers. *EBU Review – Technical*, No. 224, August 1987.

[8] UTP-3 PHY Subworking Group, ITU TSS SGXIII/WP3 discussion on sub-rate broadband service. ATM Forum, 1993.

[9] Falk, G. *et al.*, Integration of voice and data in the wideband packet satellite network. *JSAC*, **SAC-1** (6), 1076–1083, 1983.

[10] Xie, H., Narasimhan, P., Yuan, R. and Raychaudhuri, D., Data link control protocols for wireless ATM access channels. In *Proceedings of ICUPC*, November 1995.

[11] Yuan, R., Biswas, S. K. and Raychaudhuri, D., An architecture for mobility support in an ATM network. In *Proceedings of 5th WINLAB Workshop on Third Generation Wireless Networks*, April 1995.

10

From Wireless Data to Mobile Multimedia: R&D Perspectives in Europe

J. Pereira, J. Schwarz da Silva, B. Arroyo-Fernández, B. Barani and D. Ikonomou[1]

---

[1]The views expressed herein are those of the authors, and do not necessarily reflect the views of the European Commission. Any mention of companies and products is solely for illustration purposes and should not be construed as any sort of endorsement.

Insights into Mobile Multimedia Communication
ISBN 0-12-140310-6

10.1. INTRODUCTION

Mobile and personal communications are recognised as a major driving force of socio-economic progress and are crucial for fostering European industrial competitiveness and for sustained economic growth, and balanced social and cultural development. The impact of telecommunications extends well beyond the industries directly involved, in fact enabling a totally new way of life and a wealth of new ways of working and doing business.

The emergence of the Global System for Mobile Communications (GSM) as a world standard shows the potential of a concerted action at European level. The unprecedented growth of world-wide mobile and wireless markets, coupled with advances in communications technology and the accelerated development of fixed network services, points towards an urgent need for the introduction of a flexible and cost-effective third generation mobile communications system. In this context, the Universal Mobile Telecommunications System (UMTS) has been the subject of extensive research carried out primarily in the context of European Community Research and Development (R&D) programmes such as Research in Advanced Communications in Europe (RACE) and Advanced Communications Technologies and Services (ACTS).

In addition to the much needed and anticipated integration of fixed, mobile, terrestrial and satellite systems, a dramatic paradigm shift is envisioned with the adoption of *soft telecommunications* concepts, of which the software radio is just one example. At the same time, research in basic technology will insure that the necessary developments are in place (smart batteries, flat panel displays, mobility-aware protocols, communications-oriented operating systems, intelligent agents, etc.) to allow for true mass market telecommunication products. With those, a myriad of personalised telecommunication services will emerge, requiring the development of new service engineering tools and the extension to the telecommunications arena of "just-in-time" service definition and provision. Future products will offer a more user-centric perspective with personalised, "just-what-the-customer-wants" services. These new products will provide the user with simple, user-friendly interfaces to the service providers.

This chapter begins with a review of the growth of voice and data communications throughout Europe, North America and Japan. Future GSM services and enhancements are discussed, together with existing and emerging radio technologies such as DECT, HIPERLAN and UMTS. To support future broadband services, both wireless Ethernet and wireless ATM technology is considered and the ATM vs IP debate is covered in section 10.5. The chapter ends with an overview of the UMTS concept and a glimpse of the multimedia capabilities offered by Europe's third generation networks.

10.2. THE GROWTH OF MOBILE COMMUNICATIONS

All over the world, mobile communication systems have recently enjoyed tremendous growth rates, capturing the imagination of the public and becoming an essential part of our every day lives. The success of voice telephony is in stark contrast to the poor market "take-up" of wireless data products. This trend is particularly evident in Europe, where many still believe it is not yet worth the trouble nor cost.

10.2.1. Wireless in Europe – the GSM Explosion

In recent years Europe has witnessed a massive growth in mobile communications, ranging from the more traditional analogue-based systems to the current generation of digital systems such as the Global System for Mobile Communications (GSM), the Digital Communication System (DCS1800) and to a lesser extent, the Digital European Cordless Telephone (DECT). The GSM family of products (GSM and DCS1800) represents the first large-scale deployment of commercial digital cellular systems and has enjoyed world-wide success. GSM has been adopted by over 230 operators in more than 110 countries (by the year 2000, approximately 350 GSM networks in over 130 countries are expected, serving approximately 250 million users).

At current growth rates, it is envisaged that the total number of subscribers could reach 120 million in Europe alone by the turn of the century.[2] Full liberalisation of the telecommunications market, scheduled for 1998 in the majority of the EU member states, will certainly accelerate the trend. It should also be noted that while the average telephone density for fixed lines is not expected to exceed 50% (i.e., at most one phone for every two persons), mobile and personal communications promises to reach nearly 80% of Europe's population.

The very success of second generation systems in becoming more cost-effective and increasingly cost-attractive raises the prospect that it will reach an early capacity and service saturation in Europe's major conurbations. At the same time, the rapid advance of component technology, the pressure to integrate fixed and mobile networks, the developments in the domains of service engineering, network management and intelligent networks, the desire to have multi-application hand-held terminals, and above all the increasing scope and sophistication of the multimedia services expected by the customer, all demand performance advances beyond the capability of second generation technology. These pressures will lead to the emergence of third generation systems which represent a major opportunity for expansion of the global mobile market-place, rather than a threat to current systems and products.

---

[2]The projections vary from as low as 100 million according to the Solomon Brothers (Source: FT) to 115 million from the Analysis/Intercai study [1] and 120 million from the UMTS Forum [2] which reviewed the Analysis/Intercai figures.

10.2.2. Wireless in the US

In the United States, the wireless growth of approximately 10 million subscribers in 1997 is far fewer that that observed in Europe, which exceeded 18 million in the same period. As a result, at the end of 1997 Europe was already almost on a par with the US in terms of the number of subscribers and is predicted to surpass the US in January 1998. In terms of penetration, Europe still trails the US (14% vs. 20%).

US companies are already positioning themselves to provide an evolutionary path from current standards such as IS-95 to a fully IMT-2000 compliant, Wideband CDMA system (Wideband-cdmaOne). A number of major companies have jointly announced that they intend to work together with the CDMA Development Group (CDG) to develop specifications for next generation wireless communications standards (while also supporting the development of other third generation standards around the globe).

10.2.3. Wireless in Japan

While in Europe the need for a fully fledged third generation system is anticipated for 2005, the situation in Japan is such that the Japanese Ministry of Posts and Telecommunications (MPT) felt compelled to accelerate the standardisation process to have a new, high-capacity system deployed around the year 2001.

Users of wireless communication systems, both cellular (analogue, TACS, and digital, PDC) and public cordless (PHS, Personal Handyphone System), have increased dramatically in Japan, coinciding with the first tentative steps at deregulation: by the end of March 1997 there were already almost 27 million users, of which around 6 million were PHS users.

The growth was initially helped by the huge success of PHS, and the combined growth rate exceeded one million new subscribers per month. However, by the end of November 1997 there were only 34.7 million wireless subscribers, of which 7 million were PHS subscribers, reflecting a slow down mainly due to the lacklustre performance of PHS. As a result, the increase in the number of wireless subscribers in Japan is not expected to reach 11.5 million in 1997 – better than the US, but worse than Europe.

Digital overtook analogue at the beginning of 1996, coinciding with the peak of analogue (exactly as in Europe). Since then, analogue has been losing subscribers. In an effort to meet the increasing demand for mobile communication systems, measures for promoting more efficient utilisation of frequencies are being studied and implemented, and a significant impetus has been given to the early introduction of third generation systems.

In 1996 the MPT created a study group on the Next-Generation Mobile Communication Systems, and tasked it to prepare a concrete vision of next generation systems, and examine related technical issues including the current status and trends, the identification of technological requirements, and the measures to be taken towards implementation. The report has recently been published [3]: it calls for a concerted effort towards wideband-CDMA, although it stops short of an exclusive endorsement of such a solution. Interest groups will continue to study other solutions like wideband-TDMA and OFDMA. In the meantime, NTT DoCoMo has openly supported its own Wideband-CDMA system, which is planned to be deployed by the year 2001. For that purpose they have called upon a world-wide partnership with European and US companies.

10.3. THE MOBILE COMMUNICATIONS MARKET – VOICE VERSUS DATA

The numbers reviewed previously identify a fast growing industry, even when growth problems are clearly present. As can be seen from Figure 10.1, mobile revenues are significant and ever increasing.

Although Europe, the US and Japan still lead, the perspective is that the rest of the world, and namely the Asian market outside Japan will dominate in the not so distant future: the shear number of potential subscribers makes this inevitable. Closer to home, the emerging Eastern European mobile market has a tremendous growth potential. In both cases, in absence of an extensive or modern telecommunications infrastructure, wireless communications are being introduced as a leapfrog technology, helping to provide phone services at much lower costs (and much faster) than the wireline alternative.

One needs to understand that at present the vast majority of this revenue arises from voice services. This situation is bound to change in the near future: at present we live in the Personal Communications era (dominated by voice); however, the future Mobile Multimedia era already beckons and this will be dominated by data.

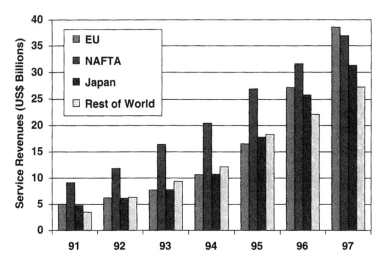

Figure 10.1 The Global Mobile Communications Market [Source: Omsync].

10.3.1. Wireless data

Relative to the huge success of mobile telephony, the take-up of wireless data is still only a promise. The current problems are mainly due to the lack of user-friendly, plug and play (P&P) applications and devices. The use of wireless data still involves a disproportionate effort: the user has to endure painful installations and in most cases has to work with cumbersome devices which are difficult to operate, and has to fight with inappropriately designed applications that take little account of the constraints of mobile usage. High costs, as well as the abundance of proprietary solutions, are also critical aspects. As a result, wireless data has stayed as a niche product.

Cellular Data

A recent Dataquest study [4] showed that the European take-up of mobile data is slow. Europe, with 11.7 million Mobile Analytical Professionals versus 10.6 million in the US and 3.5 million in Japan, had only 2.1 million mobile personal computer shipments in 1996 versus 2.5 million in Japan and 5.0 million in the US.

Moreover, with a clear lead in digital cellular telephone subscribers (20.3 million in Europe versus 9.5 in Japan and 2.9 in the US at the end of 1996), the amount of data in the digital network is negligible: only approximately 400 000 of the GSM subscribers had data adapters ($\sim 2\%$) versus 460 000 CDPD[3] users in the US (2.3% of covered subscribers); no figures are available for the number of PDC and PHS[4] data adapters in Japan. The resulting GSM data revenues correspond to only 0.5% of the total GSM revenues.

A recent ERO report [5] confirmed that European operators can still (in practice) ignore data when planning and optimising their GSM networks: the observed traffic loads, as well as the traffic characterisation, are not visibly affected by data.

In its report, Dataquest identifies some main reasons for the slow adoption of data:

- limitations of available communications technologies (e.g., GSM's basic 9.6 kbps and the call dropouts during handoffs);
- users require applications, not just a technology platform (namely GSM data);
- relevant software has been slow to materialise (e.g., GSM data support came gradually, well after the adapters had been made available).

Three more factors also seem to be relevant. The first is the *lack of plug and play terminals and of user-friendly, efficient applications*. GSM data products and services are matter-of-factly provided by industry, these have evolved separately, and inter-operability and ease of use need to be improved; furthermore, in many cases term-

---

[3]Cellular Digital Packet Data (CDPD) is a 19.2 kbps digital overlay on top of analogue AMPS.
[4]PDC offers an 11.2 kbps data service. PHS started with a 2.4–9.6 kbps data service, but recently a 32 kbps data protocol (PIAFS–PHS Internet Access Forum) has been standardised, and the service has already started (April 1997).

inals and/or applications are not tailored for the wireless environment. Without ease of use, the users will not be convinced to switch to or use wireless data. This issue was identified in the US [6], but has more acuity in Europe.

The second factor is the lack of standardisation of air interfaces and devices, and the diversity of software solutions, reflecting in many cases the focus on specialised vertical markets. This has led to a *proliferation of approaches, many of them proprietary*, further fragmenting the market, and thus thwarting the uptake.

Last but not least, the third factor is *cost*, which at current levels is effectively keeping the users away. Users in general do not perceive wireless data to be *worth* the bother. They balk at the idea of having to carry cables, plugs and adapters, and connect the laptop to the phone just to send a simple e-mail. The fact that nowadays, over GSM, apart from the Short Messaging Service (SMS) only circuit switched data is available, makes it almost prohibitively expensive for an individual to use the services. Call dropouts do not help; however, here there is some hope: the upcoming packet data services, and the development of applications that work in the background using communication resources only when necessary will help to reduce costs. However, innovative pricing policies are essential if users are to be convinced that data is finally worth while.

Dataquest's study mentions other barriers to widespread data adoption. Mainly due to the success of GSM's mobile telephony, few Europeans realise that products and services supporting mobile data communications over the GSM network exist today.

Cultural factors are also influential. European businesses have been slower than US businesses to utilise most forms of electronic data communications. For instance, over 60% of large US businesses make use of e-mail, compared with only 31% of German businesses. Attitudes however are changing, and the insistence on face-to-face meetings and telephone calls is giving way to the appropriate use of "virtual meetings" over video-conferencing, while e-mail becomes a commodity.

Cost and security definitely play a role. End-user research suggests that mobile data is viewed as a high-cost solution in terms of product acquisition, deployment and network access charges. Communication and IT managers also worry about data security and possible obsolescence. Industry cooperation is needed to create complete, secure and cost-effective solutions.

To conclude, the major stumbling block for wireless multimedia and data applications is the perceived *lack of a killer application* (see Figure 10.2) – the three top data applications are presently Internet access, database access and e-mail. Although mobile Internet is seen by many as that missing piece of the puzzle, others believe it to be the mobile office (and enhanced field service/field sales applications), while others believe it to be voice.[5] The direct result of this lack of consensus is a different emphasis on the nature and types of services and applications proposed.

---

[5]For example, Logica, the UK consultancy, believed voice-mail would be the killer application of 1997, ahead of the anticipated success of mobile Internet access. In this context, voice-mail includes, besides the traditional offerings, e-mail based alternatives as well, made more interesting through reduced tariffs and the personalisation of the voice-mail boxes. As an example, Microsoft Exchange can carry voice-mail messages as *.WAV files.

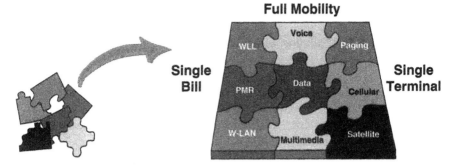

Figure 10.2 Killer application: from cumbersome, piecemeal provision of services, to integrated, user-friendly, personalised services across competing, complementary networks.

It is our opinion that the success of wireless data hinges upon a flexible solution that combines the use of wireless data as a *smart* pipe[6] to the Internet (and Intranets) with a set of tightly integrated services that complement the voice service and can serve as a differentiator in the market a federation of services on complementary and competitive network domains.

Wireless Local Area Networks (W-LAN)

The global W-LAN market is currently dominated by the US, with proprietary systems from Proxim, Symbol, DEC, AT&T/Lucent Technologies, IBM, Xircom and Motorola, among others. DASA is an honourable European exception, providing top of the range systems. All these systems operate in the 2 GHz ISM band. Unfortunately, information concerning the size of the market is not easily available, although it is thought to be quite small, a result not only of the market fragmentation stemming from the abundance of proprietary solutions, but also of the success of LANs in general and of Ethernet in particular [7].

Nowadays, most of the computers available in the market are sold with an integrated LAN interface, or have this as an option. The Ethernet solution became a *de facto* standard (while evolving from 10 Mbps to 100 Mbps and now moving towards 1 Gbps). As a result, other solutions, namely wireless, have had a hard time penetrating the market.

To have an idea of the size of the W-LAN market, consider, in absence of other information, the LAN market, making the safe assumption that it will be a small fraction of the latter. Figure 10.3 shows the size of the Western European LAN market, and its perspectives in the short term (data taken from an IDC study). An obvious conclusion is that the LAN market is comparatively small, and certainly has

---

[6]No doubt the Internet has a wealth of content. The problem is to find in reasonable time the information of interest to the user and then personalise it, getting it to the mobile device in the form best adapted to the terminal's capabilities. The most intelligent way to deal with the mobile user is to *anticipate* the information they need, use intelligent agents to search the Web and *mine* the information, post-process it to *reduce entropy* (striping graphics, etc.), and then present it to the user in *menu* format – all this with the necessary warnings for important updates.

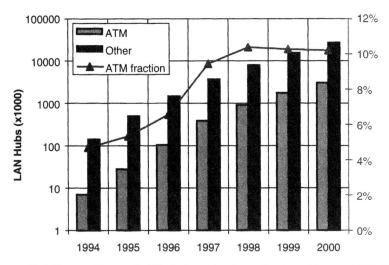

Figure 10.3 Western Europe LANs hub market by technology [Source: IDC, 1996].

to do with the number of companies, big and small, operating in the economy: if the Small Office/Home Office (SOHO) is a growing market segment, few households are expected to need a LAN. Furthermore, the cost-effectiveness of a *dedicated* wireless solution, when compared with that of a cheap and ever more powerful Ethernet alternative, is expected to keep private users away for the foreseeable future. Consequently, the W-LAN market in Europe could remain stagnant for quite some time.

10.3.2. Mobile Data and Multimedia – a Bright Future?

At present around 1% of traffic is due to data, 20 years from now it is believed that more than 90% of all communications will be in the form of data; some, like the Advisory Council of the US National Electronics Consortium, are predicting this figure could be as high as 99%.

This does not mean that voice will lose its importance. On the contrary, voice contact, possibly enhanced by images (video-telephony) and/or shared white boards (video-conference and group work), will remain an essential part of the business. In fact, the combination of voice and data capabilities confers two advantages: the operator does not need to repay the investment cost purely on the back of its data services, and costs can be shared between services and tariffs set accordingly. Users perceive and indeed expect this benefit. In fact, 75% of potential mobile data users would prefer a service which offers both voice and data.

The Internet Telephony Consortium, the Voice over IP Forum, the Voice on Net Coalition, among others, together with standardisation activities like the ETSI Project Telecommunications and IP Harmonisation Over Networks

(TIPHON), reflect the growing interest in integrating voice in data, aiming at achieving interoperability of presently proprietary interfaces with each other and the PSTN.

How much of that data (and voice as well) will be wireless? – more than we can imagine. The main driver for this renewed thrust into wireless communications is competition [8]. The technical driver is the continued dramatic increase in inexpensive processing power and storage.

To the consumer, competition means *free* choice – a refreshingly new perspective in the telecommunications industry. We have all grown up in a monopolistic (or duopolistic) telephony environment where the consumer has very limited choices. True competition, which is deemed necessary to revamp in particular the wireless data market, exists only with unrestricted entry. We are getting closer: January 1st 1998 marked the full opening of the telecommunications market to competition in most of the EU member states.

In 2005, by which time full-scale deployment of UMTS is expected, and depending on the mass market uptake, the European Mobile Multimedia *services market* is expected to grow to between 7.2 and 20.8 billion ECU in the network-centred scenario, and to between 9.8 and 17.6 billion ECU in the terminal-centred case. The size of the *terminal market* is estimated between 10 and 20 billion ECU, depending on mass market uptake.

These numbers, impressive as they are, are considered understated by some. In its recent report to the European Commission [9], the UMTS Forum indicates that higher forecast of mobile users due to continuing strong market growth, greater reductions in mobile tariffs than those considered by Analysys, and greater reductions in terminal costs will make the 2005 figures considerably higher.

For the terminal-centred scenario, considered more likely, with a developed mass market, the UMTS Forum numbers are 32 million subscribers against Analysys' 20 million, and 24 billion ECU from services against 17.6 billion. On the other hand, a conservative 10 billion ECU from terminal revenues is advanced, against 18.5 billion from Analysys.

To understand the importance of the mobile multimedia market, the projections above have to be related to the total mobile market (including voice, low data rate, SMS and messaging). The UMTS Forum estimates that 16% of all mobile users will use mobile multimedia, corresponding to 23% of the total service revenues, and 60% of the overall traffic generated.

Not all the predictions are this optimistic. Ovum suggests [10] that the take-up for high data rate services above 144 kbps may be close to zero even by 2005, while the demand for 64–144 kbps traffic will amount to less than 10% of the global mobile market.

In any case, the perspectives for the mobile multimedia market are encouraging. However, the slow uptake of data over GSM points to the dire consequences of missed opportunities, both from the regulatory and political side, if failing to provide the optimum conditions for market development, and from the marketing/competitiveness side, if failing to encourage consumer uptake through lower prices and product innovation/differentiation/customisation.

10.3.3. Wireless Application Protocol

In an effort to create a standard for bringing advanced services and applications, as well as Internet content to personal digital assistants (PDAs), digital mobile phones, and other devices, Ericsson Motorola, Nokia, and Unwired Planet launched (June 26, 1997) an initiative to cooperate in defining the new Wireless Application Protocol (WAP), an open standard for interactive wireless applications. A common standard means the potential for realising economies of scale, encouraging cellular phone manufacturers to invest in developing compatible products, and cellular network carriers to develop new differentiated service offerings as a way of attracting new subscribers. Consumers benefit through more and varied choice in advanced mobile communications applications and services.

The purpose of WAP is to provide operators, infrastructure and terminal manufacturers, and content developers with a common environment that will enable development of value-added services for mobile phones. The four founding members aim to create, together with other industry partners, a global wireless service specification, which will be independent of the network infrastructure system in place, to be adopted by appropriate standards bodies. All the applications of the protocol will be scaleable regardless of transport options and device types. The goal is to produce a license-free protocol which will be independent of the underlying air interface standard. The protocol, initially targeted at GSM-technology networks, will let compliant devices work regardless of whether a carrier uses a CDMA, GSM, or other TDMA system.

WAP will include specifications for transport and session layers as well as security features. Over and above these network layers, the protocol will define an application environment including a micro-browser, scripting, telephony value-added services and content formats. WAP will be scaleable so applications are able to make the best use of available display and network data transport capabilities across a broad range of terminal types. Services can be created from single-line text displays in standard digital mobile phones to highly sophisticated smart phone displays. The architecture of WAP (Figure 10.4) was made public on September 15th 1997 on the World Wide Web for public review and comment.

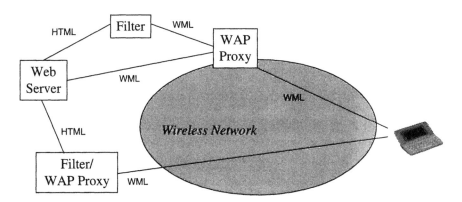

Figure 10.4 WAP Architecture.

The protocol will be designed to take advantage of the best characteristics of several data-handling approaches already in use. WAP will integrate the Handheld Device Markup Language (HDML) subset of HTML and the Handheld Device Transport Protocol developed by Unwired Planet, as well as Nokia's Smart Messaging specification and the Intelligent Terminal Transfer Protocol (ITTP) standard developed by Ericsson.

Using the existing Narrowband Sockets specification for wireless communication, WAP will accommodate the different memory, display, and keypad capabilities of various manufacturers' devices. But digital phones are not the only devices expected to take advantage of the new protocol. Almost any portable, data-capable device is a candidate for using the new protocol.

A few products already take advantage of parts of the standard. The PocketNet phone marketed by AT&T Wireless Services, for example, incorporates software written to WAP standards that lets users retrieve data from Intranet servers, e-mail, and information encoded using HDML and ITTP.

The WAP Forum was recently created to provide the means to allow terminal and infrastructure manufacturers, operators/carriers and service providers, software houses, content providers and companies developing services and applications for mobile devices to participate in the definition and development of the protocol.

10.4. MULTIMEDIA AND DATA OVER GSM

Short Message Service (SMS) and circuit switched data at 9.6 kbps (allowing e-mail and fax), are enjoying considerable success in spite of the inherent limitations. Complete solutions are however needed that match the diverse needs of mobile professionals (see Chapter 2 for a discussion of Wireless Multimedia using GSM Transport).

From this incipient start, the data capabilities of GSM are being enhanced, starting with direct connection to the ISDN network. With high-speed circuit switched data (HSCSD) and the introduction of packet switched data (GPRS), all with multi-slot capability, data over GSM will accommodate a considerable range of applications, even if those systems are not expected to be integrated in the immediate future. Finally, through higher level modulation, EDGE will further extend data rates for high link quality conditions, building upon GPRS and HSCSD.

Short Message Services

SMS is an inherent capability of GSM; it functions much like two-way paging,[7] but has the potential to rapidly evolve into an electronic messaging system. When GSM phones with PDA-like functionality appear on the market, SMS will take off as a

---

[7]SMS was originally used to let people know that voice mail was waiting, but it can also be used as an indicator of e-mail and faxes. For example, in the UK, Pipex is considering the integration of Internet access and SMS.

platform for communication applications with low data-transfer requirements. Some projections [11] in fact suggest that SMS might account for more than 10% of all cellular network revenues by 2002, representing, at that time, more than half of the expected revenues from all data services.

Work on SMS is ongoing in the European Telecommunications Standards Institute (ETSI) SMG, with SMS Interworking Extensions, a second SMS broadcast channel and concurrent SMS/CB and data transfer included in GSM 2+ Release 96. Simultaneously, new applications are being devised to make use of the SMS capabilities.

High-speed Circuit Switched Data (HSCSD)

Higher data rates will come as a result of compression, including compression at the application level.[8] But multi-slot services can take it all the way to 64 kbps (although not steady-state, as we will see immediately) and thereby facilitate inter-networking with ISDN.

The recently standardised HSCSD (GSM Release 96, March 97), allows for the combination of multiple time slots. By using up to four time slots in each direction (uplink *and* downlink), the channels can be multiplexed together to offer a *raw* data rate of up to 64 kbps (38.4 kbps *user* data rate, or up to 153.6 kbps with data compression). However, because each time slot could carry a conventional conversation, the use of multiple slots restricts the capacity for speech traffic, forcing the user to specify a minimum acceptable data rate and a preferred (and usually higher) data rate. The network will then attempt to provide as much bandwidth as required, without compromising the capacity for voice traffic. The only limitation will be the price subscribers will pay for the extra bandwidth.

HSCSD is expected to be commercially available in the 1998–1999 time-frame,[9] initially offering 19.2 kbps (two slots), the targeted service being fax and file transfer at 14.4 kbps[10] (up to 56 kbps with data compression). It will prove particularly useful for applications with high-speed data requirements, such as large file transfers, advanced fax services and mobile video communications.

GSM Circuit Switched Data

By virtue of being a digital system, GSM was conceived from the start for wireless data – all the way to ISDN rates. However, only in March 1994, when Nokia demonstrated a notebook PC linked to a GSM phone via a PCMCIA card (now PC Card), was data first sent over GSM. Today, several vendors offer PC Cards that provide a 9.6 kbps interface to the GSM network. A communications application on

---

[8]We can foresee intelligent agents and other technologies being used to automate, simplify, and otherwise enhance the mobile communications process, by filtering out unnecessary information.
[9]A first demonstration of HSCSD was performed by Ericsson in cooperation with Telia Mobitel and Telia Research in May 96.
[10]GSM Release 96 also defined a 14.4. kbps user data rate per time slot channel available under *ideal* circumstances. The effective user data rate might in fact fall below 9.6 kbps except under very good channel conditions (and almost certainly so near the cell boundaries).

the notebook/mobile computer sees the card as if it were a regular data/fax modem, enabling e-mail, file transfer, and faxing to be done with regular Windows programs. Unfortunately, all the current notebook-GSM phone interfaces are proprietary, but vendors are working their way toward a standard (see Figure 10.5).

The PC Card provides a communication link to the GSM network, but the modulation/demodulation facility is part of the network infrastructure. The Mobile Switching Centre (MSC) of the network interfaces the Public Switched Telephone Network (PSTN), and the modem pool of the GSM network operator pumps the data over the phone network. Over the air, data is conveyed by the GSM radio link protocol (RLP) that handles traffic from the data card to the MSC, while V.42-compliant communication takes over for the modem-to-modem part over the PSTN. What was needed, and is already available, is a standard that implements V.42bis on GSM networks. It has already been approved by the ETSI but has not yet been implemented by every network operator, although V.42bis enabled PC Cards are already available.

The advantages of having an all-digital mobile ISDN link are significant, particularly for short communications. Modems need time to negotiate with each other (i.e., find the highest common speed as well as the compression and error-correction capabilities): it takes 30 to 40 seconds to set up a connection over GSM. This set-up time imposes a pretty steep penalty when sending just short e-mail messages. A mobile ISDN connection, however, takes only 5 to 7 seconds to set up. Together with the high data throughput, this solution makes a number of heavy-duty applications more cost-effective.[11] A wireless ISDN user could connect, send a short message and disconnect in less time than it takes to simply set up a connection via a traditional circuit switched channel.

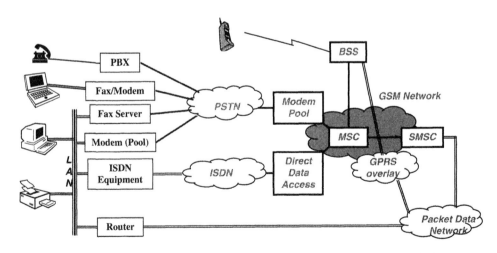

Figure 10.5 How the GSM Network now connects/will connect to a LAN.

---

[11]The cost and time savings are obvious once you consider that voice and data connections are billed in some countries (e.g., UK) in one-second increments.

Higher data throughput via V.42bis compression, coupled with the fast set-up times of all-digital connections, will boost applications such as mobile access to the Internet, particularly when this data network is used to access a corporate LAN or host. For example, UUNET Pipex in Cambridge, UK, is now offering as a commercial service access to the Internet through a direct connection between the GSM switch and its point-of-presence.

The difference between ISDN-enabled applications and systems employing the PSTN is the usage of the communication protocol transport layers. Regular AT-commands send data to the PSTN modem in the GSM infrastructure. In an all-digital scenario, a wireless AT-command extension directs the call over ISDN.

A more elegant solution would use CAPI and so-called service indicators. ETSI is currently in the process of extending CAPI to include GSM specifications. The service indicators are control instructions, which are sent over a separate channel. In this case, data and instructions are independent and can be sent simultaneously. Service indicators also enable much quicker set-up times. Thus, when a CAPI-compliant application is loaded, file transfer can start almost immediately. If the instruction is to use ISDN but the called party has an analogue connection, then the modem pool in the digital network is used. The set-up time will still be quicker, but the end-to-end transfer rate will be determined by the weakest link in the chain – the analogue modem. The separate control channel enables GSM to send a fax and receive, for example, a short message via SMS at the same time.

The availability of compression, making for faster effective transmission, and of direct links from the GSM network to data applications, reducing significantly the handshake period, and the development of software that allows an application appear to reside in the background while the GSM call is actually disconnected, are expected, according to Logica, to double the circuit switched data take-up from the end of 1996 value of under 2% to about 4% by the end of 1997.

10.4.1. General Packet Radio Service (GPRS)

As its name suggests, GPRS is based on the transportation and routing of packetised data,[12] reducing greatly the time spent setting up and taking down connections. Capacity limitation is hence in terms of the amount of data being transmitted rather than the number of connections. At the same time, charges will be based on the amount of data actually transmitted and no longer on the connection time. Multi-slot services will likely be marketed as a bandwidth-on-demand facility, making it ideal for high-speed file transfers and mobile video communications.

GPRS will work with public data networks using Internet Protocol and with legacy X.25 networks, and it is likely to prove very successful in "bursty" applications such as e-mail, traffic telematics, telemetry and broadcast services. Perhaps its greatest asset, however, is that it offers a perfect medium for Internet web-browsing. The requirements for short bursts of high-speed transmission (downloading) between longer periods of inactivity (perusing) are better suited to this connectionless

---

[12]GPRS will be available over GSM-networks but may also be supported as part of other standards, such as DECT and possibly UMTS.

approach than a circuit switched environment in which precious airtime must be charged for the entire connection time.

GPRS requires significant modifications to the GSM system architecture, with the need to introduce a data overlay. A first specification of GPRS will be available early in 1998 as part of GSM Release 97, focusing on point-to-point connections, and based upon a frame relay (FR) implementation of the overlay network. Release 98 is expected to include the specification of the ATM implementation, and full IP/IPv6 capabilities. As a result, GPRS may only be commercially launched in the year 2000.

10.4.2. Enhanced Data Rates for GSM Evolution (EDGE)

EDGE, also known as Evolved GSM (beyond GSM 2+ and to be developed in parallel with it), will enable higher data rates using the same frequency bands in use today (900 and 1800 MHz in Europe, and 1900 MHz in the US) with relatively small additional hardware and software upgrades, and keeping full GSM backward compatibility.

This will be achieved through higher level modulation: instead of the conventional GMSK of GSM, EDGE will rely on O-16QAM. As a result, instead of today's 14.4 kbps of GSM, 48 kbps per time slot will be possible. One has to understand that for the higher data rate (which can in principle be taken as high as 64 kbps per time slot in indoor environments) higher radio signal quality is required than for GSM. The system, therefore, will have to automatically adapt to radio conditions (link adaptation), dropping back to GSM data rates when necessary.

Furthermore, channels with EDGE functionality will be able to operate in either GSM or EDGE mode, allowing the two types of channels to co-exist in the same network, facilitating the step-by-step introduction of EDGE.

EDGE will make it possible to explore the full advantages of GPRS: fast set-up, higher data rates, and the fact that many users can share the same channel will result in a highly improved utilisation of the network, especially for bursty applications.

Table 10.1 GSM versus EDGE

	GSM	EDGE
Modulation	GMSK	16-QAM
Carrier spacing	200 kHz	200 kHz
Modulation bit rate	270 kbps	640 kbps
User data rate per time slot	9.6/14.4 kbps	48 kbps
Total user data rate*	76.8/115.2 kbps	384 kbps

*Total data rate available to all users, not necessarily that available to a single user.

10.5. UMTS AND WIRELESS BROADBAND COMMUNICATIONS

The never-ending appetite for higher and higher data rates even while on the move will not be satisfied with the 2 Mbps made possible by UMTS. Higher data rates, up to 155 Mbps, are being pursued to provide truly broadband wireless access on a par with broadband wired access. The flexibility and mobility offered by wireless solutions, not forgetting the significant cost savings in wiring and re-wiring buildings, plus the dropping costs of portable computers and the associated proliferation of portable devices, are motivating serious interest towards the deployment of private and public wireless broadband networks, even if the market perspectives are not that reassuring at present.

10.5.1. Mobile Broadband Systems (MBS)

For an outdoor, cellular scenario, Mobile Broadband Systems (MBS), a concept born in the RACE Programme, are under consideration, targeting data rates up to 155 Mbps with high mobility. Here again, the solution adopted in Europe is wireless ATM-based. No standardisation activity has so far been foreseen, although the ERO has considered the spectrum needs of this new service [10].

Possibly related, an interesting development in Japan is now attracting attention: the top three carriers (NTT, KDD and DDI) in cooperation with NEC, Fujitsu and Nippon Motorola will trial PHS-based connections at 25 Mbps as early as March 1998. New switches and dedicated all-optic fibre network, rather than the ISDN-based network now in operation, will provide the backbone for the new system. The upgrade of the radio system is already well under way, although no information is currently available as to its structure.

10.5.2. Wireless ATM Versus Wireless IP

The title of this section is an obvious oversimplification of the dilemma that seems to have captured a whole generation of engineers, but it certainly captures the essence of it. We discuss here the role of ATM and IP in the ongoing attempts to reach the mobile user providing access to multimedia. The risks of concentrating on wireless ATM solutions (see Chapter 9) is discussed in the light of the fast pace of IP "standardisation" and Ethernet evolution.

The approach in Europe seems to be quite distinct when discussing UMTS or beyond. While the third generation developments seem to be all geared towards providing mobile multimedia over IP at up to 2 Mbps, on top of an evolved GSM core network that builds upon the GPRS overlay, wireless ATM is currently the only solution under consideration for data rates in excess of 25 Mbps. Not so strangely, in Japan there seems to be more coherence, with a much more wireless ATM-oriented approach to third generation, reflecting their (more conventional) B-ISDN orientation.

It is true that in the fixed network, ATM offers bandwidth on demand and is capable of simultaneously supporting voice, data and video, which is important at a time when operators, carriers and ISPs are all vying to deploy an efficient, economic infrastructure, allowing them to bundle services, and be able to carry good quality traffic. These capabilities,[13] along with the required mobility management functionalities, are attractive also for wireless service provision [2].

However, against the more traditional "circuit switched" wireless ATM approach, one cannot ignore the ongoing efforts to extend the Ethernet-like IEEE 802.11 standard to higher data rates, nor the fast paced development at IETF of real time/time constrained protocols (RTP, RSVP), in parallel with the development of Mobile IP and IPv6, as well as of versions of TCP suitable for mobile environments (e.g., M-TCP [12]) and of concepts such as IP Switching [13].

The same way wireless ATM is seen as a natural extension of the ATM backbone, Ethernet-like W-LAN protocols fit naturally with the ubiquitous, ever faster (from 10 Mbps to 100 Mbps to 1 Gbps) Ethernet LANs. And here, as already pointed out in [7], the trend is not favourable to wireless ATM.

Although ATM offers clear and appealing advantages from a technological point of view, its commercialisation still faces many problems. Ovum sees a scenario where the first barrier lies in investments already carried out by network operators to build other high-speed data transmission networks which they would like to recover before boosting alternative infrastructures. A further issue concerns billing: a fair and appealing method has not yet been agreed upon. Until network operators are willing to offer ATM at attractive tariffs, network managers will also be reluctant to move to ATM in the LAN. If there is little or no likelihood of connecting individual LANs running ATM, ATM LANs just cannot compete with other LAN technologies.

All the above, combined with recent inroads in IP switching, which anticipate the deployment of fully IP-based networks, implies that IP-based W-LAN solutions, well matched to Ethernet protocols, will most certainly dominate any wireless ATM alternatives. This makes it even more urgent to redress the balance of research: the observed lack of coverage of non-wireless ATM alternatives, especially in Europe, constitutes a major handicap.

In a recent study, OVUM establishes the victory of IP over ATM [14] based upon the unstoppable growth of the Internet, its pervasive character and its continued evolution to allow the transmission of real-time multimedia. They predict that as early as 2002 the usage of IP will become dominant, surpassing that of other (non-IP and non-ATM) protocols, and certainly above ATM. ATM usage will grow in importance but less rapidly than IP, even in Europe and Japan where the influence of traditional operators is greater. At the anticipated rates of growth, the authors foresee that even before 2005 more than half of the world traffic will be carried by IP, a phenomenon that is bound to translate itself onto the mobile/wireless arena – even if the time scale turns out to be less aggressive than anticipated, more attention should clearly be paid to IP in what concerns mobile multimedia.

---

[13]They include as well as service type selection, efficient multiplexing of traffic from bursty sources, end-to-end provisioning of broadband services and suitability of available ATM switching equipment for inter-cell switching.

In the US considerable effort is also being put into alternative, more IP-oriented solutions, under the aegis of IEEE 802.11. In Europe, the single exception to the wireless ATM trend is the ETSI HIPERLAN Type 1 standard (see Chapters 37 and 38 for more details).

10.5.3. Wireless LAN Technology in Europe and North America

The wireless LAN market is currently dominated by proprietary systems operating in the 2 GHz ISM band, with data rates ranging from 1 to 16 Mbps. In order to improve customer acceptance and market penetration, interoperability becomes strategically important. The IEEE 802.11 standard defines an air interface that facilitates interoperability between wireless LAN products from many different suppliers. The Ethernet-like approach of IEEE 802.11 enables operation in two modes: independent (allowing for *ad hoc* networking) and infrastructure-based. Data rates are 1 or 2 Mbps, depending on the PHY implemented (frequency hopping spread spectrum radio, direct sequence spread spectrum radio, infrared). Recently, a new study group has been established to address higher data rates (up to 25 Mbps).

Two organisations have been launched by leading W-LAN equipment suppliers in order to promote interoperability and raise awareness. The Wireless LAN Interoperability Forum (WLIF) has been founded to promote the growth of the W-LAN market through interoperable products and services at all levels of the value chain. The Forum publishes an RF standard specification, allowing independent parties to develop compatible products. Perhaps less ambitious, the Wireless LAN Alliance (WLANA) seeks to promote increased awareness and knowledge of W-LANs among potential customers, independent software vendors and systems integrators, but without aiming for interoperability of products from different suppliers.

In opposition to this "pragmatic" approach, the attention of other standardisation bodies seems to focus now solely on wireless ATM. The ATM Forum and the ETSI Broadband Radio Access Networks (BRAN) project[14] lead, and in fact coordinate, the work in the wireless ATM area. Activities in Japan are coordinated by the Mobile Multimedia Access Communication (MMAC) organisation. MMAC services are targeted for launching around 2002, with an Advanced MMAC version anticipated for 2010 aiming at 156 Mbps even during high speed travel.

A whole family of wireless-ATM-based protocols (HIPERLAN Types 2 to 4) is under preparation for W-LAN, remote access/WLL and point-to-point usage, covering data rates from 25 up to 155 Mbps. Interestingly enough, the MMAC concept consists of two systems: Ultra-High Speed Radio LAN (fixed, up to 156 Mbps) and High-Speed Wireless Access (pedestrian speeds, up to 25 Mbps).

HIPERLAN Type 1 has already been standardised, with first products expected to come to market in 1998. The HIPERLAN Type 1 protocol, contrary to the others,

---

[14]The BRAN project was launched in early 1997 by ETSI. The aim of this project is to define the standards for service independent broadband radio access networks and systems having a peak rate of at least 25 Mbps at the user network interface. It is expected to have a set of base standards and service profiles on the radio subsystem until mid 1999. Completion of the project is targeted for 2005.

is Ethernet-like, reflecting its computer communications background (Apple, INRIA, Symbionics). The standard only defines *part* of the lower two layers of the OSI model (Physical and DLC). Within the DLC layer, only the MAC sublayer (CSMA/CA[15]) is specific to HIPERLAN. The organisation of the MAC sublayer provides a fully decentralised subsystem which does not require any central control point to operate, enabling *ad hoc* networking.

HIPERLAN nodes can communicate even when they are out of range, using other nodes to relay the messages (see Figure 10.6). This functionality is called "forwarding" or "multi-hop communication". When the final destination is not within direct reach of the transmitter, a forwarder relays the packets onto their final destination.

10.6. THE UNIVERSAL MOBILE TELECOMMUNICATIONS SYSTEM (UMTS)

The vision of UMTS, as has emerged from work undertaken within RACE [15], calls for UMTS to support all those services, facilities and applications which customers presently enjoy, and to have the potential to accommodate yet undefined broadband multimedia services and applications with quality levels commensurate to those of

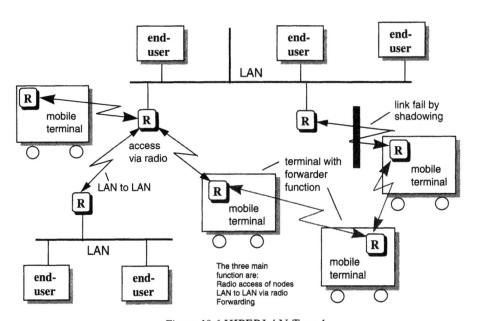

Figure 10.6 HIPERLAN Type 1.

---

[15]Carrier Sense Multiple Access with Collision Avoidance is a variation of CSMA/CD, CSMA with Collision Detection, or IEEE 802.3, usually referred to as the Ethernet protocol, although some minor differences exist.

the fixed IBC networks. In this context, projects will identify the cardinal services that UMTS must support, the "future-proofing" UMTS bearer requirements in macro-, micro- and picocell environments, and the applications likely to be supported by UMTS. A considerable effort will be devoted to determining how best to ensure that UMTS will be designed so as to be perceived by the customers as a broadband service evolution of second generation technologies, while ensuring a competitive service provision in a multi-operator environment.

UMTS represents a new generation of mobile communications systems in a world where personal services will be based on a combination of fixed and wireless/mobile services providing a seamless end-to-end service to the user. Bringing this about will require unified offering of services to the user in wireless and wired environments, mobile technology that supports a very broad mix of communication services and applications, flexible, on-demand, bandwidth allocation for a wide variety of applications, and standardisation that allows full roaming and inter-working capability, where needed, while remaining responsive to proprietary, innovative and niche markets. In particular, UMTS will support novel telematics applications such as dynamic route guidance, fleet management, freight control and travel/tourism information, specifically for road transport telematics (RTT, also known as intelligent transportation systems, ITS) and high-speed train communications. For applications where there is a very large degree of asymmetry in the downstream and upstream traffic channels, UMTS in combination with digital audio broadcasting (DAB) techniques, can provide cost-effective solutions.

With UMTS fully exploiting its capabilities as the integral mobile-access part of B-ISDN, telecommunications will make a major leap forward towards the provision of a technically integrated, comprehensive, consistent, and seamless personal communication system supported by both fixed and mobile terminals. As a result, mobile access networks will begin to offer services that have traditionally been provided by fixed networks, including wideband services up to 2 Mbps. UMTS will also function as a stand-alone network implementation.

At the time UMTS reaches service, ATM will be an established transmission technique; hence the UMTS environment should also support ATM-cell transmission through to the user's terminal. This compatibility will enable service providers to offer a homogeneous network, where users can receive variable bit-rate services regardless of their access media (mobile or fixed, including wireless local loop). The same flexibility is anticipated over IP networks, as discussed previously, so IP support is also considered essential, and it is in fact expected to be the preferred mode of operation.

UMTS will require a revolution in terms of radio air-interface design, and continued evolution of intelligent network (IN) principles. The arrival of a fully capable UMTS does not preclude the extension of such developments into those bands currently open to second generation technology. The resulting parallel process of UMTS design and second generation enhancement will call for careful market management and co-existence between UMTS and second generation services to ensure a smooth, customer sensitive transition. Indeed, multi-mode/multi-band transceiver technology may be used to provide multi-standard terminal equipment, particularly between UMTS, GSM/DCS-1800, and DECT.

10.6.1. Data Over UMTS

What is unique about data over UMTS is that in the same network, as a function of the environment and the user/terminal mobility, the maximum peak data rate available to the user will vary from 144 kbps to 2048 kbps. This will imply the need for dynamic adjustment of the maximum data rate available to the applications as the user moves around the network. Moreover, as UMTS will be progressively deployed and is not expected to provide full coverage, not even in urban areas, from day one, the user will be required to drop back to GSM data when outside coverage.[16] Then, a more complicated handover mechanism beyond rate adaptation will be required, especially if it needs to be seamless.

The final implementation of data over UMTS will certainly result from the same quest for improved efficiency relying upon data rate (i.e. bandwidth) on demand, dynamic resource allocation, and QoS negotiations based upon user profile. It is also clear that both circuit- and packet-switched data will have to be provided over a single radio, allowing the user to select one of the modes at will, unlike the case now for GSM, where GPRS and HSCSD each requires a dedicated radio implementation.

At this point in time, there seems to be consensus in that the UMTS backbone will evolve from the GSM backbone and its GPRS overlay. In fact, the GMM Report [16] singles out GPRS as a key link of GSM to UMTS, both in terms of the packetised transfer made possible, and the mere existence of an overlay. Where the perspectives differ is in the way that evolution is expected to occur.

Figure 10.7 shows the approach proposed by the ACTS project RAINBOW (see section 10.4), where the interaction of the new UMTS Radio Access Network (RAN)

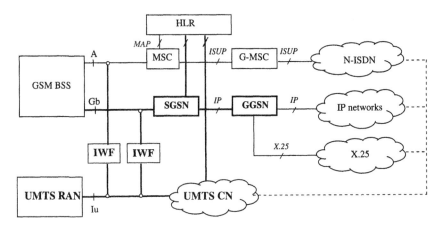

Figure 10.7 UMTS as an overlay on the evolved GSM network [Source: ACTS Project RAINBOW].

---

[16]In the beginning, a few small UMTS islands will stand out in the middle of a large GSM ocean covering the whole of Europe: GSM will be the fallback network, providing service outside UMTS coverage. This fact alone makes it essential that roaming be possible between those UMTS islands, as well as with the pervasive GSM networks, if a critical market is to be established.

and Core Network (CN) with the existing GSM network is done through interworking function (IWF) units. It is obvious from the figure that the UMTS network can thus be seen as yet another overlay on the GSM network. This conceptual image does not however spell out the exact implementation of the UMTS Core Network (possibly IP-based), nor its connection with existing networks.

The UMTS CN, if certainly IP-based,[17] be it Mobile-IP and/or IPv6, could come about in many formats: e.g., IP over ATM, native IP or a hybrid of the two. Furthermore, the overlay nature of this network will make it possible for operators (especially new entrants) even to skip the deployment of the GPRS overlay, jumping directly into UMTS.

10.7. MOBILE MULTIMEDIA IN ACTS

In the context of the ACTS programme, R&D in advanced mobile and personal communications services and networks is called upon to play an essential role. The specific objectives that are addressed by the current ACTS projects in what is called the mobile domain include the development of third generation platforms for the cost-effective transport of broadband services and applications, aiming at responding to the needs of seamless services provision across various radio environments and under different operational conditions. Since the scope of future mobile communications encompasses multimedia, far beyond the capabilities of current mobile/wireless communication systems, the objective is to progressively extend mobile communications to include multimedia and high performance services, and enable their integration and inter-working with future wired networks.

Third generation mobile communication systems aiming also at integrating all the different services of second generation systems, provide a unique opportunity for competitive service provision to over 50% of the population, and cover a wide range of broadband services (voice, data, video, multimedia) consistent and compatible with the technology developments taking place within the fixed telecommunications networks. The progressive migration from second to third generation systems, expected to start early in the next century, will therefore encourage new customers, while ensuring that existing users will perceive a service evolution that is relatively seamless, beneficial, attractive and natural.

Figure 10.8 portrays the technological capabilities of third generation systems, measured in terms of terminal mobility and required bit-rates as compared to those of second generation platforms such as GSM.

UMTS is conceived as a multi-function, multi-service, multi-application digital mobile system that will provide personal communications at rates ranging from 144 kbps up to 2 Mbps according to the specific environment, will support universal roaming, and will provide for broadband multimedia services. UMTS is designed to have a terrestrial and a satellite component with a suitable degree of commonality between them, including the radio interfaces. The R&D effort concentrates on the

---

[17]Other protocols (e.g., for video transmission) will also be supported.

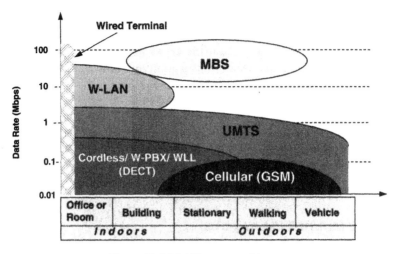

Figure 10.8 Mobility versus bit-rates.

development of technical guidelines regarding, in particular, the level of UMTS support of ATM as well as of IP, the compatibility of UMTS and fixed-network architecture, the allocation of intelligent functionality (universal personal telecommunications – UTM, and Intelligent Network – IN), the level of integration of the satellite component of UMTS, and the multi-service convergence philosophy of the UMTS radio interface.

Mobile Broadband Systems (MBS) and their Wireless Local Area Network (W-LAN) counterpart, are an extension to the wired B-ISDN, with the ability to provide radio coverage restricted to a small area (e.g., sports arenas, factories, television studios, etc.) allowing communication between mobile terminals and with terminals directly connected to the B-ISDN at rates up to 155 Mbps. Mobility and the proliferation of portable and laptop computers together with potential cost savings in the wiring and re-wiring of buildings are also driving forces in the introduction of broadband wireless customer premises networks in a picocell environment supporting the requirements for high-speed local data communication up to and exceeding 155 Mbps.

Figure 10.9 illustrates the range of service environments, from in-building to global, in which the third generation of personal mobile communication systems will be deployed. Appropriately positioned are all the present projects in the mobile domain, with some relevant ACTS projects from other domains indicated in italics.

10.7.1. Enabling Technologies

At the core of the future mobile communication networks are a series of enabling technologies whose role is essential in permitting such networks to meet the capacity and quality of service requirements at a cost/performance level attractive to both operators, service providers and users. These enabling technologies range from those usually related to hardware issues (e.g., antennas, adaptive wideband radio front-

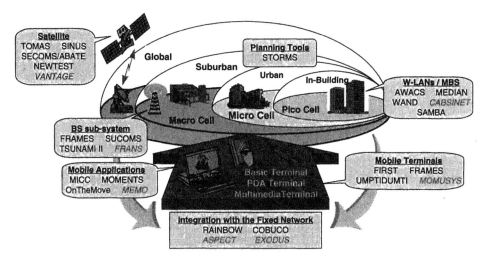

Figure 10.9 ACTS projects relative to the different system environments.

ends) to those related to software/algorithmic issues (e.g., authentication, network planning and control).

Project *TSUNAMI II* seeks to further develop the technology for smart antennas and aims to demonstrate that it is feasible and cost-effective to deploy adaptive antennas within the infrastructure of third generation mobile systems such as UMTS (see Chapter 31 for a discussion on the application of adaptive antennas for personal communications).

Design and development of sustainable transmission systems with very high capacity is a major challenge for the realisation of universal personal telecommunication (UPT). *Adaptive antennas* such as the ones developed by *TSUNAMI II* are a new technology that have the potential to provide large increases in capacity. The top level design of the field trial equipment has been completed, and the project is in the process of constructing a single fully adaptive base station (BS) that will be deployed and used in a DCS 1800 network.

The concept of software radio [17–19] has recently started to attract the interest of both manufacturers and operators (see Chapter 35 for more details). It is widely recognised that in particular software radio will have far reaching implications for all players in the mobile communications field, namely:

- customers (more service choice and better control over service provision);
- service providers (more freedom to innovate and differentiate services);
- operators (may dynamically optimise air interface parameters, re-apportion spectrum, customise terminals, offer variations of standards, change base station configurations on the fly);
- manufacturers (may reduce system/service specific product ranges);
- system designers (need to study the implications of transferring software over the air interface versus other alternatives, e.g., SIMs, *vis-à-vis* its consequences in terms of network management, and particularly signalling load);

- standardisation bodies (standards will evolve towards radio-assembly language based specifications);
- spectrum regulators (offered the possibility of more efficient utilisation of spectrum, with its implications in terms of spectrum management);
- licensing authorities (need to understand the impact on licensing and competition policies).

The aim of project *FIRST* is to demonstrate that it is feasible and cost-effective to develop and deploy intelligent *multi-mode* terminals (IMT) capable of operation with UMTS, as well as with multiple second generation standards, and with the ability to deliver multimedia services to mobile users. The work concentrates on multi-mode radio frequency (RF) and digital signal processing (DSP) sub-systems, and will identify architectures and technologies which meet the functional power and cost targets of actual third generation handsets and base station (BS) implementations. The project has contributed to the MMITS Forum.

Two third call proposals were retained in the software radio area. The *SORT* proposal aims to demonstrate the feasibility of flexible, modular, *multi-mode and multi-band* radio systems which fit with the generic radio access network (GRAN) approach. *SORT* will address issues like the identification of the air-interface functionalities, classification of functions amongst critical (real-time) and non-critical, software portability, which is particularly critical for real-time functions, and hardware re-configurability (software re-configurable hardware). It will implement channelisation, one of the critical functionalities, by means of a prototype, and will focus on baseband signal processing using FPGA. The SORT consortium brings a unique perspective of integrating terrestrial and satellite components.

The *SUNBEAM* proposal aims at developing innovative *BS array processing* architectures and algorithms, building upon the work performed in the project *TSUNAMI II*. The emphasis will be on the broadband RF array processing required to accommodate different air interfaces under realistic spatio-temporal propagation conditions. It will investigate more efficient DSP architectures which will allow as many of the array processing functions as possible to be performed using software radio techniques.

10.7.2. Mobile/Wireless Platforms

The essential platforms of relevance to mobile communications are those meeting the requirements in a variety of environments (public, private, rural, office) and at different levels of service and availability. These can be broadly classified as public cellular networks (UMTS), private wireless local area networks and public cellular mobile broadband systems (W-LANs and MBS) and global low mobility networks (satellites).

10.7.3. UMTS

The vision of UMTS, as it has emerged from work undertaken within RACE [20], calls for UMTS to support all those services, facilities and applications which customers presently enjoy, and to have the potential to accommodate yet undefined broadband multimedia services and applications with quality levels commensurate to those of the fixed IBC networks. In this context, projects will identify the cardinal services that UMTS must support, the "future-proofing" UMTS bearer requirements in macro-, micro- and picocell environments, and the applications likely to be supported by UMTS. A considerable effort will be devoted to determining how best to ensure that UMTS will be designed so as to be perceived by the customers as a broadband service evolution of second generation technologies, while ensuring a competitive service provision in a multi-operator environment.

With UMTS fully exploiting its capabilities as the integral mobile-access part of B-ISDN, telecommunications will make a major leap forward towards the provision of a technically integrated, comprehensive, consistent, and seamless personal communication system supported by both fixed and mobile terminals. As a result, mobile access networks will begin to offer services that have traditionally been provided by fixed networks, including wideband services up to 2 Mbps. UMTS will also function as a stand-alone network implementation.

At the time UMTS reaches service, ATM will be an established transmission technique; hence the UMTS environment should also support ATM-cell transmission through to the user's terminal. This compatibility will enable service providers to offer a homogeneous network, where users can receive variable bit-rate services regardless of their access media (mobile or fixed, including wireless local loop). The same flexibility is anticipated over IP networks, as discussed previously, so IP support is also considered essential, and it is in fact expected to be the preferred mode of operation.

Service provision across Europe, in a multi-operator environment, demands close attention to spectrum-sharing techniques, charging, billing and accounting, numbering, network security, privacy, etc., all of which may have regulatory implications. To these issues, one must add the degree of compatibility between UMTS and fixed network functionality, and the form of the multi-service capability at the radio interface.

UMTS will require a revolution in terms of radio air-interface design, and continued evolution of intelligent network (IN) principles. The arrival of a fully capable UMTS does not preclude the extension of such developments into those bands currently open to second generation technology. The resulting parallel process of UMTS design and second generation enhancement will call for careful market management and co-existence between UMTS and second generation services to ensure a smooth, customer sensitive transition. Indeed, multi-mode/multi-band transceiver technology may be used to provide multi-standard terminal equipment, particularly between UMTS, GSM/DCS-1800, and DECT.

UMTS Platforms

The main distinction of UMTS relative to second generation systems is the hierarchical cell structure designed for gradated support of a wide range of multimedia broadband services within the various cell layers by use of advanced transmission and protocol technologies. The three-dimensional hierarchical cell structure in UMTS aims to overcome second generation problems by overlaying, discontinuously, pico- and micro-cells over the macro-cell structure with wide area coverage. Global satellite cells provide coverage where macro-cells are not economical due to low teledensity, and support long-distance traffic.

The choice of an air-interface parameter set corresponding to a multiple access scheme is a critical issue as far as spectral efficiency is concerned. Within the framework of the ACTS Programme, project *FRAMES* is in charge of defining a hybrid multiple access scheme based on the adaptive air interface concepts that were developed earlier in RACE. After a comparative assessment of a dozen candidate techniques, the FRAMES Multiple Access (FMA) system has been proposed to ETSI for the process of definition of the UMTS air interface. In fact, three out of the five proposals initially under discussion in ETSI were based on FMA Mode 1 (Wideband TDMA with or without spreading) and FMA Mode 2 (Wideband CDMA), with small differences due to the input of other supporters. Two of them were the only real contenders for the final selection.[18] The dual mode FMA concept lends itself to the introduction of novel technologies, such as joint detection, multi-user detection for interference rejection, advanced coding schemes (e.g., turbo codes) and adaptive antennas which contribute to more efficient utilisation of the UMTS frequency bands.

The introduction of third generation mobile systems require an effective software tool able to assist and support operators to design and plan the UMTS network. The tool being designed by *STORMS* is conceived as an open software platform where external or newly developed modules can be easily plugged. A Topographical and Geographical Information System (TGIS) has been developed. A radio *coverage optimisation* module which allows the optimisation of the number of base stations necessary to ensure a given coverage with a certain confidence degree has also been developed. A fixed infrastructure (links between the UMTS network nodes) *resources optimisation* module has also been included.

The study of a "generic" UMTS access infrastructure, able to cope with different radio access techniques (long-term view) and, at the same time, to guarantee a smooth migration from second to third generation systems (short-term view), is addressed by *RAINBOW*. On the issues of network integration *RAINBOW* proposed was a reference configuration for the UMTS system introducing a *single generic functional interface* between the UMTS access part and the core network and a separation of the *radio technology dependent* parts from the *radio technology independent* parts. These two new concepts have been presented and promoted in ETSI SMG and ITU-R TG8/1 with the purpose of obtaining a unified view of the evolution of the standards towards UMTS.

---

[18]An indicative vote held in an ETSI SMG2 meeting in Madrid on 16 December 1997 saw FMA 1 with spreading (CD-TDMA) and FMA 2 (W-CDMA), proposals δ and α, respectively, split the vote: no other proposal received any votes. A combined $\alpha + \delta$ proposal was later agreed upon and met with consensus in Paris [3]. The details of the UMTS standard will now be developed by ETSI, with a conclusion expected by the end of 1999.

10.7.4. Mobile Broadband Systems (MBS) and Wireless LANs (W-LAN)

The strategic importance of mobile broadband communications systems catering for different mobility requirements ranging from stationary (wireless local loop – WLL), through quasi-stationary (office and industrial environments), to full mobility was recognised at an early stage in the context of RACE. The objectives of the work were namely to develop a quasi-mobile wireless system for bit-rates of up to 155 Mbps throughput (in the 40 or 60 GHz bands), and to create the industrial capacity to produce the necessary system components (RF, IF and baseband systems, antennas and terminals). In the context of MBS applications, the investigation and definition of system aspects, radio access schemes, network management issues, integration with IBC, among others, is critical. MBS systems will cater for novel mobile multi-media applications, including those appropriate to broadband W-LAN and WLL systems.

As wireless terminal extensions to the B-ISDN, MBS/W-LAN system concepts are being actively researched in the context of ACTS with emphasis on specific objectives including the demonstration of mobile broadband applications, video distribution, interactive video, audio and data communication service at bit-rates up to 155 Mbps on a mobile terminal connected to an IBC network; the demonstration of ATM compatibility between mobile and fixed terminals with implementation of the necessary mobility management functions (especially for handover) and the required signalling, control and service-provision protocols; and finally the validation of quality-of-service parameters corresponding to the evaluated applications.

ERO is already looking into spectrum allocation for MBS [21]. As with UMTS, the satellite component of MBS is being investigated to determine the optimum frequency band (Ka: 20–30 GHz [22], or 60 GHz). In the meantime, ERO has already made available for HIPERLAN Type I 100 (+50) MHz at 5 GHz (in contrast with the 300 MHz made available in the US) plus 200 MHz at 17 GHz.

Proliferation of portable and laptop computers, wide acceptance of mobility, and the potential cost savings in avoiding the wiring or re-wiring of buildings are driving forces for broadband wireless-access in an in-building environment. Consequently, third generation mobile systems must include W-LAN capabilities to maintain "universality". Application areas include mobile systems for offices, industrial automation, financial services, emergency and medical systems, education and training, with network connection for portable computers and personal digital assistants as well as *ad hoc* networking. The specific nature of each of the above related environments do, however, influence security, range, defined working area, transmission rate, re-using of frequencies, cost, maintenance, penetration potential, etc.

MBS/W-LAN Platforms

In creating a high-speed (up to 155 Mbps) local data communication link, significant research is required to identify a suitably reliable system and associated air interface. Important issues include frequency allocation and selection, choice of bandwidth, efficient coding schemes, specification of medium access procedures, definition of link control protocols, as well as connectivity aspects related to connection to other wired or wireless communications networks.

Figure 10.10 positions the various ACTS W-LAN/MBS projects with respect to the target environment and mobility. Wireless ATM is the common approach that will allow users to transmit and receive data at data rates (> 20 Mbps) and controlled service levels that match those of the wired ATM world.

The project *The Magic WAND* aims to develop and evaluate in realistic user environments a wireless ATM transmission facility that expands the reach of ATM technology to the premises communications networks. User trials will be carried out with selected user groups and the feasibility of a wireless ATM access system will be assessed. The results will be used to promote standardisation of wireless ATM access, notably in ETSI. The 5 GHz frequency band is targeted by the demonstrator, providing 20 Mbps data rate, but studies on higher bit rate operation (> 50 Mbps) in the 17 GHz frequency band will be carried out as well.

The main objective of the *MEDIAN* project is to evaluate and implement a high-speed wireless Customer Premises Local Area Network (WCPN/WLAN) pilot system for multimedia applications and to demonstrate it in real-user trials. The pilot system at 60 GHz relies on a multi-carrier modulation scheme (OFDM) which adjusts to the transmitted data rates (up to 155 Mbps) and channel characteristics, and on wireless ATM network extension. ATM-compatibility is achieved by transparent transmission of ATM cells. For an analysis of antenna and propagation issues for indoor wireless networks in the 60 GHz band, see Chapter 40.

The *AWACS* project is aimed at the development of a low-mobility system operating at 19 GHz and offering user bit-rates of up to 34 Mbps with a radio transmission range of up to 100 m. The demonstrator will provide propagation data, bit error rate (BER) and ATM performance, allowing the investigation of spectrum and power efficient radio access technologies. The trials are expected to indicate the

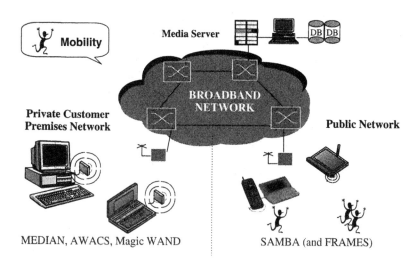

Figure 10.10 ACTS wireless broadband projects by target environment (adapted from [23]).

capacity of the system in a real user environment. For more information on the AWACS project, see Chapter 36.

The *SAMBA* project is developing an MBS trial platform operating in the 40 GHz frequency band offering transparent bearer services of up to 34 Mbps. It aims to demonstrate the feasibility of the MBS concept by offering all the essential functionalities expected from MBS. The trials performed on the *SAMBA* trial system will demonstrate the provision of multimedia services (such as high-resolution video), the integration of MBS to a fixed broadband network via standard ATM interfaces and validate the enabling technologies developed for operation in the 40 GHz frequency band.

So far the user requirements are defined and the target quality of service parameters set. The trial platform is specified. This includes the air interface, the cell shapes and the antennas, the mm-wave transceiver, baseband processing unit, control unit as well as the ATM switch and ATM mobility server. The development of ASICs, MMICs and modules as well as protocol software has started.

Figure 10.11 describes the different approaches of the wireless ATM projects to enable high bit-rate support, while Table 10.2 systematises the technical characteristics of the systems under consideration.

10.7.5. Advanced Mobile/Wireless Services and Applications

The essential key question with which the mobile communications sector as a whole will be confronted is that of which are the services and applications that will be the key drivers of the new technological generation. While it is widely accepted that voice will certainly continue to be one of the key requirements to be met by a mobile/wireless system, the debate is still wide open regarding the nature, type and main characteristics of future advanced multimedia mobile services, as well as its likely date of "take-off".

In conjunction to the above system platform concepts and enabling technologies projects, there are a number of application development projects, expected to provide the "proof of concept" as well as an insight on what the users should expect from future generations of mobile communications networks. When feasible, trials are based on platforms developed by the system projects of the domain.

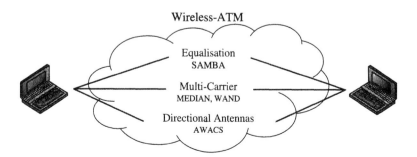

Figure 10.11 Approaches to support high bit-rate wireless ATM.

Table 10.2 Technical characteristics of the W-LAN/MBS projects

	The Magic WAND	AWACS	MEDIAN	SAMBA
Frequency (GHz)	5	19	60	40
Data rate (Mbps)	20	34	155	34
Coverage (m)	~30	~30	~10	~100
Modulation	16 point OFDM	O-QPSK with coherent detection	256 point OFDM	O-QPSK
Multiple access	TDMA	TDMA	TDMA	TDMA/FDMA
Domain	WCPN	WCPN	WCPN/W-LAN	Public Network
Mobility	portable/low	portable/low	portable/low	high, seamless handover

The distribution of advanced multimedia products by wireless means is the main subject of *MOMENTS*, which also covers the related billing and charging issues. So far the configuration for the service platform has been defined, and the hardware and the software for the basic service platform and terminals for 100 end-users have been delivered and installed to each trial network. Contributions to standardisation bodies (ITU-T and MPEG4) have been submitted regarding the video coding schemes.

The construction sector will benefit from the work of *MICC* which will bring to construction site workers completely new communications capabilities through mobile communications. *MICC* will publish a set of tested European wide recommendations.

Every user of telecommunications services has individual special needs; by taking account of the needs of users with disabilities the usability requirements of the population as a whole can be accommodated. This is the basic theme of *UMPTIDUMPTI* whose primary objective is to verify that emerging broadband and mobile services and equipment can be used by everyone, including people with special needs. Within ETSI TC Human Factors the project has initiated two new work items concerned with standardisation of text-telephony (a key service for the deaf) and the further analysis of requirements for mobile video telephony.

To facilitate and promote the development of a wide spectrum of mobile multi-media applications, a standardised mobile application programme interface (API) is being developed by *OnTheMove*. The project has succeeded in defining the mobile applications support environment (MASE) middleware architecture and its services offered to mobility-aware applications. These include among others location aware-ness, messaging and multi-party services, cost control, security, disconnected opera-tion, terminal adaptation, profile management and support for distributed environments. The application designed and implemented for this experiment included stock portfolio information, news feeds, travel information and access to news magazines as well as a dashboard providing feedback to the users about their location and available bandwidth.

10.8. CONCLUSIONS

The whole idea of third generation systems revolves around the provision of mobile multimedia. Supported by telecommunications operators, equipment manufacturers, service providers, research institutes, universities and leading-edge users, a number of EU-funded R&D projects have proved to be instrumental in the development of third generation systems. The participation of representative users reflects the policy that R&D must be demand-driven to ensure that the developed technologies, services and applications are responsive to market requirements.

The case is also made for the urgent need for enhanced, more efficient mobile communication systems, providing access to mobile multimedia services on a par with the fixed networks. R&D in mobile and personal communications is believed to be essential, due to its multiplicative, enabling effect in all areas of the information society.

By building upon the capabilities of existing second generation systems and creating a market of mobile-aware if not mobile-oriented multimedia applications, in the right regulatory context and with true competition at the network and service provider levels, and by intelligently exploiting the wealth of information available, mobile multimedia will certainly become a mass-market commodity, as has been long anticipated.

ACKNOWLEDGEMENTS

The authors wish to express their appreciation to all participants in the ACTS mobile projects and all those who, throughout the duration of the programme, have generously contributed to the discussion regarding the future direction of R&D in Mobile and Personal Communications.

Particular thanks are due to Ermano Berruto, manager of the project RAINBOW, for his availability and patience while discussing the likely evolution of the GSM network, and to Larry Taylor for his passionate perspectives on HIPERLAN standardisation.

REFERENCES

[1] Mobile broadband systems (MBS). Final Report, ERO, July 1997.
[2] Raychauduri, D. and Wilson, D., ATM-based transport architecture for multiservices wireless personal communication networks. *IEEE Journal on Selected Areas in Communications*, October 1994.
[3] Next-generation mobile communications systems. MPT Study Group, June 1997.
[4] Mobile data communications in Europe: a market perspective. Dataquest, October 1996.
[5] Traffic loading, preliminary report. 1998, ERO.
[6] Pereira, J. M. *et al.*, Communications Document, National ITS Architecture Study. GTE Laboratories Inc., November 1995.

[7] Pereira, J. M., Indoor wireless broadband communications: R&D perspectives in Europe. In *Proceedings of the Colloquium on Indoor Communications*, TU Delft, The Netherlands, October 1997.

[8] Cooper, M., Wireless in the 21st Century. Cellular & Mobile International, January 1998.

[9] A regulatory framework for UMTS. UMTS Forum, June 1997.

[10] Third generation mobile systems. White paper, OVUM, 1997.

[11] The European and US markets for wireless data services. Strategy Analytics, November 1997.

[12] Brown, K., and Singh, S., M-TCP: TCP for mobile cellular networks. *ACM Computer Comm.* Review, October 1997.

[13] IP Switching: the intelligence of routing, the performance of switching. White paper, Ipsilon Networks.

[14] The future of broadband networking: ATM vs. IP, white paper, OVUM, 1997.

[15] RACE vision of UMTS. In *Proceedings of the Workshop on Third Generation Mobile Systems*, DG XIII-B4, European Commission, Brussels, January 1995.

[16] A standardisation framework. GMM Report. ETSI SMG 3GIG, June 1996.

[17] Mitola, J., Special issue on Software Radio, *IEEE Communications Magazine*, May 1995.

[18] *Proceedings of Software Radio Workshop*, European Commission, Brussels, DG XIII-B4, May 1997.

[19] *Proceedings of ACTS Mobile Telecommunications Summit*, European Commission, DG XIII-B4, Aalborg, Denmark, October 1997.

[20] RACE vision of UMTS. In *Proceedings of the Workshop on Third Generation Mobile Systems*, DG XIII-B4, European Commission, Brussels, January 1995.

[21] Mobile broadband systems (MBS). Final Report, ERO, July 1997.

[22] 20/30 GHz, Draft Final Report, ERO, June 1997.

[23] Mikkonen, J., Wireless-ATM: enabling technology for multimedia services. In *Proceedings of the Wireless Broadband Communications Workshop*. European Commission, Brussels, DG XIII-B4, September 1997.

11

Security in Future Mobile Multimedia Networks

Chris J. Mitchell and Liqun Chen

11.1. INTRODUCTION

Current digital mobile networks, e.g. those based on the ETSI GSM standards, provide a robust set of security facilities to protect communications across the air interface. The main GSM security services are confidentiality of user and signalling data (across the air interface), user authentication to a base station, and user identity confidentiality (across the air interface). Because of their universal nature and the extra requirements of high data rate multimedia traffic, standards for future networks will need to support a larger range of security services. Possible new services

Insights into Mobile Multimedia Communication
ISBN 0-12-140310-6

include: end-to-end data confidentiality and integrity, incontestable charging, and a more robust user identity confidentiality.

There is also much to be gained by standardising management aspects of security provision. In GSM, although the management security requirements are clear, the exact way in which user key information is generated, stored and accessed is left to network operators (NOs) and equipment providers to arrange. This can make security service provision costly for all concerned, since every NO may arrange security management differently. In future mobile networks, possibly operating in a rather more deregulated environment than at present, standardised support for security management will be a very important feature. Without such standards, the required co-operation between the likely large numbers of competing NOs and service providers (SPs) could become impossibly complex to arrange.

In this chapter we examine some of the security provisions in the emerging ETSI Universal Mobile Telecommunications System (UMTS) and ITU Future Public Land Mobile Telecommunications System (FPLMTS) standards for future mobile telecommunications networks. After a brief review of some of the most significant areas for the provision of security services, we focus our attention on the simultaneous provision of identity and location privacy for the mobile user and mutual authentication between mobile user and base station. Much of the work presented here has emerged from the DTI/EPSRC-funded LINK project "Third Generation System Security Studies" (3GS3) [1].

11.2. THIRD-GENERATION MOBILE SYSTEMS

The term "third generation" refers to mobile systems which will follow existing digital networks such as GSM, DCS1800 and DECT; such systems are currently being standardised by ITU (FPLMTS) and ETSI (UMTS). They have the following characteristics: multiple operators, multiple environments (residential cordless, mobile, satellite, etc.), multi-vendor and standardised interfaces, use of the WARC-assigned FPLMTS band, higher bit-rates (up to 2 Mb/s), and migration from existing systems.

Current GSM systems support security features such as confidentiality of user and signalling data on the air interface, authentication of users, and user identity confidentiality. There are areas where security can be enhanced in third generation systems, partly based on lessons learnt from second generation systems, but mostly derived from the new characteristics noted above.

In our discussion of security in mobile communications we use a simple model with four roles: users, NOs, SPs and intruders, which are defined fully in Clause 3.2 of [1]. Briefly:

- a *user* is an entity authorised to use particular network services;
- a *network operator (NO)* is an entity providing network capabilities to support particular services, and which allows users to access the network to use the services;

- a *service provider (SP)* is an entity responsible for the provision of particular services, and will typically do so by means of contractual relationships with NOs;
- an *intruder* is an entity that abuses the network infrastructure or services on the network.

11.3. SECURITY FEATURES FOR FUTURE NETWORKS

Initial studies in 3GS3 identified the likely security threats to future mobile networks in the context of role and functional models [1]. Security features necessary to address these threats were identified and classified and include the following:

- *Entity authentication.* Entity authentication between a user and network operators and/or service providers is a key feature. A number of mechanisms, based on various cryptographic methods, were examined, classified, and tested (formally and informally). As a result, an entity authentication mechanism was proposed to both UMTS and FPLMTS, and subsequently was incorporated into both sets of draft standards. This mechanism, briefly described in [2], is considered in section 11.5.1; a further mechanism is considered in section 11.5.2. Problems arising when some of the "authentication servers" within a system may be unreliable, [3,4], and the effect of the properties of the underlying components of an authentication mechanism on its design, [2], were also considered.
- *Novel techniques for key distribution.* Maurer has shown, [5], how Wyner's "Wire tap channel" concept, [6], can be used much more widely than originally envisaged. The idea makes use of the universal presence of noise in communications channels to help two users agree a secret key using only "public" channels. The practicality of this idea was investigated, and new theoretical results were discovered, [7].
- *End-to-end encipherment, and warranted interception facilities.* Multimedia terminals will place demanding bandwidth requirements on the mobile network. These requirements have relatively little direct effect on security feature provision, except that any directly data-related security features, such as the provision of data confidentiality, must be implemented using methods which can handle high-bandwidth data. In practice this means that air-interface encryption methods must be able to handle high throughput rates. However, this should not be too difficult since multimedia terminals will not be low cost items, and the provision of processing capabilities to handle high-data-rate encipherment should not add significantly to the overall cost of such devices.

  More significant to the design of security features are the likely needs of users of these multimedia services; these needs are potentially very different from those of "voice" users of existing networks. Of particular importance are likely to be issues such as end-to-end integrity and confidentiality, albeit that existing networks do not support integrity, and only provide encryption for the air interface. Of all the end-to-end security features, end-to-end confidentiality raises most problems. The problems are mainly political rather than technical, and arise from the need of law enforcement agencies for access to certain communications paths, when a warrant exists. Such access is valuable in combating criminal activity, but also needs to be carefully controlled because of the civil liberties issues. This issue has given rise to a public debate on "key escrow" schemes, starting with the US *Clipper* scheme; see, for example [8]. There is a growing consensus that trusted third parties (TTPs) offer a means of supporting warranted access at the same time as meeting legitimate user needs for confidentiality. A TTP-based scheme for warranted access has been developed, offering considerable advantages over its competitors [9].

- *Identity and location privacy.* In mobile telecommunications systems, each user must let its SP know where he/she is so that its call route can be maintained by the system. This is achieved by the registration and location update mechanisms which a user employs to inform its SP (via the NO) of its current location. This has the side effect that anyone wanting to track this particular user can do so by monitoring the identity and location messages transmitted during the registration and location update processes.

  Users of public telecommunications networks are likely to regard the possibility of their location being revealed by these mechanisms as an unacceptable breach of personal privacy. Thus, in order to prevent users' identity and location information being disclosed to unauthorised parties, *an Identity and Location Privacy* (ILP) mechanism is needed; such a mechanism protects users against tracing of their physical location by illegal means.

  Current GSM networks provide a level of user identity confidentiality, but the mechanism used is less appropriate for future networks, not least because of the multi-operator environment likely to prevail. New mechanisms, based on both public key and "conventional" cryptographic techniques, have been examined, and are the focus of the remainder of this chapter.

- *Simultaneous multiple access channel coding and encipherment.* The claim that CDMA, a likely multiple access method for future mobile networks, is inherently secure has been considered and rejected. Options for using CDMA sequences for encipherment have been examined by Brown [10].

- *Terminal-related security.* Current networks enable black-listing of stolen terminals, and detection of non-type-approved terminals. The need for such facilities in future was reviewed, given that most mobile terminals are likely to be relatively low-cost. Whether a universal scheme is adopted, or a scheme only applying to valuable (e.g. multimedia) terminals, remains a topic for debate.

11.4. IDENTITY AND LOCATION PRIVACY

The remainder of this chapter is concerned with two particularly important security services for future mobile multimedia networks: identity and location privacy (ILP) for the mobile user, and mutual authentication between mobile user and base station. We start by considering in detail the provision of ILP services. Subsequently we consider two mutual authentication mechanisms also providing ILP.

11.4.1. The GSM Approach

In GSM, ILP is achieved by using temporary identities (TIs) over the air interface instead of real identities (RIs).[1] The TI is chosen by an NO[2] and is valid only in a given location area. The SP[3] maintains a database of current TI/RI relationships and can therefore determine the real identity of a user, i.e. it can determine the RI from the TI. TIs are changed on each location update and on certain other network-defined occasions.

In more detail, the user identifies himself by sending the old TI during each location update process (this occurs prior to authentication, and the TI must therefore be sent unencrypted). The new TI is returned after authentication is complete and a new session key has been generated, and hence the new TI can be, and is, encrypted when sent to the user. This prevents an interceptor from linking one TI to the next, and blocks tracing of user movements by linking TIs. If a TI is unavailable or invalid, e.g. if during the initial location registration the old NO is unreachable or the old TI is unknown, [11], then a user has to identify itself using its RI. In this event a new TI is allocated and returned encrypted.

11.4.2. Possible Threats to the GSM Approach

In this section we consider seven possible threats to the GSM ILP scheme.

T1. *Intercepting communications between user and NO.* An intruder can obtain an RI from the GSM air interface whenever an RI is sent in clear text, i.e. in the following cases: initial location registration, "old visitor location register unreachable", and "no old TI available".

T2. *Impersonating a user.* In a mobile telecommunications environment an intruder may be able to fabricate and/or interfere with a user's messages to an NO. An intruder could modify the user's TI and/or the location area identifier, both of which are sent from user to NO in clear text. This will mean that the NO fails to recognise the user (or is unable to contact the "old" NO), causing the NO to ask the user to send its RI unencrypted over the air interface. Such a procedure could be repeated, enabling an intruder to track a user.

T3. *Impersonating an NO.* In GSM, user authentication is unilateral, i.e. the NO verifies the user's identity, but the user does not verify the NO's identity. Hence an intruder could impersonate an NO and instruct a user to send its RI unencrypted over the air interface. As is the case for threat T2, such a procedure could be repeated as often as required, enabling an intruder to track a user.

T4. *Intercepting channels between NOs and SPs.* If an intruder could monitor the channel between NO and SP, it could observe a user's identity and location

---

[1] In GSM TIs and RIs are called temporary mobile subscriber identities (TMSIs) and international mobile subscriber identities (IMSIs) respectively.
[2] In GSM an NO is a base station subsystem, mobile switching centre and visitor location register (BSS/MSC/VLR).
[3] In GSM an SP is an authentication centre and home location register (AuC/HLR).

information, and hence track a user, because each updated location message is sent from an NO to an SP, possibly in clear text.

T5. *Malicious NOs.* It is possible for a malicious NO to track a user because TIs are chosen by NOs, and hence NOs have access to a user's RI.

T6. *Impersonating an SP to an NO.* In GSM the SP verifies the user's identity during the user authentication process, but no mechanisms are provided for the NO and/or the user to verify the SP's identity. In practice where such a threat exists proprietary techniques are used to protect SP/NO communications, and hence (indirectly) protect the user against an intruder impersonating an SP. However, if the NO does not authenticate the SP, then an intruder could impersonate an SP to an NO to obtain the user's identity and location information (and thereby track the user).

T7. *Malicious SPs.* A user's physical location could be disclosed to an intruder if an SP abuses the user's identity and location information. However, it is essential that the SP knows the user's identity and location since the user has a contractual/charging relationship with its SP. Hence SPs will need to protect their users against breaches of privacy, and utilise secure access control and audit mechanisms for their user databases.

11.4.3. Requirements for an ILP Mechanism

We now list general requirements for ILP mechanisms, based on our analysis of GSM.

- The user's RI should never be transmitted unprotected across the air interface (hence addressing threats T1, T2 and T3).
- The user's RI should never be transmitted unprotected between network entities (NOs and/or SPs), unless the communications path is inherently secure (hence addressing threat T4).
- The user's RI should only be given to parties needing it for correct network operation; in the limit this could mean that the user's SP is the only entity knowing the user's RI (addressing threat T5). For service provision, only the user's SP needs to know the user's RI, since when an NO provides service to a user it only needs to know the user's TI and who the user's SP is, so that the NO can subsequently charge the SP for service provided to the user (the SP will also need to keep the TI so that the charge can be matched against the user's RI).
- Third parties should be unable to track users by impersonating an SP to an NO, an NO to a user, a user to an NO, or an NO to an SP (hence addressing threats T2, T3 and T6).

Not all these requirements can always be met in a practical system, although at least the first requirement should always be met (unlike in GSM).

11.4.4. General Approaches for Providing ILP

We now discuss two general approaches for providing ILP, which typically occurs in combination with entity authentication. Section 11.5 contains examples of the two approaches.

The fundamental problem is to meet the first identified requirement, i.e. to avoid transmission of users' RIs on the air interface. Note that the reason why addresses of some kind need to be sent across the air interface is because it is a broadcast medium; NOs need to have a means of distinguishing between users, and users need to have a way of deciding which communications are meant for them.

In the first approach, where *symmetric encipherment* is used, addresses cannot be enciphered. This is because the NO needs to know which key to use to decipher an address, i.e. the NO needs to read the address *before* deciphering it. Similarly, a user needs to read an address embedded in an enciphered data string before deciding whether it should attempt to decipher it. Of course, these problems disappear if all entities use the same key, but this is very insecure and we do not consider this approach further here. This has led to the use of temporary identities (as in GSM) where the RI is not used as an address, and instead a "temporary" address (TI) is used to identify a user, and this TI changes at regular intervals. The new TI is chosen by the NO and sent to the user in enciphered form, thus preventing an intruder from linking old TIs to new ones. The problem with the GSM approach is the need to use the user's RI prior to setting up an initial TI; this problem can be avoided by using two levels of TIs, as in the approach of section 11.5.1. Apart from this example, another scheme using TIs has been proposed by Mu and Varadharajan, [12], who refer to *subliminal identities* instead of TIs.

In the second approach, where *asymmetric encipherment* is used, it is possible to encipher addresses, at least on the "up link", i.e. in communications between mobile users and an NO. This is because users can encipher data sent to the NO using the NO's public encipherment key. Protecting the "down link" is rather more problematic, and still requires the use of some form of TI. However the "set up" problems associated with GSM can probably be avoided by using this approach. The only remaining problem is to ensure that a user knows which NO it is sending to (and hence can use the right public key), and possesses reliable copies of public encipherment keys for all NOs it may wish to use. An example of such a scheme is given in section 11.5.2 and another scheme of this type is reported in section 9.4 of reference [13].

In addition, Beller *et al.*, [14], give one symmetric-based and three asymmetric-based authentication protocols for use in mobile systems. Whilst the symmetric-based mechanism does not provide ILP services, the asymmetric-based protocols provide a level of ILP by encrypting the user identity using the public key of an entity roughly corresponding to our NO, thus ensuring that only the NO knows the user's true identity. Carlsen, [15], proposed some enhancements to the protocols in [14], although the ILP mechanisms remain the same. Federrath *et al.*, [16], proposed an ILP scheme for mobile systems which prevents a user's SP from tracking a user's movements, and Jackson, [17], in the same proceedings, proposed a very similar scheme to prevent "management" from spying on users. In these schemes, a mobile user needs to know the entire route from himself to his SP (consisting of a number of

NOs), all these NOs' public keys, and also has to compute asymmetric encryptions several times (one for each NO in the route) during every location update process. These requirements are probably unrealistic for the real mobile user with limited computational power and memory.

11.4.5. Legal and Operational Limitations on ILP

In the discussion of ILP requirements in section 11.4.3, we ignored the domain management requirements applying to NOs. There are two issues, applying in some domains, affecting the provision of ILP.

- The *calling line identifier (CLI)* requirement necessitates that called entities are provided with the CLI (which typically means the telephone number) of the party calling them.
- The *warranted interception* requirement means that law enforcement agencies must be given access to certain calls starting or terminating within their domain, typically when an interception warrant has been issued. In principle this requirement could also be applied to all calls routed through a domain, even if they do not start or terminate within that domain, although this is unlikely (see [8]). For details of evolving European rules see [18].

This means that some NOs may need to know the RIs of users sending and/or receiving calls within their network. However, this does not mean it is essential for the ILP scheme used to always transfer a user's RI to an NO. It may be more appropriate to have RIs routinely transferred from SPs to NOs only when NOs need them for legal and/or operational reasons. Thus one could envisage a situation where some NOs will (by law) not provide service to a user unless the user's SP is prepared to provide the user's RI to the NO, whilst some users may be so concerned about privacy that they refuse to use their mobile telephone in networks where their RI has to be divulged. Hence, if the ILP mechanism can avoid the need for the user's RI to be distributed outside the SP, a whole range of privacy options become possible, giving both users and government agencies the maximum flexibility to manage ILP.

11.5. MECHANISMS FOR MUTUAL AUTHENTICATION PROVIDING ILP

In GSM networks it is theoretically possible for an intruder to masquerade as an NO by imitating a base station, as GSM only provides *unilateral* authentication of a user to an NO. For GSM it is hard to see how the intruder could gain much from doing this; however, in third generation systems it is likely that NOs will have much more over-the-air control of users. For instance, they may be able to disable faulty terminals directly, or write billing data direct to the user identity module (UIM, the UMTS equivalent of a subscriber identity module or SIM). Thus *mutual* (two-way) entity authentication is necessary.

In both mechanisms described, the NO is not automatically given the user's RI; if required for legal or operational reasons, the RI can be sent from SP to NO in addition to the specified information. Also in both mechanisms the SP acts as a TTP to help provide authentication and key establishment.

11.5.1. A Mechanism Based on Symmetric Cryptography

Background

This mechanism was previously outlined in [2]. It has the advantage that it establishes a temporary user-NO key, i.e. there is no need for NO-SP communication once a user has registered with an NO. This contrasts with the GSM scheme, which needs regular NO-SP communications to transfer challenge–response pairs. It combines the provision of ILP, entity authentication and session key generation in a single mechanism, and also conforms to the relevant ISO/IEC standard, [19].

The mechanism provides the following security features:

1. mutual entity authentication between user and NO;
2. user identity confidentiality over the communications path between user and NO;
3. session key establishment between user and NO for use in providing other security features, e.g. for confidentiality and/or integrity for data passed between user and NO.

The mechanism makes use of the following types of cryptographic key:

- *User-SP key*: K_{SU}, a secret key known only to a user and its SP, and which remains fixed for long periods of time.
- *User-NO key*: K_{NU}, a secret key known only to a user, its SP and its "current" NO. These keys may remain fixed while a user is registered with an NO. Associated with every such key is a key offset (KO), which is used in conjunction with the user-SP key K_{SU} to generate K_{NU}.
- *Session key*: K_S, a secret key known only to a user and its current NO, i.e. the NO with whom the user is registered. A new session key, for use in data encipherment and/or other security features, is generated as a result of every use of the authentication mechanism.

The mechanism makes use of the following cryptographic algorithms:

- *User authentication algorithm*: A_U, which takes as input a secret key and data string and outputs a check value RES.
- *SP authentication algorithm*: A_S, which takes as input a secret key and data string and outputs a check value RES. This algorithm may be the same as A_U.
- *Identity hiding algorithm*: C_U, which takes as input a secret key and data string and outputs a string CIPH used to conceal a user identity.
- *Session key generation algorithm*: A_K, which takes as input a secret key and data string and outputs a session key K_S.

- *User-NO key generation algorithm:* A_N, which takes as input a secret key and data string and outputs a user-NO secret key K_{NU}. This algorithm may be the same as A_K.

The mechanism makes use of the following types of temporary identifiers:

- *Temporary user identity for NO:* TI_N, an identity used to identify a user to the NO with which they are currently registered. It is known to the user and the current NO.
- *Temporary user identity for SP:* TI_S, an identity used to identify a user to its SP. It is known to the user and its SP.

There are two versions of the mechanism, depending on whether or not the user is currently registered with the NO; we consider them separately, although they are closely related. In the description, as throughout, $X \| Y$ denotes the concatenation of data items X and Y.

Current Registrations

We first consider the case where the user is already registered with the NO, so that the user and NO share a valid temporary identity TI_N and secret key K_{NU}. The mechanism for this case consists of three messages exchanged between user and NO (the SP is not involved):

1. **user → NO:** TI_N, RND_U
2. **NO → user:** $RND_N, TI'_N \oplus CIPH_N, RES_N$
3. **user → NO:** RES_U

RND_U and RND_N are random "challenges" generated by user and NO respectively. RES_U and RES_N are "challenge responses" generated by user and NO respectively, where $RES_N = A_U(K_{NU}, RND_N \| RND_U \| TI'_N)$, and $RES_U = A_U(K_{NU}, RND_U \| RND_N)$. TI'_N is the "new" user TI for use with the NO, and will replace the current value TI_N. $CIPH_N$ is a string of bits used to conceal TI'_N whilst in transit between NO and user, where $CIPH_N = C_U(K_{NU}, RND_U)$. The user and NO can compute a session key K_S as $K_S = A_K(K_{NU}, RND_U \| RND_N \| TI'_N)$.

New Registrations

We then consider the case where the user is not registered with the NO, and so user and NO do not share any information. The mechanism for this case consists of five messages exchanged between user, NO, and the user's SP.

1. **user → NO:** TI_S, RND_U
2. **NO → SP:** TI_S, RND_U
3. **SP → NO:** $TI'_S \oplus CIPH_S, KO, K_{NU}, RES_S$
4. **NO → user:** $TI'_S \oplus CIPH_S, KO, RES_S, RND_N, TI'_N \oplus CIPH_N, RES_N$
5. **user → NO:** RES_U

First note that we assume that a secure channel is available for exchanging messages 2 and 3 between NO and SP. As previously, RND_U and RND_N are random "challenges" generated by user and NO respectively, and RES_U, RES_N, and RES_S are "challenge responses" generated by user, NO, and SP respectively. RES_N and RES_U are calculated as in section 11.5.2, and $RES_S = A_S(K_{SU}, RND_U || KO || TI'_S)$. TI'_S is the "new" user TI for use with the SP, and will replace the current value TI_S. As previously, TI'_N is the "new" user TI for use with the NO. $CIPH_S$ is a string of bits used to conceal TI'_S whilst in transit between SP and user, where $CIPH_S = C_U(K_{SU}, RND_U)$. $CIPH_N$ (computed as previously) is a bit-string used to conceal the new TI TI'_N whilst in transit between NO and user. On receipt of message 4, the user can compute the NO secret key $K_{NU} = A_N(K_{SU}, KO || NOID)$, where $NOID$ is the NO's identifier; the same calculation is done by the SP on receipt of message 2. As previously, user and NO can compute session key $K_S = A_K(K_{NU}, RND_U || RND_N || TI'_N)$. As a result of the mechanism, user and NO will share a secret key K_{NU} and a TI, TI'_N.

11.5.2. A Mechanism Based on Asymmetric Cryptography

Requirements

This mechanism is based on a combination of public key encipherment and symmetric cryptographic techniques. Nonces are used for checking timeliness. The following cryptographic functions are used.

- A *public key encipherment function E* (which the user and SP must implement). We use $E_{K+}[X]$ to denote public key encipherment of data X using public encipherment key $K+$.
- A *cryptographic check function f* (which the user, NO and SP must implement). We use $f_K(X)$ to denote the (check-value) output of f given input data X and key K.
- A *symmetric encipherment function e* (which user, NO and SP must implement). We use $e_K(X)$ to denote the output of e given input data X and key K. This encipherment algorithm must provide integrity and origin authentication (cf. requirements (a), (b) in Clause 4 of ISO/IEC 11770-2, [20]). If necessary, the encipherment algorithms used by the two pairs: user/SP, and NO/SP, can be distinct; we have assumed that a single algorithm is used to simplify the presentation.

The following keys need to be in place:

- The SP needs to generate a public key/private key pair for the public key encipherment algorithm. The user must have a reliable copy of the SP's public encipherment key, K_{S+}.
- The user and SP must share a secret key K'_{US} for the cryptographic check function f.
- The two entity pairs: user/SP, and NO/SP, both need to share a secret key for the symmetric encipherment algorithm, denoted by K_{US} and K_{NS} respectively.

In addition the user, NO and SP must be able to generate non-repeating nonces, the user must be able to generate temporary identities, and the SP must be able to generate session keys.

The Protocol

The following protocol (partly) conforms to Key Establishment Mechanism 9, specified in Clause 6.3 of ISO/IEC 11770-2, [20]. One point at which it significantly diverges from the standard is that the user's identity U is never sent in clear text and is known only to the SP and itself (the standard protocol would require U to be sent in clear text in message **M2**).

M1. user $\rightarrow$ **NO**: $R_U\|S\|N\|E_{K_{S+}}[U\|T_U\|f_{K'_{US}}(R_U\|U\|N\|T_U)]$

M2. NO $\rightarrow$ **SP**: $R_N\|R_U\|S\|N\|E_{K_{S+}}[U\|T_U\|f_{K'_{US}}(R_U\|U\|N\|T_U)]$

M3. SP $\rightarrow$ **NO**: $e_{K_{NS}}(R_N\|K_{UN}\|T_U)\|e_{K_{US}}(R_U\|K_{UN}\|N\|T_U)$

M4. NO $\rightarrow$ **user**: $T_U\|R'_N\|e_{K_{US}}(R_U\|K_{UN}\|N\|T_U)\|f_{K_{UN}}(R'_N\|R_U\|T_U)$

M5. user $\rightarrow$ **NO**: $f_{K_{UN}}(R_U\|R'_N\|N)$

The protocol procedure is as follows; if at any point a check fails, then the protocol is aborted.

1. The user generates and stores a nonce R_U, and generates a new temporary identity, T_U. The user then sends the NO an authentication request **M1**, in which it lets the NO know its SP is S. The user's real identity (U) is enciphered using K_{S+} so that only the SP can read it.

2. In **M2** the NO forwards the user's request to the SP, appending (and storing) a nonce R_N.

3. On receipt of **M2** the SP deciphers the enciphered string using its private key. The SP then checks the output of f using its copy of K'_{US}. The SP retrieves the temporary identity T_U, generates a session key K_{UN} for use by user and NO, and distributes them in **M3**. The SP maintains a database of relationships between users and temporary identities.

4. On receipt of **M3**, the NO deciphers (and simultaneously integrity checks) the first part of the message. The NO then checks that the nonce it contains is correct, and also uses the nonce to link the message with the correct "transaction". The NO then retrieves the new temporary identity T_U and session key K_{UN}, and uses the latter to generate the check-value in message **M4** which is a function of a second nonce, R'_N, which the NO also stores. Note that, when using broadcast channels, the user's address must be embedded in any message sent to it. Thus message **M4** is prefixed with T_U, to indicate that, if necessary, T_U can be used as the broadcast address for user U without compromising user U's anonymity.

5. On receipt of **M4**, the user deciphers (and integrity checks) the enciphered part. The user checks that the nonce it contains is correct, and retrieves the new session key K_{UN}, which is then used to verify the check-value in the message and to generate message **M5**.

6. On receipt of **M5**, the NO verifies the check-value by recomputing it.

The NO is not given the user's RI, and can only identify a user by the temporary identity T_U supplied by the SP. The NO will use the temporary identity T_U when communicating with SP in order to be recompensed for the cost of providing service to the user.

11.6. CONCLUSIONS

The two protocols described here have the following advantages over the GSM approach mentioned in section 11.4.1.

1. User RIs are never transmitted in clear text in the mobile radio path (or, for the 2nd mechanism, in the NO-SP channel).
2. NOs are not given access to a user's RI.
3. Authentication of both NO and SP is implicitly included.

The protocols can prevent threats T1–T6 in section 11.4.2. Threat T7, i.e. that an SP abuses user identity and location information, can only be prevented by internal management controls imposed by an SP.

A variant of SVO logic, [21], has been used to verify the mechanisms' correctness; in fact logical analysis revealed a subtle flaw in a previous version of the first mechanism which has now been corrected.

The cost of the second mechanism as compared with conventional protocols, for example that presented in section 11.5.1, is as follows. Each SP must have a public key known to all its users and keep a corresponding private key secret, and each user has to compute $E_{K_{S_+}}[U||T_U||f_{K'_{US}}(R_U||U||N||T_U)]$, which has then to be checked by the SP. Note that, for the RSA algorithm, an encryption operation can be made significantly more efficient than a signature operation, since a relatively small public exponent can be chosen. Moreover, transmission of a user-computed signature could also potentially compromise the confidentiality of a user, if the user's public verification key is widely known.

Finally it should be noted that both protocols rely on the shared key K_{US} remaining secret long term; other slightly more complex versions of the mechanisms can be devised which do not have this requirement. Also, variants of the second protocol can be devised to deal with various location update requirements, including a three-message scheme corresponding to the current registration case of section 11.5.1.

ACKNOWLEDGEMENTS

The work described in this chapter has been performed under the DTI/EPSRC-funded LINK project "Third Generation System Security Studies" (3GS3). This project was performed in collaboration with Vodafone Ltd, GPT Ltd. The authors would like to acknowledge the invaluable support and advice of colleagues in 3GS3, without which this chapter could not have been written.

REFERENCES

[1] Security features for third generation systems. UK DTI/EPSRC LINK PCP 3GS3 Technical Report 1. Vodafone Ltd, GPT Ltd, Royal Holloway, University of London. Final version, February 1996.

[2] Chen, L., Gollmann, D. and Mitchell, C., Tailoring authentication protocols to match underlying mechanisms. In Pieprzyk, J. and Seberry, J. (eds), *Information Security and Privacy*, Springer-Verlag LNCS 1172, 121–133, 1996.

[3] Chen, L., Gollmann, D. and Mitchell, C.J., Distributing trust amongst multiple authentication servers. *Journal of Computer Security*, **3**, 255–267, 1994/95.

[4] Chen, L, Gollmann, D. and Mitchell, C. J., Authentication using minimally trusted servers. *ACM Operating Systems Review*, **31** (3), 16–28, 1997.

[5] Maurer, U. M., Secret key agreement by public discussion from common information. *IEEE Transactions on Information Theory*, **39**, 733–742, 1993.

[6] Wyner, A. D., The wire-tap channel. *Bell System Technical Journal*, **35**, 1355–1387, 1975.

[7] Mitchell, C. J., A storage complexity based analogue of Maurer key establishment using public channels. In Boyd, C. (ed.), *Cryptography and Coding – Proceedings 5th IMA Conference, Cirencester*, Springer-Verlag LNCS 1025, 84–93, 1995.

[8] Hoyle, M. P. and Mitchell, C. J., On solutions to the key escrow problem. In *Proceedings of State of the Art and Evolution of Computer Security and Industrial Cryptography*, Leuven, June 1997.

[9] Jefferies, N., Mitchell, C. and Walker, M., A proposed architecture for trusted third party services. In Dawson, E. and Golic, J. (eds), *Cryptography: Policy and Algorithms*, Springer-Verlag LNCS 1029, 98–104, 1996.

[10] Brown, J., Combined multiple access and encryption for CDMA systems. In *Proceedings of the 3rd International Symposium on Communication Theory and Applications*, Ambleside, UK, July 1995.

[11] Security related network functions. European Telecommunications Standards Institute, ETSI/PT12 GSM-03.20, August 1992.

[12] Mu, Y. and Varadharajan, V., On the design of security protocols for mobile communications. In Pieprzyk, J. and Seberry, J. (eds), *Information Security and Privacy*, Springer-Verlag LNCS 1172, 134–145, 1996.

[13] Security mechanisms for third generation systems. UK DTI/EPSRC LINK PCP 3GS3 Technical Report 2. Vodafone Ltd, GPT Ltd, Royal Holloway, University of London, May 1996.

[14] Beller, M. J., Chang, L. and Yacobi, Y., Privacy and authentication on a portable communications system, *IEEE J. on Selected Areas in Comms*, **11**, 821–829, 1993.

[15] Carlsen, U., Optimal privacy and authentication on a portable communication system. *ACM Operating Systems Review*, **28** (3), 16–23, 1994.

[16] Federrath, H., Jerichow, A. and Pfitzmann, A., MIXes in mobile communication systems: Location management with privacy. In Anderson, R. (ed.), *Information Hiding*, Springer-Verlag LNCS 1174, 121–135, 1996.

[17] Jackson, I. W., Anonymous addresses and confidentiality of location. In Anderson, R. (ed.), *Information Hiding*, Springer-Verlag LNCS 1174, 115–120, 1996.

[18] International requirements for the lawful interception of communications, European Union Council Resolution, January 1995.

[19] Information technology – Security techniques – Entity authentication – Part 4: Mechanisms using a cryptographic check function. ISO/IEC 9798-4, International Organization for Standardization, Geneva, Switzerland, 1995.

[20] Information technology – Security techniques – Key management – Part 2: Mechanisms using symmetric techniques. ISO/IEC 11770-2. International Organization for Standardization, Geneva, Switzerland, 1996.

[21] Syverson, P. and van Oorschot, P. C., On unifying some cryptographic protocol logics. *Proceedings of the 1994 IEEE Computer Society Symposium on Research in Security and Privacy*. IEEE Computer Society Press, 14–28, 1994.

Part 4

Source Coding: Speech and Audio

There is a great deal of interest in developing high-quality, low-bit-rate speech coding schemes for mobile multimedia applications. In recent years there has been a lot of interest in providing real-time speech over the Internet and the next generation of cellular system. These have resulted in a number of ITU and ETSI standards with bit rates as low as 2.4 kb/s. It is also widely accepted that the quality in many multimedia applications is determined primarily by the speech coding algorithms. Hence there is a lot of interest in developing high-quality speech coding algorithms suitable for such applications. The chapters in this section show the different speech and audio coding schemes for mobile multimedia applications.

In Chapter 12 the author provides an excellent review of the speech coding schemes from ADPCM to more advanced VS-CELP coders. The CELP-based technology has been widely used in many of the speech coding standards developed by ITU due to its superior performance at low bit rates. The chapter also deals with quality measures and provides a good comparison between the different speech coding standards. Some of the more advanced speech coding techniques based on sinusoidal modelling and waveform interpolation are presented in Chapter 13. The authors list four different sinusoidal transform coding (STC) schemes for low-bit-rate speech coding. They highlight some of the practical as well as the theoretical aspects of STC and conclude with a performance comparison of different STC schemes. The performance of the speech codec will be affected by other sub-systems in the transceiver. Chapter 14 investigates ten different combinations of sub-systems for modulation, channel coding and investigates the performance of two speech coding schemes, 32 kb/s ADPCM and 13 kb/s GSM speech codec, for wireless applications. The authors show that the performance of the speech coding will be determined by the components in the transceiver and the channel characteristics. They conclude that for Gaussian channels ADPCM can provide better-quality speech with low complexity while the GSM codec is more robust and performed equally well. In Chapter 15 the author proceeds to show how a dual-mode adaptive speech codec can be employed for wireless applications. The bit rate of the dual mode codec is controlled by the network and will invoke higher bit rates when the channel SNR is very high and switch to a lower rate for low channel SNRs. The error sensitivity of the different speech parameters was investigated and an unequal error protection scheme was proposed to improve the performance in hostile channel conditions.

Chapter 16 proposes a scalable audio coding scheme for mobile multimedia applications. The authors review the MPEG-2 audio coding and propose a wavelet scheme for low-bit-rate audio coding. The performance of the wavelet coder was shown to produce transparent quality audio at 32 kb/s and was considered to be superior to MPEG-2 audio. In Chapter 17 the authors propose a high-quality low-delay speech codec based on the CELP algorithm. By using a QMF filter bank the wideband signal is split into two components and coded separately to achieve better performance. The authors show the improvement in speech quality and found the performance of the proposed scheme at 16 kb/s to be comparable to G.722 speech coding standard at 48 kb/s.

12

Speech Coding for Mobile Telecommunications

C. I. Parris

12.1. INTRODUCTION

Low-bit-rate speech coding is an important requirement in mobile telecommunications systems. In recent years speech coding has gone through a revolutionary development culminating in a number of ITU and ETSI standards with bit rates as low as 2.4 kb/s. This chapter reviews the speech coding technology employed in existing cellular mobile communication systems. Without exception all current systems employ linear predictive coding techniques. The basics of linear predictive coding

Insights into Mobile Multimedia Communication
ISBN 0-12-140310-6

are presented here. Code Excited Linear Prediction (CELP) represents the state of the art in speech coding for near transparent coding of clean speech at bit rates as low as 4 kb/s. The CELP cornerstones of vector quantisation, analysis-by-synthesis and perceptual filtering are described. Power consumption, which translates directly to implementation complexity, is a key concern in mobile applications and is presented here along with other key performance measures such as the processor and memory requirements.

12.2. LINEAR PREDICTIVE (LP) CODING

Speech waveforms are highly correlated and thus exhibit a considerable amount of redundancy. In linear predictive coding (LP) each speech sample is predicted from a linear combination of previous speech samples. Typically speech codecs utilise two predictors, a short-term predictor which operates on the most recent speech samples and a long-term predictor which operates on samples one pitch period previous. Long-term predictors are commonly referred to as pitch predictors.

12.2.1. Short-term Prediction

Figure 12.1 illustrates the short-term prediction process for an eighth-order predictor as used in the full-rate GSM system. As might be expected, the higher the prediction order the greater the compression that can be achieved. Nevertheless the performance increase tends to level off when the predictor order reaches sixteen for male speech and twenty for female speech.

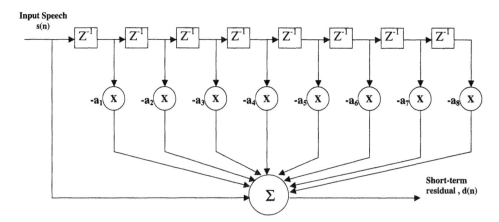

Figure 12.1 Short-term linear prediction.

For the predictor in Figure 12.1, the transfer function is given by

$$A(z) = 1 - \sum_{i=1}^{8} a_i z^{-i} \tag{12.1}$$

$A(z)$ is referred to as the LPC analysis filter and the coefficients a_i as the linear predictive coefficients. There are a number of methods to perform the calculation of the LP coefficients, each of which produces slightly different results based on different assumptions on the windowing applied [1]. The short-term predictor models the gross spectral shape of the speech or the *formants*. The short-term residual signal $d(n)$ has a flattened spectrum which is noise-like. However, in voiced segments of speech, some periodicity remains in $d(n)$ due to the regular excitation in the glottis of the speaker. This periodic redundancy can be modelled by long-term prediction.

12.2.2. Long-term Prediction (LTP)

Long-term prediction is in principle similar to short prediction but the predictor delay line now spans a much wider range, which corresponds to the pitch periods of human voice, typically varying from 20 to 120 samples. If LPC analysis is performed on voiced speech with such a high-order predictor, it is found that the majority of coefficients are near zero, with a cluster of significantly sized coefficients around the delay corresponding to the pitch of the speech. In many applications, long-term predictors are restricted to have only one non-zero tap (first order) at the delay (or lag) which is specified as the pitch period. Long-term prediction is also called pitch prediction. The pitch predictor transfer function is given by

$$B(z) = 1 - b_c z^{-N_c} \tag{12.2}$$

where b_c is the linear predictive coefficient (pitch gain) and N_c is the index of the peak non-zero coefficient (pitch).

Figure 12.2 illustrates the basic LP coder structure. Here the parameters of the two predictors are calculated at the encoder, quantised and transmitted to the decoder. The long-term residual signal $e(n)$ has a flat spectrum with little or no harmonic structure. However, it still contains important noise-like features of the speech, especially for unvoiced regions. It is noted that this structure is *forward adaptive* because the predictor coefficients are derived from the input speech and then explicitly relayed as side information. The alternative *backward adaptive* form derives the predictor coefficients from the output signal estimate and no side information is used.

12.2.3. Latency and Backward Adaptive Coders

Forward adaptive coders operate by optimising the predictor parameters over a fixed window of speech. The short-term predictor is updated at the frame rate which is of the order of 10 to 30 ms with 20 ms most commonly used for mobile applications. The pitch predictor is usually updated at the sub-frame rate which is typically every 3

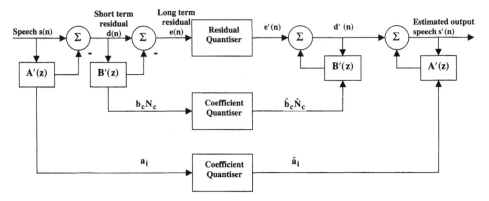

Figure 12.2 Forward adaptive linear predictive codec.

to 10 ms. The analysis process requires a full frame of speech at the input buffer before processing can commence and this introduces significant latency into the system. Most mobile communication systems operate with an end-to-end delay of three to four times the algorithmic latency of the encoder which may sometimes dictate the need for acoustic echo cancellation. There are also mobile communications systems such as DECT which utilise low-delay speech compression. Backward adaptive LP coders are used in such systems.

As noted above in backward adaptive LP coders, the predictor parameters are derived from the past synthetic speech output rather than the actual speech signal. In order to keep the encoder and decoder in the same state, the predictor parameters are derived locally at both the encoder and decoder. The optimisation window is restricted to the past synthetic speech samples; however, the update rate may be anything up to the speech sample rate (typically 8 kHz). Backward adaption has the additional advantage that no bandwidth is required to transmit these parameters, allowing higher prediction orders. In the absence of channel errors the synthetic speech at the decoder will be identical to that which can be locally derived in the encoder, this is essential for the encoder and decoder to track. Currently backward adaptive LP coders require around twice the bit rate for comparable speech quality compared to their forward adaptive counterparts. This is illustrated in the next section in the comparison of the G.728 [2] and G.729 speech compression standards. They are less robust to channel bit errors. The DECT system uses a low-delay speech codec G.721.

12.2.4. Classification of LP Coders

The LP coders can be classified according to how the long-term residual is modelled and quantised. The major classes are

- adaptive differential pulse code modulation (ADPCM);
- residual pulse excitation (RPE);
- multi-pulse excitation (MPE);
- codebook excitation (CELP).

ADPCM

In ADPCM the residual is quantised using an adaptive scalar quantiser. An example of such a codec is the G.721 speech codec used in DECT. G.721 scalar quantises each sample of the residual using a 4-bit adaptive quantiser. The range of the quantiser is adapted to the rate of change of the residual signal. Since G.721 is a backward adaptive coder, only the residual is quantised leading to a bit rate of 32 kbit/s (the sample rate of telephony speech being 8 kHz). Each sample is quantised in isolation and so the algorithmic delay is one sample period, 0.125 ms. The G.726 standard encapsulates G.721 to extend to ADPCM codecs operating at 16, 24 and 48 kbit/s. At 32 kbit/s and above, the ADPCM codec is considered capable of transparent speech quality (toll quality).

RPE

RPE coding is currently only used in forward adaptive coders. RPE coders model the residual as a regular pulse train where the pulse spacing is preset by a decimation ratio. The pulse grid or phase is updated at the sub-frame rate. Each pulse is scalar quantised individually but for quantisation efficiency it is also normalised to the largest pulse in a sub-frame hence each sub-frame of residual $e(n)$ is treated as a vector e.

An example of an RPE codec is the full-rate GSM (FR-GSM) speech coder [3]. It is also called the residual pulse excitation – long-term predictive (RPE–LTP) coder. The FR-GSM coder operates with a frame size of 20 ms and a sub-frame size of 5 ms. The long-term residual signal e for an entire sub-frame is treated as a vector and quantised as a block. In this coder, the compression of e is achieved by first decimating by three to reduce the number of samples in the vector from 40 to 13. The decimation is adaptive in that all four possible phases are considered. The phase with the highest energy is then quantised using adaptive pulse code modulation (APCM). A two-bit index M_c is used to indicate the phase selected. Prior to decimation a decimation filter, or weighting filter, is applied to prevent excessive aliasing. APCM as used in the GSM system involves quantising logarithmically the largest sample in the vector of 13 using six bits (x_{max}) and then all 13 samples are normalised to this value before quantisation at three bits per sample. FR-GSM operates at 13 kbit/s and offers near transparent speech quality. Out of the 13 k bit/s, 9.4 kbit/s are used for the quantisation of the residual.

Multi-pulse Excitation

Multi-pulse excitation is an extension of the RPE scheme where the pulses are not restricted to be regularly spaced. The position and amplitude of each pulse are quantised separately. Compared to the RPE model, multi-pulse excitation requires fewer pulses to achieve the same quality at a lower bit rate. An example is the British Telecom Sky-Phone codec which operates at 9.6 kbit/s, which offers near transparent speech quality.

CELP

Codebook excited linear predictive coders (CELP) vector quantise the residual. Typically a gain-shape vector quantiser is used where the codebook contains normalised vectors. The transmitted parameters are the codebook indices and the scaling gains. The vector dimension is typically the sub-frame size. Figure 12.3 illustrates the CELP synthesis model. The adaptive codebook is a closed-loop long-term predictor and the post-filtering attempts to enhance the synthetic speech quality.

Many acronyms exist for CELP coders which utilise codebooks which exhibit particular properties; the common acronyms are given below:

- SELP for stochastic excited linear predictive coder;
- VSELP for vector sum excited linear predictive coder [4];
- ACELP for algebraic codebook excited linear predictive coder;
- CS-CELP for conjugate structure CELP;
- PSI-CELP for pitch synchronous innovation CELP;
- QCELP for Qualcomm CELP.

Examples of standardised CELP codecs for mobile applications include:

- GSM (half-rate) – VSELP;
- GSM (enhanced full rate) – ACELP;
- IS54 (full rate) - VSELP;
- IS95 – QCELP;
- IS136 (enhanced full rate) – ACELP;
- PCS1900 – ACELP;
- PDC (full-rate) – VSELP;
- PDC (half-rate) – PSI-CELP.

12.3. FULL-RATE GSM CODEC OVERVIEW

Table 12.1 summarises the parameters of the RPE-LTP coder. The log area ratio (LAR) parameters are a transformation of the LPC parameters since the LPCs are

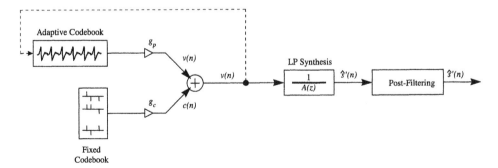

Figure 12.3 CELP synthesis model.

Table 12.1 Parameters of the full-rate GSM coder

Parameter	Update interval	Number of bits	Number of bits per frame
LAR_1		6	6
LAR_2		6	6
LAR_3		5	5
LAR_4	160 samples	5	5
LAR_5		4	4
LAR_6		4	4
LAR_7		3	3
LAR_8		3	3
pitch period N_c		7	28
pitch gain, b_c		2	8
grid index, M_c		2	8
block amplitude, X_{maxc}	40 samples	6	24
APCM sample $X_c(0)$		3	12
$X_c(1)$		3	12
:		:	:
$X_c(12)$		3	12
			260

never quantised directly due to their uneven error sensitivity. Other well-known transformation sets include the reflection coefficients and line spectral frequencies. With the exception of these eight LAR parameters, all other parameters are updated every sub-frame (40 samples). The LARs are updated every frame (160 samples), but are also interpolated during the first sub-frame to ensure smooth evolution of the synthetic speech spectrum.

12.4. OPEN-LOOP AND CLOSED-LOOP ANALYSIS

Two fundamental approaches exist in the calculation and quantisation of the parameters used to model speech waveforms: open-loop analysis and closed-loop analysis. In speech coding, closed-loop analysis is often referred to as analysis by synthesis.

12.4.1. Open-loop Analysis

In open-loop analysis the various speech parameters are derived directly from the input speech. The analysis involves correlation of delayed versions of waveforms derived from filtering the speech or its residuals. Quantisation is typically performed directly on the residual waveforms (as in ADPCM), on the correlation coefficients (such as pitch periods and pitch gains), or on a parameter set derived from the correlation coefficients (such as LARs).

12.4.2. Closed-loop Analysis

In closed-loop analysis the various speech parameters are derived by generating what would be the synthetic speech and comparing it against the original speech. The comparison usually involves the calculation of an objective metric such as the mean squared error or the signal-to-noise ratio between the original and synthetic speech. Ideally all possible synthetic waveforms should be generated and evaluated by taking all possible combinations of the quantiser levels and codebooks in all parameters. However, this is impractical in real-time implementations and so in practice each parameter is optimised and quantised sequentially.

12.4.3. Perceptual Filtering

Clearly the performance of the closed-loop analysis scheme is heavily dependent on the distortion metric used. One commonly used metric is the perceptually weighted signal-to-noise ratio. The perceptual weighting network is a filter, $W(z)$, derived from the short-term predictor.

$$W(z) = \frac{1 - \sum_k a_k z^{-k}}{1 - \sum_k a_k \alpha^k z^{-k}}$$

Figure 12.4 illustrates the quantisation noise spectrum obtained by using such a filter. The quantisation noise appears in the formant regions where it is masked by the higher energy. Noise outside the formant regions is suppressed, thereby improving the perceived speech quality.

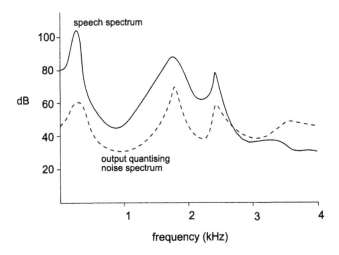

Figure 12.4 The effect of applying a weighted perceptual filter.

12.5. STANDARD CODECS

A number of standard speech coding techniques have been developed by the ITU and others in recent years. These are summarised in Table 12.2 (p. 202).

12.6. PERFORMANCE MEASURES

12.6.1. Quality

Measures of speech quality are primarily divided into two categories: subjective and objective. The ultimate measure of quality must be the satisfaction of the human user; therefore properly controlled subjective testing is very important. As the test population has to be large for the results to be statistically significant, subjective tests are expensive and hence used sparingly. They are however usually compulsory in the selection of standard coders when the quality has to be guaranteed in a public network.

The mean opinion score (MOS) is a subjective measure specified by the ITU-T for the evaluation of speech transmission quality in telephone networks [5]. The scale of scoring is given in Table 12.3. When applied to narrowband speech (200 to 3400 Hz) a score of 4 to 5 is referred to as network or toll quality speech, with 64 kbit/s G.711 PCM having a representative MOS of 4.53 [6]. An MOS score of 3.5 to 4.0 is referred to as communication quality. Such quality is generally held to be acceptable in applications such as voice mail and mobile radio. MOS scores of 2.5 to 3.5 are referred to as synthetic quality.

Table 12.4 illustrates the mean opinion scores for the some of codecs given in Table 12.2. It should be noted that since the various MOS scores in Table 12.4 were obtained during different evaluation programmes they are not directly comparable. For example most people would agree that the G.721 far exceeds the performance of G.729A particularly when background noise is present and that G.729 has superior speech quality to G.723.

Other subjective quality measures include the Quality Distortion Unit (QDU) and the Diagnostic Rhyme Test (DRT) [7]. The QDU is defined in ITU-T G.113 as the distortion introduced by one 64 kbit/s G.711 PCM codec pair. The DRT is a measure of intelligibility where the listener has to recognise one of two possible stimuli in each of 96 rhyming pairs. For most standard speech coders operating at above 4 kbit/s, the DRT scores are above 90%. This leaves little room to differentiate the performance of one coder against another. A high DRT score should be considered a prerequisite to any good quality speech coder.

Objective measures are cost-effective and repeatable but care must be taken to select the correct measure for a given type of speech coder. Examples of objective speech metrics include:

- signal to noise ratio (SNR) and perceptually weighted SNR as described above;
- articulation index (AI) [8];
- log spectral distortion (LSD) and log cepstral distance (LCD) [9];
- bark spectral distortion [10].

Table 12.2 Summary of standard codecs

Coder	Standard	Predictor type	Bit rate (kbit/s)	Frame rate (ms)	Sub-frame rate (ms)	STP order	LTP analysis	Residual quantisation	Residual modelling
G.726	ITU, DECT	backward	16/24/32/64	0.125	–	2^1	–	ADPCM	open-loop
FR	GSM, DCS1800, PCS1900	forward	13	20	5	8	open-loop	RPE	open-loop
HR	GSM	forward	5.6	20	5	10	open and closed-loop	CELP	closed-loop
G.728	ITU	backward	16	0.625	–	50	–	CELP	closed-loop
G.729	ITU	forward	8	10	5	10	open and closed-loop	ACELP	closed-loop
G.729A	ITU	forward	8	10	5	10	open and closed-loop	ACELP	closed-loop
G.723	ITU	forward	5.3/6.3	30	7.5	10	open and closed-loop	ACELP	closed-loop
EFR	GSM, DCS1800, PCS1900	forward	12.2	20	5	10	open and closed-loop	ACELP	closed-loop
EFR	IS136	forward	7.95	20	5	10	open and closed-loop	ACELP	closed-loop
FR	PDC, PHS	forward	6.9	20	5	10	closed-loop	VSELP	closed-loop
HR	PDC, PHS	forward	3.45	40	5	10	open and closed-loop	PSI-CELP	closed-loop
FR	IS54	forward	8	20	5	10	closed-loop	VSELP	closed-loop

[1]It is actually a combined second order auto-regressive predictor and a sixth order moving average predictor.

Table 12.3 The mean opinion score scale

Number scores	Quality scale	Impairment scale
5	Excellent	imperceptible
4	Good	(just) perceptible but not annoying
3	Fair	(perceptible and) slightly annoying
2	Poor	annoying (but not objectionable)
1	Unsatisfactory	very annoying (objectionable)

Table 12.4 The mean opinion scores for selected codecs

Coder	Bit rate (kbit/s)	MOS
G.721	32	3.8
FR-GSM	13	3.6
HR-GSM	5.6	3.2
G.728	16	3.8
G.729	8	3.9
G.729A	8	3.8
G.723	5.3/6.3	3.9
IS54	8	3.4

12.6.2. Complexity of Implementation

The most common methods of describing the complexity of an algorithm are millions of instructions per second (MIPS) and millions of operations per second (MOPS). Neither of these figures adequately defines the complexity of an algorithm. They are only meaningful in terms of the specific instruction set of a particular digital signal processor (DSP). Comparison of the same algorithm on different DSP devices cannot be quantitative and the machine cycle time should always be taken into account. What is really important is the total cost and power consumption of an implementation which are not dictated solely by the MIPS/MOPS figure and memory requirements. There is usually a trade-off between the requirements of an algorithm in terms of MIPS or MOPS and its memory requirement. Measurements of speech coding algorithm complexity based on combinations of MIPS/MOPS and memory requirements have been investigated but no single formula is available. Table 12.5 compares the processor requirements for implementations of the codecs given in Table 12.2. The MIPS quoted are for the TMS320C50 processor.

Implementation cost includes processor hardware, memory and firmware. For high-volume applications such as mobile communications, DSP vendors usually supply the applications software free of charge to promote their devices. For most of those applications, only a single DSP is required and the users' choice is a trade-off between device cost and power consumption. It is in this market place that application specific integrated circuits (ASICs) excel, offering the best price–performance

Table 12.5 The complexity for selected codecs

Coder	Bit rate (kbit/s)	MIPS
G.721	32	11
FR-GSM	13	4
HR-GSM	5.6	25
G.728	16	37
G.729	8	22
G.723	5.3/5.6	18
IS54	8	18

compromise. Previous experience has shown that an eight-channel ASIC implementation of the speech codecs in a DECT base station requires less power than a single-channel implementation on a TMS320C50 processor.

12.7. SUMMARY

In this chapter, an overview has been given to the background of linear predictive coding and a number of its applications in standard codecs. Associated issues including the speech quality, computational complexity and channel coding are also covered. For further reading, the following articles are recommended: [11,12,13].

REFERENCES

[1] Rabiner, L. R. and Schafer, R. W., *Digital Processing of Speech Signals*, Chapter 8, 396–455, Prentice-Hall, Englewood Cliffs, NJ, 1978.
[2] Chen, J-H, Cox, R. V., Lin, Y.-C., Jayant, N. and Melchner, M. J., A low-delay CELP coder for the CCITT 16 kbit/s speech coding standard. *IEEE Journal on Selected Areas in Communications*, 10 (5), 830–849, 1992.
[3] ETSI, Recommendations GSM 5 series, Groupe Speciale Mobile, European Conference of European Post and Telecommunications Administrations.
[4] Gerson, I. A. and Jasiuk, M. A., Vector sum excited linear prediction (VSELP). *Advances in Speech Coding*, Kluwer Academic Publishers, 1991.
[5] CCITT Recommendation P.77, Method for evaluation of service from the standpoint of speech transmission quality. CCITT Red Book vol. V, Geneva, 1985.
[6] Daumer, W. R., Subjective evaluation of pre- and post-filtering in PAM, PCM and DPCM voice communication systems. *IEEE Trans. Comm.*, April 1982.
[7] Voiers, W. D., Diagnostic evaluation of speech intelligibility. In Hawley, M. (ed.), *Speech Intelligibility and Speaker Recognition*, Dowden Hutchinson Ross, Stroudsburg, PA, 1977.
[8] Steeneken, H. J. M. and Houtgast, T., A physical method for measuring speech-transmission quality. *J. Acoust. Soc. Am.*, 67, 318–326, 1980.
[9] Kitawaki, N., Nagabuchi, H. and Itoh, K., Objective quality evaluation for low-bit-rate speech coding systems. *IEEE J. Selected Areas in Comms.*, 6, 242–248, 1988.
[10] Wang, S., Sekey, A. and Gersho, A., An objective measure for predicting subjective quality of speech coders. *IEEE J. Selected Areas in Comms.*, 10 (5), 819–829, June 1992.

[11]   Xydeas, C., An overview of speech coding techniques. *IEE Colloquium on Speech Coding – Techniques and Applications*, Digest No 1992/090, April 1992.
[12]   Boyd, I., Speech coding for telecommunications. *Electronics & Communication Engineering Journal*, 273–283, October 1992.
[13]   Demolitas, S., Standardising speech-coding technology for network applications. *IEEE Comms. Mag.*, **31** (11), 26–33, 1993.

13

Advanced Speech Coding Techniques

X. Q. Sun and B. M. G. Cheetham

13.1. INTRODUCTION

In recent years, many actively studied techniques for telephone band width speech coding at very low bit-rates (i.e. at bit-rates of 4.8 kb/s and below) have been based on sinusoidal modelling and waveform interpolation (WI). Among these techniques are *sinusoidal transform coding* (STC) [1,2,3], *multi-band excitation* coding (MBE) [4,5,6,7], *prototype waveform interpolation* (PWI) [8,9,10,11,12] and *time frequency interpolation* (TFI) [13,14]. A version of MBE has already been adopted for a commercial mobile communication system [6], and the others are being actively researched as strong candidates for future mobile applications. The aim of this chapter is to give some insight into the theory of some of the more recent sinusoidal modelling and WI techniques.

Insights into Mobile Multimedia Communication
ISBN 0-12-140310-6

13.2. SINUSOIDAL MODELLING AND WAVEFORM INTERPOLATION (WI)

These techniques characterise speech by analysing short representative segments extracted at suitable time intervals. Each analysis is considered to produce an *instantaneous* measurement of the speech waveform in the vicinity of an *update-point* normally in the centre of the segment. The results of the analysis may be used to define a set of sinusoids which, when summed, produce a close approximation to the segment. Values of amplitude, frequency and phase for each sinusoid are required for the approximation, and these values may be quantised and efficiently encoded to constitute a low bit-rate speech representation. To reconstruct speech at the decoder, frames of speech may be synthesised as sums of sinusoids with smoothly changing amplitudes, frequencies and phases obtained by interpolating between the decoded parameters. Synthesis frames begin and end at update-points and, in general, the lengths of analysis segments and synthesis frames will be different. The interpolation process is intended to approximate the changes that occur in the original speech, from one update-point to the next.

Fundamental differences between the coding techniques mentioned in the introduction lie in the way the spectral analysis is performed, the width of the analysis window, the use or non-use of an LPC synthesis filter, the treatment of voicing decisions and the interpolation techniques used at the decoder. First the general features of each technique will be outlined.

13.2.1. Sinusoidal Transform Coding (STC)

Sinusoidal transform coding (STC) is based on a method [2] of representing speech as the sum of sinusoids whose amplitudes, frequencies and phases vary with time and are specified at regular update-points. The model was first applied to low bit-rate (8 kb/s) speech coding [1] in 1985, and has subsequently been modified in various ways for use at lower bit-rates (4.8 kb/s and below) [3].

At the analysis stage of STC, segments of the windowed speech, each of two and a half pitch-periods duration and centred on an update-point, are zero padded and spectrally analysed via a Fast Fourier transform (FFT). For voiced segments, the magnitude spectra will have peaks, in principle at harmonics of the fundamental frequency as illustrated in Figure 13.1. Unvoiced segments, which are given a fixed duration, will have randomly distributed peaks, as illustrated in Figure 13.2. The frequencies of the peaks may be identified by applying a peak-picking algorithm and the corresponding magnitudes and phases are then obtained. These measurements become the frequencies, magnitudes and phases of the fundamental sinusoidal model. It has been found that speech synthesised by the model can be made essentially indistinguishable from the original speech.

For low bit-rate coding it is not possible to encode the parameters of the sinusoidal model directly with sufficient accuracy. An indirect method is adopted whereby an STC coder is represented by a small number of parameters. These parameters which are encoded at each update-point are:

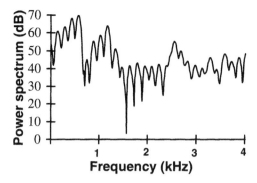

Figure 13.1 Voiced power spectrum.

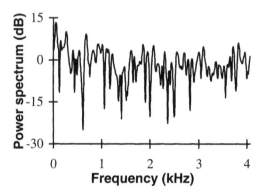

Figure 13.2 Unvoiced power spectrum.

- a measurement of the pitch-period in the vicinity of the update-point;
- a spectral envelope represented by a set of cepstral or LPC coefficients;
- a *voicing probability* frequency.

The spectral envelope is obtained by applying cubic spline interpolation to produce a smooth curve through the spectral peaks, and this envelope may be further processed to produce a high order all-pole envelope efficiently parameterised by LPC coefficients. The voicing probability frequency caters for partially and fully unvoiced speech by dividing the spectrum into two bands; a lower band, below the specified frequency, considered voiced and an upper band considered unvoiced.

The STC decoder receives a set of parameters as listed above for each update-point. Below the voicing probability frequency, the amplitudes required for a sinusoidal model are obtained by sampling the decoded envelope at the pitch-frequency harmonics. The phase of each sinusoid is not available directly from the decoded parameters, but is deduced from the spectral envelope on the assumption that the envelope is the gain response of a minimum phase transfer function. The phase spectrum is therefore derived via a discrete Hilbert transform and this is sampled at the pitch frequency harmonics as for the magnitude spectrum. STC models speech above the voicing probability frequency by sinusoids closely spaced in frequency (say 100 Hz apart) synthesised with random phase. Again the amplitudes of these sinusoids are obtained by sampling the decoded envelope. The process of generating frames of speech between each update-point requires natural and normally smooth evolution of the amplitudes and instantaneous phases of the modulated sinusoids constituting the sinusoidal model. This is an interpolation process, the details of which will be discussed later.

13.2.2. Multi-band Excitation (MBE)

The multi-band excitation coding technique was proposed by Griffin and Lim in 1988 [4] for 8 kb/s coding and was later improved and adapted to 4.8 kb/s by Hardwick and Lim in 1989 [5]. This improved version, referred to as "IMBE", is the basis of the Inmarsat-M voice codec for 6.4 kb/s speech transmission, with error protection, for a satellite mobile communication application [6].

The IMBE coder extracts parameters at update-points every 20 ms (160 samples with sampling rate 8 kHz). At the encoder, a two-stage pitch-period extraction procedure is applied. The first stage achieves an accuracy to within one half of a sampling interval, and the second stage refines the result of the first stage to an accuracy of one quarter of a sampling interval. Both stages are based on the same principle of "analysis-by-synthesis" whereby a set of pitch-period candidates, from 21 to 114 sampling intervals, are identified and an "error function" is calculated for each candidate. This error function measures the difference between the original speech segment and the same segment synthesised with the given pitch period. The technique is made more reliable and robust to noise affected speech by "pitch tracking" which aims to eliminate abrupt changes between successive frames. This smoothing procedure uses the pitch-period estimates for the previous two update-points and "look-ahead" estimates for the following two update-points. Looking ahead requires a buffer which contributes considerably to the overall delay of about 100 ms. In contrast to the variable 2.5 pitch-period spectral analysis window used by STC, the spectral analysis window used by IMBE is fixed and in most cases will be considerably larger.

From the more accurate pitch-period estimate, non-overlapping frequency bands are defined, each of bandwidth equal to a specified number of pitch-frequency harmonics, typically three. The number of bands is therefore pitch-period dependent and may lie between 3 and 12. A separate voiced/unvoiced decision is now made for each of these bands in the original speech spectrum. This is done for each band by measuring the similarity between the original speech in that band and the closest approximation that can be synthesised using the (normally three) estimated pitch frequency harmonics that lie within the band. When the similarity is close, the band is declared voiced; otherwise it is declared unvoiced.

Once the voiced/unvoiced decision has been made for each band, three "spectral amplitudes" are determined for each band. For a voiced band, these spectral amplitudes are simply the amplitudes of the pitch frequency harmonics within the band. For an unvoiced band, the spectral amplitudes are determined by the root mean square of a "frequency bin" centred at the specified harmonic frequency and with bandwidth equal to the calculated fundamental frequency. Note that even unvoiced speech frames are assumed to have a fundamental frequency which will be determined by any small amounts of correlation within the unvoiced frame and, perhaps more importantly, correlation in the previous and next frames as examined by the pitch tracking procedure. The input speech is now assumed to be characterised at each update-point by a pitch-period estimate (from which the number of frequency bands may be determined) and, for each of these bands, a voiced/unvoiced decision and three spectral amplitudes.

At the decoder, the synthesised speech for each frame is a combination of voiced and unvoiced bands. Voiced speech is produced as the sum of sinusoids whose amplitudes, frequencies and phases vary across the frame. Unvoiced bands are generated by extracting appropriately band-limited portions from the DFT spectrum of a white pseudo-random sequence. Although earlier versions of MBE encoded phase information, the improved version (IMBE) does not. For voiced bands, the phases at the update-points are chosen simply to maintain continuity with the waveform synthesised in the previous frame. Therefore there is no attempt to preserve the same phase relationship between the harmonics as existed in the original speech.

13.2.3. Prototype Waveform Interpolation (PWI)

Prototype waveform interpolation (PWI) was proposed [8, 10] in 1991 for speech coding at 3 to 4 kb/s. A prototype waveform is a segment extracted from a pseudo-periodic signal, i.e. a signal for which there is strong correlation, viewed over a suitably short time window, between itself and a truly periodic signal. The extracted segment must be centred on a given point in time and be of length equal to the instantaneous period at that point in time.

The term "prototype waveform" may be applied to segments extracted directly from voiced speech waveforms or to segments extracted from residual signals obtained from an LPC analysis filter. A prototype waveform may start at any point within a pitch-cycle, i.e. not necessarily at a vocal tract excitation point. The original idea of PWI [8] was to encode separately the lengths and shapes of prototype residual waveforms extracted, as illustrated in Figure 13.3, at update-points which lie within voiced portions of the speech. It was proposed that a voiced/unvoiced decision should be made for each update-point, and that unvoiced segments should be

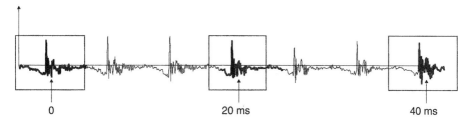

Figure 13.3 Prototype waveforms exracted from voiced speech.

encoded by switching to a form of CELP coder. The prototype residual waveform is analysed to produce a Fourier series:

$$e(t) = \sum_{\ell=1}^{P/2} [A_\ell \cos([2\pi/P]\ell t) + B_\ell \sin([2\pi/P]\ell t)] \tag{13.1}$$

whose period P is the instantaneous pitch-period. Over the analysis interval, i.e. an interval of length P centred on the update-point, $e(t)$ is equal to the prototype waveform residual. "Time-alignment" is then applied by delaying $e(t)$ to make the Fourier series coefficients A_ℓ and B_ℓ as close as possible to the $A_{\ell-1}$ and $B_{\ell-1}$ coefficients of the previous update-point. The resulting A_ℓ, B_ℓ and LPC coefficients, the pitch period P, and a one-bit voicing decision are quantised and efficiently encoded for voiced speech.

At the decoder, as Fourier series sine as well as cosine amplitudes are received, the correct phase relationships between pitch-frequency harmonics may be preserved, thus maintaining the true shapes of prototype waveforms. However, pitch synchronism with the original speech is not maintained because of the time alignment at the encoder carried out for efficient quantisation. Further time alignment is needed at the decoder to maintain the continuity of the synthesised waveform across frame boundaries. This is achieved by equating the phases at the update-point to the instantaneous phases attained at the end of the previous synthesis frame. A synthesised waveform similar to the original, but with an imperceptible time shift, which varies from frame to frame, is characteristic of PWI-encoded speech waveforms.

Characteristic Waveforms (CW)

Recently [16,17,12], the concept of a prototype waveform has been generalised to include arbitrary length segments of unvoiced speech or residual. The term "characteristic waveform" is now used. Further, instead of extracting a single characteristic waveform at each update-point, a sequence of about ten are placed at regular intervals between consecutive update-points. When these waveforms are time-aligned as illustrated in Figure 13.4, the changes that occur to their corresponding

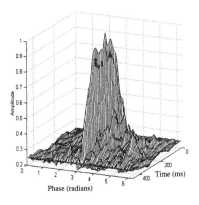

Figure 13.4 Evolution of prototype waveform.

Fourier series coefficients are indicative of the nature of the speech. Rapid changes occur for unvoiced speech, slow changes for voiced. High-pass and low-pass digital filtering is applied to separate the effect of these changes and the resulting filtered Fourier series coefficients characterise a *slowly evolving waveform* (SEW) and a *rapidly evolving waveform* (REW) which sum to form the true characteristic waveform coefficients. The SEW is down-sampled to one waveform per update-point, rapid fluctuations having been eliminated by the low-pass filtering. The REW cannot be accurately represented at a low bit-rate. Fortunately it may be replaced at the decoder by a random waveform with similar spectral shape. The parameters for this generalisation of PWI, i.e. LPC coefficients, pitch-period and characteristics of the SEW and REW, may be encoded at 2.4 kb/s.

13.2.4. Time-frequency Interpolation (TFI)

The concept of time-frequency interpolation (TFI) [13,14] can be viewed as a generalisation of a range of waveform interpolation techniques. The term was used by Shoham [13] to describe an approach to speech coding which defines an "instantaneous" short term spectrum for each sample $s[n]$ of a speech signal. These spectra may be assumed to evolve slowly from speech sample to speech sample. Samples of the evolving spectrum may be obtained by DFT analysis at suitable intervals. The aim is to efficiently encode these sampled spectra and to reconstruct speech by interpolating between them at the decoder. Particular forms of TFI are investigated by Shoham [13,14] which are similar to PWI but have distinguishable features.

The concept of TFI is to associate each speech or residual sample $x[n]$ with a sequence of $M[n]$ samples which are spectrally analysed to produce a short term DFT spectrum $\{X_n[k]\}$. If such spectra are computed at regular intervals of N samples, the "rate" of the TFI scheme is said to be $1/N$ spectra per sampling interval. The value of $M[n]$ may be variable and is often, but not necessarily, related to the instantaneous pitch period.

Two forms of TFI are referred to by Shoham: "low rate" and "high rate". Low rate TFI (LR-TFI) has $N > M[n]$ which means that the sampled spectra are relatively far apart and the interpolation process must generate a time waveform between pairs of spectrum sampling points, whose instantaneous spectrum evolves in a suitable way. When $M[n]$ is made equal to the instantaneous pitch-period for voiced speech, as in PWI, the interpolation is achieved by means of the inherent periodicity of the inverse DFT. Outside the DFT time window, the IDFT waveform repeats with period $M[n]$ and will therefore tend to match the pseudo-period voiced waveform. When $M[n]$ is larger than the pitch-period, as with STC, a similar effect can be achieved by determining the pitch-frequency harmonics by peak picking, and interpolating between the parameters of these harmonics. Having $M[n]$ smaller than the pitch-period raises considerable difficulties and is not considered

High rate TFI (HR-TFI) has $N > M[n]$ which means that the spectra obtained at the spectral sampling points will be more similar to each other than with LR-TFI, though there will be more of them over a given time span. In most cases the spectral analysis windows will overlap. The periodicity of the inverse DFT plays a less critical part in the interpolation process, and in principle, a smoother and more accurate description of the signal can be obtained, though encoding the spectra at low bit-rates becomes a more challenging problem. The use of a differential vector quantisation scheme to encode the difference between successive spectra, as used with PWI for example, is still viable. The greater similarity between successive spectra with HR-TFI will enhance the effectiveness of the differential aspect of the scheme, thus compensating for the fact that there are more spectra to quantise in a given time span. Whereas the original PWI concept is similar to LR-TFI, the later work by Kleijn [12] on characteristic waveform interpolation has aspects in common with HR-TFI.

As well as defining the general concept of TFI, Shoham [13, 14] also proposes a low bit-rate coder based on HR-TFI. This coder has two versions for 4.05 kb/s and 2.4 kb/s operation. The HR-TFI procedure is applied in the residual domain with $N = 40$ samples and DFT window length $M[n]$ said to be approximately [13] or exactly [14] equal to the instantaneous pitch-period at time n. The LPC coefficients are updated less frequently than with PWI, for example (every 60 ms for the 2.4 kb/s version), and are block-interpolated at intervals of N samples. A predictive vector quantisation process is used, the differences between successive spectra being quantised to a trained code book. Test results were reported [13, 14] to show that the 2.4 kb/s TFI coder performed very similarly to full rate (13 kb/s) GSM and 7.95 kb/s IS54 when coding IRS filtered telephone bandwidth speech.

13.3. RECONSTRUCTING THE SPEECH

Waveform interpolation and sinusoidal coding techniques, including those referred to in this chapter, re-synthesise frames of speech at the decoding stage by interpolating spectral information from one update-point to the next as illustrated in Figure 13.5.

The synthesis procedure may be described in terms of a generalisation of the concept of a Fourier series, where the amplitudes and frequencies of the harmonics may vary with time. Such a generalised Fourier series is:

$$x(t) = \sum_{\ell=1}^{L} [a_\ell(t) \cos(\theta_\ell(t)) + b_\ell(t) \sin(\theta_\ell(t))] \qquad (13.2)$$

where $x(t)$ is a speech or LPC residual signal, $a_\ell(t)$ and $b_\ell(t)$ are the instantaneous amplitudes and $\theta_\ell(t)$ is the instantaneous phase of the ℓth sine and cosine term of the series. Equation (13.2) can always be re-expressed in the following form:

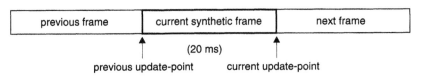

Figure 13.5 Relationship of synthesis frames and update-points.

$$x(t) = \sum_{\ell=1}^{L} [A_\ell(t) \cos(\sigma_\ell(t))] \qquad (13.3)$$

where $A_\ell(t) = \sqrt{a_\ell(t)^2 + b_\ell(t)^2}$ and $\sigma_\ell(t) = \theta_\ell(t) + \tan^{-1}(b_\ell(t)/a_\ell(t))$.

Both forms of this generalised Fourier series are seen in sinusoidal and WI coders; the first is used in the synthesis procedure proposed for PWI, and the second is used by STC, IMBE and TFI.

Let $t = 0$ and $t = NT$ correspond to successive update-points. The objective is to synthesise a frame of speech or residual for values of t between 0 and NT. Consider the simpler all-cosine Fourier series first. Instantaneous amplitude measurements, $A_\ell(0)$ and $A_\ell(NT)$, will be known at $t = 0$ and $t = NT$. To obtain intermediate values across the frame, a straightforward linear interpolation scheme may be used:

$$A_\ell(t) = A_\ell(0) + (A_\ell(NT) - A_\ell(0))t/NT \qquad (13.4)$$

Deriving formulae for the instantaneous phases $\sigma_\ell(t)$ is a little more complicated and there are important differences in the approaches adopted by different sinusoidal coding techniques. It must be arranged that the instantaneous phases of all sinusoids smoothly increase across the synthesis frame and that discontinuities do not occur at synthesis frame boundaries. The time-derivative of the instantaneous phase $\sigma_\ell(t)$ of a sinusoid is the instantaneous frequency denoted $\omega_\ell(t)$.

For each sinusoid, $\omega_\ell(t)$ will be known at each update-point and must also be arranged to change smoothly across the synthesis frame. The instantaneous frequency of each sinusoid at an update-point will be determined by the encoded pitch-period. If the pitch-period is denoted by $P(0)$ seconds at $t = 0$ and $P(NT)$ seconds at $t = NT$, and the modulated sinusoids are assumed to remain harmonically related; i.e. $\omega_\ell(t) = \ell\omega_1(t)$ for all t, it follows that $\omega_\ell(0) = 2\pi\ell/P(0)$ and $\omega_\ell(NT) = 2\pi\ell/P(NT)$ for each value of ℓ. The continuity requirement means that the value of instantaneous phase $\sigma_\ell(0)$ for each sinusoid at the beginning of a synthesis frame must be equal to $\sigma_\ell^{prev}(NT)$ where this denotes the value reached at $t = NT$ by the formula for $\sigma_\ell(t)$ used in the previous synthesis frame. For each value of ℓ, the three conditions:

(i) $\omega_\ell(0) = 2\pi\ell/P(0) = d\sigma_\ell(t)/dt$ at $t = 0$
(ii) $\omega_\ell(NT) = 2\pi\ell/P(NT) = d\sigma_\ell(t)/dt$ at $t = NT$
(iii) $\sigma_\ell(0) = \sigma_\ell^{prev}(NT)$

may be satisfied by a quadratic polynomial:

$$\sigma_\ell(t) = \sigma_{\ell 0} + \omega_{\ell 0}t + \frac{\omega_{\ell 1} - \omega_{\ell 0}}{2N}t^2 \qquad (13.5)$$

where $\sigma_{\ell 0} = \sigma_\ell^{prev}(NT)$, $\omega_{\ell 0} = 2\pi\ell/P(0)$ and $\omega_{\ell 1} = 2\pi\ell/P(NT)$.

With this quadratic interpolation formula used for each $\sigma_\ell(t)$, the sinusoids in equation (13.3) will remain exactly harmonically related throughout each synthesis frame. However, the instantaneous phases may drift over time with respect to each other due to the accumulation of rounding errors. This is essentially the approach used by IMBE.

The technique above has no provision for making the phase relationships between the sinusoids correspond to the best possible model of the phase spectrum of the speech or residual waveform. It is well known that the resulting phase distortion introduces a degree of unnaturalness and loss of speech quality. Various coding techniques, including STC, attempt to make the instantaneous phases at the update-points correspond to the true phase spectrum. Considering again the three conditions on $\sigma_\ell(t)$ given above, this requirement modifies the third condition to:

(iii) $\sigma_\ell(0) = \phi_{\ell 0}$

and introduces a fourth condition:

(iv) $\sigma_\ell(NT) = \phi_{\ell 1} + 2\pi M_\ell$

where $\phi_{\ell 0}$ and $\phi_{\ell 1}$ are the instantaneous phases, specified in the range 0 to 2π, of the ℓth sinusoid at the update-points $t = 0$ and $t = NT$ respectively. M_ℓ must be an integer. The $2\pi M_\ell$ term must be included because $\sigma_\ell(t)$ may pass through 2π several times as t goes from 0 to NT. To satisfy these four conditions, the following cubic polynomial may be employed:

$$\sigma_\ell(t) = \phi_{\ell 0} + \omega_{\ell 0}t + \alpha_\ell t^2 + \beta_\ell t^3 \tag{13.6}$$

where α_ℓ and β_ℓ must be chosen to ensure that conditions (ii) and (iv) are satisfied, i.e.

$$\omega_{\ell 0} + 2\alpha_\ell NT + 3\beta_\ell(NT)^2 = \omega_{\ell 1} \tag{13.7}$$

$$\phi_{\ell 0} + \omega_{\ell 0}NT + \alpha_\ell(NT)^2 + \beta_\ell(NT)^3 = \phi_{\ell 1} + 2\pi M_\ell \tag{13.8}$$

For a given value of M_ℓ, these simultaneous equations are easily solved for α_ℓ and β_ℓ. The integer M_ℓ has a critical effect on the change in instantaneous frequency that occurs across the current synthesis frame. In all cases the amplitude and instantaneous frequency will be correct at the beginning and the end of the frame. However, for inappropriate choices of M_ℓ, the change in frequency will be unnecessarily large across the frame. It may be shown [2] that the best value of M_ℓ is that which minimises $|\beta_\ell|$. This may be found by setting $\beta_\ell = 0$ in equations (13.7) and (13.8), solving for α_ℓ and M_ℓ, and rounding the value obtained for M_ℓ to the nearest integer. Equations (13.7) and (13.8) are then solved again for the true values of α_ℓ and β_ℓ given the required integer value of M_ℓ.

A reasonable approximation to the effect of interpolation, at much less computational expense, is sometimes obtained by a technique referred to as "overlap-and-add". This technique is used by IMBE for unvoiced to voiced transitions and elsewhere. For each update-point, a frame is synthesised using sinusoids and/or random signals with specified parameters which now remain constant instead of being modulated. This frame is then merged with a frame produced similarly for the previous update point. The merging of the frames is achieved by multiplying the previous frame by a decaying window function, multiplying the current frame by a growing window function, and then adding the two frames together.

13.3.1. Some Practical Considerations

Cubic interpolation was originally used by STC to reconstruct speech waveforms with the phase spectrum derived by discrete Hilbert transform applied to the parameterised magnitude spectrum. More recently an overlap-add technique has been used to reduce computational complexity.

In IMBE, equation (13.5) is used directly for sinusoids up to about 1 kHz. For higher frequencies it is modified to avoid the "buzziness" associated with too much phase coherence. The modification is achieved by replacing $\omega_{\ell 0}$ by $\omega_{\ell 0} + \Delta\omega_\ell$ where, for each ℓ, $\Delta\omega_\ell$ is a small offset which varies randomly from frame to frame. This slightly perturbs the exact harmonic frequency relationship between the sinusoids and therefore, over time, unlocks the relationship between the instantaneous phases. The instantaneous phases attained at the end of the frame are affected, and consequently the phase offsets for the next frame must be adjusted to maintain continuity. With IMBE, quadratic interpolation is used for harmonics within bands declared voiced when the difference between the pitch-frequencies is relatively small. Otherwise an overlap-and-add technique is used. With quadratic interpolation, the phases of each sinusoid will drift freely in relation to the phases of the original speech and the phase relationships between individual sinusoids will drift due to accumulated rounding error and the injection of random phase. These effects will cause the reconstructed speech waveshape not to resemble the original, though the phases changes are found to be, to a considerable extent, inaudible.

PWI employs linear interpolation of the pitch-period rather than the pitch-frequency, to determine a formula for the instantaneous pitch-frequency. Other instantaneous frequencies are assumed to be harmonics of the pitch-frequency and these are integrated in the normal way to obtain formulae for the instantaneous phases $\theta_\ell(t)$ of the Fourier series (13.2). Equation (13.2) is used to synthesise an LPC residual $x(t)$ for $t = 0, T, 2T, \ldots, (N-1)T$, with $a_\ell(nT)$ and $b_\ell(nT)$ coefficients which are linearly interpolated between known values at $t = 0$ and $t = NT$. This approach is similar to quadratic interpolation, except that the use of sine as well as cosine terms in the Fourier series preserves the original phase relationships between harmonics.

Relating equations (13.2) and (13.3), it may be seen that the PWI phase interpolation technique effectively separates the instantaneous phase, $\sigma_\ell(t)$ of the ℓth sinusoid into the sum of two components, $\theta_\ell(t)$ and $\phi_\ell(t) = \tan^{-1}(b_\ell(t)/a_\ell(t))$. The first component, $\sigma_\ell(t)$, forces the ℓth sine and the ℓth cosine term to remain harmonics of the interpolated fundamental and maintains their continuity at frame boundaries. $a_\ell(t)$ and $\phi_\ell(t)$ introduce the spectral colouration which preserves the naturalness of the excitation signal. Since the instantaneous frequency of the ℓth sinusoid is the time-derivative of $(\theta_\ell(t) + \phi_\ell(t))$ rather than $\theta_\ell(t)$, the time-derivative of $\phi_\ell(t)$ should, strictly, be made zero at frame boundaries. This is not guaranteed when $a_\ell(t)$ and $b_\ell(t)$ are linearly interpolated, as in PWI, though the resulting frequency changes and discontinuities are not normally serious as changes in the coefficients are small from one frame to the next. The discrepancy may be eliminated by cubic interpolation of $a_\ell(t)$ and $b_\ell(t)$, thus allowing their time-derivatives to be constrained to zero at $t = 0$ and $t = NT$.

13.4. COMPARISONS OF SINUSOIDAL CODING TECHNIQUES

Now that four sinusoidal coding techniques have been introduced and their recon-struction techniques at the receiver have been discussed in some detail, a more general comparison can be made between them.

Important differences lie in the lengths of the window used for short-term spectral analysis at the encoder. For MBE, the relatively large analysis segment gives good spectral resolution and noise immunity though the effects of non-stationarities, due to pitch-frequency variation for example, will be apparent in the resulting spectra. These effects are accommodated to a considerable extent by the multi-band approach which can declare a band unvoiced when frequency spreading causes it to appear not to be harmonically related to the estimated pitch frequency. A PWI analysis segment, being of length equal to a single pitch-period, may be expected to give a more accurate spectral representation of voiced speech at an update-point and be less affected by pitch-frequency variation and other non-stationarities. The vari-able length of the analysis segment adopted by STC is not a critical factor and is chosen to allow an accurate spectral envelope to be determined without serious distortion caused by non-stationarity.

Also of considerable importance is the fact that STC and IMBE directly encode the speech signal whereas PWI and TFI, as originally proposed, apply sinusoidal coding and interpolation to an LPC residual. This demarcation has been trans-gressed with STC-based coders designed to operate in the residual domain [18] and there is probably no fundamental reason why PWI should not be applied directly to a speech waveform. However, the intricate details of each coder have been specifically adapted to the intended domain of operation, and would require careful modification if applied in a different domain.

IMBE, STC and TFI have the advantage of not including any phase information in the encoded output, whereas PWI attempts to encode the phase of the residual by sending Fourier series sine and cosine coefficients. IMBE and TFI disregard the phase spectrum entirely producing speech or residual waveforms that are not expected to resemble the original. The reconstruction procedure used by STC uses a minimum phase assumption to regenerate a phase spectrum from encoded magni-tude-only information. The addition of judicial amounts of random noise into the phase spectra of re-synthesised speech, usually in higher frequency bands, has been found to subjectively improve speech quality. Most sinusoidal and WI coders make some use of this fact.

At the decoder, there are important differences in the way each of the techniques adapt the model to unvoiced frames. IMBE applies band-limited portions from the DFT spectrum of a white pseudo-random sequence to unvoiced bands; PWI switches to CELP for unvoiced frame; STC generates a set of sinusoids whose frequencies are 100 Hz apart and phases are random between $-\pi$ to π to the unvoiced frequency range.

Among these sinusoidal and WI coding techniques IMBE is the only coder that has been adopted for commercial use, which is 6.4 kb/s Inmarsat-M voice codec including error correction. PWI, TFI and STC are still in a stage of continual development. Variations of these coding techniques have been proposed, such as a version of PWI with the interpolation applied in the time-domain [20] rather than frequency-domain. New lower bit-rates are being achieved. Both IMBE and STC based 2.4 kb/s coders were submitted as candidates for the latest DOD standard. It was also reported [12] that a 2.4 kb/s WI coder has equivalent performance to the 4.8 kb/s FS1016 standard.

13.5. CONCLUSIONS

The use of sinusoidal and waveform interpolation coding techniques has enabled great advances to be made in the field of low bit-rate coding for telephone bandwidth speech. They are clearly techniques that will be increasingly important in the field of mobile telecommunications. The computational complexity of the techniques described here is quite high, but generally within the capacity of modern single chip DSP microprocessors. IMBE has been implemented on such processors, for speech coding at 6.4 kb/s (with error coding). Lower bit-rate versions based on IMBE have been proposed. The development of the best possible 4 kb/s speech coder for mobile and other applications remains a topic for active research, and new ideas are being reported all the time. At present it seems possible that there is much potential for refinement and cross-fertilisation of the ideas behind the different coding techniques reported here.

REFERENCES

[1]   McAulay, R. J. and Quatieri, T. F., Mid-rate coding based on a sinusoidal representation of speech, *Proceedings of the IEEE Int. Conf. ASSP*, 945–948, Tampa, FL, 1985.

[2]   McAulay, R. J. and Quatieri, T. F., Speech analysis/synthesis based on a sinusoidal representation, *IEEE Trans. ASSP-34*, no. 4, 744–754, August 1986.

[3]   McAulay, R. J. and Quatieri, T. F., Low-rate speech coding based on the sinusoidal-model. In *Advances in Speech Signal Processing*, Furui, S. and Sondhi, M. M. (eds), Marcel Dekker, New York, 165–208, 1992.

[4]   Griffin, D. W. and Lim, J. S., "Multi-band excitation vocoder", *IEEE Trans. ASSP-36*, no. 8, 1223-1235, Aug 1988.

[5]   Hardwick, J. S. and Lim, J. S., A 4800 bps improved multi-band excitation speech coder, *Proceedings of the IEEE Workshop on Speech Coding for Telecoms*, Vancouver, Canada, 1989.

[6]   Inmarsat M Voice Codec, version 3.0, Maritime System Development Implementation, 1991.

[7]   Hardwick, J. S. and Lim, J. S., The application of the IMBE speech coder to mobile communications. In *Proceedings of the IEEE ICASSP-91*, 249–252, 1991.

[8]   Kleijn, W. B., Continuous representations in linear predictive coding. In *Proceedings of the IEEE, ICASSP-91*, **s1**, 201–204, May 1991.

[9]   Kleijn, W. B., Analysis-by synthesis speech coding based on relaxed waveform-matching constraints. Ph.D. dissertation, Delft University of Technology, Delft, The Netherlands, 55–62, 1991.

[10] Kleijn, W. B. and Granzow, W., Methods for waveform interpolation in speech coding, *Digital Signal Processing*, **1**, 215–230, 1991.

[11] Kleijn, W. B., Encoding speech using prototype waveforms. *IEEE Trans. SAP-1* no. 4, 386–399, 1993.

[12] Kleijn, W. B. and Haagen, J., A speech coder based on decomposition of characteristic waveforms. In *Proceedings of the IEEE ICASSP-95*, 508–511, 1995.

[13] Shoham, Y., High-quality speech coding at 2.4 to 4.0 kbps based on time-frequency interpolation. *IEEE Trans ASSP*, **2**, 167–170, 1993.

[14] Shoham, Y., High-quality speech coding at 2.4 kbps based on time-frequency interpolation. In *Proceedings of the European Conference on Speech Communication and Technology*, Berlin, Germany, **2**, 741–744, 1993.

[15] Cheetham, B. M. G., Sun, X. Q. and Wong, W. T. K., Spectral envelope estimation for low bit-rate sinusoidal speech coders. *Proc. Eurospeech'95*, Madrid, Spain, 693–696, September 1995.

[16] Kleijn, W. B. and Haagen, J., A general waveform-interpolation structure for speech coding; Signal processing VII: theories and applications. *Proc. EUSIPCO-94*, **III**, 1665–1668, Edinburgh, UK.

[17] Kleijn, W. B. and Haagen, J., Transformation and decomposition of the speech signal for coding. *IEEE Signal Processing Letters.* **1** (9), 136–138, 1994.

[18] Yeldener, S., Kondoz, A. M. and Evans, B. G., Sine-wave excited linear predictive coding of speech. In *Proceedings of the International Conference on Spoken Language Processing*, Kobe, Japan, 4.2.1–4.2.4, November 1990.

[19] Makhoul, J., Roucos, S. and Gish, H., Vector quantization in speech coding. *Proc. IEEE*, **73** (11), 1551–1588, 1985.

[20] Li, H. and Lockhart, G. B. Non-linear interpolation in prototype waveform interpolation (PWI) encoders. In *Proceedings of the IEE Colloquium*, 3/1–3/5, 1994.

14

Low-complexity Wireless Speech Communications Schemes*

L. Hanzo, J. E. B. Williams and R. Steele

14.1. INTRODUCTION

There is much activity world-wide in attempting to define the third generation personal communications network (PCN). In Europe the Research in advance communications equipment (RACE) project comparatively studied an advanced time division multiple access (ATDMA) and a code division multiple access (CDMA) scheme. This research culminated in defining a hybrid framework, accommodating both TDMA and CDMA schemes under the auspices of the European Advanced Communications Technologies and Services (ACTS) project [1]. In North America and Japan there are also intensive activities in both of these areas. In designing a

---

*This chapter is partially based on Williams, J., Hanzo, L., Steele, R. and Cheung, J. C. S., A comparative study of microcellular speech transmission schemes. *IEEE Tr. on Veh. Technology*, **43** (4), 909–925, 1994.

third generation system cognisance has to be given to the second generation systems. Indeed, we may anticipate that some of the sub-systems of the Global System of Mobile communications known as GSM [2] and Digital European Cordless Telephone (DECT) scheme [2,3] may find their way into PCNs either as a primary sub-system or as a component to achieve backward compatibility with systems in the field. This approach may result in hand-held transceivers that are intelligent multi-mode terminals, able to communicate with existing networks, while having more advanced and adaptive features that we would expect to see in the next generation of PCNs.

In this chapter we present the results of a series of experiments we conducted. We concerned ourselves with the radio link and introduced a "toolbox" of system components that includes some second generation sub-systems for backward compatibility. We also included 16-level quadrature amplitude modulation (16-QAM) [4], since it exhibits a high bandwidth efficiency in environments where low interference levels prevail. We assumed that the system components, such as the speech- and channel codecs as well as the modulation schemes, can be adaptively reconfigured in order to meet the prevalent system optimisation criteria and documented the expected system performance trade-offs. In our deliberations we restricted ourselves to non-dispersive, low-interference microcellular propagation environments [2,4].

This chapter is organised as follows. In sections 14.2, 14.3 and 14.4 we consider radio modems, speech codecs and the associated channel codecs, respectively. Armed with a set of these sub-systems we construct a "toolbox" of speech transmission systems in section 14.5. Section 14.6 discusses the simulation results of our candidate speech links, while in section 14.7 we provide some conclusions.

14.2. MODEM SCHEMES

The choice of modem is based on the interplay of equipment complexity, power consumption, spectral efficiency, robustness against channel errors, co-channel and adjacent channel interference, as well as the propagation phenomena, which depends on the cell size [2]. Equally important are the associated issues of linear or non-linear amplification and filtering, the applicability of non-coherent, differential detection, soft-decision detection, equalisation and so forth [4].

In our experiments we used three different modems, namely GMSK [2], $\frac{\pi}{4}-$DQPSK and 16-StQAM [4], each with a low- and a high-complexity detector. Here we refrain from detailing the operation of these modems for reasons of space economy; for an in-depth treatment the interested reader is referred to the above references. Suffice to say here that the Global System of Mobile Communications known as GSM [2], which is the most widespread mobile system world-wide at the time of writing, employs the above-mentioned constant-envelope GMSK scheme, partly due to its resilience to power-efficient non-linear class-C amplification. As in GSM, in our GMSK modem a normalised bandwidth bit-timing product of $B_T = 0.3$ was favoured. The typical bandwidth efficiency of GMSK is about 1.35 bit/s/Hz, as seen in Systems A & B of Table 14.1. This table will be

Table 14.1 System comparison (© IEEE, 1994, Williams et al. [5])

System	Modulator	Detector	FEC	Speech codec	Complexity order	Baud rate (kBd)	User bandwidth (kHz)	Min CSNR (dB) AWGN	Min CSNR (dB) Rayleigh
A	GMSK	Viterbi	No	ADPCM	2	32	23.7	7	∞
B	GMSK	Freq. Discr.	No	ADPCM	1	32	23.7	21	31
C	$\frac{\pi}{4}$-DQPSK	MLH-CR	No	ADPCM	4	16	19.8	10	28
D	$\frac{\pi}{4}$-DQPSK	Differential	No	ADPCM	3	16	19.8	10	28
E	16-SQAM	MLH-CR	No	ADPCM	6	8	13.3	20	∞
F	16-SQAM	Differential	No	ADPCM	5	8	13.3	21	31
G	GMSK	Viterbi	BCH	RPE-LTP	8	24.8	18.4	1	15
H	GMSK	Freq. Discr.	BCH	RPE-LTP	7	24.8	18.4	8	18
I	$\frac{\pi}{4}$-DQPSK	MLH-CR	BCH	RPE-LTP	10	12.4	15.3	5	20
J	$\frac{\pi}{4}$-DQPSK	Differential	BCH	RPE-LTP	9	12.4	15.3	6	18
K	16-SQAM	MLH-CR	BCH	RPE-LTP	12	6.2	10.3	13	25
L	16-SQAM	Differential	BCH	RPE-LTP	11	6.2	10.3	16	24

referred to throughout our further discourse. A low-complexity frequency discrimi-nator and a higher-complexity Viterbi detector were deployed [2].

A $\frac{\pi}{4}$ – DQPSK modulation scheme was favoured by the Americans in their IS-54 D-AMPS network and by the Japanese digital cellular system, which is a strong justification for including these schemes in our experiments. In these modems we decided to compare the performance of a low-complexity differential detector to a more complex maximum likelihood correlation receiver (MLH-CR), as shown for Systems C & D of Table 14.1. We used square root raised cosine Nyquist filters with a roll-off factor of 0.35 [4] and achieved an approximate modem bandwidth effi-ciency of 1.64 bit/Hz. The differentially coded 16-level twin-ring star QAM (16-StQAM) arrangement [4] included in our studies was shown to work well in micro-cellular environments, where high SNRs and low dispersion are generally the norm and the high traffic density requirements justify the use of linear amplification. The low-complexity non-coherent differential detector's performance [6] was compared to that of an MLH-CR, as seen in Systems E & F of Table 14.1. We also note that although in the linearly amplified modems we opted for low-complexity differentially detected schemes, their error resilience can be further improved by about 6 dB, when opting for pilot-assisted coherently detected schemes [4]. When using square root raised cosine Nyquist filters with a roll-off factor of unity and requiring a spectral attenuation of 24 dB at the edge of the adjacent channel, the modem bandwidth efficiency was 2.4 bit/s/Hz. However, this value can be increased up to values of 3–3.2 bits/s/Hz, when using lower excess bandwidths, which translates directly to a higher number of users supported. In this work we considered modems which are reconfigurable on a medium- to long-term basis.

The most recent research trend in modulation is to improve the transceiver per-formance by facilitating so-called burst-by-burst based reconfiguration of TDMA systems, where the number of bits per modulation symbol is adjusted on the basis of the instantaneous signal-to-noise ratio (SNR), while maintaining a constant band-width requirement. Steele and Webb [6,4] proposed adaptive differentially encoded, non-coherently detected star-constellation quadrature amplitude modulation [4] (Star-QAM) for exploiting the time-variant Shannonian channel capacity of fading channels. Their work has stimulated further pioneering work, in particular by Kamio, Sampei, Sasaoka, Morinaga, Morimoto, Harada, Okada, Komaki and Otsuki at Osaka University and the Ministry of Post in Japan, [8]–[11], as well as by Chua and Goldsmith [12] at CalTech in the USA or by Pearce, Burr and Tozer [13] in the UK. Further contributions in the field of adaptive modulation are due to Torrance et al. [14]–[18]. Having briefly considered the associated mod-ulation aspects, let us now focus our attention on the issues of speech coding.

14.3. SPEECH CODECS

In recent years both speech coding research and IC-technology went through a revolutionary development, culminating in a number of standards with bit rates as low as 2.4 kb/s. For an in-depth overview of speech coding the interested reader is referred to Chapter 12 in this book. The selection of the speech codec for mobile applications is based on the appropriate combination of parameters such as speech quality, computational complexity related to power consumption, bit rate, delay and robustness against channel errors.

The two speech codecs included in our experiments are the CCITT G. 721 32 kb/s adaptive differential pulse code modulation (ADPCM) codec used in the DECT system, and the 13 kb/s GSM and DCS 1800 regular pulse excited (RPE), long term predictor (LTP) assisted codec [2]. The ADPCM codec used in our schemes fully complies with the CCITT recommendation, hence we will not describe its operation here.

Our RPE-LTP codec is a slightly modified version of the standard GSM scheme [19], transmitting 268 bits per 20 ms frame and hence the overall bit rate is increased from 13 kb/s to 13.4 kb/s. For robustness against channel errors some speech bits are better protected than others. To identify the effect of bit errors on speech quality in our modified GSM-like codec we performed a bit sensitivity analysis based on a combination of segmental SNR (SSNR) and cepstral distance (CD) degradation. Our findings are presented in reference [19], which suggested a sensitivity order very similar to those found by exhaustive subjective testing and documented in the GSM recommendation, verifying our approach. For the sake of low complexity a simple twin-class error protection scheme is used, where the more important bits are termed as class 1 (C1) bits and the less sensitive ones as class 2 (C2) bits.

For the sake of comparison we mention that the 32 kb/s ADPCM waveform codec has a segmental SNR (SSNR) of about 28 dB, while the 13 kb/s analysis-by-synthesis (ABS) RPE-LTP codec has a lower SSNR of about 16 dB, associated with similar subjective quality rated as a mean opinion score (MOS) of about four. This discrepancy in SSNR values is because the RPE-LTP codec utilises perceptual error weighting, which degrades the objective speech quality in terms of both SSNR and CD, but improves the subjective speech quality. The cost of the RPE-LTP codec's significantly lower bit rate and higher robustness compared to ADPCM is its increased complexity and encoding delay.

14.4. CHANNEL CODING AND BIT-MAPPING

The cordless telecommunications (CT) schemes [3] considered here are based on the ADPCM codec. No error correction coding is used, because it is not necessary in the microcellular environments at the bit rates employed. In the higher complexity, more robust, lower bit rate speech transmission systems based on the RPE-LTP codec we favour binary Bose–Chaudhuri–Hocquenghem (BCH) block codes [2]. They combine low computational complexity with high error correcting power and have reliable error detection properties, which are used by our speech post-enhancement scheme [20], when the received speech is corrupted due to an FEC decoding failure. The channel codec may also be employed to assist in initiating fast handovers, which are vitally important in small microcells, where they may occur frequently and fast [21]. In our twin-class FEC scheme [19] the 116 C1 bits of the RPE-LTP codec are protected by a shortened binary BCH (62,29,6) code yielding 248 bits, while the 152 C2 bits are coded by a shortened binary BCH (62,38,4) code which also yields 248 bits. Four 62-bit codewords are used in each class, and are rectangularly interleaved to curtail error propagation across 20 ms speech frames. The 496 bits in one speech frame are transmitted during 20 ms, yielding a total transmission bit rate of 24.8 kbits/s.

14.5. SPEECH TRANSMISSION SYSTEMS

In our simulations we used the GMSK, $\frac{\pi}{4}$ – DQPSK and 16-StQAM modems combined with both the unprotected low-complexity 32 kb/s ADPCM codec (as in DECT and CT2) and the 13 kb/s RPE-LTP codec protected by its FEC. Each modem had the option of either a low or a high complexity demodulator, as seen in Table 14.1. Synchronous transmissions and perfect channel estimation were assumed in evaluating the relative performances of the systems listed in Table 14.1. Our results represent performance upper bounds, allowing relative performance comparisons under identical circumstances. The propagation environment was assumed to be Gaussian or Rayleigh [2]. We presumed that the microcells were sufficiently small for the transmitted symbol rate, considering that dispersion in the radio channels was a rare event.

Returning to Table 14.1, the first column shows the system classifiers A–L, the next the modulation used, the third the demodulation scheme employed, the fourth the FEC scheme and the fifth the speech codec invoked. The sixth column gives the estimated relative order of the complexity of the schemes, where the most complex one having a complexity parameter of 12 is the 16-StQAM, MLH-CR, BCH, RPE-LTP arrangement, System K. The single-user speech Baud rate, taking into account the corresponding 1, 2 and 4 bits/symbol capacity of the modems used and the TDMA user bandwidth, are given next. Observe that in the latter column the DQPSK and 16-StQAM baud rate figures have been moderated by the actual bandwidth requirements computed by taking into account their filtering requirements. Explicitly, the 1 bit/symbol GMSK, 2 bit/symbol $\frac{\pi}{4}$-DQPSK and 4 bit/symbol 16-StQAM modems have moderated relative spectral efficiency figures of 1.35, 1.62 and 2.4 bits/Hz, respectively. Hence the single-user bandwidths are yielded upon dividing the speech codec rate, in case of the more robust schemes the FEC-coded rate, by the above modem spectral efficiencies. A signalling rate of 400 kBd was chosen for all our experiments, irrespective of the number of modulation levels, to provide a fair comparison for all the systems under identical propagation conditions. The corresponding 400 kBd systems have a total bandwidth of $400/1.35 = 296$ kHz, $2 \times 400/1.62 = 494$ kHz and $4 \times 400/2.4 = 667$ kHz, respectively. The last two columns of Table 14.1 list the minimum channel SNR values in dB required for the system to guarantee nearly unimpaired speech quality through Gaussian and Rayleigh channels. These are characteristics closely related to the system's robustness, will be discussed in section 14.6.2. A range of higher complexity transceivers were proposed for example in References [22,23,24].

14.6. SPEECH PERFORMANCE

The system performances apply to microcellular conditions. We used a carrier frequency of 2 GHz, a signalling rate of 400 kBd and a mobile speed 15 m/s. At 400 kBd in microcells the fading is flat and usually Rician. The transmission rate in DECT is 1152 kBd and even at this high rate it is not necessary to use FEC codecs or equalisers. The best and worst Rician channels are the Gaussian and Rayleigh fading channels, respectively, and we performed our simulations for these channels to obtain upper and lower bound performances. Our experimental conditions of 2 GHz, 400 kBd and 15 m/s are equivalent to a fading profile that can be obtained for a variety of different conditions, for example, for the GSM-like scenario of 900 MHz, 271 kBd and 23 m/s. Our goal is to compare the objective speech performances of the systems defined in Table 14.1, when operating according to our standard conditions. In evaluating the speech performance we used both SSNR and CD, hence our conclusions are based on both. However, due to lack of space here we only present SSNR results. The interested reader is referred to reference [5] for full details.

14.6.1. Cordless Telecommunication Schemes

Cordless telecommunication (CT) systems [3] are typically used in office environments, where benign channel conditions prevail. For the schemes A–F in Table 14.1 we employed the low-complexity ADPCM codec without the use of FEC coding. Let us consider systems A and B, employing GMSK with $B_T = 0.3$ and the ADPCM speech codec. In error-free conditions this speech codec achieved a segmental SNR (SSNR) of 29 dB with our test speech signal. Figure 14.1 shows that unimpaired speech quality was achieved for the Gaussian channel with the Viterbi detector (System A) for channel SNRs (CSNR) in excess of about 6 dB, whereas the discriminator (System B) required about 20 dB. For transmissions over the Rayleigh fading channel, the Viterbi detector never reached the error-free SSNR = 29 dB, since the modem had an irreducible error rate at high SNRs. Despite its poorer performance at low SNRs, System B did not suffer from the problem of irreducible error rate. Consequently, above 25 dB channel SNR its speech SSNR performance was better than that achieved by System A, when transmitting over Rayleigh fading channels. The associated minimum required CSNR values are listed in Table 14.1 for the various systems.

The objective speech quality of schemes C and D is shown in Figure 14.2 in terms of SSNR versus CSNR. The more complex MLH-CR of system C required some 2 dB lower channel SNR to achieve the same SSNR performance as the differential detector of system D for transmissions over the Gaussian channel, a trade-off not necessarily worth while. Over Rayleigh fading channels the two detectors performed similarly, requiring about 25 dB channel SNR for unimpaired speech quality.

The performance of Systems E and F evaluated in terms of SSNR versus channel SNR is given in Figure 14.3. Over the Gaussian channel system E

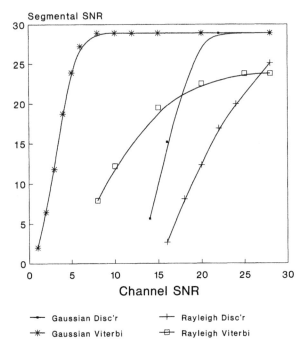

Figure 14.1 Speech performance of systems A and B: SSNR (dB) versus channel SNR (dB) (© IEEE, 1994, Williams *et al.* [5]).

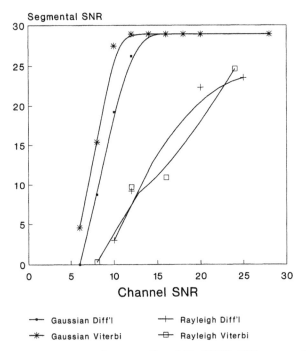

Figure 14.2 Speech performance of systems C and D: SSNR (dB) versus channel SNR (dB) (© IEEE, 1994, Williams *et al.* [5]).

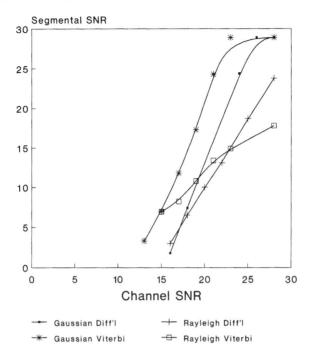

Figure 14.3 Speech performance of systems E and F: SSNR (dB) versus channel SNR (dB) (© IEEE, 1994, Williams *et al.* [5]).

requires about 22 dB channel SNR for unperturbed speech quality, while system F is some 3 dB less robust in terms of channel SNR. For transmissions over the Rayleigh channel the lower-complexity system F recovers faster from low-channel SNR conditions, as the SNR improves, reaching almost error-free conditions for SNRs of about 30 dB.

From our assessment of systems using the ADPCM speech codec, we conclude that for the Gaussian channel the GMSK system offers the best performance in terms of SNR, but it also has the lowest throughput. Doubling the number of bits per symbol, by moving to the DQPSK scheme, sacrifices 5 dB in channel SNR. Changing to Star QAM, with four bits per symbol, results in a further 12 dB CSNR penalty. For high-channel SNR values the lower-complexity frequency discriminator and differential decoder schemes are preferable to the more complex MLH-CR. This is because they have similar or, for high-channel SNR values, even superior BER and SSNR performances. However, if the channel was fading, the less complex discriminator provided better speech quality. For the multi-level modems the low-complexity differential approach was preferred. A minimum of about 25 dB channel SNR was required for unprotected, unimpaired speech quality. Again, we note that upon using pilot symbol assisted 16QAM [4] nearly 6 dB lower CSNRs would suffice in comparison to the lower-complexity star 16QAM employed in our experiments. Our findings are summarised in the last two columns of Table 14.1.

14.6.2. Robust Systems

We now consider the more sophisticated schemes G-L using FEC coding and the RPE-LTP speech coder. Again, the GMSK, $\frac{\pi}{4}$-DQPSK and 16-StQAM modems were used, each with either a simple or a complex detector.

Systems G and H consisted of a BCH-coded RPE-LTP speech codec in conjunction with a GMSK modem using either Viterbi detection or a frequency discriminator. Since the FEC now removed the modem's residual BER, the SSNR reached the error-free unimpaired condition for all the scenarios, as shown in Figure 14.4. The poor performance of the frequency discriminator was apparent. The Viterbi detector required some 10 dB less channel SNR than the frequency discriminator for both Gaussian and Rayleigh channels for an unimpaired SSNR of about 16 dB. The discriminator operating over the Gaussian channel appeared to have an extremely inconvenient systematic distribution of errors, causing the FEC to be frequently overloaded. Surprisingly, this meant that the bursty errors over the Rayleigh channel were more easily accommodated. Thus the Viterbi detector was the most appropriate demodulator for this scheme.

Systems I and J incorporated RPE-LTP speech coding, BCH FEC and DQPSK with either MLH-CR or differential detection. Figure 14.5 shows that the SSNR objective speech performance of the two schemes was very similar, hence the less complex differential detector was favoured. For Gaussian channels an SNR of about

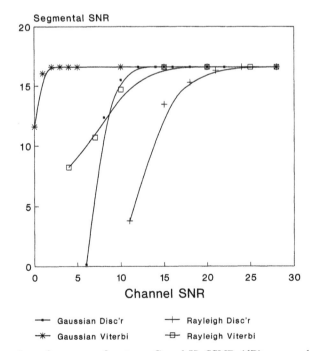

Figure 14.4 Speech performance of systems G and H: SSNR (dB) versus channel SNR (dB) (© IEEE, 1994, Williams *et al.* [5]).

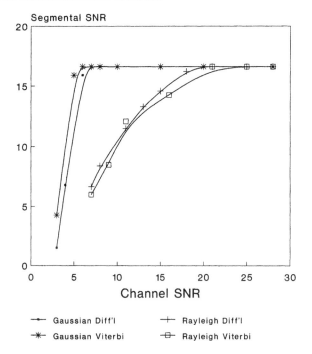

Figure 14.5 Speech performance of systems I and J: SSNR (dB) versus channel SNR (dB) (© IEEE, 1994, Williams *et al.* [5]).

5 dB is sufficient for error-free speech reception, while the Rayleigh channel requires about 20 dB channel SNR.

The SSNR versus channel SNR performances of systems K and L are shown in Figure 14.6. The systems with differential or MLH-CR detection had similar overall SSNR performances for both Gaussian and Rayleigh channels, requiring some 15 and 25 dB channel SNRs to achieve unperturbed speech quality, respectively. However, the complexity advantage of the differential scheme made it more attractive.

As anticipated, the BCH-protected RPE-LTP schemes were more robust. For the GMSK modem the Viterbi detector was very attractive for both the channels we investigated, requiring a mere 10 dB channel SNR for unimpaired speech quality in the presence of Rayleigh fading. For the multi-level modems the simpler differential detector was preferred. These BCH-protected systems removed the modems' residual BER, yielding essentially unimpaired, error-free speech quality, which was rarely achieved without FEC coding via fading channels for unprotected CTs. The more robust, higher-quality speech transmissions are attainable at a channel coded rate of 24.8 kb/s, instead of the unprotected and less complex 32 kb/s ADPCM transmissions, a trade-off available to system designers. Over non-fading Gaussian channels, typically 5 dB lower channel SNR was sufficient for similar speech quality, when using the FEC-protected robust schemes, assuming identical modems. Again, these results are summarised in the last two columns of Table 14.1.

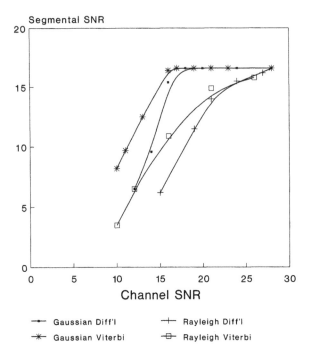

Figure 14.6 Speech performance of systems K and L: SSNR (dB) versus channel SNR (dB)
(© IEEE, 1994, Williams *et al.* [5]).

14.7. CONCLUSIONS

In our discourse we have examined a number of radio links formed with different
sub-systems, some of which are to be found in the second generation cordless telecommunication (CT) systems of CT2 and DECT, and in the cellular systems of
GSM, DCS 1800 and IS54. We consider that third generation systems will need
backward compatibility with these second generation systems, and therefore third
generation transceivers will contain in firmware formats sub-systems that can be
reconfigured under software control to emulate the second generation transceivers
as well as those of the more advanced third generation PCNs [7].

We envisage an adaptive transceiver storing the software of a wide range of sub-
systems, such as the low-complexity schemes listed in Table 14.1, plus a variety of
more complex arrangements [22,23,24] in order to provide services matching the
predominant optimisation criteria. The number of modulation levels and detection
algorithm, the activation of the BCH channel codec and loading of the appropriate
speech codec algorithm and other transceiver features are programmable. As channel
conditions worsen, or as the offered teletraffic increases, the transceiver reconfigures
itself to match the optimisation criterion, which might be minimum power consumption, maximum robustness against channel errors or minimum transmission bandwidth requirement. The definition of these optimisation algorithms requires further
research.

ACKNOWLEDGEMENT

The financial support of the following organisations is gratefully acknowledged: Radiocommunications Agency, UK; Motorola ECID, Swindon, UK; European Community, Brussels, Belgium; Engineering and Physical Sciences Research Council, Swindon, UK; Mobile Virtual Centre of Excellence, UK.

REFERENCES

[1]   Advanced Communications Technologies and Services (ACTS), Workplan, DGXIII-B-RA946043-WP, Commission of the European Community, Brussels, 1994.

[2]   Steele, R., *Mobile Radio Communications*, Pentech Press, London, 1992.

[3]   Tuttlebee, W. H. W., (ed.), *Cordless Telecommunications in Europe*, Springer-Verlag, London, 1990.

[4]   Webb, W. T. and Hanzo, L., *Modern quadrature amplitude modulation: Principles and applications for fixed and wireless channels*, IEEE Press-Pentech Press, ISBN 0-7273-1701-6, 557, 1994.

[5]   Williams, J., Hanzo, L., Steele, R., and Cheung, J. C. S., A comparative study of microcellular speech transmission schemes. *IEEE Transactions on Vehicle Technology*, **43** (4), 909–925, 1994.

[6]   Webb, W. T., Hanzo, L. and Steele, R., Bandwidth efficient QAM schemes for Rayleigh fading channels. *IEE Proc*, Part I, **138**, (3), 169–175, 1991.

[7]   Steele, R. and Webb, W. T., Variable rate QAM for data transmission over Rayleigh fading channels. In *Proceedings of Wireless '91*, Calgary, Alberta, IEEE, 1–14, 1991.

[8]   Kamio, Y., Sampei, S., Sasaoka, H. and Morinaga, N., Performance of modulation-level-control adaptive-modulation under limited transmission delay time for land mobile communications. In *Proceedings of the 45th Vehicular Technology Conference*, IEEE, 221–225, 1995.

[9]   Sampei, S., Komaki, S. and Morinaga, N., Adaptive modulation/TDMA scheme for large capacity personal multi-media communication systems. *IEICE Transactions on Communications*, **77** (9), 1096–1103, 1994.

[10] Morimoto, M., Harada, H., Okada, M. and Komaki, S., A study on power assignment of hierarchical modulation schemes for digital broadcasting. *IEICE Transactions on Communications*, **77**, (12), 1495–1500, 1994.

[11] Otsuki, S., Sampei, S. and Morinaga, N. Square-QAM adaptive modulation TDMA/TDD systems using modulation level estimation with Walsh function. *Electronics Letters*, November, 169–171, 1995.

[12] Chua, S.-G. and Goldsmith, A., Variable-rate variable-power mqam for fading channels. In *Proceedings of the 46th Vehicular Technology Conference*, IEEE, Atlanta, USA, 815–819, 1996.

[13] Pearce, D. A., Burr, A. G. and Tozer, T. C., Comparison of counter-measures against slow Rayleigh fading for TDMA systems. In *Proceedings of the Colloquium on Advanced TDMA Techniques and Applications*, IEEE, 9/1–9/6, 1996.

[14] Torrance, J. M. and Hanzo, L., Upper bound performance of adaptive modulation in a slow Rayleigh fading channel. *Electronics Letters*, April, 169–171, 1996.

[15] Torrance, J. M. and Hanzo, L., Adaptive modulation in a slow Rayleigh fading channel. In *Proceedings of the 7th Personal, Indoor and Mobile Radio Communications (PIMRC) Conference*, IEEE, 497–501, 1996.

[16] Torrance, J. M. and Hanzo, L., Optimisation of switching levels for adaptive modulation in a slow Rayleigh fading channel. *Electronics Letters*, June, 1167–1169, 1996.

[17] Torrance, J. M. and Hanzo, L., Demodulation level selection in adaptive modulation. *Electronics Letters*, **32** (19), 1751–1752, 1996.

[18] Torrance, J. and Hanzo, L., Latency considerations for adaptive modulation in slow Rayleigh fading. In *Proceedings of IEEE VTC'97*, Phoenix, USA, 1204–1209, 1997.

[19] Webb, W., Hanzo, L., Salami, R. A. and Steele, R. Does 16-QAM Provide an Alternative to a Half-Rate GSM Speech Codec? In *Proceedings of the IEEE Vehicular Technology Conference*, St Louis, USA, 511–516, May 1991.

[20] Hanzo, L., Steele, R. and Fortune, P. M., A subband coding, BCH Coding and 16-QAM system for mobile radio communication. *IEEE Transactions on VT* **39** (4), 327–340, 1990.

[21] Steele, R., Twelves, D. and Hanzo, L., Effect of cochannel interference on handover in microcellular mobile radio. *Electronics Letters*, **25** (20), 1329–1330, 1989.

[22] Hanzo, L. and Woodard, J. P., An intelligent multimode voice communications system for indoors communications. *IEEE Transactions on Vehicle Technology*, **44** (4), 735–749, ISSN 0018-9545, 1995.

[23] Hanzo, L., Salami, R., Steele, R. and Fortune, P. M., Transmission of digitally encoded speech at 1.2 KBd for PCN. *IEE Proc-I*, **139** (4), 427–447, 1992.

[24] Hanzo, L. and Woodard, J. P. *Modern Voice Communications: Principles and applications for fixed and wireless channels*, IEEE Press, in preparation. For detailed contents please refer to http://www-mobile.ecs.soton.ac.uk.

15

An Intelligent Dual-mode Wireless Speech Transceiver*

J. P. Woodard and L. Hanzo

15.1. INTRODUCTION

The near future is likely to witness the emergence of adaptive mobile speech communicators that can adapt their parameters to rapidly changing propagation environments and maintain a near-optimum combination of transceiver parameters in various scenarios. Therefore it is beneficial, if the transceiver can drop its speech source rate for example from 6.5 kbits/s to 4.7 kbits/s and invoke 16-level quadrature amplitude modulation (16-QAM) instead of 64-QAM, while maintaining the same bandwidth. In this chapter the underlying trade-offs of using such a dual-rate algebraic code excited linear predictive (ACELP) speech codec in conjunction with a

*This chapter is partially based on Hanzo, L. and Woodard, J. P., An intelligent multimode voice communications system for indoors communications. *IEEE Tr. on Veh. Technology*, **44** (4), 735–749, 1995.

diversity- and pilot-assisted coherent, re-configurable, unequal error protection 16-QAM/64-QAM modem are investigated. Section 15.2 briefly describes the re-configurable transceiver scheme, section 15.3 details the proposed dual-rate ACELP codec, followed by a short discussion on bit sensitivity issues in section 15.4 and source-matched embedded error protection in section 15.5. Packet reservation multiple access (PRMA) is considered in section 15.6. Finally, before concluding, the system performance is characterised in section 7.

15.2. THE TRANSCEIVER SCHEME

The schematic diagram of the proposed re-configurable transceiver is portrayed in Figure 15.1. A voice activity detector (VAD) similar to that of the Pan-European GSM system [1] enables or disables the ACELP encoder [2] and queues the active speech frames in the PRMA [3] contention buffer for transmission to the base station (BS). The 4.7 or 6.5 kbps ACELP coded active speech frames are mapped according to their error sensitivities to *n* protection classes by the bit mapper, as shown in Figure 15.1, and source sensitivity-matched binary Bose–Chaudhuri–Hocquenghem (BCH) encoded [4]. The "Map and PRMA slot allocator" block converts the binary bit stream to 4- or 6-bit symbols, injects pilot symbols and ramp symbols and allows the packets to contend for a PRMA slot reservation. After BCH encoding the 4.7

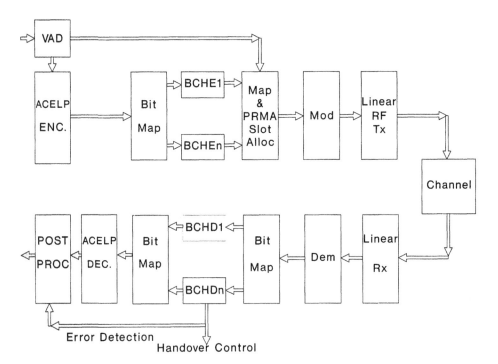

Figure 15.1 Transceiver schematic.

and 6.5 kbps speech bits they are converted to 4- or 6-bit symbols, which modulate a re-configurable 16- or 64-level quadrature amplitude modulation (QAM) scheme [5]. However, both operational modes will have the same signalling rate and bandwidth requirement. Therefore this transmission scheme can provide higher speech quality, if high channel signal-to-noise ratios (SNR) and signal-to-interference ratios (SIR) prevail, while it can be reconfigured under network control to deliver lower, but unimpaired speech quality amongst lower SNR and SIR conditions. The various components of the transceiver will be considered in more depth during our forthcoming discourse. Let us commence with the characterisation of the speech codec.

15.3. THE DUAL-RATE ACELP CODEC

The structure of our dual rate speech encoder is shown in Figure 15.2. The transfer function $A(z)$ represents an all zero filter of order ten, which is used to model the short term correlation in the speech signal $s(n)$. The filter coefficients are determined for each 30 ms speech frame by minimising the variance of the prediction residual $r(n)$, using the autocorrelation method [7]. The filter $1/A(z)$, referred to as the inverse or synthesis filter, is used in the decoder to produce the reconstructed speech signal from the excitation signal $u(n)$. The filter coefficients, as delivered by the autocorrelation method, are not suitable for quantization due to stability problems of the all pole synthesis filter $1/A(z)$. Therefore they are converted to line spectrum frequencies (LSFs) [7], which are then scalar quantized with a total of 34 bits per frame.

The excitation signal $u(n)$ of Figure 15.2 has two components, both of which are determined for each 5 or 7.5 ms subsegment of a 30 ms speech frame, depending on the targeted output bit rate. The first component is an entry from the adaptive codebook Figure 15.2, which is used to model the long term periodicity of the speech signal. The other excitation component of Figure 15.2 is derived from a fixed code-

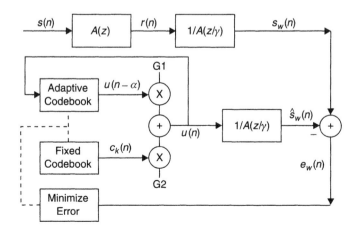

Figure 15.2 ACELP encoder structure (© IEEE, 1995, Hanzo and Woodard [6]).

book, which models the random, Gaussian-like prediction residual of the speech signal after both its long- and short-term correlations have been removed. The excitation is described in terms of the following parameters of Figure 15.2, which are also summarised in Table 15.1:

- The adaptive code-book delay α that can take any integer value between 20 and 147 and hence it is represented using 7 bits.
- The adaptive code-book gain G_1 which is non-uniformly quantised with 3 bits.
- The index of the optimum fixed code-book entry $c_k(n)$, which is one of 4096 entries and therefore it is represented with 12 bits.
- The fixed code-book gain G_2, the magnitude of which is quantised with a 4-bit logarithmic quantiser and its sign is assigned an additional bit.

Thus a total of 27 bits are needed to represent the excitation signal $u(n)$. In the low- and high-rate mode of the ACELP codec each 30 ms frame is split into four and six sub-frames, respectively, for each of which the excitation information is optimised and quantised using 27 bits. Thus for the low-rate mode we have a total of $(34 + 4 \cdot 27) = 142$ bits per 30 ms frame, or a rate of about 4.7 kbits/s, while in the high rate mode the encoder generates $(34 + 6 \cdot 27) = 196$ bits per 30 ms frame, yielding a data rate of just over 6.5 kbits/s.

In the encoder of Figure 15.2 the excitation is passed through the so-called weighted synthesis filter $1/A(z/\gamma)$. Similarly, the prediction residual $r(n)$ is passed through an identical filter to produce a weighted version $s_w(n)$ of the input speech signal $s(n)$. It is then the weighted error $e_w(n)$ between the original and reconstructed speech, which the encoder attempts to minimise when choosing the excitation parameters. This weighting process de-emphasises the energy of the error signal in the formant regions, where the speech has high energy. The factor γ determines the extent of the weighting, and was set to 0.8 for the 6.5 kbits/s codec, and 0.9 for the 4.7 kbits/s codec.

The excitation parameters listed above are determined using a closed loop analysis-by-synthesis procedure. This significantly improves the codec's performance, at the expense of a major increase in complexity. First the effect of passing each possible adaptive code-book word through the weighted synthesis filter is calculated. Then the parameters G_1 and α, which minimise the total squared error between the output of the filter and the original weighted speech, are chosen. Finally, given the adaptive code-book parameters, the fixed code-book parameters are calculated in a similar way.

Table 15.1 Bit allocation scheme (© IEEE, 1995, Hanzo and Woodard [6])

Bit Index	Parameter
1 to 34	LSFs
35 to 41	Adaptive Code-book Delay
42 to 44	Adaptive Code-book Gain
45 to 56	Fixed Code-book Index
57	Fixed Code-book Gain Sign
58 to 61	Fixed Code-book Gain

Traditionally the major part of a CELP codec's complexity accrues from the fixed code-book search. For a twelve bit code-book the effect of filtering 4096 codewords through the weighted synthesis filter must be determined. In our codec we opted for an algebraic code excited linear predictive (ACELP) structure [2,4], in which each 40 or 60 samples long excitation pattern has only four non-zero excitation pulses given by $1, -1, 1, -1$. Each pulse has eight legitimate positions, and the optimum positions can be determined efficiently using a series of four nested loops. This dramatically reduces the complexity of the code-book search, and also removes the need for a fixed code-book to be stored both at the encoder and decoder. Let us now focus our attention on the sensitivity of the various speech bits.

15.4. ERROR SENSITIVITY ISSUES

In order to achieve high robustness over fading mobile channels, in this section ways of quantifying and improving the error sensitivity of our codec are discussed. It has been noted [8,9] that the spectral parameters in CELP codecs are particularly sensitive to errors. This impediment can be mitigated by exploiting the ordering property of the LSFs [7]. When a non-ordered set of LSFs is received, we examine each LSF bit to check whether inverting this bit would eliminate the error. If several alternative bit inversions are found, which correct the cross-over, then the one that produces LSFs as close as possible to those in the previous frame is used. With this technique we found that when the LSFs were corrupted such that a cross-over occurred, which happened about 30% of the time, the corrupted LSF was correctly identified in more than 90% of the cases, and the corrupted bit was pinpointed about 80% of the time. When this occurs, the effect of the bit error is completely removed and hence 25% of all LSF errors can be entirely removed by the decoder. When un-equal error protection is used, the situation can be further improved because the algorithm will be aware of which bits are more likely to be corrupted, and hence can attempt correcting the LSF crossover by inverting these bits first.

The algebraic code-book structure used in our codec is inherently quite robust to channel errors. This is because if one of the code-book index bits is corrupted, the code-book entry selected at the decoder will differ from that used in the encoder only in the position of one of the four non-zero pulses. Thus the corrupted code-book entry will be similar to the original. This is in contrast to traditional CELP codecs which use a non-structured, randomly filled code-book. In such codecs when a bit of the code-book index is corrupted, a different code-book address is decoded and the code-book entry used is entirely different from the original, inflicting typically a segmental signal-to-noise ratio (SEGSNR) degradation of about 8 dB. In contrast, our codec exhibits degradation of only about 4 dB, when a code-book index bit is corrupted in every frame.

The magnitude of the fixed code-book gain tends to vary quite smoothly from one sub-frame to the next. Therefore errors in the code-book gain can be spotted using a smoother, indicating on the basis of neighbouring gains, what range of values the present gain is expected to lie within. If a code-book gain is found, which is not in this range, then it is assumed to be corrupted, and replaced with an estimated gain.

After careful investigation we implemented a smoother, which was able to detect almost 90% of the errors in the most significant bit of the gain level, and reduced the error sensitivity of this bit by a factor of five. The fixed code-book gain sign shows erratic behaviour and hence does not lend itself to smoothing. This bit has a high error sensitivity and has to be well protected by the channel codec.

Seven bits per sub-frame are used to encode the adaptive code-book delay, and most of these are extremely sensitive to channel errors. An error in one of these bits produces a large SNR degradation not only in the frame in which the error occurred, but also in subsequent frames, and generally it takes more than ten frames before the effect of the error dies out. Therefore these bits should, wherever possible, be strongly protected. The pitch gain is much less smooth than the fixed code-book gain, and hence it is not amenable to smoothing. However, its error sensitivity can be reduced by Gray-coding the quantiser levels.

Clearly, some bits are much more sensitive to channel errors than others, and the sensitive bits must be more heavily protected by the channel codec. However, it is not obvious how the sensitivity of different bits should be quantified. A commonly used approach is, for a given bit, to invert this bit in every frame and evaluate the SEGSNR degradation. The error sensitivity for our 4.7 kbits ACELP codec measured in this way is shown in Figure 15.3 for the 34 LSF bits and the 27 first sub-frame bits. Again, the bit allocation scheme used is shown in Table 15.1.

The problem with this approach of quantifying the error sensitivity is that it does not take adequate account of the different error propagation properties of different bits. This means that if, instead of corrupting a bit in every frame it is corrupted randomly with some error probability, then the relative sensitivity of different bits will change. Hence we propose a new measure of error sensitivity. For each bit we find the average SNR degradation due to a single bit error both in the frame in which the error occurs and in consecutive frames, until the degradation decays down. The total SNR degradation is then found by adding or integrating the degradations

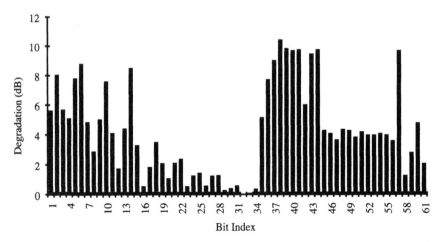

Figure 15.3 SEGSNR degradations of the 4.7 kbps codec due to single errors corrupted with 100% probability in each frame (© IEEE, 1995, Hanzo and Woodard [6]).

caused in the erroneous frame and in all the following frames, which are affected by the error. Figure 15.4 demonstrates the total SNR degradations evaluated in this way. The graph is significantly different from that in Figure 15.3. In particular, the importance of the adaptive code-book delay bits due to their memory propagation becomes more explicit.

Our final error sensitivity measure is based on the total SEGSNR degradations described above and on a similar measure for the total so-called *cepstral distance* degradation. The two sets of degradation figures are given equal weight and combined to produce an overall sensitivity measure, which was invoked to assign the speech bits to various bit protection classes. We are now ready to consider the aspects of bit-sensitivity matched error protection in the next section.

15.5. EMBEDDED ERROR PROTECTION

15.5.1 Low-quality Mode

In this section we highlight our code design approach using the 4.7 kbps codec and note that similar principles were followed in case of the 6.5 kbps codec. The sensitivity of the 4.7 kbps ACELP source bits was evaluated in the previous section in Figures 15.3 and 15.4, but the number of bit protection classes n still remains to be resolved. Intuitively, one would expect that the more closely the FEC protection power is matched to the source sensitivity, the higher the robustness. In order to limit the system's complexity and the variety of candidate schemes, in case of the 4.7 kbps ACELP codec we have experimented with a full-class BCH codec, a twin-class and a quad-class scheme, while maintaining the same coding rate.

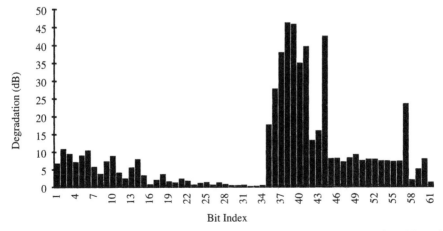

Figure 15.4 SNR degradation of the 4.7 kbps codec due to single errors in various bits, when taking into account their prolonged effect over consecutive frames (© IEEE, 1995, Hanzo and Woodard [6]).

Non-coherent QAM modems are less complex to implement, but typically require up to 6dB higher SNR and SIR values than their coherent-detection pilot-assisted counterparts. Hence in our proposed scheme second-order switched-diversity assisted coherent pilot symbol assisted modulation (PSAM) using the maximum–minimum-distance square QAM constellation is preferred [5]. Our propagation conditions are characterised by a pedestrian speed of 3 m.p.h., propagation frequency of 1.8 GHz, pilot symbol spacing of $P = 10$ and a signalling rate of 100 kBd, which will fit in a bandwidth of 200 kHz, when using a unity roll-off factor. For a channel SNR of about 20 dB this modem provides two independent QAM subchannels exhibiting different bit error rates (BERs). The BER is about a factor three to four times lower for the higher integrity path referred to as the Class 1 (C1) subchannel than for the C2 subchannel over Rayleigh-fading channels [10]. We capitalise on this feature in order to provide unequal source sensitivity-matched error protection combined with different BCH codecs for our ACELP codecs.

If the ratio of the BERs of these QAM subchannels does not match the sensitivity constraints of the ACELP codec, it can be "fine-tuned" by the help of different BCH codecs [4], while maintaining the same number of BCH-coded bits in both subchannels. However, the increased number of redundancy bits of stronger BCH codecs requires that a higher number of sensitive bits are directed to the lower integrity C2 subchannel, whose coding power must be concurrently reduced in order to accommodate more source bits. This non-linear optimisation problem can only be solved experimentally, assuming a certain subdivision of the source bits, which would match a given pair of BCH codecs.

For the full-class system we have decided to use the approximately half-rate BCH (127,71,9) codec in both subchannels, which can correct 9 errors in each 127-bit block, while encoding 71 primary information bits. The coding rate is $R = 71/127 \approx 0.56$ and the error correction capability is about 7%. When splitting the ACELP source bits into two classes each hosting 71 bits, the more sensitive bits require almost an order of magnitude lower BER than the more robust bits, in order to inflict a similar SEGSNR penalty. Hence both classes must be protected by different codes, and after some experimentation we found that the BCH (127,57,11) and BCH (127,85,6) codes employed in the C1 and C2 16-QAM subchannels provide the required integrity. The SEGSNR degradation caused by a certain fixed BER assuming randomly distributed errors is portrayed in Figure 15.5 for both the full-class and the above twin-class system, where the number of ACELP bits in these protection classes is 57 and 85, respectively. Note that the overall coding rate of this system is the same as that of the full-class scheme, namely $(57 + 85)/(2 \cdot 127) \approx 0.56$ and each 142-bit ACELP frame is encoded by two BCH codewords, yielding $2 \cdot 127 = 254$ encoded bits, hence curtailing error propagation at the block boundaries. Nonetheless, there is error propagation through the speech codec's memory. The FEC-coded bit rate became ≈ 8.5 kbps.

With the incentive of perfectly matching the FEC coding power and the number of bits in the distinct protection classes to the ACELP source sensitivity requirements we also designed a quad-class system, while maintaining the same coding rate. We used the BCH(63,24,7), BCH(63,30,6), BCH(63,36,5) and BCH(63,51,2) codes and transmitted the most sensitive bits over the C1 16-QAM subchannel using the two

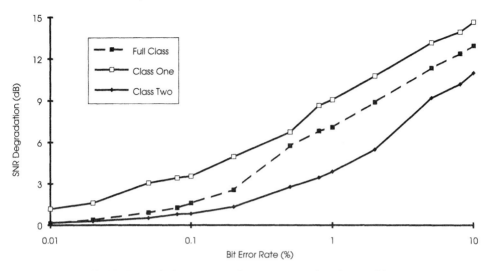

Figure 15.5 SEGSNR Degradation versus Bit Error Rate for the 4.7 kbps ACELP codec (© IEEE, 1995, Hanzo and Woodard [6]).

strongest codes and relegated the rest of them to the C2 subchannel, protected by the two weaker codes.

The PRMA control header was allocated a BCH(63,24,7) code in case of all schemes and hence the total PRMA framelength became 317 bits, representing 30 ms speech and yielding a bit rate of ≈ 10.57 kbps. The 317 bits constitute 80 16-QAM symbols, and 9 pilot symbols as well as $2+2=4$ ramp symbols must be added, resulting in a PRMA framelength of 93 symbols per 30 ms slot. Hence the signalling rate becomes 3.1 kBd. Using a PRMA bandwidth of 200 kHz, similarly to the Pan-European GSM system, and a filtering excess bandwidth of 100% allowed us to accommodate 100 kBd/3.1 kBd ≈ 32 PRMA slots. This efficiency can be further improved upon using an excess bandwidth of 50%, allowing us to create 42 time slots and hence serving up to an additional 20 users at the cost of increased system complexity.

15.5.2. High-quality Mode

Following the approach proposed in the previous subsection we designed a triple-class source-matched protection scheme for the 6.5 kbps ACELP codec. The reason for using three protection classes this time is that the 6.5 kbps ACELP codec's higher bit rate must be accommodated by a 64-level QAM constellation, which inherently provides three different subchannels. When using second-order switched-diversity and pilot-symbol assisted coherent square-constellation 64-QAM [5] amongst our previously stipulated propagation conditions with a pilot-spacing of $P = 5$ and channel SNR of about 25 dB, the C1, C2 and C3 subchannels have BERs of about 10^{-3}, 10^{-2} and $2 \cdot 10^{-2}$, respectively [9].

The source sensitivity-matched codes for the C1, C2 and C3 subchannels are BCH (126,49,13), BCH (126,63,10) and BCH (126,84,6), while the packet header was allocated again a BCH (63,24,7) code. The total number of BCH-coded bits becomes $3 \cdot 126 + 63 = 441/30$ ms, yielding a bit rate of 14.7 kbps. The resulting 74 64-QAM symbols are amalgamated with 15 pilot and 4 ramp symbols, giving 93 symbols/30 ms, which is equivalent to a signalling rate of 3.1 kBd, as in case of the low-quality mode of operation. Again, 32 PRMA slots can be created, as for the low-quality system.

15.6. PACKET RESERVATION MULTIPLE ACCESS

PRMA was designed for conveying speech signals on a flexible demand basis via time division multiple access (TDMA) systems [3]. In our system a voice activity detector (VAD) similar to that of the GSM system [1] queues the active speech spurts to contend for an up-link TDMA time-slot for transmission to the BS. Inactive users' TDMA time slots are offered by the BS to other users, who become active and are allowed to contend for the un-used time slots with a given permission probability P_{perm}. In order to prevent users from consistently colliding in their further attempts to attain a time-slot reservation we have $P_{perm} < 1$. If several users attempt to transmit their packets in a previously free slot, they collide and none of them will attain a reservation. In contrast, if the BS receives a packet from a single user, or succeeds to decode an uncorrupted packet despite a simultaneous transmission attempt, then a reservation is granted. When the system is heavily loaded, the collision probability is increased and hence a speech packet might have to keep contending in vain, until its life-span expires due to the imminence of a new speech packet's arrival after 30 ms. In this case the speech packet must be dropped, but the packet dropping probability must be kept below 1%. Since packet dropping is typically encountered at the beginning of a new speech spurt, its subjective effects are perceptually insignificant.

Our transceiver used a signalling rate of 100 kBd, in order for the modulated signal to fit in a 200 kHz GSM channel slot, when using a QAM excess bandwidth of 100%. The number of time-slots created became TRUNC(100 kBd/3.1 kBd) = 32, where TRUNC represents truncation to the nearest integer, while the slot duration was 30 ms/32 = 0.9375 ms. One of the PRMA users was transmitting speech signals recorded during a telephone conversation, while all the other users generated negative exponentially distributed speech spurts and speech gaps with mean durations of 1 and 1.35 s. These PRMA parameters are summarised in Table 15.2.

In conventional time division multiple access (TDMA) systems the reception quality degrades due to speech impairments caused by call blocking, hand-over failures and corrupted speech frames due to noise, as well as co- and adjacent-channel interference. In PRMA systems calls are not blocked due to the lack of an idle time-slot. Instead, the number of contending users is increased by one, slightly inconveniencing all other users, but the packet dropping probability is increased only gracefully. Hand-overs will be performed in the form of contention for an uninterfered idle time slot provided by the specific BS offering the highest signal quality amongst the potential target BSs.

Table 15.2 Summary of PRMA parameters (© IEEE, 1995, Hanzo and Woodard [6])

PRMA parameters	
Channel rate	100 kBd
Source rate	3.1 kBd
Frame duration	30 ms
No. of slots	32
Slot duration	0.9375 ms
Header length	63 bits
Maximum packet delay	30 ms
Permission probability	0.2

If the link degrades before the next active spurt is due for transmission, the subsequent contention phase is likely to establish a link with another BS. Hence this process will have a favourable effect on the channel's quality, effectively simulating a diversity system having independent fading channels and limiting the time spent by the MS in deep fades, thereby avoiding channels with high noise or interference.

This attractive PRMA feature can be capitalised upon in order to train the channel segregation scheme proposed in reference [11]. Accordingly, each BS evaluates and ranks the quality of its idle physical channels constituted by the unused time slots on a frame-by-frame basis and identifies a certain number of slots, N, with the highest quality i.e. lowest noise and interference. The slot-status is broadcast by the BS to the portable stations (PSs) and top-grade slots are contended for using the less robust, high speech quality 64-QAM mode of operation, while lower quality slots attract contention using the lower speech quality, more robust 16-QAM mode of operation. Lastly, the lowest quality idle slots currently impaired by noise and interference can be temporarily disabled. When using this algorithm, the BS is likely to receive a signal benefiting from high SNR and SIR values, minimising the probability of packet corruption due to interference and noise. However, due to disabling the lowest-SNR and -SIR slots the probability of packet dropping due to collision is increased, reducing the number of users supported. When a successful, uncontended reservation takes place using the high speech quality 64-QAM mode, the BS promotes the highest quality second-grade time slot to the set of top-grade slots, unless its quality is unacceptably low. Similarly, the best temporarily disabled slot can be promoted to the second-grade set in order to minimise the collision probability, if its quality is adequate for 16-QAM transmissions.

With the system elements of Figure 15.1 described we now focus our attention on the the performance of the re-configurable transceiver proposed.

15.7. SYSTEM PERFORMANCE AND CONCLUSIONS

The number of speech users supported by the 32-slot PRMA system becomes explicit from Figure 15.6, where the packet dropping probability versus number of users is displayed. Observe that more than 55 users can be served with a

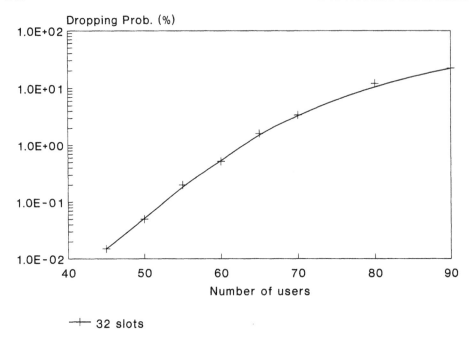

Figure 15.6 Packet dropping probability versus number of users for 32-slot PRMA (© IEEE, 1995, Hanzo and Woodard [6]).

dropping probability below 1%. The effect of various packet dropping probabilities on the objective speech SEGSNR quality measure is portrayed in Figure 15.7 for both the 4.7 kbps and the 6.5 kbps mode of operation. This Figure implies that packet dropping due to PRMA collisions is more detrimental in case of the higher quality 6.5 kbps codec, since it has an originally higher SEGSNR, despite having a shorter, 20 ms, rather than 30 ms framelength. In order to restrict the subjective effects of PRMA-imposed packet dropping, according to Figure 15.6 the number of users must be below 60. However, in generating Figure 15.7 packets were dropped on a random basis and the same 1% dropping probability associated with initial talk-spurt clipping only, imposes much less subjective annoyance or speech quality penalty than intra-spurt packet loss would. As a comparative basis, it is worth noting that the 8 kbps CCITT/ITU ACELP candidate codec's target was to inflict less than 0.5 mean opinion score (MOS) degradation in case of a speech frame error rate of 3% [12].

   The overall SEGSNR versus channel SNR performance of the proposed speech transceiver is displayed in Figure 15.8 for the various systems studied, where no packets were dropped, as in a TDMA system supporting 32 subscribers. Observe that the source sensitivity-matched twin-class and quad-class 4.7 kbps ACELP-based 16-QAM systems have a virtually identical performance, suggesting that using two appropriately matched protection classes provides adequate system performance, while maintaining a lower complexity than the quad-class scheme. The full-class 4.7 kbps/16-QAM system was outperformed by both source-

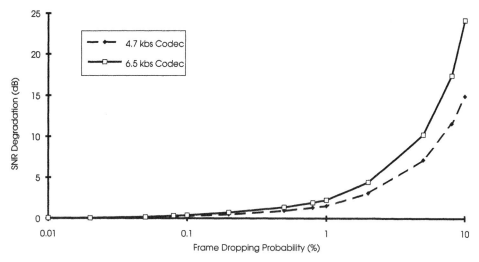

Figure 15.7 Speech SEGSNR degradation versus packet dropping probability for the 4.7 and 6.5 kbits/s ACELP codecs (© IEEE, 1995, Hanzo and Woodard [6]).

matched schemes by about 4 dB in terms of channel SNR, the latter systems requiring an SNR in excess of about 15 dB for nearly unimpaired speech quality over our pedestrian Rayleigh-fading channel. When the channel SNR was in excess of about 25 dB, the 6.5 kbps/64-QAM system outperformed the 4.7/16-QAM scheme in terms of both objective and subjective speech quality. When the proportion of corrupted speech frames due to channel-induced impairments and

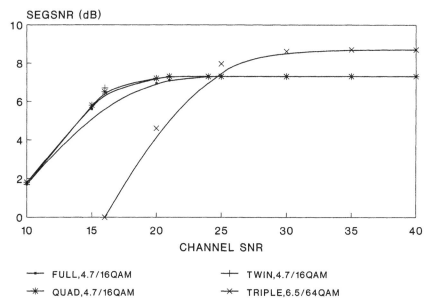

Figure 15.8 SEGSNR versus channel SNR performance of the proposed 100 kBd transceiver using 32-slot TDMA (© IEEE, 1995, Hanzo and Woodard [6]).

due to random packet dropping as in Figure 15.7 was identical, similar objective and subjective speech degradations were experienced. Furthermore, at around a 25 dB channel SNR, where the 16-QAM and 64-QAM SEGSNR curves cross each other in Figure 15.8, it is preferable to use the inherently lower quality but unimpaired mode of operation. When supporting more than 32 users, as in our PRMA-assisted system, speech quality degradation is experienced due to packet corruption caused by channel impairments and packet dropping caused by collisions. These impairments yield different subjective perceptual degradation, which we will attempt to compare in terms of the objective SEGSNR degradation. Quantifying these speech imperfections in relative terms in contrast to each other will allow system designers to adequately split the tolerable overall speech degradation between packet dropping and packet corruption. The corresponding SEGSNR versus channel SNR curves for the twin-class 4.7 kbps/16-QAM and the triple-class 6.5 kbps/64-QAM operational modes are shown in Figure 15.9 for various numbers of users between 1 and 60. Observe that the rate of change of the SEGSNR curves is more dramatic due to packet corruption caused by low-SNR channel conditions than due to increasing the number of users. As long as the number of users does not significantly exceed 50, the subjective effects of PRMA packet dropping show an even more benign speech quality penalty than that suggested by the objective SEGSNR degradation, because frames are typically dropped at the beginning of a speech spurt due to a failed contention.

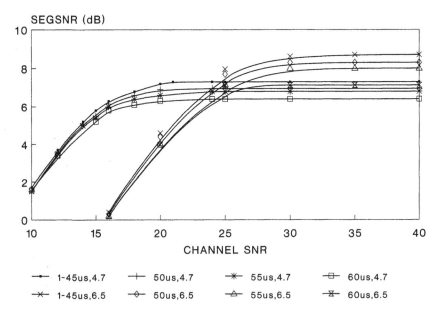

Figure 15.9 SEGSNR versus channel SNR performance of the re-configurable 100 kBd transceiver using 32-slot PRMA for different number of conversations (© IEEE, 1995, Hanzo and Woodard [6]).

Table 15.3 Transceiver parameters (© IEEE, 1995, Hanzo and Woodard [6])

Parameter	Low/high quality mode
Speech codec	4.7/6.5 kbps ACELP
FEC	Twin-/triple-class binary BCH
FEC-coded rate	8.5/12.6 kbps
Modulation	Square 16-QAM/64-QAM
Demodulation	Coherent diversity PSAM-assisted
Equaliser	No
User's signalling rate	3.1 kBd
VAD	GSM-like [1]
Multiple access	32-slot PRMA
Speech frame length	30 ms
Slot length	0.9375 ms
Channel rate	100 kBd
System bandwidth	200 kHz
No. of users	> 50
Equivalent user bandwidth	4 kHz
Min. channel SNR	15/25 dB

In conclusion, our re-configurable transceiver has a single-user rate of 3.1 kBd, and can accommodate 32 PRMA slots at a PRMA rate of 100 kBd in a bandwidth of 200 kHz. The number of users supported is in excess of 50 and the minimum channel SNR for the lower speech quality mode is about 15 dB, while for the higher quality mode about 25 dB. The number of time slots can be further increased to 42, when opting for a modulation access bandwidth of 50%, accommodating a signalling rate of 133 kBd within the 200 kHz system bandwidth. This will inflict a slight bit error rate penalty, but pay dividends in terms of increasing the number of PRMA users by about 20. The parameters of the proposed transceiver are summarised in Table 15.3. In order to minimise packet corruption due to interference, the employment of a time-slot quality ranking algorithm is essential for invoking the appropriate mode of operation. When serving 50 users, the effective user bandwidth becomes 4 kHz, which guarantees the convenience of wireless digital speech communication in a bandwidth similar to conventional analogue telephone channels.

ACKNOWLEDGEMENT

The financial support of the following organisations is gratefully acknowledged: Motorola ECID, Swindon, UK; European Community, Brussels, Belgium; Engineering and Physical Sciences Research Council, Swindon, UK; Mobile Virtual Centre of Excellence, UK.

REFERENCES

[1]   Hanzo, L. and Stefanov, J., The Pan-European Digital Cellular Mobile Radio System – known as GSM. In Steele, R. (ed.) *Mobile Radio Communications*, Pentech Press, London, Chapter 8, 677–773, 1992.

[2]   Laflamme, H. C., Adoul, J.-P. and and Morissette, S., On reducing the complexity of codebook search in CELP through the use of algebraic codes. In *Proceedings of ICASSP 1990*, 177–180, 1990.

[3]   Goodman, D. J. and Wei, S. X., Efficiency of packet reservation multiple access, *IEEE Transactions on vehicular technology*, **40**, (1), 170–176, 1991.

[4]   Wong, K. H. H. and Hanzo, L. Channel coding, Steele, R. (ed.) *Mobile Radio Communications*, Pentech Press, London, Chapter 8, 347–488, 1992.

[5]   Webb, W. T. and Hanzo, L., *Modern Quadrature Amplitude Modulation: Principles and Applications for Fixed and Wireless Communications*. IEEE Press–Pentech Press, London, 1994.

[6]   Hanzo, L. and Woodard, J. P., An intelligent multimode voice communications system for indoors communications. *IEEE Transactions on Vehicle Technology*, **44** (4), 735–749, ISSN 0018-9545, 1995.

[7]   Salami, R. A., Hanzo, L. *et al.*, Speech coding, in Steele, R. (ed), *Mobile Radio Communications*. Pentech Press, London, Chapter 3, 347–488, 1992.

[8]   Cox, R. V., Kleijn, W. and Kroon, P., Robust CELP coders for noisy backgrounds and noisy channels. In *Proceedings of ICASSP 1989*, 739–742, 1989.

[9]   Hanzo, L., Salami, R., Steele, R. and Fortune, P. M., Transmission of digitally encoded speech at 1.2 Kbaud for PCN. *IEE Proceedings-I*, **139** (4), 437–447, 1992.

[10] Stedman, R., Gharavi, H., Hanzo, L. and Steele, R. Transmission of subband-coded images via mobile channels. *IEEE Transactions on Circuits and Systems for Video Technology*, **3** (1), 15–27, 1993.

[11] Frullone, M., Riva, G., Grazioso, P. and Carciofi, C., Investigation on dynamic channel allocation strategies suitable for PRMA schemes *Proceedings of the 1993 IEEE International Symposium on Circuits and Systems*, Chicago, 2216–2219, May 1993.

[12] Salami, R. A., Laflamme, C., Adoul, J.-P. and Massaloux, D., A toll quality 8 Kb/s speech codec for the personal communications system (PCS) *IEEE Transactions on Vehicle Techology*, **43** (3), 808–816, 1994.

16

High Quality Audio Coding for Mobile Multimedia Communications

M. B. Sandler, A. J. Magrath and P. Kudumakis

Insights into Mobile Multimedia Communication
ISBN 0-12-140310-6

16.1. INTRODUCTION

Audio is perhaps the forgotten element of multimedia. If not true now, it was almost certainly true until quite recently and still seems the case that it is considered the minor partner. This belies its real importance as the front-line means of communication between humans. Of course, speech has always been an important part of telecommunications, but considerations of quality have been secondary to intelligibility.

The main body of this chapter presents results from two projects which examine high quality, low bit rate coding of audio, both music and speech. One approach uses wavelets, which offer advantages over polyphase filter banks and discrete cosine transform (DCT) in MPEG music coding at low bit rates. The other approach uses the long established linear predictive coding (LPC) technique, but in a modified guise, with orders significantly greater than 10. It also uses least mean squares techniques to fit lines in time-frequency space to the line spectral pair (LSP) representation of LPC.

The first section offers an introduction to audio coding and processing in the broadest sense as is applicable to mobile multimedia. Here it is emphasised that in a complete mobile multimedia (MMM) system, the designer needs to take account of more than just the most effective coding technique. Following this, a new approach to speech coding is described and some experimental results presented. Some attention is paid to the means by which high quality, high order LPC speech may be re-synthesised. Finally the principles of the use of wavelets in audio coding are covered, and these too are supported by results.

16.1.1. How Does Audio Fit in with Mobile Multimedia?

Audio is about much more than coding. Obviously it includes the capture and replay of audio signals (microphone and loudspeaker technology, amplification, ADC and DAC technology) and modifications to spectral distributions (filtering) but it also now includes aspects such as spatial reality and sound field synthesis.

Thus we can hope to see high quality replay of audio from a mobile multimedia station to make the most of the coding effort. We can also expect enhanced teleconferencing facilities incorporating spatial separation in the listener's environment of the various conferees: this is for both headphone and loudspeaker reproduction.

Also, of course, mobile multimedia does not necessarily involve mobile communications (e.g. it includes the use of CD-ROM from a mobile workstation), though it will always include data compression. Presumably the mobile office of the near future will not only seek to keep the business-person-on-the-move in touch with colleagues at all sorts of mobile and fixed locations, it will also seek to keep him or her amused with the replay of compact discs (CDs) and digital versatile discs (DVDs), games and Internet radio. One could envisage audio-on-demand services, similar to video-on-demand and falling somewhere between digital radio and stored audio (CD, mini-disk etc.).

16.1.2. What is Audio?

Audio is a single-sense communications mechanism. Its two most important forms are speech and music – the first normally conveying meaning and the second emotion, pleasure or similar. Both have been and continue to be important in the development of human societies. Other forms of audio can be broadly be categorised as noise, though this does not necessarily imply that it is unwanted. Consider for example the studio dubbing of speech onto a film sound track. Without what are known as room tones, the speech is dry and unrealistic: "noise" must be added. Consider also the effect of the non-coding of silence (pauses) in low bit rate speech. The total absence of sound can be somewhat distracting to the listener and again the insertion of the correct sort of noise will be beneficial. So audio is not only more than speech, it is more than speech and music, though in this chapter we will ignore the importance of noise (in the sense used above).

16.1.3. What is Mobile?

This is not a fatuous question. What is expected in a physical sense (size, power consumption and so on) of an end-user terminal will have an influence on its capabilities. A transportable system will be able to provide greater power amplification and better quality than a fully portable system. It is possible that, because of power supply limitations, the portable system will be able to devote less processing power to signal coding and processing than the transportable system. This brings in issues of scaleable coding schemes. These are covered in more detail in section 16.3.

16.1.4. What is Multimedia?

Multimedia is about communications between humans. This implies the use of human sense organs. To date only sight and hearing have been addressed as multimedia modalities. Strangely however, while all possible uses of the visual sense are explicitly mentioned (images, video, text and graphics), the use of the aural sense are covered by the single collective noun – either sound or audio. Once we start to consider that audio is made up of music, speech and noise, one immediately re-categorises radio as multimedia rather than single-medium. This is more important than mere pedantry: such an approach immediately recognises that when speech and music are mixed, for instance in a broadcast, coding each separately should bring compression advantages. This is no different from coding mixed video and audio or mixed text and graphics – in each case the two constituents are coded separately into a representation of the whole.

16.1.5. What is Mobile MM used for?

Here we must start to consider what uses MMM is to be put to in order to better understand the roles that audio is to play, and to understand what processing is appropriate. The following is a list – by no means complete – of current and obvious future applications:

- conferencing;
- entertainment;
- services (medical, police, news gathering);
- information (e.g. route finding);
- wireless office LANs;
- mobile office.

Not all of these require communications. The mobile office requires it only sporadically, typically for email and fax. Entertainment can be wholly off-line (as in CD replay) or partially off-line in the sense that, for example a book to read or a piece of music to listen to can be downloaded as a background task or from a base-station rather than on the move.

16.1.6. What is Wanted from Audio?

We are now ready to answer this key question. Still of primary importance is the requirement that audio is intelligible. Different psychoacoustic intelligibility criteria pertain for speech and music (e.g. plain old telephone system – POTS – quality speech and AM-mono music) but they do exist. However, the demand will exist for significantly improved speech transmission (wider bandwidth, improved dynamic range and SNR) and FM-stereo to CD quality music (or better). This is to say that clarity of signal is important. So too are fidelity and reality. These are the four levels of performance that are required of MMM audio sub-system performance.

The ordering of the terms clarity, fidelity and reality are somewhat arbitrary but might convey the following. Clarity implies that there is greater system bandwidth and/or SNR, but the reproduction electronics (power amplifier, speakers, screening levels) are only of moderate quality. Fidelity would imply that reproduction electronics are up to high standards (though not necessarily to hi-fi buff standards) and using only 1 or 2 channels. Reality implies multichannel reproduction of some sort. It might well involve 3D auralisation techniques to simulate room acoustics (binaural or transaural). Such an approach is clearly of interest in teleconferencing with multiple parties, where preferably each participant is separately located in a 3D listening environment. Precisely the same functionality is going to be of use in MMM for entertainment.

16.1.7. What Signal Processing is Required?

Clearly, signal compression and coding are required, and a discussion of these forms the major part of this chapter. We will see that analysis and synthesis of signals forms a core part of compression, as does pattern recognition. In fact the new MPEG4 recommendations explicitly recognise that signal coding and signal interpretation are two different views of the same requirement [1].

Less obvious as topics for attention in MMM are data conversion, amplification and transducers. In each case, size must be as small as possible and power consumption as low as possible, while still retaining, as far as possible, clarity, fidelity and reality. There is obviously no way that a laptop or palmtop computer can provide audio reproduction down to 50 or even 100 Hz, but this is clearly the ideal. Small electrostatic loudspeakers would seem to be appropriate to an MMM terminal, yet to the author's knowledge, no-one has yet combined these two technologies.

A great deal of power is dissipated in the amplifier driving loudspeakers. Work around the world over the last decade and more has focused on the use of switching (Class D) output stages driven either by a pulse width modulated (PWM) or a sigma–delta modulated (SDM) digital signal. Currently SDM looks to have significant advantages in this area. Such approaches promise power efficiencies of 90% and more [2]: conventional class B or AB output stages might only offer 50–70% power efficiency.

These days most audio and speech band analogue to digital converters (ADCs) use SDM techniques, but still convert the resulting one-bit signal to a PCM format. There is considerable mileage in investigating systems which process the signal in its one-bit format. This might be equalisation or it might be compression. In other words silicon, and therefore cost and power, efficiencies are available to an MMM terminal designer if ADC and processing are no longer seen as separate functions and instead a multifunctional approach is taken [28].

It is possible to provide a complete sound field reproduction using either headphones (binaural) or loudspeakers (transaural). Transaural is less demanding computationally but there is only a small "sweet spot" for the listener's head, at which the full effect will be perceived. Although this can be a problem for hi-fi or film reproduction, the MMM terminal user will often be in a well-defined location with respect to the transducers built into the terminal. This functionality is of practical importance in teleconferencing and of significance for leisure uses of MMM [3].

Most often an MMM terminal will be used in a more or less audio-hostile environment. While headphones can improve the situation on the replay side, advanced microphone techniques are needed for signal capture. Thus we can expect to see microphone arrays providing high directionality for the wanted signal and high attenuation to the background.

16.1.8. An Overview of Audio Coding

Most audio compression techniques fall into one of the following categories:

- waveform coding;
- transform coding;
- parametric coding.

Of these waveform coding [4] provides the least amount of compression for a desired signal quality. The best known audio implementation of waveform coding is near instantaneous companded audio multiplex (NICAM) as used in television broadcasts since the 1980s.

Both transform and parametric coding are covered each in a single embodiment in the following sections. Each has its strengths.

Transform coding, which includes sub-band coding, is the prevalent technique for music and is used in MPEG [5,6], AC-3 [7] and others [8,9,10]. The signal is partitioned into equal-sized segments each of which is separately processed. The processing can be some form of DCT or sub-band filtering followed by a quantisation process which takes account of the human auditory system by means of psycho-acoustics to provide the compression. Included in this category is the wavelet transform approach as described in section 16.3.

Parametric coding also operates on signal segments. However, it differs in that the signal is modelled as a filter (finite or infinite impulse response – FIR or IIR, but normally IIR) which when driven by an appropriate signal (whose parameters must also be found) best models the original. Typical of this category is the LPC approach described in the next section.

Thus all the compression techniques of current interest deal with the signal in segments. The primary reason for this is the need to capture the time-varying statistics of the signal. Both transform and parametric coders provide a time-frequency mapping of the signal.

In cases when audio is allied to other coded information – typically video – there is a need to determine how best to allocate the overal bit budget to each signal stream. In MPEG1 video of the 1.5 Mbps only 64 kbps are allocated to audio. There is still an open question concerning whether or not this is optimal. In tele-conferencing scenarios, where there is less bandwidth, it is unlikely that the best solution allocates a similarly low proportion of bits to the audio, especially where multi-way conferencing is taking place.

16.2. A HIGH QUALITY SPEECH CODER

Speech coding systems are used to transmit and store speech with minimal bandwidth (or bit rate). Whilst low/medium quality speech coding is acceptable for mobile and network communications, applications are arising where high quality, natural speech is desirable, for example, audio/video teleconferencing, digital audio broadcasting, virtual reality, secure voice transmission, digital storage systems, voice mail and voice response systems.

An established technique for speech coding is linear predictive coding (LPC), which derives parameters of the vocal tract filter and excitation source that model the human vocal system [11]. The digitised speech signal is initially divided into frames and the coefficients of an all-pole filter $H(z)$ of order p are obtained for each frame. The poles represent resonances, or formants, of the vocal tract, whose frequency and bandwidth vary with time. More information on LPC coding is presented in Chapter 12 of this book.

Most of the current standards use a speech bandwidth of 4 kHz, and the formants are represented by a filter of order 10 or 12. For natural sounding speech, a wider bandwidth is desirable, and to code the additional formants at high frequencies, a higher order filter is required. Higher order models are also desirable as they improve the modelling of spectral zeros during nasalised sounds [12], improve imunity to external acoustic noise [13], and can increase the medium to long-term redundancy in the residual [14]. The latter implies that the residual signal can be coded with greater efficiency and therefore reduced bit rate.

16.2.1. Line Spectral Pair Least Mean Squares (LSP–LMS) Coding

The cost of higher order coding is an increase in the number of filter parameters that need to be sent to the receiver. A solution proposed in [15] is to transform the LPC coefficients into line spectral pair (LSP) frequencies and parameterise the variations over K consecutive LPC frames.

The LSP transformation [16] involves mapping the p zeros of the inverse vocal tract filter $A(z) = 1/H(z)$ onto the unit circle through two z-transforms $P(z)$ and $Q(z)$ of order $p + 1$.

$$P(z) = A(z) + z^{-(p+1)}A(z^{-1}) \tag{16.1}$$

$$Q(z) = A(z) - z^{-(p+1)}A(z^{-1}) \tag{16.2}$$

Figures 16.1 and 16.2 show an example speech waveform, sampled at 8 kHz and the frequencies of the roots of the $P(z)$ and $Q(z)$ polynomials obtained for 22nd order LPC analysis with a frame length of 160 samples. The LSP representation produces

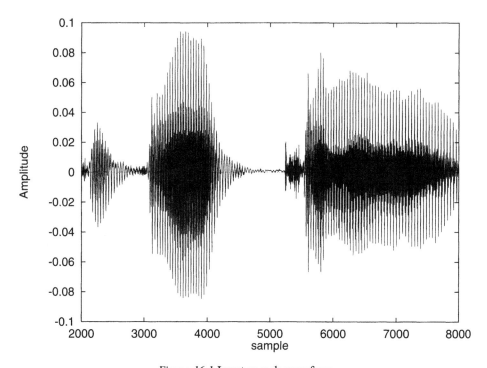

Figure 16.1 Input speech waveform.

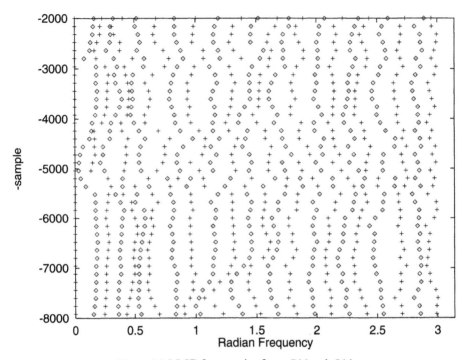

Figure 16.2 LSP frequencies from $P(z)$ and $Q(z)$.

frequency variations which can be reasonably modelled by piecewise linear approximations. This leads to a coding scheme in which the LSP trajectories over K consecutive LPC frames are modelled as straight-line segments. A method of obtaining
the parameters of the line segments using the combinatorial Hough transform has
been described in [15]. Here we describe an alternative method using a least mean
squares algorithm which has been shown [17] to have higher accuracy than the
Hough transform.

16.2.2. Least Mean Squares Algorithm

The least mean squares (LMS) algorithm finds the line of best fit through a collection
of data points. Each coding frame f is subdivided into K LPC frames. The equation
of each LSP line segment is given by

$$\theta_f^{(n)}(k) = \sigma_f^{(n)} + \frac{k}{K}\delta_f^{(n)} \tag{16.3}$$

Here $k : \{0 : K - 1\}$ is the LPC frame index, $\sigma_f^{(n)}$ is the nth LSP frequency at start
of coding frame f, $\delta_f^{(n)}$ is the rate of change of the line segment. To ensure continuity
between line segments, $\sigma_f^{(n)}$ is connected to the end point of the previous line segment,
and is therefore a constant in the LMS analysis. The aim of the analysis is to find the

value of $\delta_f^{(n)}$ which minimises the mean square deviations between the candidate line and the LSP data points.

Defining each data point as $\vartheta_f^{(n)}(k)$, the error on the frequency axis between a point on the candidate line and the data point is:

$$\xi_f^{(n)}(k) = \vartheta_f^{(n)}(k) - \theta_f^{(n)}(k) \tag{16.4}$$

The sum of the squares of this function is

$$\Gamma = \sum_{k=0}^{K-1} \xi_f^{(n)}(k)^2 \tag{16.5}$$

$$= \sum_{k=0}^{K-1} \left\{ \vartheta_f^{(n)}(k) - \sigma_f^{(n)} - \frac{k}{K} \delta_f^{(n)} \right\}^2 \tag{16.6}$$

We minimise Γ which leads to

$$\delta_f^{(n)} = \frac{K \left\{ \sum_{k=0}^{K-1} k \vartheta_f^{(n)}(k) - \sigma_f^{(n)} \sum_{k=0}^{K-1} k \right\}}{\sum_{k=0}^{K-1} k^2} \tag{16.7}$$

Equation (16.7) is calculated for each LSP data set of $P(z)$ and $Q(z)$. Figure 16.3 shows the line segments obtained by applying this analysis to the data of Figure 16.2. For stability in the vocal tract filter $H(z)$, the adjacent line segments must not intersect [16]; however, this may occur due to errors in the LMS approximation.

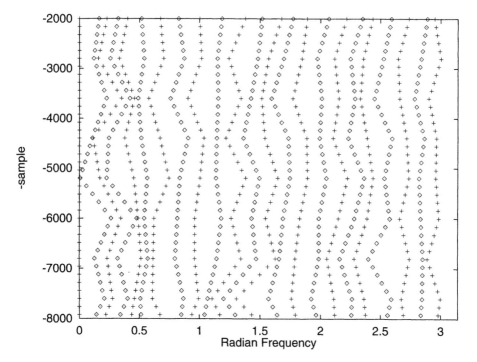

Figure 16.3 Coded LSP frequencies from $P(z)$ and $Q(z)$.

A post-processing algorithm is used to modify the equations of intersecting lines to prevent instability.

16.2.3. Coder/Decoder

In this section we propose a speech coder/decoder based upon the above technique. The coder is shown in Figure 16.4. LPC analysis is performed on K consecutive speech frames. The LSP–LMS coder produces a quantised value of $\delta_f^{(n)}$ for each LSP trajectory n. To compensate for errors in the LSP–LMS coding, the residual is obtained from the coded LPC coefficients, by means of a local decoder and inverse filter. In this implementation the residual is coded using multipulse coding [18] with long-term prediction (LTP).

As an example of the performance of the coder, Figure 16.5 shows the resynthesised speech obtained using $K = 5$, the LPC parameters of section 16.2.1 and multipulse coding with 10 pulses for each residual subframe of 40 samples and a single tap LTP. Assuming the LSP line equation parameters are quantized with the same accuracy as the standard (non-coded) LSP parameters, this coder would produce a five-fold reduction in the number of bits required to represent the vocal tract filter. Research is now directed towards improving the accuracy of the LMS model by the use of polynomial based line-fitting.

16.3. HIGH QUALITY MUSIC CODING

Compression of digital audio signals marks the second phase of the revolutionary changes to the audio world which has been brought by digital signal processing. About ten years ago, when the CD had just been introduced, the first proposals for digital data reduction systems were greeted with suspicion and disbelief. Almost nobody thought that digital compression was necessary, given the vast storage capa-

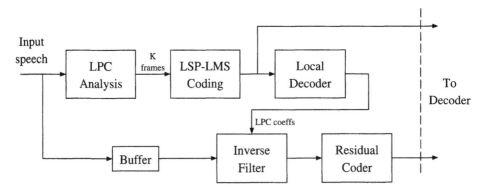

Figure 16.4 Block diagram of speech coder.

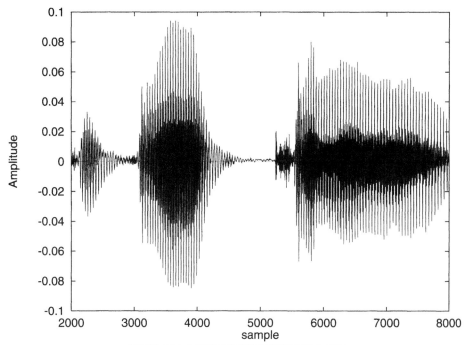

Figure 10.3 Resynthesised speech waveform.

city of CD. Everybody agreed that it would be impossible to fulfil the audio quality requirements of the "golden ears" while removing 75% or more of the digital data.

However, since 1988 collaborative work by an international committee of high-fidelity audio compression experts within the Moving Picture Experts Group (MPEG) has resulted in the first international standard for the digital compression of high-fidelity audio, the so-called MPEG-1 audio (coding of mono and stereo signals at sampling rates of 32, 44.1 and 48 kHz). The International Standards Organization and the International Electrotechnical Commission (ISO/IEC) adopted this standard at the end of 1992 [5].

Thus, these days the situation of digital compression for high-fidelity audio is completely different. Due to the widespread interest in transmitting, storing and retrieving CD-quality sound, high quality audio coding is an area attracting considerable interest within the DSP community. Current applications for high quality audio coding include digital audio broadcasting (radio and TV), portable audio units (digital compact cassette and minidisc), and also transmission of coded sound over high speed networks. Forthcoming applications where audio compression is involved include interactive mobile multimedia communication, videophone, mobile audio-visual communication, multimedia electronic mail, electronic newspapers, interactive multimedia databases, and multimedia videotex. Thus, audio compression methods are included in almost every digital audio system. In fact, in many cases digital audio is used at bit rates which are too low even by the standards of the inventors of audio coding systems. Most of this change is due to the work of the companies involved in MPEG audio. That is because wide acceptance of the MPEG audio standard will permit manufacturers to produce and sell, at reasonable cost, large numbers of MPEG audio codecs and multimedia related products.

Due to the high interest these days in high quality low bit rate audio coding, the MPEG committee also developed, at the end of 1994, the MPEG-2 audio standard (backwards compatible coding of 5.1 multichannel audio and low bit rate coding of mono and stereo signals at half sampling rates in respect to MPEG-1 audio) [6].

Moreover, backwards compatibility (BC) means that an MPEG-1 audio decoder can decode two channels of the MPEG-2 stream and an MPEG-2 audio decoder can decode an MPEG-1 audio stream as if it were an MPEG-1 audio decoder.

However, the MPEG committe today focuses on next generation audio coding standard the so-called MPEG-4. MPEG-4 will be the coding system for the multi-media applications of the future. It is designed to facilitate the growing interaction and overlap between the up-to-now separate worlds of computing, electronic mass media (TV and radio) and telecommunications. International standard status of MPEG-4 is expected in November 1998 [19].

16.3.1. MPEG: Layer I, II and III

The MPEG committee chose to recommend three compression methods and named them audio Layer I, II and III. This provides increasing quality/compression ratios with increasing complexity and demands on processing power.

Thus a wide range of trade-offs between codec complexity and compressed audio quality is offered by the three layers. The reason for recommending three layers was partly that the testers felt that none of the coders was 100% transparent to all material and partly that the best coder (layer III) was so computing-intensive that it would seriously impact the acceptance of the standard.

16.3.2. MPEG: Non-backwards Compatible (NBC) Audio Coding

BC built in MPEG-2 audio is an important service feature for many applications, such as television broadcasting. This compatibility, however, entails a degree of quality penalty that other applications need not pay. Work in this area of MPEG audio has produced a non-backwards compatible (NBC) extension to MPEG-2.

The NBC standard (part 7 of MPEG-2) is bringing down to 64 kbps virtual transparency of single-channel music which MPEG-1 audio had set at 128 kbps. It is expected that interesting performance will be obtained even at lower bit rates than 64 kbps. Since MPEG-4 will not introduce new tools for coding of audio signals at 64 kbps and above, the NBC extension of MPEG-2 is therefore already providing part of the MPEG-4 audio standard. More work, however, needs to be done in the bit rate range much lower than 64 kbps. This is an area where there is a need for a generic technology serving such different applications as satellite and cellular com-munications, mobile multimedia communications, Internet, etc. Thus in this chapter we present an alternative NBC approach to low bit rate audio coding based on the wavelet packet algorithm.

16.3.3. Wavelet Packet Codec: an Alternative Approach to Low Bitrate Audio Coding

A wavelet packet (WP) or, in other words, a five-stage 32-band uniform frequency subdivision subband coding scheme, based on a tree-structure filterbank, has been designed and implemented [20]. The frame length was set equal to 1024 samples. However, efficient signal compression results when sub-band signals are quantised with sub-band-specific bit allocation, based on the input power spectrum (fast Fourier transform – FFT) and the model of auditory perception. Thus, this filterbank is combined with dynamic bit allocation (DBA), based on psychoacoustic model-1 as adapted for use with Moving Pictures Expert Group (MPEG) layer-2 [5].

In our experiments we have not included the tonality in order to keep the overall complexity of the codec low. This also ensures that the segmental signal-to-noise ratio (SSNR) is more valid since masking phenomena have not been utilised. However, the tonality would improve the compression further if included. Finally, while the dynamic bit allocation strategy exploits some of the human hearing characteristics, further reduction in the bit rate requires removing the statistical redundancies of the signal. That is the ideal case for entropy noiseless Huffman coding. This has been embedded in the WP coder as in MPEG layer-3 [5].

The advantages of the WP coder are:

- The increasing quality/compression ratios with increasing complexity achieved by the three layers of MPEG may be achieved with the WP codec while switching to different wavelet filters. There is obviously a tremendous cost saving in software and hardware by using one instead of three layers while keeping their features.
- The high coding gain of the WP algorithm ensures that the WP codec outperforms the MPEG in the bit rate range of interest, namely much lower than 64 kbps. This can be observed from Figures 16.6–16.8 for three different music signals [21].

16.3.4. Scaleability and MPEG-4 Functionalities

Perhaps the most significant feature which has not yet been fully addressed and embedded in audio coding systems, although well known in video coding technology, is the concept of scaleability. Scaleability is the property of a coded signal that part of the coded bit stream can be decoded in isolation. The fact that a subset of the coded bits is sufficient for generating a meaningful audio or video signal is important in at least two contexts. One is where both cheap–simple and expensive–complex decoders are envisaged as being receivers of the signal: it is as if both AM and FM radio quality are available in the same transmitted signal dependent on an appropriate decoder. A second is when the transmission channel cannot guarantee the full necessary bandwidth to handle the complete bitstream, for example: Internet radio. There are several types of scaleability, in terms of bandwidth; number of channels; and the most important, the so-called SNR scaleability [22].

Scaleable systems currently require a higher bit rate to achieve the same quality as a single-stage perceptual audio codec. Since scaleability offers many advantages, a small performance penalty may be acceptable [23]. However, we have seen in [24]

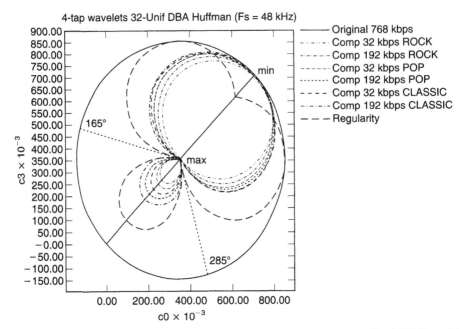

Figure 16.6a The performance of 4-tap wavelet filters, in terms of normalised SSNR and in comparison to their regularity, for low (32 kbps) and high (192 kbps) bit rates and for three different music signals: ROCK (8 s), POP (6 s), CLASSIC (28 s)

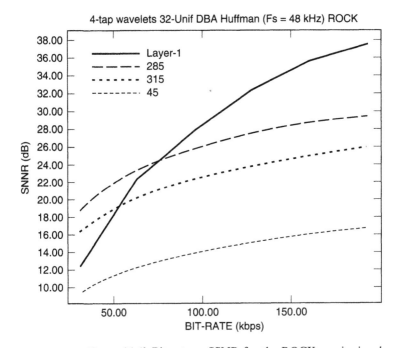

Figure 16.6b Bit rate vs SSNR for the ROCK music signal.

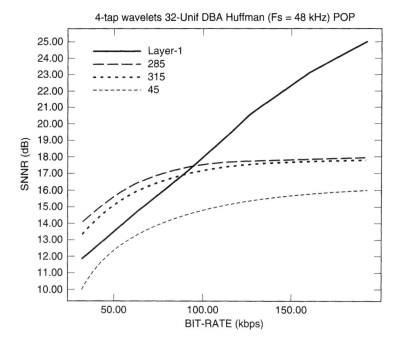

Figure 16.7 Bit rate vs SSNR for the POP music signal.

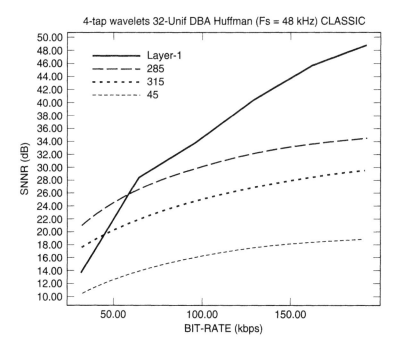

Figure 16.8 Bit rate vs SSNR for the CLASSIC music signal.

that using the wavelet packet approach this performance penalty becomes performance advantage, particularly at low bit rates. For example we have compared MPEG and wavelet-based two-stage scaleable coders and found that if a 64 kbps stream allocates 32 kbps to each of two stages, the overall SSNR is 22.10 dB for wavelets and 16.04 dB for MPEG. Even allocating all 64 kbps to MPEG (i.e. single stage standard MPEG) only achieves 22.32 dB SSNR. AT 32 kbps SSNR of 18.72 dB is achieved with wavelets, performance which is superior to a scaleable MPEG coder with double the number of bits [24].

Progress on scaleability can be expected if a true integration of speech and sound coding can be achieved. In a scaleable codec a reasonable specification requires the inner layer to provide 3.5 kHz of audio bandwidth at bit rates ranging from 3 to 16 kbps. An additional scaleability layer may result resulting in an intermediate audio bandwidth of 7 to 11 kHz at 16 to 40 kbps. Finally an additional high quality layer operating at bit rates in excess of 100 kbps per channel would be useful for studio applications [23].

Such a range of performance was one of the justifications for examining in [25] the performance of various wavelet filter families at both low ($Fs = 8$ kHz) and high ($Fs = 48$ kHz) sampling frequencies. This performance evaluation has also shown the limits of a bandwidth scaleable wavelet-packet-based codec. Transparency (CD sound quality) can be obtained at 24 kbps with $Fs = 8$ kHz, and at 64 kbps with $Fs = 48$ kHz.

All perceptual scaleable audio codecs today are based on a 1024 band MDCT filterbank, as used for MPEG-2 NBC coding [23]. This also justifies our decision at early stages of this research for the wavelet packet transform length used, instead of the 384 samples used by Layer I and 1152 samples by Layer II of MPEG-1 and MPEG-2 BC.

Since MPEG-4 audio will contain provisions to transmit synthetic speech and audio at very low bit rates, scaleability should also be considered in such systems. For very low rates in speech coding synthetic reproduction of speech is used. So far nothing has been presented for the transmission of music. MIDI operates at 32 kbps, which is a rate where true audio codecs can already operate quite well. Therefore synthetic audio must be kept in mind for later designs [23].

However, we have shown that almost transparent (AM sound quality) coding can be achieved with wavelet-packet-based audio compression systems even at bit rates as low as 32 kbps ($Fs = 48$ kHz), and in comparison to MPEG-audio standards, wavelet-packet-based systems result in better sound quality than Layer I, and are competitive with Layer II [25]. Some ways to generate synthetic audio using the wavelet packet model have also been presented in [25], e.g. reconstructing the signal from its scale factors or random-number-based algorithmic composition.

We conclude this discussion of scaleability with some thoughts motivated from our detailed study of 4-tap wavelet filters and their regularity (see Figure 16.6) [26]. There are three types of 4-tap wavelet filters in terms of scalable SSNR performance and complexity. This is clarified in Table 16.1. We may need to consider that the SSNR values are dependant on the bit-rate and signal characteristics, while in terms of complexity we could use for the two scalability stages $2+2$, $2+4$ or $4+4$ tap filters corresponding to various SSNR performances depending on the selection of the angle θ which determines the wavelet filter coefficients. Of course, we could

Table 16.1 Complexity and SSNR scaleability using wavelet filters with $N = 2$ vanishing moments. The SSNR values are given approximately and at 32 kbps with $Fs = 48$ kHz for the ROCK (8 s) music signal.

$\theta(°)$	TAPS	$SSNR_\theta$ (dB)	Reg.
45	2	9	0
(135,525,315)	2	16	0
(165,285)	4	18	0.55

extend this technique using longer than 4-tap wavelet filters with a larger number of filter selections [27].

Current research on scaleability is focused on encoding of mono signals only. Scaleable multichannel, stereo and other MPEG-4 functionalities, for example support for pitch/time-scale change and editability/mixing on the compression bit stream, are areas which require further research.

Wavelets are particularly suited to scaleable coding because they inherently embody bandwidth scalability. At low bit rates the results of this paper have already demonstrated that wavelets compete strongly with current techniques. Wavelets are also very useful in multimedia applications where pitch and/or time changes are necessary (e.g. alongside slowed-down or speeded-up video). Also of interest is the plan that MPEG-4 will include synthetic music as one of its functionalities: here again wavelets are ideally suited as [25] demonstrates. Therefore what is still missing is to demonstrate the maturity of wavelet strategies and their readiness for exploitation in digital broadcasting, multimedia, digital hi-fi audio, speech coding/telephony and in particular to cover speech, "AM", "FM" and "CD" quality music under one unified coding scheme.

16.4. CONCLUSIONS

This chapter seeks to highlight the growing importance of research into audio to the fast-moving field of mobile multimedia. Much of it has dealt with two specific projects aimed at high quality coding of audio. However, the Introduction surveyed all aspects of digital and analogue audio relevant to the context of mobile multimedia systems implementation. The coding strategy introduced in section 16.2 is in its infancy and the results are preliminary. Nevertheless it appears that significant compression of the vocal tract models is possible when formant trajectories, in the form of LSP representations, are coded. To make a complete, reduced-entropy coding scheme requires further research, including examination of improved residual coders. It is already demonstrably true that wavelets offer significant performance advantages over DCT, MDCT and the polyphase filter-banks used in MPEG at low bit rates. Also, because the wavelet transform may be simply adapted to provide the basis of any psychoacoustic model, it is more efficient than MPEG as no FFT is needed. This makes real-time software encoders possible for mid-performance workstations, and will make ASIC implementations more cost effective. We therefore propose the use of the WT and WPT in scaleable audio coders and demonstrate that again, at low bit rates, this approach has significant advantages over MPEG. Note that all this work uses MPEG 1 Layers 1 to 3 as the basis for comparison. Future stages of the research programme will compare wavelet approaches, suitably modified, to the higher performing MPEG 2 and 4 audio standards.

ACKNOWLEDGEMENT

The authors would like to acknowledge the support of EPSRC grant nos GR/L 15272 and GR/L 21914.

REFERENCES

[1] Requirements for low bit rate audio coding/MPEG-4 audio. ISO/IEC JTC1/SC29/ WG11 MPEG94/443, 1994.

[2] Magrath, A. and Sandler, M., Hybrid pulse width modulation/sigma-delta modulation power digital to analogue conversion. *IEE Procs Circuits, Devices & Systems*, **143** (3), 149–156, 1996.

[3] Hardman, V., Sasse, M. A. and Kouvelas, K., Successful multi-party audio communication over the Internet. *Communications of the ACM*, **41** (5), 1997.

[4] Jayant, N. S. and Noll, P., *Digital Coding of Waveforms: Principles and Applications to Speech and Video*. Prentice Hall, Englewood Cliffs, NJ, 1984.

[5] Information technology – Coding of moving pictures and associated audio for digital storage media at up to about 1,5 Mbit/s – (Part 3: Audio), ISO/IEC 11172-3, August 1993.

[6] Information technology – Generic coding of moving pictures and associated audio – (Part 3: Audio), ISO/IEC 13818-3, November 1994.

[7] Todd, C. C., Davidson, G. A., Davis, M. F., Fielder, L. D., Link, B. D. and Vernon, S., AC-3: flexible perceptual coding for audio transmission and storage. In *Proceedings of the 96th AES Convention*, Amsterdam, February 26–March 1, 1994.

[8] Smyth, S. M. F., Smith, W. P., Smyth, M. H. C., Yan, M. and Jung, T., DTS coherent acoustics. Delivering high quality multichannel sound to the consumer. In *Proceedings of the 100th AES Convention*, Copenhagen, 11–14 May, 1996.

[9] Tsutsui, K., Suzuki, H. and Shimoyoshi, O., ATRAC: adaptive transform acoustic coding for MiniDisc. In *Proceedings of the 93rd AES Convention*, San Francisco, October 1–4, 1992.

[10] Tan, R. K. C. and Hawksford, M. O. J., Multi-pulse adaptive sub-band coding (MASC) using psychoacoustic optimization algorithm. In *Proceedings of the 93rd AES Convention*, San Francisco, October 1–4, 1992.

[11] Markel J. D. and Gray, A. H. Jr., *Linear Prediction of Speech*. Springer-Verlag, New York, 1976.

[12] O'Shaughnessy, D. *Speech Communication. Human and Machine*. Addison-Wesley Series in Electrical Engineering: Digital Signal Processing, Addison-Wesley, Reading, MA, 1987.

[13] Liu, Y. J., A robust 400-bps speech coder against background noise. In *Proceedings of the IEEE International Conference on Acoustics, Speech and Signal Processing*, 601–604, 1991.

[14] Chen, J.-H., Cox, R. V., Lin, Y.-C., Jayant, N. and Melchner, M. J., A low-delay CELP coder for the CCITT 16 kb/s speech coding standard. *IEEE Journal on Selected Areas in Communications*, **10** (5), 601–604, 1992.

[15] Magrath, A. J., Linear predictive coding of speech at high order using the Hough transform. In *Proceedings of the 2nd IEEE UK Symposium on Applications of Time-Frequency and Time-Scale Methods*, August 1997.

[16] Soong, F. K. and Juang, B.-H., Line spectral pair (LSP) and speech data compression. In *Proceedings of the IEEE International Conference on Acoustics, Speech and Signal Processing*, 1984.

[17] Magrath, A. J., Linear predictive coding of speech at high order using the Hough transform and least-mean squares algorithm. Internal Report of the Signals, Circuits and Systems Research Group, King's College London No. 117/SCS/97, ISBN 1-898-783-08-X, March 1993.

[18] Berouti, M., Garten, H., Kabal, P. and Mermelstein, P., Efficient computation and encoding of the multipulse excitation for LPC. In *Proceedings of the IEEE International Conference on Acoustics, Speech and Signal Processing*, 1984.

[19] Chiariglione, L., MPEG and multimedia communications. http://www.cselt.stet.it/ufv/leonardo/paper/isce96.htm.

[20] Kudumakis, P. and Sandler, M. On the performance of wavelets for low bit rate coding of audio signals. In *Proceedings of the IEEE ICASSP'95*, **5**, 3087–3090, Detroit, MI. May 8–12, 1995.

[21] Eveleigh, ?? and Wilkinson, ??, GUD Toons (Good Tunes), in Kinti, K. The future music, CD Vol. 4, demo no. 6, in *Future Music Magazine* No. 14, December 1993.

[22] Brandenburg, K. and Grill, B., First ideas on scalable audio coding. In *Proceedings of the 97th AES Convention*, San Francisco, November 10–13, 1994.

[23] Grill, B. and Brandenburg, K., A two- or three-stage bit rate scalable audio coding system. In *Proceedings of the 99th AES Convention*, New York, October 6–9, 1995.

[24] Kudumakis, P. and Sandler, M., Wavelet packet based scalable audio coding. In *Proceedings of IEEE ISCAS'96*, **2**, 41–44, Atlanta, May 1996.

[25] Kudumakis, P., Synthesis and coding of audio signals using wavelet transforms for multimedia applications. Ph.D. Thesis, King's College University of London, 1997.

[26] Kudumakis, P. and Sandler, M., On the compression obtainable with 4-tap wavelets. *IEEE Signal Processing Letters*, **3** (8), 231–233, 1996.

[27] Kudumakis, P. and Sandler, M., On the usage of short wavelets for scalable audio coding. In *Proceedings of SPIE'97*, San Diego, CA, USA, 27 July–1 August, 1997.

[28] Anderson, M., Summerfield, S., Kershaw, S. and Sandler, M., Realization and implementation of sigma delta bitstream FIR filter. *IEE Procs Circuits Devices & Systems*, **143** (5), 267–273, 1996.

17

High Quality Low Delay Wideband Speech Coding at 16 kb/s

A. W. Black and A. M. Kondoz

17.1. INTRODUCTION

Until recently most speech coding applications have required that the signal is band limited from 300 Hz to 3.4 kHz prior to digital sampling. This then allows the signal to be transmitted over the PSTN network, a requirement which is necessary for most telephony-based speech communication. However, with the emergence of the Integrated Digital Service Network (ISDN) there is no such stipulation. This has paved the way for high quality multimedia communication services, such as high quality telephony and audio-video teleconferencing, which can be transmitted over these bandwidth-rich environments. One way of achieving this desired quality is to use wideband speech which is assigned the band from 50 Hz to 7 kHz and sampled at 16 kHz for subsequent digital sampling. The extra low frequency components increase the voice naturalness, whereas the added high frequencies make the speech sound crisper, more intelligible and "brighter" when compared to narrowband speech. The quality produced by wideband speech is said to be equivalent to FM

Insights into Mobile Multimedia Communication
ISBN 0-12-140310-6

radio, with a richness notably greater than telephone bandwidth speech together with a very high intelligibility and naturalness. Wideband speech can be thought of as consisting of a base band which contains approximately 80% of the perceptually important speech spectral information [1], and a high band which contributes to the overall perceived intelligibility of the speech. Typically during voiced speech there is a large spectral contrast between the two bands with most of the signal energy distributed across the lower frequencies. A wideband coding algorithm must, therefore, accurately code the perceptually important lower frequency components, and yet retain enough of the higher frequency information such that the perceived richness and fidelity of the original speech is preserved. Another major consideration for network and multimedia-based applications is coding delay, i.e. the amount of buffering that the encoder requires such that the correlations present in the waveform can be exploited for economical digital representation. The encoder buffering necessary for parameter analysis in linear-prediction-based coders can be several tens of milliseconds. Such delays can have important ramifications for speech coders used in networks such as the ISDN where the one way encoder–decoder delay is a very important criterion for service quality. Delay may necessitate the use of echo cancellation, which, in some applications, is detrimental to non-voice services. In some circumstances, the delay may become so large that it remains an impairment even after echo cancellation has been performed. It is therefore vital to keep the transmission delay as low as possible in such networks. One important method of controlling the overall delay is to ensure that the buffering required by the speech codec is kept to a minimum.

17.2. BACKWARD LPC PREDICTION FOR WIDEBAND SPEECH

The conventional CELP coder uses forward prediction to determine the LPC parameters for transmission to the receiver. The process of forward predictive coding relies on a block of future samples, typically 20 to 30 ms long, to calculate the LPC coefficients. This results in a long coding delay which is unacceptable in many applications. Low delay-CELP (LD-CELP) [2] avoids this excessive coding delay by performing the LPC analysis in a backward mode. This is where the above analysis is performed on past quantised speech, thus eliminating the need for a long look-ahead data frame. Consequently, as this speech is present both at the encoder and decoder, no LPC information needs to be transmitted. This then allows all the transmission bits to be assigned to the excitation vectors. In order to compensate for the inaccuracies incurred due to backward LPC prediction it is then necessary to have very accurate modelling of the reference vector. This is achieved by keeping the speech vector size as small as possible, which in turn results in a low algorithmic buffering delay since the coder encodes the speech on a block-by-block basis. However, this does have the disadvantage that the excitation parameter set has to be updated more frequently than in a conventional forward predictive LPC system, often resulting in higher bit rates. The original narrowband 16 kb/s LD-CELP uses a short data frame of just 0.625 ms (5 samples at 8 kHz) [2]. In order to keep the bit rate to a minimum the coder dispenses with the need for a long-term predictor

(LTP), instead opting for a 50th-order LPC predictor. Since the coefficients are backward predicted no additional bits are needed. A Barnwell window [3] is used to window the past quantised speech since this allows the autocorrelation coefficients to be calculated efficiently and also improves speech quality. The inclusion of the 50th-order LPC also had several other advantages: namely the algorithm is less speech specific, since it does not assume speech quasi-periodicity in the form of an LTP. As a consequence, all the transmission bits are assigned to the fixed excitation codebook.

Since speech signals are assumed to be stationary over very short time, frequent update rates of backward LPC coefficients can provide prediction gains close to forward adaptive schemes. Table 17.1 shows the prediction gain of a 16th-order predictor for various update rates of forward and backward schemes. The LPC analysis is based on a Barnwell window of 320 samples of quantised wideband speech, and the excitation is modelled using a single tap forward LTP and an 8-bit Gaussian secondary codebook. The excitation vector size is fixed at 0.62 ms (10 samples at 16 kHz sampling).

From the values given in Table 17.1 it can be seen that a decrease in predictor update rate is accompanied by a drop in backward LPC prediction performance, when compared to the equivalent forward predictive schemes. However, since the length of the excitation vector is kept relatively short, this results in an almost negligible degradation in speech quality. This is primarily due to the fact that the excitation sources are able to accurately match the reference vector in the analysis by synthesis (AbS) loop, and therefore, compensate for any discrepancies incurred by backward LPC prediction over the limited number of samples. Table 17.2 shows the effect on the prediction gain when the excitation vector size increases with the update rate of the backward LPC predictor.

Clearly Table 17.2 shows that the prediction gain of the backward LPC predictor starts to deteriorate rapidly as the length of the excitation vector exceeds 10 samples. This is accompanied by a rapid drop in speech quality, especially over the higher

Table 17.1 Prediction gains (dB) comparison of forward and backward 16th-order LPC prediction for a fixed excitation vector length of 10 samples

Scheme	Update rate (samples)			
	10	20	30	40
Forward	17.32	17.32	17.33	17.34
Backward	15.36	14.96	14.90	14.77

Table 17.2 Prediction gains (dB) for different size excitation vector

Scheme	Excitation vector size (samples)			
	10	20	30	40
Backward	15.16	12.98	11.34	10.58

frequencies. However, previous studies for narrowband speech [4] have shown that for update rates and excitation vector sizes of up to 2 ms rates (which corresponds to a vector size of 32 samples at 16 kHz sampling) backward prediction can provide prediction gains close to that achieved using forward schemes. The poor performance of wideband speech for large excitation vector sizes (above 10 samples) can be explained from Figure 17.1 which shows the effect of increasing the size of the excitation vector on the backward predictive LPC spectrum for a single frame of speech. As the vector dimension is increased it becomes more difficult for the excitation source to match the reference vector, and hence the backward LPC analysis is performed over a noisy signal. This is a problem for both wideband and narrowband speech, although this effect is exacerbated in wideband speech. This is due to the fact that the signal has a greater proportion of higher frequency components which when compared to the magnitude of the lower frequencies are typically lower in energy. Thus, as a consequence, during the minimum squared error (MSE) search process the excitation vector will be chosen primarily to model the higher energy lower

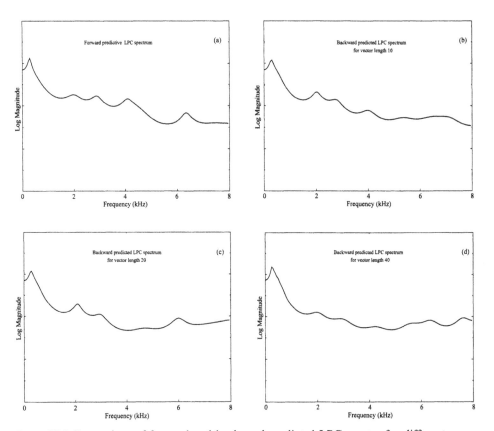

Figure 17.1 Comparison of forward and backward predicted LPC spectra for different excitation vector lengths: (a) forward predicted spectrum; (b) backward predicted with 10 sample excitation; (c) backward predicted with 20 sample excitation; (d) backward predicted with 40 sample excitation.

frequency components. This inevitably leads to less accurate modelling over the higher frequencies and therefore the coded speech signal becomes noisy. As the noise component increases, the signal at the higher frequencies will become immersed first since they are lower in energy. The backward predicted LPC spectrum will then become distorted over these frequencies, which results in the smearing of formant structure, when compared to the forward predicted spectrum as seen in Figure 17.1(a). As the vector length increases then the noise component becomes larger in magnitude which ultimately affects a wider range of frequencies. This effect can be alleviated by reducing the wideband coder's excitation vector length to a level such that the secondary excitation and LTP contributions compensate for the errors incurred due to the backward prediction scheme. This update length limit was found to be about 15 samples or 0.94 ms. However, this resulted in a higher overall bit rate of typically 32 kb/s when utilising both LTP and secondary codebooks. The quality produced by such systems is high and tests have shown that wideband coders utilising update rates of up to 15 samples can have a perceptual performance which is equivalent to the standard G.722 at 64 kb/s [5].

17.3. SPLIT BAND APPROACH

In order to reduce the bit rate further it is necessary to increase the excitation vector size which is far larger than experienced in equivalent narrowband systems. However, as we have just seen this results in a severe degradation in quality. Alternatively, a quadrature mirror filter (QMF) [6] bank can be used to split the speech signal into two individual bands. The split band structure offers benefits over the full band scheme which more than compensates for the additional complexity incurred. One advantage is that any distortion due to the quantisation process is confined to the band where it is produced. This then allows the low band information to be treated separately from the troublesome higher frequencies, a requirement which is deemed necessary when using backward LPC prediction over wideband speech. In this particular application, the signal is split into its respective bands by using the same QMF bank as stipulated in the G.722 wideband audio coding standard [7]. In this standard the QMF is implemented as an FIR filter which has the advantage of a linear phase response.

The input speech signal is split into two 8 kHz sampled signals, one representing the lower sub-band information (0–4 kHz) which is perceptually more important as it contains most of the speech spectral information. The other 8 kHz sampled signal represents the higher sub-band (4–8 kHz) information which primarily contributes to the overall quality and fidelity of the perceived speech. The contrast between higher and lower sub-band information can be clearly seen in Figure 17.2, where most of the speech information signal is contained in the lower band. Dividing the signal into two equal bands allows longer update rates to be applied for backward LPC prediction, especially over the lower band as the signal is now effectively a narrowband signal. The higher frequencies can be treated separately.

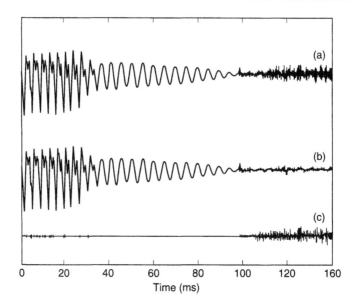

Figure 17.2 Split band approach: (a) full-band signal; (b) lower sub-band signal; (c) higher sub-band signal.

17.3.1. Coding of the Lower Sub-band

The lower sub-band signal is quantised using a backward LPC predicted CELP coder. The 10 LPC coefficients are updated every 1.75 ms (14 samples at 8 kHz sampling) by performing the LPC analysis over a Barnwell windowed 20 ms of past quantised speech. The backward analysis is now confined over the 4 kHz spectrum, which eliminates most of the inaccuracies by predicting over the full band signal. This then allows the excitation vector length to be also extended to 14 samples (or equivalently 28 samples at 16 kHz sampling). Previous studies [4] have shown that for low delay narrowband coders using backward LPC analysis, with an excitation vector size greater than five samples it is necessary to incorporate a forward pre-dictive LTP in addition to the secondary codebook contribution. This is because the excitation contribution from a fixed secondary codebook is not enough to compen-sate for the discrepancies incurred by backward LPC analysis. The fixed secondary codebook consists of an 8-bit PRELP excitation. The structure of the lower sub-band coder is shown in Figure 17.3.

17.3.2. Coding of the Higher Sub-band

Although from Figure 17.2 the upper band signal appears to be noise-like there is still a reasonable amount of sample to sample correlation which can be modelled using a low order LPC predictor. Backward LPC prediction over the upper band signal proved to be unsuitable. This was primarily due to the fact that the analysis was performed over a noisy synthetic signal. Therefore to achieve better quality

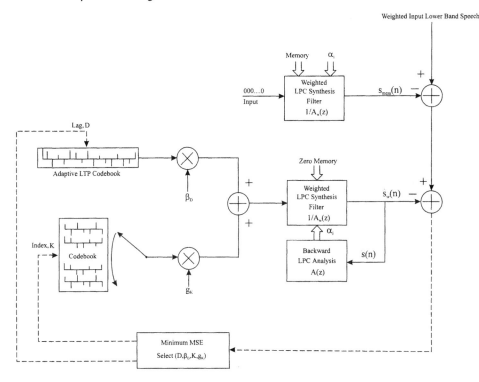

Figure 17.3 Lower sub-band PRELP coder.

using backward prediction the upper band signal needed to be modelled more accurately. However, this would require a short excitation vector length and close matching of the random-like signal in order to keep the noise component to a minimum. This ultimately would require a large expenditure of bits, and since the upper band contains only 10–20% of the information in the speech signal it is a rather inefficient method of representing this signal. Instead, it was decided to represent the higher sub-band spectral envelope using a forward predicted LPC. Figure 17.4 shows the LPC filter prediction gain against filter order for the higher sub-band signal. It can be seen that the prediction gain saturates at an order of 6. Thus the spectral envelope of the signal contained within this band can be modelled using a 6th-order LPC filter. The coefficients are forward predicted over 20 ms of Barnwell windowed speech where the update length of the filter is set at four times the lower sub-band frame length, which obviously increases the delay of the coder.

As a result of performing forward LPC prediction over the higher sub-band, the excitation signal can now be more coarsely quantised. In fact due to the random nature of the excitation signal in the upper band, it was decided not to match the signal on a sample by sample basis. Instead, the upper band signal at the decoder was represented by simply exciting the LPC synthesis filter with a randomly generated vector drawn from a unit variance, zero mean Gaussian source. The length of this vector was set to be the same as the LPC parameter update rate of 56 samples. This proved to be sufficient for modelling the higher sub-band information. However, one

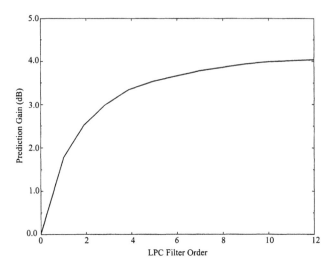

Figure 17.4 LPC prediction gain versus LPC order for upper band speech signal.

problem with this method was finding the correct gain for the vector. Initially, an open loop technique was tried which involved finding the RMS energy of the inverse filtered signal. Perceptual listening tests indicated a satisfactory performance if this value was attenuated by 0.6. Occasionally, however, during some fricative sounds there appeared to be perceptual disturbances. In order to alleviate this effect it was decided that a tighter control was needed on the magnitude of this term. Thus, a closed loop technique was adopted where the magnitude was calculated such that the energy of the excited synthesis filter was the same as the energy of the upper band speech signal.

The total energy of the synthesis filter consists of the energy contribution of the memory response, $s_{mem}(n)$, as well as the energy induced by exciting the filter with a Gaussian vector. Let x_u be a randomly generated vector from the Gaussian source. Then the synthetic output of the analysis filter due to x_u can be expressed as

$$y_u(n) = \sum_{i=0}^{n} h(i)x_u(n-i) \qquad 0 \le n \le N-1 \tag{17.1}$$

where $h(n)$ is the impulse response of the 6th-order synthesis filter. The gain, g_u, must be chosen such that the total energy of the filter is equal to the energy of the upper band signal, $s_u(n)$

$$\sum_{n=0}^{N-1}(s_{mem}(n) + g_u y_u(n))^2 = \sum_{n=0}^{N-1} s_u^2(n) \tag{17.2}$$

Writing equation (17.2) in terms of g_u

$$g_u^2 \sum_{n=0}^{N-1} y_u^2(n) + 2g_u \sum_{n=0}^{N-1} y_u(n)s_{mem}(n) + \left(\sum_{n=0}^{N-1} s_{mem}^2(n) - \sum_{n=0}^{N-1} s_u^2(n)\right) = 0 \tag{17.3}$$

g_u is found by solving the quadratic given in equation (17.3).

$$g_u = \frac{-b \pm \sqrt{b^2 - 4ac}}{2a}$$

$$b = 2 \sum_{n=0}^{N-1} y_u(n) s_{mem}(n)$$

$$a = \sum_{n=0}^{N-1} y_u^2(n) \tag{17.4}$$

$$c = \sum_{n=0}^{N-1} s_{mem}^2(n) - \sum_{n=0}^{N-1} s_u^2(n)$$

When the roots of equation (17.4) are real the root nearest to zero is chosen. This ensures that the minimum excitation energy is used which proved to be subjectively more acceptable. However, complex roots will occur if the energy from the memory response is greater than the energy of the upper band signal, and there is insufficient correlation between the memory response and the filtered excitation to be able to remove part of the memory contribution and thereby attain the desired energy level. In this case the energy cannot be matched. For this situation the RMS energy of the inverse filtered upper band speech signal is used.

As the gain is principally calculated in a closed loop manner it is necessary that both the excitation source and filter memories at the encoder and decoder are synchronised. Thus the speech is synthesised both at the encoder and decoder as shown in Figure 17.5.

17.4. PARAMETER QUANTISATION

Backward prediction requires that the analysis frame and excitation vectors are updated regularly. In the case of the lower sub-band coder, this implies that a

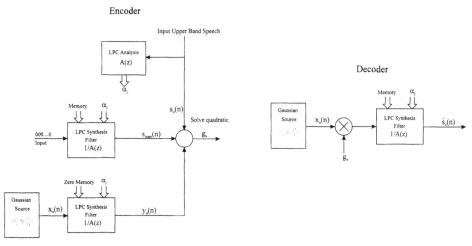

Figure 17.5 Higher sub-band speech codec.

large amount of the total channel capacity will be utilised by the LTP and secondary codebook gains. Therefore, in order to keep the overall bit rate to a minimum it is advantageous to represent these two parameters with as few bits as possible, but without degrading the quality of the synthetic speech.

Due to the high update rate there is a strong correlation between successive parameters. This fact can be used to reduce the variance of the secondary codebook gain. In the original LD-CELP at 16 kb/s [2] only three bits are assigned for quantising the gains. The technique uses linear prediction on the previous energies of the excitation signals in the logarithmic domain. However, as the excitation vector sizes increases above 10 samples, then the prediction gain drops dramatically. This is mainly due to irrational variations in the gain magnitudes. One simple solution to overcome this problem is to use the very simple technique of backward average mean smoothing [7]. The predicted gain, G_p, is generated by finding the mean of the previous three absolute quantised gains:

$$G_p = \frac{1}{3} \sum_{i=1}^{3} |g_{-i}| \tag{17.5}$$

This predicted value is then used as a normalisation factor for the current secondary codebook gain. The normalised gain, g_n, can then be vector quantised with the pitch gain, g_p, in a closed loop manner by incorporating the lower sub-band coder into the minimisation process. The vector index, j, is chosen from an 8-bit codebook to minimise the following MSE distortion measure.

$$D = \min \|s_w - Y\Phi g_j\|^2 \tag{17.6}$$

where s_w is the perceptually weighted speech vector with the memory contribution subtracted. Φ is a diagonal two by two matrix with the leading diagonal containing the predicted gain, G_p, for the secondary codebook gain and unity for the pitch gain. g_j is the jth code vector in the gain codebook, and Y is a matrix whose column vectors are the AbS-selected filtered excitation vectors from the LTP and the fixed secondary codebook. The technique of closed loop training the codebook contents is used to optimise the performance of the quantiser. This technique was first reported by Chen [2], who used it to train the contents of a secondary excitation codebook. It is based on minimising the perceptually weighted error criterion of the coder where the encoder is used for coding the entire training database.

In training the gain codebook the index j is chosen to minimise the weighted MSE distortion measure given by equation (17.6). So, let N_j be the set of training base indices for which g_j is selected as the best gain vector while encoding the training set. By starting with an initial open loop trained codebook, the low band coder is used to code the entire training database of original speech. The total accumulated distortion due to the jth cluster corresponding to g_j is given by

$$D_j = \sum_{n \in N_j} \| s_w - Y\Phi g_j \|^2 \tag{17.7}$$

To minimise the distortion due to the jth cluster the above equation is differentiated with respect to g_j and the result is set to zero.

$$\frac{\partial D_j}{\partial g_j} = -2 \sum_{n \in N_j} \Phi^T Y^T s_w + 2 \sum_{n \in N_j} \Phi^T Y^T Y \Phi g_j = 0 \qquad (17.8)$$

Thus the centroid due to the jth cluster which minimises D_j satisfies the following normal equation

$$g_j \sum_{n \in N_j} [\Phi^T Y^T Y \Phi] = \sum_{n \in N_j} \Phi^T Y^T s_w \qquad (17.9)$$

The two summations on both sides of this normal equation are separately accumulated for each of the codebook indices j. After the entire training set has been encoded the resulting normal equations are then solved for each of the indices to obtain a new set of centroids. These centroids are used to replace the old codebook and the training data-base is re-encoded using the newly obtained codebook. As pointed out by Chen, unlike the open loop LBG training algorithm which is used for direct VQ, the overall distortion and updated codebook is not guaranteed to converge to a local optimum. Therefore the codebook values are selected from the training iteration which gives a maximum improvement in objective quality.

For the upper sub-band, the excitation gain, g_u, is scalar quantised using four bits. The six LPC coefficients are coded using line spectral frequencies (LSF) and then split vector quantised using 16 bits. Table 17.3 shows the configuration and bit allocations for each of the bands in the coder. As indicated in Section 17.3.2 the upper sub-band coder requires a look-ahead of four times the basic frame length of the lower sub-band coder. Therefore, the overall coding delay including the inherent delay of the QMF bank is 8.4 ms.

17.5. SUBJECTIVE PERFORMANCE OF THE LOW DELAY WIDEBAND CODER

The performance of the LD-wideband coder was assessed by comparing it against the G.722 7 kHz audio standard operating at 48 kb/s and 64 kb/s [5]. The speech material used in the trial consisted of 20 s of male and female speech. Each utterance was coded in turn by G.722 operating at the above rates and the LD-wideband coder

Table 17.3 Bit allocation for LD-wideband coder

Parameter	Bits	Update rate (Hz)	Bits/s
		Lower sub-band coder	
LTP lag	7	571.4	3999.8
CB index	8	571.4	4571.4
VQ gains index	8	571.4	4571.4
		Higher sub-band coder	
LSFs	16	142.8	2284.8
Gain	4	142.8	571.4
Total			16000

as described in Table 17.3. The coded utterances were then grouped into pairs. Each pair consisted of the LD-wideband coder and G.722 operating at either 48 kb/s or 64 kb/s. A pairwise comparative listening test was then performed by 15 subjects. Where for each pair (9 in all) the subject was asked to compare the quality of the second utterance to the first by using the following scale: 5 = Better, 4 = Slightly Better, 3 = Same, 2 = Slightly Worse and 1 = Worse. From the results of the test it was found that the LD-wideband scores an average value of 2.0 and 1.9 for male and female speech respectively when compared to G.722 at 64 kb/s. This indicated that in a pairwise comparison test the subjective quality of the LD-wideband coder was judged to be slightly worse than the reference coder (G.722 at 64 kb/s). However, when compared against G.722 at 48 kb/s the LD-wideband coder scored an average of 3.1 and 2.9 for male and female speakers respectively. The perceived quality was therefore judged to be the same as G.722 operating at 48 kb/s.

17.6. CONCLUSIONS

The technique of backward LPC prediction was investigated over wideband speech. It was found that for relatively short excitation vector lengths the technique can produce high quality speech with little or no distortion. Unfortunately, this typically resulted in a high overall bit rate which was primarily due to frequent updating of the excitation parameter set. In order to reduce the overall bit rate it was necessary that the length of the excitation vector was increased. However, this resulted in less accurate modelling of the reference signal and as a consequence the backward LPC prediction was performed over a noisy synthetic signal. This caused the pre-dicted spectrum to be distorted, especially over the higher frequencies. This was primarily caused by the noise level immersing these lower magnitude frequency components, i.e. the high frequencies. As a result of this spectral distortion the processed speech became distorted resulting in a quality which was unacceptable for a high quality wideband system. This distortion was controlled by splitting the wideband speech into two narrowband signals. The lower sub-band signal was still coded using backward LPC prediction. However since the signal was limited to 4 kHz a longer vector length could be used without incurring gross distortions in the backward predicted spectrum. The upper sub-band signal on the other hand was coded using a forward predicted LPC system. The main reason for this was that backward LPC prediction would require an accurate estimation of the signal in order to maintain the noise component at a minimum level. This in turn would require a significant coding rate and since this band only contained approximately 20% of the speech information it was deemed an inefficient use of available bandwidth. Instead, the band was modelled by a spectrally shaped noise source whose energy was closed loop optimised. This resulted in a 16 kb/s wideband coder whose subjective quality was comparable to G.722 operating at 48 kb/s operating with an algorithmic coding delay of 8.4 ms.

REFERENCES

[1] O'Shaughnessey, D., *Speech Communication Human and Machine*. Addison-Wesley, Reading, MA, 1987.

[2] Chen, J.-H., High-quality 16 kb/s speech coding with a one-way delay less than 2 ms. In *Proceedings of International Conference on Acoustics, Speech and Signal Processing*, 453–456, Albuquerque, April 1990.

[3] Barnwell, T. P., Recursive windowing for generating autocorrelation coefficients for LPC analysis. In *IEEE Transactions on Acoustics, Speech and Signal Processing*, ASSP-29(5), 1062–1066, October 1991.

[4] Soheili, R., Kondoz, A. and Evans, B., Design and implementation of a 4 msec LD-CELP coder at 8 kb/s. In *Proceedings of the International Symposium on Signals, Systems and Electronics*, 233–236, Paris, September 1992.

[5] Ordentlich, E. and Shoham, Y., Low-delay code-excited linear-predictive coding of wideband speech at 32 kb/s. In *Proceedings of the International Conference on Acoustics, Speech and Signal Processing*, 9–12, Glasgow, April 1990.

[6] Crochiere, R. and Rabiner, L., *Multirate Digital Signal Processing*, Signal Processing Series, Prentice Hall, Englewood Cliffs, NJ, 1983.

[7] CCITT, 7kHz audio coding within 64 kb/s. In *Recommendation G.722*, vol. Fascicle III.4 of Blue Book, 269–341, Melbourne 1988.

Part 5

Source Coding: Video and Multimedia

Research into video coding and compression has been active for many years and significant advances have been made in the past two decades. The aim of such techniques has been simply to optimise algorithm performance for a given data rate. However, in order to support mobility, additional criteria must also be considered. Future coding methods will need to offer functionalities such as high performance compression, scalability, content-based interactivity, error robustness and universal accessibility. Some of these issues are covered in Part 5.

For example, despite progress in VLSI technology, implementation complexity is still a key issue. Secondly, techniques which can offer acceptable performance at very low bit rates are needed to support existing systems such as GSM. An overview of such algorithms is presented in Chapter 18, covering waveform-based, model-based and fractal-based methods.

With the increasing globalisation of communication technology, the need for interoperability has also become paramount. This latter point has led to a number of standardisation activities, most notably ISO MPEG-4, which is targeted at bit rates between 10 kb/s and 2 Mb/s with features relevant to both mobile and broadcast applications. Talluri, in Chapter 19, emphasises the functionalities needed for low bit rate coding and overviews the standards activities in H.263 and MPEG-4. Chapter 23 discusses another important feature of next generation video coding schemes – scalability.

A further important aspect of low bit rate multimedia coding for mobile applications relates to the integration of audio and video coding tools. This is covered in Chapters 20 and 24. Hanzo *et al.* in Chapter 20, address coding and modulation issues while Eryurtlu *et al.* in Chapter 24 focus on coding trade-offs.

One of the problems of image and video coding at high compression ratios is the distortion introduced by the coding process. Conventional approaches based on a hybrid motion compensated DCT suffer blocking artefacts and alternative methods can yield superior results. Wilkinson in Chapter 21 addresses issues associated with an alternative coding method based on the use of wavelets. A second approach, based on a morphological segmentation of the displaced frame difference signal is presented by Czerepiński and Bull in Chapter 22.

18

New Trends in Very Low Bit Rate Video Coding – An Overview

T. Ebrahimi, M. Kunt, O. Egger and E. Reusens

18.1. INTRODUCTION

The importance of visual communications has increased tremendously in the few last decades. The progress in microelectronics and computer technology together with the creation of networks operating with various channel capacities provides the basis of an infrastructure for a new era of telecommunications. New applications are bringing about a revolution in everyday life. Emerging applications such as video-conferencing [1], cellular videophones [2] and multimedia [3,4,2] will have a great impact on professional life, education and entertainment. The digital representation of visual information in its canonic form generates a huge amount of data. In order to meet the requirements of these new applications, powerful image sequence compression techniques are needed to drastically reduce the global bit rate.

Insights into Mobile Multimedia Communication
ISBN 0-12-140310-6

A number of standards have been defined for the compression of visual information. The JPEG [5] still image compressor was proposed by the Joint Photographic Expert Group and is a general-purpose image compression standard. The MPEG (Moving Picture Expert Group) standards address the compression of video signals. MPEG-1 [6] operates at bit rates of about 1.5 Mbit/s and targets storage and transmission over communication channels such as the integrated-services digital network (ISDN) or local area networks (LAN). MPEG-2 [7] operates at bit rates around 10 Mbit/s and is designed for the compression of higher resolution video signals. The recommendation H.261 [8] (also known under the acronym $p \times 64$) was proposed by the International Telegraph and Telephone Consultative Committee (CCITT, now known as ITU-T). Based on this standard, videoconferencing at bit rates down to 64 kbit/s has become feasible. This requires the capacity of one ISDN channel. In the near future, modern visual communications applications will be available to the general public. To meet this objective, transmission must be possible over Public Switched Telephone Networks (PSTN) or mobile channels. The transmission of video sequences at bit rates as low as 9.6 kbit/s is therefore necessary. Efforts to define new standards for these applications are still in their relative infancy. However, several expert groups have been created to pursue this objective, the major ones being ISO/MPEG-4 [9] and ITU-T/H.263.

An uncompressed video sequence for very low bit rate applications typically requires a bitstream of up to 10 Mbit/s. In order to achieve very low data rates, compression ratios of about 1000 : 1 are therefore required. Intensive research has been performed in the last decade to attain this objective [10,11]. Variations of the recommendation H.261 for very low bit rate applications have been defined as simulation models in COST 211ter (SIM3) [12] and in ITU-T H.263 (TMN) [13]. For these simulation models, severe blocking artifacts occur at very low data rates. Much ongoing research is devoted to developing methods based on philosophies which differ drastically from the existing standards developed for higher bit rates.

The aim of this chapter is to provide the reader with an overview of existing approaches which target very low bit rates and with a list of major publications in the field. Section 18.2 gives a review of high compression image coding techniques. Section 18.3 overviews video coding techniques aimed at very low bit rates. These have been classified into four classes: waveform, object-based, model-based and fractal-based coding techniques. Then, pre- and post-processing of the visual information is overviewed in section 18.4. Existing products in the market are described in section 18.5 and finally, conclusions are drawn in the last section.

18.2. HIGH COMPRESSION IMAGE CODING

High compression image coding has attracted significant interest in recent years. In this type of coding, visible distortions of the original image are accepted in order to obtain very high compression ratios. High compression image coders can be split into three distinct categories. The first category is referred to as waveform coding and consists of transform and sub-band coding. The second group, often called second-generation techniques, comprises techniques which attempt to describe an image in terms of visually meaningful primitives (contour and texture, for example). The third category is based on fractal theory.

A waveform-based coding scheme typically comprises the following steps: (i) decomposition/transformation of the image data, (ii) quantization of the transform coefficients, and (iii) source coding of the quantized coefficients. The first step transforms the image into an alternative representation, where most of the energy is compacted into a small number of coefficients. As a general approach, the sub-band analysis/synthesis system was first introduced for 1D data by Crosier [14] in 1976. Smith and Barnwell proposed solutions possessing the property of perfect reconstruction [15] and the extension to 2D signals was reported by Vetterli [16]. Only later was the method applied to image compression by Woods and O'Neil [17]. In sub-band coding, an image is split into a set of sub-band images by using a group of bandpass filters [18] followed by critical subsampling. A special case of this is transform coding which includes the discrete cosine transform (DCT) coding technique as employed in the still image compression standard JPEG [5]. At compression factors of about 30 to 40 this technique begins to produce visible *blocking artifacts*. All transform coders suffer from this distortion and unfortunately, the human eye is very sensitive to it. Hence block coders are generally considered inappropriate for low bit rate image coding. Sub-band coding of images has been the subject of intensive research in recent years [19,20,21,22]. The main artifact which occurs at high compression factors (around 50) is due to the Gibbs phenomenon. This is associated with the use of linear filters and produces a *ringing effect* around image edges. For low bit rate sub-band coding [23] (compression ratios higher than 50) it is of major importance to exploit the existence of zero correlation across the sub-bands as proposed in [24,25,26] in order to maintain a good quality. A scheme combining vector quantization (VQ) and the prediction of insignificance across the bands has been proposed in [27,28]. Although it is possible to reduce the ringing effect by an appropriate choice of the sub-band filters [29,30,31,32], it is not possible to find linear sub-band filters which completely eliminate this effect. To avoid this artifact, morphological sub-band decompositions have been proposed [33,34,35] which lead to good quality decoded pictures at compression ratios as high as 70–80.

The second group of methods is based on second-generation techniques. These attempt to decompose the data into visual primitives such as contours and textures [36,37]. One approach is to divide the image into directional primitives as proposed in [38]. Segmentation-based coding techniques [39] extract regions from the image data which are represented by their shape and their textural content. Following similar ideas, sketch-based image coding [40] has been proposed. This is based on extracting the contours of an image, namely geometric and intensity information, resulting in the so-called sketch picture. The texture is then defined by the difference between the original and the sketch image and is coded using waveform coding techniques. An extension of this technique has been proposed by Ran and Farvardin [41,42] and is based on a three-component image model. This technique divides the image into the strong edge, texture and smooth components. The strong edge component is encoded separately whereas the texture and smooth components are encoded using waveform coding techniques. A solution which identifies the most important image features has been proposed by Mallat and Zhong [43] using multiscale edges. A double layer technique based on multiscale edges and textures has then been proposed in [44]. In general, second-generation techniques become efficient at higher compression ratios (above 50) when compared to other methods.

Iterated functions systems (IFS) theory, closely related to fractal geometry, has recently found application in the area of image compression. Barnsley [45] and Jacquin [46] pioneered this field followed by numerous contributions from other researchers [47,48]. The approach consists of expressing an image as the attractor of a contractive function system which can be retrieved simply by iterating the set of functions starting from any initial arbitrary image. The form of redundancy exploited is referred to as *piecewise self-transformability*. This refers to the property whereby each segment of an image can be properly expressed as a simple transformation of another part with higher resolution. IFS-based still image compression techniques can pretend to very good performances at high compression ratios (about 70–80) as proved by [49,50].

18.3. VERY LOW BIT RATE VIDEO CODING

In contrast to still image coding, the compression of video signals also allows exploitation of the temporal dimension. Existing compression techniques for still images can serve as basis for the development of video coding techniques. Either these can be generalized to 3D signals, or a hybrid approach based on motion compensation can be defined.

The input video data for very low bit rate applications typically comprises small images (approximately 144×176 pixels) with a frame rate of about 5–10 frames/s. The target bit rates vary from 4.8 kbit/s to 64 kbit/s depending on the desired application. In the following discussion, very low bit rate video compression techniques have been classified into four classes: waveform, object-based, model-based and fractal coding techniques.

18.3.1. Waveform-based Techniques

Viewing the temporal axis as a third dimension, all the waveform coding techniques developed for still image compression can be generalized to the compression of video signals. Only limited work has been published in the use of 3D transforms for sequence compression [51]. The blocking artifacts at low bit rates, however, make it improper to code image sequences. Three dimensional sub-band coding of video was first introduced by Karlsson and Vetterli [52]. In this work, standard sub-band filters are used for the spatial dimensions while a DCT derived filter bank is applied to the temporal dimension. Variations of this scheme have been reported elsewhere in the literature [53,54].

The drawback of 3D sub-band coding is that the temporal filtering is not performed along the direction of motion. A solution to this is to combine the temporal SBC component with motion compensation (MC) [55] as proposed in [56,57]. This scheme can be extended by the use of lattice vector quantization and MC with sub-pixel accuracy [58]. The problem of coding the resulting prediction error images or *displaced frame differences* (DFD) has been addressed by using linear transforms such as the DCT [59,60] and by using the wavelet transform [61]. Video coding standards such as MPEG2 [7] and H.261 [8] propose the use of DCT-based algorithms for coding the DFDs. A method of preprocessing the DFD using a morphological segmentation has been proposed by Li *et al.* [62,63].

The application of waveform coding algorithms to very low bit rate video coding has been proposed based on 3D sub-band coding [64,65], motion compensated sub-band coding [66,67,68,69,70,71] and motion compensated transform coding [12,13].

18.3.2. Object-based Techniques

The promising results obtained with second-generation techniques for still images has motivated their extension to image sequence compression. A straightforward solution is to extend the 2D techniques used previously in a 3D context. One approach is to perform a 3D segmentation of a sequence viewed as a 3D volume. In this way Willemin *et al.* [72,73] proposed an octree split and merge segmentation as a generalization of the quadtree segmentation previously used in still image coding [36]. Another similar technique has been introduced by Salembier and Pardàs [74], based on mathematical morphology allowing arbitrary region shapes. Along the same lines, an object-oriented scheme, in which objects are defined as regions with three associated parameters, these being shape, textural content and motion, was proposed in [75]. In this case the parameters are obtained by image analysis based on source models of either moving 2D or 3D objects [76].

All these approaches require the transmission of the object descriptions created at the encoder. The textural content of the objects can be coded efficiently using transform-based techniques similar to those used in block-based methods [77,78,79,80]. Typically, the shape information is represented by a chain code of the contour information [81], quadtree shape representation [82,83] or the medial axis transform [84,85]. Simulation results show that these representations consume a significant portion of the global bit rate. One solution to reduce this cost is to use more efficient techniques for shape representation such as the geodesic morphological skeleton as proposed by Brigger *et al.* [86] and/or perform a simplification of the contours by appropriate postprocessing operations prior to their encoding [87]. Another solution is to avoid frequent transmission of contour information by object tracking [88,89,90,91,92]. Finally, a third solution is to define objects with simple shapes which need fewer bits to be transmitted for their shape representation [93,94,95,96,97,98].

18.3.3. Model-based Techniques

All compression techniques rely to some extent on the existence of an underlying model. The term model-based coding, however, refers specifically to an approach which seeks to represent the projected 2D image of a 3D scene by a semantic model. The goal is to find an appropriate model together with its corresponding parameters.

This technique can be divided into two main steps: analysis and synthesis. Analysis is the most difficult task due to the complexity of most natural scenes. Thus to date, most effort has been concentrated on simple scenes such as head-and-shoulder sequences [99,100,101]. The synthesis block, however, is easier to realise as it can build on techniques developed for image synthesis in the field of computer graphics. Further details of model-based techniques can be found in the excellent tutorial on this subject by Pearson [102].

18.3.4. Fractal-based Techniques

The promising performance provided by fractal-based still image compression techniques has encouraged their application to video compression. Different approaches have recently been proposed. Beaumont suggested a straightforward extension of 2D approach to 3D data volumes [103]. In order to reduce the computational burden, Li *et al.* proposed a 3D approach without domain block search but with increased contractive transformation complexity [104]. Reusens worked on a scheme where sequence volume is adaptively segmented along an octree structure and 3D blocks coded either by contractive transformation or 3D temporal block matching [93]. Independently, Lazar and Bruton followed the same approach but allowed only contractive transformations [105]. Hürtgen and Buttgen introduced a 2D approach where regions classified as foreground, are coded by intraframe fractal approximation [106].

18.4. PRE- AND POST-PROCESSING

In the majority of multimedia applications, the data capture mechanism (i.e. the camera) should be sufficiently cheap to make it affordable in volume production. In addition, compact solutions (miniaturized terminals) are desirable for portable applications. However, it is generally accepted that the quality of such equipments is lower than its more expensive professional counterpart. It is therefore mandatory to use a pre-processing stage prior to coding in order to enhance the quality of the pictures, and to remove the various distortions which affect the performance of very low bit rate algorithms. Solutions have been proposed in the field of image processing to enhance the quality of images for various applications [107,108,109]. A more appropriate approach is to take into account the characteristics of the coding scheme when designing such operators.

Mobile communications is an important application for very low bit rate video coding. Terminals in such applications are usually subject to varying forms of motion including tilting and jitter, which translate into a global motion in the scene. This component of the motion can be detected and extracted by appropriate post-processing stages. Results reported in the literature show an important improvement in the coding performance when a global motion estimator is used [110,111,112,113].

It is normal to expect a certain degree of distortion in the decoded images for very low bit rate applications. An appropriate coding scheme, however, will introduce these distortions in ways that are visually less annoying for the viewer. For example, shortly after the introduction of the first block transform-based coding schemes, solutions were proposed to reduce the blocking artifacts which appear at high compression ratios [114,115,116,117,118]. The same approach has been used to improve the quality of decoded signals in other coding schemes, reducing other kinds of artifacts such as ringing, burring and mosquito noise [119,120].

18.5. PRODUCTS

A number of products currently exist in the market capable of transmitting audio, video and data at bit rates lower than 32 kbit/s. Most of these are based on the principles used in the recommendations H.261 [8,121], and can be used via modems on public switched telephone networks. The first product of this kind was the VideoPhone 2500 [122] which was introduced in 1990. This was followed by others, such as the British Telecom/Marconi Relate 2000, and COMETH Labs STU-3 Secure. All these products suffered from low image resolution in both the spatial and temporal domains.

In parallel, a number of collaborative activities led to the development of other variants of H.261. These include the SIM3 proposed by COST 211 ter[12], and TMN-4 by ITU-T [13]. The latter achieves an acceptable performance at bit rates as low as 8 kbit/s, for videotelephony sequences. A recent application of this scheme has been implemented in software on 7 DSPs, in the form of a demonstrator capable of encoding audio, video and data at bit rates between 9.6 and 28.8 kbit/s [123].

All these techniques suffer from the well-known artifacts associated with DCT-based hybrid schemes (blockiness and mosquito noise) and do not facilitate the introduction of new functionalities needed for future multimedia applications. The most recent techniques however, produce competitive results with more sophisticated pre-processing stages. This extraordinary amount of fine tuning has pushed hybrid DCT coding to the limit of its achievable performance.

18.6. CONCLUSIONS

The purpose of this chapter has been to provide the reader with a means of appreciating the wealth of techniques available for very low bit rate image and video coding. It is clear that the techniques developed for still pictures can be either extended to 3D signals, or a hybrid approach can be defined based on motion compensation. Waveform-based techniques can prove efficient when proper low bit rate oriented features are added. However, other techniques specifically designed for very low bit rate applications show significant promise, namely object-oriented, model-based and fractal-based techniques. As yet the most appropriate approach for mobile multimedia applications is difficult to identify and this is still the topic of extensive research effort.

REFERENCES

[1] Sabri, S. and Prasada, B. (invited paper). Video conferencing systems. *Proceedings of the IEEE*, **73** (4), 671–688, 1985.

[2] Special Issue on Visual Communications. *AT&T Technology*, **7** (3), 1992.

[3] Elton, M. C. J. (invited paper). Visual communication systems: trials and experiences. *Proceedings of the IEEE*, **73** (4), 700–705, 1985.

[4] Chang, S.-K. (invited paper). Image information systems. *Proceedings of the IEEE*, **73** (4), 754–764, 1985.

[5] ISO/IEC JTC1 Committee Draft, JPEG 8-R8. Digital compression and coding of continuous-tone still images. August 1990.

[6] ISO/IEC JTC1 CD 11172. Information technology – coding of moving pictures and associated audio for digital storage media up to about 1.5 Mbit/s – Part 2: coding of moving picture information. Technical report, International Organization for Standardization, 1991.

[7] ISO/IEC DIS 13818-2. Information technology – generic coding of moving pictures and associated audio information – Part 2: video. Technical report, International Organization for Standardization, 1994.

[8] CCITT SG XV. Recommendation H.261 – video codec for audiovisual services at $p \times 64$ kbit/s. Technical Report COM XV-R37-E, International Telecommunication Union, August 1990.

[9] ISO/IEC JTC1/SC29/WG11. First draft of MPEG-4 requirements. Technical Report N0711, International Standardisation Organization, March 1994.

[10] Pearson, D. E. and Robinson, J. A. (invited paper). Visual communications at very low data rates. *Proceedings of the IEEE*, **73** (4), 795–812, 1985.

[11] Ebrahimi, T., Reusens, E., Li, W. and Cicconi, P. (invited paper). New trends in very low bitrate video coding. In *Proceedings of the IEEE*, July 1995.

[12] COST 211ter Simulation Subgroup. Simulation model for very low bitrate image coding (SIM). Technical report, December 1993.

[13] ITU-T SG15 WP15/1. Video codec test model TMN4 Rev. 1. Technical report, NTR, October 1994.

[14] Croisier, A., Esteban, D. and Galand, C., Perfect channel splitting by use of interpolation, decimation, tree decomposition techniques. In *Proceedings of the International Conference on Information Sciences/Systems*, 443–446, Patras, August 1976.

[15] Smith, M. J. T. and Barnwell, T. P., Exact reconstruction techniques for tree structured subband coders. *IEEE Transactions on Acoustics, Speech, and Signal Processing*, **34** (3), 434–441, 1986.

[16] Vetterli, M., Multi-dimensional subband coding: some theory and algorithms. *Signal Processing*, **6** (2), 97–112, 1984.

[17] Woods, J. and O'Neil, S., Subband coding of images. *IEEE Transactions on Acoustics, Speech, and Signal Processing*, **34** (5), 1278–1288, 1986.

[18] Johnston, J. D., A filter family designed for use in quadrature mirror filter banks. In *Proceedings of the International Conference on Acoustics, Speech, and Signal Processing ICASSP*, 291–294, April 1980.

[19] Adelson, E. H., Simoncelli, E. and Hingorani, R., Orthogonal pyramid transforms for image coding. In *Proceedings of SPIE's Visual Communications and Image Processing VCIP*, **845**, 50–58, Cambridge, MA, October 1987.

[20] Gharavi, H. and Tabatabai, A., Subband coding of monochrome and color images. *IEEE Transactions on Circuits and Systems*, **35**, 207–214, 1988.

[21] Woods, J. W., *Subband Image Coding*. Kluwer Academic Publishers, Boston, 1991.

[22] Nicoulin, A., Mattavelli, M., Li, W. and Kunt, M. (invited paper). Subband image coding using jointly localized filter banks and entropy coding based on vector quantization. *Optical Engineering*, **32** (7), 1438–1450, 1993.

[23] Goh, K. H., Soraghan, J. and Durrani, T. S., Multi-resolution based algorithms for low bit-rate image coding. In *Proceedings of the International Conference on Image Processing ICIP*, **II**, 285–289, Austin, USA, November 1994.

[24] Shapiro, J. M., An embedded wavelet hierarchical image coder. In *Proceedings of the International Conference on Acoustics, Speech, and Signal Processing ICASSP*, **IV**, 657–660, San Francisco, USA, March 1992.

[25] Shapiro, J. M., Application of the embedded wavelet hierarchical image coder to very low bit rate image coding. In *Proceedings of the International Conference on Acoustics, Speech, and Signal Processing ICASSP*, **V**, 558–561, Minneapolis, USA, April 1993.

[26] Shapiro, J. M., Embedded image coding using zerotrees of wavelet coefficients. *IEEE Transactions on Signal Processing*, **41** (12), 3445–3462, 1993.

[27] Moccagatta, I. and Kunt, M., A pyramidal vector quantization approach to transform domain. In *Proceedings of the European Signal Processing Conference EUSIPCO*, **III**, 1365–1368, Brussels, Belgium, August 1992.

[28] Moccagatta, I. and Kunt, M., VQ and cross-band prediction for color image coding. In *Proceedings of the Picture Coding Symposium PCS*, 383–386, Sacramento, USA, September 1994.

[29] Egger, O. and Li, W., Subband coding of images using asymmetrical filter banks. *IEEE Transactions on Image Processing*, **4** (4), 478–485, 1995.

[30] Egger, O., Nicoulin, A. and Li, W., Embedded zerotree based image coding using linear and morphological filter banks. In *Proceedings of the International Conference on Acoustics, Speech, and Signal Processing ICASSP*, Detroit, USA, May 1995.

[31] Caglar, H., Liu, Y. and Akansu, A. N., Optimal PR-QMF design for subband image coding. *Journal of Visual Communications and Image Representation* **4** (4), 242–253, 1993.

[32] Akansu, A. N., Haddad, R. A. and Caglar, H., The binomial QMF-wavelet transform for multiresolution signal decomposition. *IEEE Transactions on Signal Processing*, **41** (1), 13–19, 1993.

[33] Egger, O. and Li, W., Very low bit rate image coding using morphological operators and adaptive decompositions. In *Proceedings of the International Conference on Image Processing ICIP*, **II**, 326–330, Austin, USA, November 1994.

[34] Egger, O., Li, W. and Kunt, M. (invited paper), High compression image coding using an adaptive morphological subband decomposition. *Proceedings of the IEEE*, **83** (2), 1995.

[35] Florêncio, D. A. F. and Schafer, R. W., A non-expansive pyramidal morphological image coder. In *Proceedings of the International Conference on Image Processing ICIP*, **II**, 331–335, Austin, USA, November 1994.

[36] Kunt, M., Ikonomopoulos, A. and Kocher, M. (invited paper), Second-generation image coding techniques. *Proceedings of the IEEE*, **73** (4), 549–574, 1985.

[37] Kunt, M., Bénard, M. and Leonardi, R., Recent results in high-compression image coding. *IEEE Transactions on Circuits and Systems*, **34** (11), 1306–1336, 1987.

[38] Ikonomopoulos, A. and Kunt, M., High compression image coding via directional filtering. *Signal Processing*, **8** (2), 1985.

[39] Kocher, M. and Kunt, M., Image data compression by contour texture modelling. In *Proceedings SPIE of the International Conference on the Applications of Digital Image Processing*, 131–139, Geneva, Switzerland, April 1983.

[40] Carlsson, S., Sketch-based coding of grey level images. *Signal Processing*, **15** (1), 57–83, 1988.

[41] Ran, X. and Farvardin, N., Low bit-rate image coding using a three-component image model. Technical Report TR 92–75, University of Maryland, 1992.

[42] Ran, X. and Farvardin, N., Adaptive DCT image coding on a three-component image model. In *Proceedings of the International Conference on Acoustics, Speech, and Signal Processing ICASSP*, **III**, 201–204, San Francisco, USA, March 1992.

[43] Mallat, S. G. and Zhong, S., Characterization of signals from multiscale edges. *IEEE Transactions on Pattern Analysis and Machine Intelligence*, **14** (7), 710–732, 1992.

[44] Froment, J. and Mallat, S. G., *Second Generation Compact Image Coding With Wavelets*. Academic Press, New York, January 1992.

[45] Barnsley, M. F., *Fractals Everywhere*. Academic Press, San Diego, 1988.

[46] Jacquin, A. E., Image coding based on a fractal theory of iterated contractive image transformations. *IEEE Transactions on Image Processing*, **1**, 18–30, 1992.

[47] Fisher, Y., A discussion of fractal image compression. In Saupe, D., Peitgen, H. O. and Jurgens, H. (eds), *Chaos and Fractals*, 903–919. Springer-Verlag, New York, 1992.

[48] Oien, G. E., L2-optimal attractor image coding with fast decoder convergence. PhD thesis, Trondheim, Norway, 1993.

[49] Jacobs, E. W., Fisher, Y. and Boss, R. D., Image compression: a study of iterated transform method. *Signal Processing*, **29**, 251–263, 1992.

[50] Barthel, K., Voyé and Noll, P., Improved fractal image coding. In *Proceedings of the Picture Coding Symposium PCS*, number 1.5, Lausanne, Switzerland, March 1993.

[51] Baskurt, A. and Goutte, R., 3-Dimensional image compression by discrete cosine transform. In *Proceedings of the European Signal Processing Conference EUSIPCO*, **IV**, 79–82, 1988.

[52] Karlsson, G. and Vetterli, M., Three dimensional subband coding of video. In *Proceedings of the International Conference on Acoustics, Speech, and Signal Processing ICASSP*, **II**, 1100–1103, New York City, USA, April 1988.

[53] Ebrahimi, T. and Kunt, M., Image sequence coding using a three dimensional wavelet packet and adaptive selection. In *Proceedings of SPIE's Visual Communications and Image Processing VCIP*, **1818**, 222–232, Boston, USA, November 1992.

[54] Chang, E. and Zakhor, A., Scalable video coding using 3-D subband velocity coding and multirate quantization. In *Proceedings of the International Conference on Acoustics, Speech, and Signal Processing ICASSP*, **V**, 574–577, Minneapolis, USA, April 1993.

[55] Dufaux, F. and Moscheni, F., New perspectives in motion estimation techniques for digital TV. *Proceedings of the IEEE*, July 1995.

[56] Ohm, J. R., Temporal domain subband video coding with motion compensation. In *Proceedings of the International Conference on Acoustics, Speech, and Signal Processing ICASSP*, **III**, 229–232, San Francisco, USA, March 1992.

[57] Ohm, J. R., Three-dimensional subband coding with motion compensation. *IEEE Transactions on Image Processing*, **3** (5), 559–571, 1994.

[58] Ohm, J. R., Three dimensional SBC-VQ with motion compensation. In *Proceedings of the Picture Coding Symposium PCS*, number 11.5, Lausanne, Switzerland, March 1993.

[59] Gilge, M., A high quality videophone coder using hierarchical motion estimation and structure coding of the predictive error. In *Proceedings of SPIE's Visual Communications and Image Processing VCIP*, **1001**, 864–874, Cambridge, USA, November 1988.

[60] Stiller, C. and Lappe, D., Laplacian pyramid coding of predictive error images. In *Proceedings of SPIE's Visual Communications and Image Processing VCIP*, **1605**, 47–57, Boston, USA, 1991.

[61] Zhang, Y.-Q. and Zafar, S., Motion-compensated wavelet transform coding for color video compression. In *Proceedings of SPIE's Visual Communications and Image Processing VCIP*, **1605**, 301–316, Boston, USA, 1991.

[62] Li, W. and Mateo, F., Segmentation based coding of motion compensated prediction error images. In *Proceedings of the International Conference on Acoustics, Speech, and Signal Processing ICASSP*, **V**, 357–360, Minneapolis, USA, April 1993.

[63] Li, W. and Kunt, M., Morphological segmentation applied to displaced frame difference coding. Special issue of *Signal Processing on Mathematical Morphology and its Applications to Signal Processing*, **38** (1), 45–56, 1994.

[64] Ngan, K. N. and Chooi, W. L., Very low bit rate video coding using 3D subband approach. *IEEE Transactions on Circuits and Systems*, **4** (3), 309–316, 1994.

[65] Podilchuk, C., Low bit-rate subband video coding. In *Proceedings of the International Conference on Image Processing ICIP*, **III**, 280–284, Austin, USA, 1994.

[66] Westerink, P. H., Biemond, J. and Muller, F., Subband coding of image sequences at low bit-rates. *Signal Processing: Image Communication*, **2** (4), 441–448, 1990.

[67] Ebrahimi, T. and Kunt, M., A video codec based on perceptually derived and localized wavelet transform for mobile applications. In Vandewalle, J. *et al.* (eds), *Signal Processing VI, Theories and Applications*, North-Holland, Amsterdam, 1361–1364, 1992.

[68] Thoreau, D. and Vial, J.-F., Very low bit rate subband coding. In *Proceedings of the Picture Coding Symposium PCS*, **11–10**, 304–307, Sacramento, USA, 1994.

[69] Qian, D., Block-based motion compensated subband coding at very low bit rates using a psychovisual model. In *Proceedings of the Workshop on Very Low Bitrate Video VLBV*, number 6.8, Essex, UK, 1994.

[70] Dachiku, K. *et al.*, Motion compensation subband extra/interpolative prediction coding at very low bit rate. In *Proceedings of the Workshop on Very Low Bitrate Video VLBV*, number 6.2, Essex, UK, 1994.

[71] Katto, J., Ohki, J., Nogaki, S. and Ohta, M., A wavelet codec with overlapped motion compensation for very low bit-rate environment. *IEEE Transactions on Circuits and Systems for Video Technology*, **4** (3), 328–338, 1994.

[72] Willemin, P., Reed, T. and Kunt, M., Image sequence coding at very low bit rates with a 3-D split and merge algorithm. Presented at the *Second International Workshop on 64kbit/s coding of Moving Video*, Hannover, Germany, September 4–6, 1989.

[73] Willemin, P., Reed, T. and Kunt, M., Image sequence coding by split and merge. *IEEE Transactions on Communications*, **39** (12), 1845–1855, 1991.

[74] Salembier, P. and Pardàs, M., Hierarchical morphological segmentation for image sequence coding. *IEEE Transactions on Image Processing*, **3** (5), 639–651, 1994.

[75] Hötter, M., Object-oriented analysis-synthesis coding based on moving two-dimensional objects. *Signal Processing: Image Communication*, **2** (4), 409–428, 1990.

[76] Jozawa, H. and Watanabe, H., Video coding using segment-based affine motion compensation. In *Proceedings of the Picture Coding Symposium PCS*, 238–241, Sacramento, USA, 1994.

[77] Kocher, M. and Leonardi, R., Adaptive region growing technique using polynomial functions for image approximations. *Signal Processing*, **11** (1), 47–60, 1986.

[78] Gilge, M., Engelhardt, T. and Mehlan, R., Coding of arbitrarily shaped image segments based on a generalized orthogonal transform. *Signal Processing: Image Communication*, **1** (2), 153–180, 1989.

[79] Chen, H. H., Civanlar, M. R. and Haskell, B. G., A block transform coder for arbitrarily shaped image segments. In *Proceedings of the International Conference on Image Processing ICIP*, **I**, 85–89, Austin, USA, 1994.

[80] Lavagetto, F., Cocurullo, F. and Curinga, S., Texture approximation through discrete Legendre Polynomials. In *Proceedings of the Workshop on Very Low Bitrate Video VLBV*, number 6.5, Essex, UK, 1994.

[81] Freeman, H., On the encoding of arbitrary geometric configurations. *IRE Transactions on Electronic Computers*, **EC-10**, 260–268, 1961.

[82] Dyer, C. R., Rosenfeld, A. and Samet, H., Region representation: boundary codes from quadtrees. *Commun. of the ACM*, **23** (3), 171–179, 1980.

[83] Samet, H., Region representation: quadtrees from boundary codes. *Commun. of the ACM*, **23** (3), 163–170, 1980.

[84] Blum, H. and Nagel, R. N., Shape description using weighted symmetric axis features. *Pattern Recognition*, **10**, 167–180, 1978.

[85] Rosenfeld, A. and Kak, A. C., *Digital Picture Processing*, **2**, Chapter 11: Representation. Academic Press, New York, 1982.

[86] Brigger, P., Ayer, S. and Kunt, M., Morphological shape representation of segmented images based on temporally modeled motion vectors. In *Proceedings of the International Conference on Image Processing ICIP*, **III**, 756–760, Austin, USA, November 1994.

[87] Gu, C. and Kunt, M., Contour simplifications and motion compensation for very low bit-rate video coding. In *Proceedings of the International Conference on Image Procesing ICIP*, **II**, 423–427, Austin, USA, 1994.

[88] Yokoyama, Y., Miyamoto, Y. and Ohta, M., Very low bit-rate video coding with object-based motion compensation and othogonal transform. In *Proceedings of SPIE's Visual Communications and Image Processing VCIP*, **2094**, 12–23, Cambridge, USA, November 1993.

[89] Gu, C. and Kunt, M., Contour image sequence coding by motion compensation and morphological filters. In *Proceedings of the Workshop on Very Low Bitrate Video VLBV*, number 7.1, Essex, UK, 1994.

[90] Casas, J. R. and Torres, L., Coding of details in very low bit-rate video systems. *IEEE Transactions on Circuits and Systems for Video Technology*, **4** (3), 317–327, 1994.

[91] Cicconi, P. and Nicolas, H., Efficient region-based motion estimation and symmetry oriented segmentation for image sequence coding. *IEEE Transactions on Circuits and Systems for Video Technology*, **4** (3), 357–364, 1994.

[92] Garcia-Garduno, V. and Labit, C., On the tracking of regions over time for very low bit rate image sequence coding. In *Proceedings of the Picture Coding Symposium PCS*, 257–260, Sacramento, USA, 1994.

[93] Reusens, E., Sequence coding based on the fractal theory of iterated transformations systems. In *Proceedings of SPIE's Visual Communications and Image Processing VCIP*, **1**, 132–140, Boston, 1993.

[94] Strobach, P., Tree-structured scene adaptive coder. *IEEE Transactions on Communications*, **38** (4), 477–486, 1990.

[95] Lu, L. and Pearlman, W. A., Multi-rate image sequence coding with quadtree segmentation and backward motion compensation. In *Proceedings of SPIE's Visual Communications and Image Processing VCIP*, **1818**, 606–614, Boston, USA, 1992.

[96] Ebrahimi, T., A new technique for motion field segmentation and coding for very low bitrate video coding applications. In *Proceedings of the International Conference on Image Processing ICIP*, **II**, 433–437, Austin, USA, 1994.

[97] Asai, K., Yamada, Y. and Murakami, T., Video sequence coding based on segment-model and priority control. In *Proceedings of the Picture Coding Sympsium PCS*, 325–328, Sacramento, USA, 1994.

[98] Queloz, M. P., Macq, B., A split-and-merge motion estimation/compensation technique for very-low bitrate image coding. In *Proceedings of the Workshop on Very Low Bitrate Video VLBV*, number 5.2, Essex, UK, 1994.

[99] Aizawa, K., Harashima, H., and Saito, T., Model-based analysis synthesis image coding (MBASIC) system for a person's pace. *Signal Processing: Image Communication*, **1** (2), 139–152, 1989.

[100] Huang, T. S., Reddy, S. C. and Aizawa, K., Human facial motion modeling, analysis, and synthesis for video compression. In *Proceedings of SPIE's Visual Communications and Image Processing VCIP*, **1605**, 234–241, Boston, USA, 1991.

[101] Li, H. and Forchheimer, R., Two-view facial movement estimation. *IEEE Transactions on Circuits and Systems for Video Technology*, **4** (3), 1994.

[102] Pearson, D. (invited paper). Developments in model-based video coding. *Proceedings of the IEEE*, July 1995.

[103] Beaumont, J. M., Image data compression using fractal techniques. *British Telecommunication Technology Journal*, **9** (4), 93–109, 1991.

[104] Li, H., Novak, M. and Forchheimer, R., Fractal-based image sequence compression scheme. *Optical Engineering*, **32** (7), 1588–1595, July 1993.

[105] Lazar, M. S. and Bruton, L. T., Fractal-based image sequence compression scheme. *IEEE Transactions on Circuits and Systems for Video Technology*, **4** (3), 297–308, 1994.

[106] Hürtgen, B. and Buttgen, P., Fractal approach to low rate video coding. In *Visual Communications and Image Processing '93*, **1**, 120–131, Boston, 1993.

[107] Ngan, K. N., Lin, D. W. and Liou, M. L., Enhancement of image quality for low bit-rate video coding. *IEEE Transactions on Circuits and Systems*, **38** (10), 1221–1225, 1991.

[108] Sezan, M. I., Ozkan, M. K. and Fogel, S. V., Temporally adaptive filtering of noisy image sequences using a robust motion estimation algorithm. In *Proceedings of the International Conference on Acoustics, Speech, and Signal Processing ICASSP*, **4**, 2429–2432, Toronto, Canada, 1991.

[109] Mattavelli, M. and Nicoulin, A., Pre and post processing for very low bit-rate video coding. In Chiariglione, L. (ed.), *Proceedings of the International Workshop on HDTV*, Torino, Italy, Springer-Verlag, Berlin, October 1994.

[110] Hoetter, M., Differential estimation of the global motion parameters zoom and pan. *Signal Processing*, **16**, 249–265, 1989.

[111] Wu, S. F. and Kittler, J. A differential method for simultaneous estimation of rotation, change of scale and translation. *Signal Processing: Image Communication*, **2**, 69–80, 1990.

[112] Tse, Y. T. and Baker, R. L., Global zoom/pan estimation and compensation for video compression. In *Proceedings of the International Conference on Acoustics, Speech, and Signal Processing ICASSP*, **4**, 2725–2728, Toronto, Canada, May 1991.

[113] Horne, C., Improving block based motion estimation by the use of global motion. In *Proceedings of SPIE's Visual Communications and Image Processing VCIP*, **2094**, 576–587, 1993.

[114] Reeve, H. C., and Lim, J. S., Reduction of blocking effects in image coding. *Optical Engineering*, **23** (1), 34–37, January/February 1984.

[115] Ramamurthi, B. and Gersho, A., Nonlinear space-variant post-processing of block coded images. *IEEE Transactions on Acoustics, Speech, and Signal Processing*, **34** (5), 1258–1268, 1986.

[116] Stevenson, R. L., Reduction of coding artifacts in transform image coding. In *Proceedings of the International Conference on Acoustics, Speech, and Signal Processing ICASSP*, **V**, 401–404, Minneapolis, USA, 1993.

[117] Yang, Y., Galatsanos, N. P., Katsaggelos, A. K., Iterative projection algorithms for removing the blocking artifacts of block-DCT compressed images. In *Proceedings of the International Conference in Acoustics, Speech, and Signal Processing ICASSP*, **V**, 405–408, Minneapolis, USA, 1993.

[118] Macq, B., Mattavelli, M., Van Calster, O., van der Plancke, E., Comes, S. and Li, W., Image visual quality restoration by cancellation of the unmasked noise. In *Proceedings of the International Conference on Acoustics, Speech, and Signal Processing ICASSP*, Adelaide, Australia, **5**, 53–56, 1994.

[119] Chen, T., Elimination of subband, coding artifacts using the dithering technique. In *Proceedings of the International Conference on Image Processing ICIP*, **II**, 874–877, Austin, Texas, 1994.

[120] Li, W., Egger, O. and Kunt, M., Efficient quantization noise reduction device for subband image coding schemes. In *Procedings of the International Conference on Acoustics, Speech, and Signal Processing ICASSP*, Detroit, USA, May 1995.

[121] CCITT SG XV. Description of Ref. Model 8 (RM8). Technical report, June 1989.

[122] Early, S. H., Kuzma, A., and Dorsey, E., The videophone 2500 – video telephony on the public switched telephone network. *AT&T Technical Journal*, **72** (1), 22–32, January/February 1993.

[123] Eude, G., Schmitt, C. and Loret, B., Videotelephony communication on personal computer at very low bit rate. In *Proceedings of the Picture Coding Symposium PCS*, 292–295, Sacramento, USA, September 1994.

19

Algorithms for Low Bit Rate Video Coding

Raj Talluri

Insights into Mobile Multimedia Communication
ISBN 0-12-140310-6

19.1. INTRODUCTION

In the past few years, various technological advances have substantially increased the prevalence of digital video in our everyday life. Video coding algorithms have made significant progress, and it is now possible to compress video to a few kbps with acceptable quality for many applications. Modem speeds have also steadily increased with 56 kbps now available. The penetration of ISDN to home and office environments has substantially increased the effective channel bandwidths, ranging from 64 kbps to 384 kbps. The clock speeds and processing power of the PC and DSPs has also increased significantly and the cost of memory has decreased considerably. The explosion of the Internet and World Wide Web in the last few years has also made it possible for users to incorporate video into their web pages. The increased penetration of wireless networks and end equipment has added yet another avenue for communication of video data. With all these advances, there is a significantly increased activity in the area of low bit rate video coding. In this chapter, we will first describe some of the applications of low bit rate video, followed by a discussion of the requirements of these applications. We will then present an overview of current technology in low bit rate video coding and highlight the relative merits of each of the approaches. International standards play a huge role in video coding, being the key to enabling interoperability. The aim of this chapter is to introduce some applications and methods for low bit rate coding and to examine how these have been adopted in recent standardization activities.

19.2. APPLICATIONS

There are many useful applications that have and will be enabled by low bit rate video coding. We discuss below some of the more important applications.

19.2.1. Videoconferencing

Desktop videoconferencing over local area networks within an office network, site to site communication using dedicated room videoconferencing systems and Internet videoconferencing over long distances (possibly over analogue telephone lines) are some of the popular incarnations of videoconferencing systems. It is possible to achieve reasonably high bandwidths (up to 10 Mbps) for desktop videoconferencing systems in an office LAN. However, these are not dedicated bandwidths and the effective bandwidth fluctuates with the network load. Also, establishing good synchronization with the audio and the video data, and maintaining low latency, are some of the issues that need to be addressed. Room videoconferencing systems typically operate over high bandwidth T1 lines and now more commonly over ISDN lines. The ISDN lines can achieve effective bandwidths ranging from 64 kbps to 384 kbps. Internet video conferencing systems operate over widely varying bandwidth ranges with the users connecting on analogue phone lines typically in the

28.8 kbps range. Internet connections with increased bandwidth are discussed further in Chapter 8 of this book.

19.2.2. Videophones

The distinction between videoconferencing systems and videophones is somewhat vague. In general, it is possible to imagine a wired videophone as a stand-alone handheld or wall mounted unit that operates over analogue phone lines. In some nations such as Germany, ISDN videophones are becoming increasingly popular. A wireless videophone on the other hand operates over wireless channels such as provided by a cellular phone system. The requirements for very low bit rate coding are significant in videophone applications. Typical wired videophones only provide bandwidths of up to 28.8 kbps for audio, video and systems data. The wireless videophones are even more stringent and may need to operate in the sub 15 kbps range.

19.2.3. Internet Video

With the increased popularity of the World Wide Web, Internet video is gaining popularity. It is not uncommon to find web pages with embedded video. Until recently, to play a video clip in a web page it was necessary to first download the entire clip before the decoder could play it (e.g. Apple QuickTime). This typically resulted in a significant delay. Nowadays, with the advent of streaming video technology, it is possible to start playing the video as the decoder receives it, hence making the service more appealing (e.g. Microsoft Netshow). A significant number of people connect to the Internet on low bandwidth channels such as provided by analogue modems from home. Low bit rate video coding is thus an essential component in making the Internet-based video attractive and usable.

19.2.4. Surveillance and Monitoring

There are many requirements for video surveillance and monitoring of remote sites and facilities via various combinations of telephone and Internet connections. For example, small business owners can dial up from their homes to visually inspect business premises, without travelling to the remote location. Improvements in very low bit rate video communications capabilities coupled with the advances in low cost, good quality video cameras are making these applications commonplace. Low cost video communications can also enable a variety of tele-operation applications. A basic example is the remote control of camera functions such as pan, tilt, and zoom, where real-time video feedback is important for rapidly reaching the desired view. Applications such as these require provisions for two-way data communications to be supported in the systems layer of video compression standards.

19.2.5. Desktop Multimedia

Desktop multimedia is an application scenario where low bit rate coding is essential to be able to deal with the tremendous bandwidth requirements of the raw video data. CDROMs and the newer Digital Versatile Disk (DVD) bring to the end user the ability to create, playback, and embed, compelling video, audio and graphics clip art into existing and new applications. Multimedia titles of interactive games, and other education and entertainment software that utilize compressed video are also becoming increasingly popular.

19.3. REQUIREMENTS

19.3.1. Coding Efficiency

One of the primary requirements of low bit rate video coding continues to be coding efficiency. Video compression will play a crucial role in making the applications of digital video pervasive. The amount of information contained in a single image is substantial: a colour TV quality image typically contains about 720×480 pixels, with each pixel requiring 16 bits of resolution (8 for luminance and 8 for chrominance), giving a total of 5.5 megabits. For broadcast quality TV, to portray full motion video, the individual images must be presented at a frame rate of 30 frames per second. Hence, TV resolution digital video contains 166 megabits of information per second. A 10 second clip of raw video therefore requires 207 megabytes of storage space. It is clear that, in order to store and transmit digital video data effectively, substantial compression is needed.

Current video compression techniques are extremely effective in reducing the bandwidth requirements of digital video. They employ methods to exploit the temporal and spatial redundancies present in the video signal in order to reduce the effective video bandwidth to manageable levels. Current digital TV standards such as MPEG-2 typically compress the above 166 Mbps signal to about 6 Mbps – a 28:1 compression ratio.

For a number of video applications it is not necessary to have the same high spatial and temporal resolutions as required for broadcast TV (720×480 at 30 frames/s). Video information coded at temporal resolutions of 10–15 frames/s with spatial resolutions of 352×288 or sometimes even as low as 176×144 prove to be quite adequate for many low bit rate applications such as videophones, videoconferencing and surveillance. In such applications the video content is typically "simpler" (e.g. a stationary person talking in front of a computer or a static scene with occasional motion) thus making it possible to achieve even higher compression ratios. Typically in these applications it is possible to compress the video by 100–200 times and still provide adequate video quality. However, the goal of almost all video compression techniques continues to be to improve the "quality" of the compressed video at a given bandwidth. Recent advances have made it possible to provide significantly improved quality at low bit rates and research continues to improve the coding efficiency of the video coding algorithms.

19.3.2. Perceptual Quality

In order to achieve the extreme bandwidth reductions required, video coding algorithms are typically lossy (i.e., the input video data is not faithfully reproduced at the decoder). This loss results in a degradation of the video quality manifest as coding artifacts such as blocking, blurring, ringing and mosquito noise.

In almost all digital video applications the decoded video is for human consumption. Hence perceptual quality is an important factor which should be optimised by the coding technique. It is, however, difficult to derive an objective measure of this perceived video quality which encapsulates the effect of the coding artifacts as perceived by the end user. There is a significant amount of ongoing research in this area, attempting to derive perceptual metrics for video quality. Typically most video coding algorithms have a number of parameters that are controlled during the encoding process. One compelling advantage of having a perceptual quality metric is that the video coding algorithm can utilize this metric to optimize its parameters to achieve the highest video quality. Another advantage of having such a metric is that it can be used to compare different video coding algorithms that achieve the same coding efficiency but introduce different types of coding artifact.

One popular metric that is currently used to rank the distortions introduced by video coding algorithms is the average peak signal to noise ratio or PSNR. This is defined as the logarithm of the mean squared error between the original and the compressed video data averaged over the entire sequence. While PSNR is not a perceptual quality metric, it does give some indication of the fidelity of the compressed video data to the original data.

There is ongoing research attempting to produce models of the human visual system (HVS) and to develop a video quality metric by utilizing these models [1]. Results which use this in the optimization of a coder have also been reported [2]. However, this work is still in the development stage and not yet widely applied in the design of current video coding algorithms.

19.3.3. Error Resilience

Current video compression techniques used in real-time video communications achieve efficient compression by using predictive coding techniques such as motion compensation together with variable length entropy coding techniques such as those based on Huffman codes. When the compressed video stream is transmitted over a real communication channels such as an analogue phone line or a wireless link, it is often corrupted by channel noise and multipath reflections. Variable length coding schemes are highly susceptible to such errors. As a result, the decoder can lose synchronization with the encoder. Predictive coding techniques make matters much worse since the errors in one video frame quickly propagate across the entire video sequence, rapidly degrading the decoded video quality and rendering it totally unusable. Hence, unless the video decoder takes proper remedial steps, the video communication system totally breaks down.

Error resilience can be described as the ability of the video compression scheme to tolerate errors introduced into the compressed bit stream and still maintain acceptable video quality. This property is particularly important in wireless channels where bit error rates (BER) in excess of 10^{-3} are possible. Current video compression techniques apply a number of different approaches to combat channel degradations and these are discussed in more detail in Chapter 25.

19.3.4. Scalability

Scalability can be defined as one embedded bit stream with many available rates. The video encoder encodes the input video once and generates a coded bit stream. If the encoder is scalable then it is possible to decode/transmit a part of this scalable bit stream and the decoder will generate a reduced quality video. The reduced quality can be due to either reduced spatial resolution (spatial scalability), temporal resolution (temporal scalability) or reduced PSNR (quality scalability).

One of the advantages of a scalable bit stream is that the encoder can make optimal use of the varying channel bandwidth by extracting a subset of the bit stream to match the available channel bandwidth. In a multicast environment, scalability is a particularly useful property. If the different receivers are connected by communication channels of widely varying channel bandwidths, the encoder can achieve efficient transmission from the same coded bit stream without the need for re-encoding at different rates. Also if the decoder is connected to the encoder by a channel that has a time varying bandwidth, then scalability makes progressive transmission possible. That is, as the channel bandwidth increases the decoder receives more of the bit stream and the video quality progressively increases.

Current video compression techniques support varying levels of scalability depending on the technique used. Typically, in order to achieve scalability properties, the encoder has to sacrifice some amount of coding efficiency. However, some approaches such as those based on wavelets enable seamless scalable properties possible with very little overhead [3]. See Chapter 23 for further details.

19.3.5. Complexity

The sheer quantity of the data to be processed makes video coding a highly computing-intensive task. Depending on the nature of the application the video coding/decoding techniques can be symmetric or asymmetric in the amount of computation required. Broadcast video is an example of an asymmetric system where the video coder can be computationally more intensive than the decoder since the encoding process can often be performed off-line. The decoder has more severe requirements imposed on it and it has to run in real time without noticeable delay. Internet video – where video is coded and stored on the web pages to be repeatedly played by the user is also such an example of asymmetric video coding, and video stored on CD ROMs is yet another example. Videoconferencing and videophone are more symmetric applications where both the encoder and the decoder have to run in real time. Here the computational load must be balanced on both ends of the video codec.

With the increasing advances in IC technology, much faster processors with clock speeds in the hundreds of megaherz range are now possible. Large memories are also becoming increasingly affordable. All these advances are making what was once considered unrealizable, now possible in real time on inexpensive computing platforms. Typical video coding algorithms are designed such that they perform similar operations such as discrete cosine transforms (DCTs) on small local neighbourhoods repeatedly across the entire image. These kinds of operations are ideally suited for digital signal processors (DSPs). Hence, we see increased penetration of these devices in video coding applications.

19.3.6. Interactivity

In most of the traditional applications of digital video such as broadcast TV, video-conferencing and videophone, the end user is a passive consumer of the video and does not generally attempt to interact with the video in any sense. In some of the newer applications such as Internet video, DVD, and video games, interactivity, is becoming a key requirement. The user typically likes to (at least) have VCR-like controls such as pause, play, stop and rewind. In addition, the user would also like to be able to select a portion of the video and edit it or add it to another application not unlike a multimedia-authoring tool. Individual objects within video segments may be selected and combined with other objects, and a standard video representation is needed to support this. Future video compression systems will thus need to embrace these interactive requirements and support selective video encoding in real time. For example, users may specify objects to view without the background, or a background scene without objects in the foreground. Other multimedia needs include selective embedding of video in graphics, efficient overlays of graphics on video, and seamless compatibility of video and graphics to support augmented reality and virtual reality applications. The newer video coding standards such as MPEG4, with the advent of video coding techniques such as object-based coding, support such functionalities.

19.4. ELEMENTS OF VIDEO COMPRESSION

The goal of all video compression schemes is to reduce the amount of data required to represent the video signal. This is accomplished by performing some or all of the following three steps: (1) redundancy removal, (2) quantization and (3) lossless coding.

19.4.1. Redundancy Removal

Prediction can be used to reduce redundancy both spatially (within a frame of the sequence) and also temporally (between one frame and the next). A simple example of spatial redundancy removal techniques is differential pulse code modulation (DPCM). In this technique, for each pixel in the image, a prediction based on the neighbouring pixel values is formed. Instead of coding the raw pixel values at each location the difference between the raw pixel value and the prediction is then encoded. Since in most images there is a high degree of spatial similarity between pixels in a local neighbourhood, this technique proves to be quite effective. Transform techniques such as the discrete cosine transform and wavelet transform also help remove spatial redundancy by transforming the image intensity values into a transform domain where the redundancy can be better isolated and removed. Motion compensation is a technique for reducing the temporal redundancy. This technique uses information in the previous frame to predict the current frame, so that the overall information required is reduced.

19.4.2. Quantization

This is the lossy stage of the compression process. This stage attempts to reduce the dynamic range of the data to be encoded by mapping the input values into a reduced set of output values. For example, if the input data has a resolution of 8 bits per pixel (256 possible values), a quantizer can reduce the dynamic range by mapping all the input values that lie between 1 to 16 to a particular output value, the value from 16 to 32 to another output value and so on. Hence, the quantized output has a reduced dynamic range of only 4 bits (16 possible values) – a 4:1 compression. This process introduces error in the input signal. Quantizers are designed to reduce the visible distortion introduced by this error. The range of the quantizer, the step size, and the distribution of the steps are all important choices in designing a quantizer that minimizes the visible distortion introduced.

19.4.3. Lossless Encoding

Most compression schemes incorporate a lossless stage which exploits the probability distribution of the input data. Huffman coding is a lossless coding scheme that exploits this property by assigning short codewords to the more frequently occurring input values and longer codewords to the less frequently occurring ones. Run length coding is a lossless coding scheme that achieves compression by detecting and efficiently representing long runs of the same values in the input sequence. Arithmetic coding and LZW codes are also examples of lossless coding schemes.

Current compression schemes such as MPEG-1, MPEG-2, H.261 and H.263 all employ the above three steps in some form or the other. Later sections describe the architecture of typical schemes and illustrate how the above ideas are used.

19.5. VIDEO COMPRESSION TECHNIQUES

In this section we describe some of the popular video compression techniques, illustrate the basic principles behind them and discuss their relative merits. We consider (1) block-based coders, (2) object-based coders and (3) model-based coders.

19.5.1. Block-based Coding

This is by far the most popular of the current video coding techniques. Most of the existing video coding standards such as MPEG-1 [4], MPEG-2 [5], H.261 [6] and H.263 [7,8] are essentially block-based coders. Block-based coders make very few assumptions about the input data and hence work very well across a wide variety of input video data and across a wide range of bit rates. This is one of the reasons for their popularity. There has been considerable amount of work in these coders and they are well optimized. The computational model of block-based coders is also well understood and well designed. They mostly make only local references to data (in terms of blocks) and hence are amenable to fast implementations without the need for large amounts of memory. A generic block-based video coding structure is shown in Figure 19.1.

Typically the input video data to these video compression schemes is in YUV colour space in what is called a 4:2:0 format. In this format, the colour representation for each pixel consists of three components: a luminance, Y, and two chrominance components, Cb and Cr. The human visual system (HVS) is most sensitive to the resolution of an image's luminance component, so the Y pixel values are encoded at a higher resolution than the chrominance pixel values. The chrominance values are subsampled by 2 in both the horizontal and vertical directions, which reduces the amount of information to be coded by 4 without significant degradation in visual quality.

All block-based coders have four essential stages of processing: (1) motion estimation/compensation, (2) DCT coding, (3) quantization and (4) entropy coding.

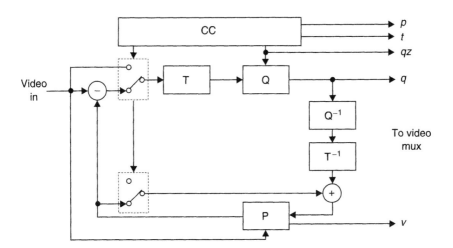

Figure 19.1 Block-based video coder architecture (T: transform (DCT), Q: quantizer, P: picture memory with motion compensated variable delay, CC: coding control p: flag for intra/inter, t: flag for transmitted or not, qz: quantizer indication, q: quantizing index for transform coefficients, v: motion vector.)

Motion Compensation

Motion compensation is the stage of the video coding process that reduces the temporal redundancy of the video signal. Each frame is first divided into rectangular units of 16×16 pixels. Each of these 16×16 units is known as a macroblock. Each macroblock is made up of six 8×8 blocks – 4 luminance blocks and 2 chrominance blocks (since the chrominance signal is subsampled by 2 in each direction). In the motion estimation stage, for each of the macroblocks in the current image reference macroblock) the video encoder searches the previous frame to find a macroblock location (target macroblock) that minimizes the pixel intensity differences between the reference macroblock and the target macroblock. The two-dimensional location of the target macroblock is known as the motion vector of the current macroblock. Thus, the motion estimation stage computes a motion vector for each macroblock of the current frame. The motion compensation stage applies these motion vectors to the previous (reference image) and generates a motion compensated image. This motion compensated image (along with the associated motion vectors) thus represents an approximation or prediction of the current image based on the previous (reference) image.

Discrete Cosine Transform

After the motion compensation has removed as much of the temporal redundancy in the signal as possible, a discrete cosine transform (DCT) is applied to remove the spatial redundancies. First the difference between the original image to be compressed and the predicted image (after motion compensation) is computed. This is referred to as the residual image or displaced frame difference signal (DFD). The DCT is applied to all the 8×8 blocks of the residual image. The DCT converts an 8×8 block of pixel values to an 8×8 matrix of horizontal and vertical spatial frequency coefficients. The DCT decorrelates the spatial information by transforming it into the frequency domain. Most of the energy is then concentrated in the low frequency coefficients. The higher order AC coefficients can now be represented with much less accuracy without significant loss in the quality of the reconstructed image.

Quantization

After applying the DCT to the residual image, quantization is applied to compress the input data. Quantization maps a range of input values into a single output value in the range. The quantized range can be concisely represented as an integer code, which can be used to recover the quantized value during decoding. The difference between the actual value and the quantized value is called the quantization noise. Under some circumstances, the HVS is less sensitive to quantization noise so such noise can be allowed to be large, thus increasing coding efficiency. Each array of 8×8 coefficients produced by the DCT is quantized to produce an 8×8 array of quantized coefficients. Normally, the number of non-zero quantized coefficients is quite small, and this is one of the main reasons why the compression scheme works as well as it does. Typically, the coefficients are quantized with a uniform quantizer.

The value of the coefficient is divided by the quantizer step size and rounded to the nearest whole number to produce the quantized coefficient. The quantizer step size is derived from the quantization matrix and the quantizer scale. It can thus be different for different coefficients, and may change between macroblocks. The only exception is the DC coefficient, which is treated differently. The eye is quite sensitive to large area luminance errors, and so the accuracy of coding the DC value is fixed.

Entropy Coding

As mentioned above, after quantization a number of the DCT coefficients are zero valued. Considerable compression gain can therefore be achieved by using a run length encoding scheme. A zig-zag scanning is performed to order the DCT coefficients in a way that clusters most of the non-zero coefficients together. This ordering concentrates the highest spatial frequencies at the end of the scan. These long runs of zeros and the following non-zero DCT coefficients are then encoding using a variable length coding (VLC) technique. This VLC technique is a lossless coding scheme that assigns short codewords to more probable events and long codewords to the less probable ones. On average, the more frequent shorter codewords dominate such that the code string is shorter than the original data.

19.5.2. Object-based Coding

Coder Architecture

There has been considerable research in object-based coding for very low bit rate video compression [9,10,11]. These techniques achieve efficient compression by separating coherently moving objects from stationary background and compactly representing their shape, motion and content. Some of the advantages of these approaches include a more natural and accurate rendition of the moving object motion and efficient utilization of the available bit rate by focusing on the moving objects. In addition to these compression advantages, if suitably constructed the object-based coding techniques can also support content-based functionalities such as the ability to selectively code, decode and manipulate specific objects in a video stream. The ability to scale the bit stream at an object level can also be supported by these techniques. These functionalities can prove useful in some PC multimedia authoring tools and real-time video conferencing system operating over low bandwidth channels. In light of these advantages it is expected that the object-based coding methods will be incorporated into international standards such as ISO MPEG-4 [12].

Object-based video compression techniques operate by estimating moving objects and coding compact representations of the object's shape, motion and the content [13]. A representative class of object-based coders is shown in Figure 19.2 [14,15], which shows the overall block diagram of the approach and also some of the typical intermediate images. The first frame of the video sequence is coded as an intraframe using the block-based DCT coding scheme. Subsequent frames are coded in a modular five-step process described below.

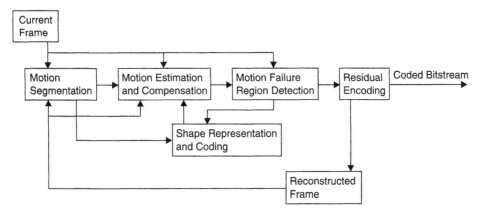

Figure 19.2 The overall block diagram of an object-based coder.

We illustrate these steps with examples for the *mother and daughter* video data from which two frames are reproduced in Figure 19.3.

Moving Object Detection

The first stage isolates moving objects in the video sequence from the stationary background using a simple, fast and effective motion segmentation technique. The intensity difference between the current image and the previously coded image is first computed. By thresholding the intensity differences, regions of significant changes in intensity are isolated. Morphological image processing operations, namely *close* (*dilate* and then *erode*) followed by *open* (*erode* and then *dilate*) with a small circular kernel, are applied to clean up the raw intensity differences and cluster the moving objects regions. Connected component analysis is performed on the output of this morphological segmentation to consistently label at the segmented regions and also

Figure 19.3 Two consecutive frames of the Mother and Daughter video sequence.

eliminate small insignificant regions. Further details on morphological coding can be found in Chapter 22.

Figure 19.4 shows examples of image regions detected as moving objects for two frames of the *mother and daughter* video. This segmentation does not exactly follow the boundary between the foreground and background, but it is still a useful approximation for efficiently identifying and encoding the change areas in a video sequence. If the entire scene is moving or if the camera is panning, a single large region may be detected; in this case compression will still succeed as a result of subsequent steps. Alternatively, global motion compensation techniques [16] may be incorporated to enable better performance for moving camera situations, but the advantages might be offset for some applications by considerations such as computational expense.

Shape Coding

The task of shape representation is to efficiently code the bounding contours of image regions that were produced in the moving object detection stage. One way to accomplish this is by first computing the bounding rectangle for each of the moving objects. The rectangle is then tiled with 16×16 pixel blocks. A bitmap with a 1 representing the blocks that lie inside the object and a 0 representing the blocks that lie outside the object is then formed. This bit map is efficiently coded using a run length encoding scheme. In order to exploit the frame to frame temporal redundancies of the shape, at each stage the coding technique searches in the previously encoded and transmitted shapes to try and identify a shape that is *closest* to the current shape. If such a shape exists, only changes from the previous shape are coded along with a pointer to the previous shape. Further a more accurate shape representation is achieved by coding the border blocks at a finer resolution.

As an alternative, a spline-based shape representation strategy can also be used. In this approach the encoder detects corner points [17] along the contours and sends these to the decoder using a differential coding technique; the coder and decoder

Figure 19.4 Image regions produced by moving object detection.

both approximate the contour by fitting Catmull–Romm splines to the corner points, so that the coder and decoder agree precisely. The spline-based shape representation is more accurate but consumes more bandwidth than the block-based representation. Hence, a trade-off can be made between accuracy of representation and bandwidth, depending on the application. In general, if the automatic motion segmentation strategy does not result in pixel-accurate boundaries of the moving object, there is no need to apply an accurate shape representation such as Catmull–Romm splines and a block-level representation will suffice. However, if the segmentation is performed off-line using more precise techniques such as blue-screening, this representation can prove to be very useful.

Motion Estimation

In this stage, the goal is to remove temporal redundancies of the video signal. Motion estimation is used to predict the contents of segmented regions from the previous images. A block-based motion estimation technique can be used. For each 16×16 block that lies inside a moving object, the algorithm searches to find a corresponding location in the previous frame such that the sum of absolute differences between the current and previous blocks is a minimum. To obtain a good match the search is conducted at half-pixel resolution, similar to the motion estimation strategy of block-based coders. The motion vector corresponding to the location that produces the minimum error is then efficiently encoded. Some object-based coders use higher order motion models to capture the motion of each of the moving objects.

Motion Failure Region Detection

After motion compensation, there is still a significant amount of residual information in some areas of the image where the motion compensation alone was not adequate. Motion segmentation is then re-applied to the compensated image and the original image to isolate these *motion failure* regions of high prediction error. Examples of motion failure regions detected in the *mother and daughter* video are shown in Figure 19.5.

Residual Error Encoding

The goal of this final step is to encode the residual information in the motion failure regions using a block-based DCT scheme. After the DCT transform, the DCT coefficients are quantized, scanned in a zig-zag fashion, run length and Huffman coded as for the block-based coder described above. Other object-based coders have also used simpler texture coding schemes such a polynomial approximations or vector quantization (VQ) methods.

This overall coding scheme maintains the high coding efficiency at very low bit rates with reduced artifacts. It also enables the coder to support content-based functionalities such as object scalability and object manipulation. Another advantage of the object-based coder is that efficient error concealment and error correction

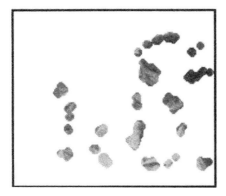

Figure 19.5 Motion failure regions after motion compensation.

strategies can be applied to the overall coding scheme to enable robust transmission of the compressed video data over noisy communication channels [13]. The object-oriented structure of the coder and the associated syntax enables them to contain the effects of errors within the boundaries of each coded object and hence mitigate the propagation of errors throughout the image sequence.

19.5.3. Model-based Coding

The model-based approach to video compression is a process of analysis and synthesis that uses parametric models of objects in the scene. This approach can enable effective visual communications at rates of 2 kbps and lower for some purposes. Model parameters are obtained by analysing object appearance and motion in the video scene. Then the parameters are transmitted to a remote location where a video display of the object is synthesized using models stored in the receiver. In principle, only a few parameters are needed to communicate changes in complex objects, thus enabling the required compression ratios. Several exploratory research efforts have demonstrated promising variations on model-based techniques [18,19].

A fanciful illustration of model-based visual communications is shown in Figure 19.6. The man at the left is unkempt, but his synthetic rendition in the video display at the right is tidy and neat. A primary challenge for model-based video compression in this example is facial analysis, to produce concise facial model parameters that are sufficiently expressive. Another challenge is facial synthesis, to produce a display that is a useful if not faithful reproduction of actual appearance.

Examples of facial analysis-synthesis techniques are embodied in the FACS + + capability [19] in which various facial expressions are classified from video data and synthesized using an animated three-dimensional face model. In other face modelling research, head models have been constructed using three-dimensional imaging systems [20]. Facial models such as these can then be animated to simulate other facial expressions, head motions, and views from arbitrary positions.

Figure 19.6 Model-based video compression.

19.6. STANDARDS

International standards play a very important role the development of digital video compression. The standards development process provides an open forum where many experts from different companies and organizations can legally engage in collaborative research and development. From the United States alone, at least 600 different experts from over 100 different organizations have contributed to the work over the last seven years. Combined with about an equal number from other countries, this represents an enormous amount of R&D effort.

The nature of today's market opportunities for consumer electronics makes it very attractive to build products that can be sold all over the world with only minor customizations. International standards enable interoperability and hence foster a healthy competition in the market place that brings down the cost of the end product and hence enables volume markets. There are two important international standards bodies that have developed (or are in the process of developing) low bit rate video compression standards: the International Telecommunications Union (ITU) and the Moving Pictures Experts Group (MPEG) committee of International Standardization Organization (ISO).

The International Telecommunications Union has developed two related standards for the particular needs of low bit rate video coding for videoconferencing and videophone applications. The initial video coding standard, H.261, was approved in 1990, and was intended for use on ISDN lines at some integer multiple of 64 kbps. The overall system standard, which uses the H.261 video codec, is H.320.

Work started in 1992 to enhance and modify the codec to be useful at bit rates lower than 64 kbps, with the goal of being able to operate over the Public Switched Telephone Network (PSTN), that is, ordinary consumer telephone lines. The new video codec (H.263) was finalized in November 1995 together with the new overall system standard, H.324. H.263 is a very impressive standard for low bit rate video coding. At similar bit rates, H.263 provides roughly twice the quality of H.261. Although it is currently used only in the low bit rate videophone market, it is expected that the H.263 will soon replace H.261 in the videoconferencing market also.

Both H.261 and H.263 use essentially similar underlying technology. They are both block-based coders and use block motion compensation followed by 8×8 DCT. H.263 uses an advanced motion compensation technique called overlapped block motion compensation (OBMC) to achieve even better performance at low bit rates [21]. H.263 also has several other options such as the ability to code 8×8 or 16×16 motion vectors, allow motion vectors to point outside the frame boundaries, a syntax-based arithmetic coder for the entropy coding, a three dimensional VLC table for coding the DCT coefficients, both forward and backward predicted coding of frames (PB-frames) and also a slightly different quantization scheme. All these options combine to enable H.263 to outperform H.261 at the same bit rate [22].

ISO MPEG-4 [23] is very similar to ITU H.263 and is in fact based on it. For very low bit rate coding H.263 and MPEG-4 are almost identical. However, the MPEG-4 also has a number of other useful properties that include much improved error resilience for transmission over noisy communication channels, the ability to also operate efficiently at much higher bit rates (tested up to 6 MB), the ability to code arbitrary shaped objects (not just rectangular frames) and provisions to code computer generated video in addition to natural video [12]. In addition, the MPEG-4 coder can also handle interlaced video material for high bit rate broadcast applications.

19.7. CONCLUSIONS

Recent technological advances have enabled a number of compelling applications for digital video and there is an increasing demand for technology capable of coding good quality video at low bit rates. Current video compression techniques and standards (MPEG-4, H.263, and H.261) have improved significantly and provide methods of achieving good video quality at very low bit rates (below 28 kbps). Object-based coding techniques have also matured end enable a number of new functionalities such as content-based manipulation and scalability. The object-based coding techniques have slowly found their way into the international standards such as MPEG-4. Error resilience aspects of video coding have also improved making it possible to transmit compressed video over noisy communication channels such as wireless links. Model-based communication is a future direction of research in very low bit rate video coding that promises to take us to even lower bit rates (in the 2–3 kbps range).

REFERENCES

[1]   Watson, A. B. (ed.), *Digital Images and Human Vision*. MIT Press, Cambridge, MA, 1993.
[2]   Watson, A. B., DCT quantization matrices visually optimized for individual images. In Rogowit, B. (ed.), *Human Vision, Visual Processing, and Digital Display IV*, Proc. SPIE 1913–14, 1993.

[3]   Shapiro, J. M., Embedded image coding using zerotrees of wavelet coefficients, *IEEE Trans. Signal Processing*, **41** (12), 3445–3462, 1993.

[4]   ISO/IEC/JTC1/SC29/WG11 CD 11172, *Information technology – Coding of moving pictures and associated audio for digital storage media up to about 1.5 Mbps – Part 2: Coding of moving picture information*, International Organization for Standardization, 1991.

[5]   ISO/IEC/JTC1/SC29/WG11 DIS 13818-2, *Information technology – Generic coding of moving pictures and associated audio information – Part 2: Video*, International Organization for Standardization, 1994.

[6]   ITU-T Recommendation H.261, *Video Codec for audiovisual services at 64-1920 Kbps*, 1993.

[7]   Rijkse, K., ITU standardization of very low bitrate video coding algorithms. *Signal Processing: Image Communications*, 7, 553–565, 1995.

[8]   Draft ITU-T Recommendation H.263, *Video coding for low bitrate communication*, May 1996.

[9]   Musmann, H. G., A layered coding scheme for very low bitrate video coding. *Signal Processing: Image Communication*, 7, 267–278, 1995.

[10] Gerken, P., Object-based analysis-synthesis coding of image sequences at very low bit rates. *IEEE Trans. Circuits and Systems for Video Technology*, 4, 228–235, June 1994.

[11] Ebrahimi, T., Reusens, M. and Li, W., New trends in very low bit rate video coding. *Proc. IEEE*, **83**, 877–891, 1995.

[12] Talluri, R., MPEG4 – status and directions. In *SPIE Proceedings* **CR60**, *Standards and Common Interfaces for Information Systems*, Philadelphia, PA, October 25–26, 1995.

[13] Talluri, R. and Flinchbaugh, B., Video coding below twenty kilobits per second. *Proc. European Conference on Multimedia Applications, Services and Techniques, ECMAST '96*, Louvain-la-Neuve, Belgium, May 1996.

[14] Talluri, R., Oehler, K., Bannon, T., Courtney, J., Das, A and Liao, J., A robust, scaleable, object-based video coding technique for very low bitrate video coding. *IEEE Transactions on Circuits and Systems for Video Technology*, 7 (1), 221–233, 1997.

[15] Talluri, R., A hybrid object-based video compression technique. *Proc. International Conference on Image Processing, ICIP '96*, Lausanne, 387–390, September 1996.

[16] Irani, M., Ruso, B. and Peleg, S., Computing occluding and transparent motions. *International Journal of Computer Vision*, **12**, 5–16, 1994.

[17] Medioni, G. and Yasumoto, Y., Corner detection and curve representation using cubic b-splines. *Computer Vision, Graphics and Image Processing*, 39, 267–278, 1987.

[18] Aizawa, K. and Huang, T., Model-based image coding: Advanced video coding techniques for very low bitrate applications. *Proc. IEEE*, **83** (2), 259–271, 1995.

[19] Essa, I. A. and Pentland, A. P., A vision system for observing and extracting facial action parameters. In *Proceedings of the Computer Vision and Pattern Recognition Conference*, 76–83, 1994.

[20] Terzopoulos, D. and Waters, K., Physically based facial modeling, analysis, and animation. *Journal of Visualization and Computer Animation*, **1** (2), 73–80, 1990.

[21] Orchard, M. T. and Sullivan, G. J., Overlapped block motion compensation: an estimation-theoretic approach. *IEEE Trans. Image Processing*, **3** (5), 693–699, 1994.

[22] Girod, B., Steinbach, E. and Farber, N., Comparison of H.623 and H.261 video compression standards. *SPIE Proceedings Vol. CR60, Standards and Common Interfaces for Information Systems*, Philadelphia, PA, October 25–26, 1995.

[23] ISO/IEC/JTC1/SC29/WG11 MPEG97/N1642, MPEG4 video verification Model Version 7.0, Bristol, April 1997.

20

A Narrowband Mobile Multimedia System*

L. Hanzo, J. Streit, R. A. Salami and W. Webb

20.1. INTRODUCTION

While the second generation digital mobile radio systems are being deployed in Europe, throughout the Pacific Rim and the United States, researchers turned their attention towards the true personal communication system (PCS) of the near future. This chapter is devoted to specific algorithmic and system aspects of a

---

*This chapter is partially based on L. Hanzo *et al.*: A low-rate multi-level voice/video transceiver for personal communications, *Wireless Personal Communications*, Kluwer Academic Publishers, **2** (3), 217–234, 1995.

Insights into Mobile Multimedia Communication
ISBN 0-12-140310-6

complete voice/video phone transceiver suitable for the future third generation PCS. The system bandwidth was assumed to be 30 kHz, as in the American IS-54 standard system, which allowed us to assess the potential of the proposed scheme in comparison to a well known benchmarker.

Sections 20.2 and 20.3 address speech and video compression issues, while section 20.4 is focused on the choice of modulation, in particular on 16-level quadrature amplitude modulation (16-QAM) [1]. Forward error correction coding (FEC) is considered in section 20.5. The proposed transceiver scheme is described in section 20.6 and the system performance is analysed in section 20.7, before some conclusions are drawn in section 20.8.

20.2. SPEECH CODING ISSUES

20.2.1. Advances in Speech Compression

Let us commence our discourse with a rudimentary overview of the recent speech compression literature [2]–[24]. For an in-depth discourse the interested reader is referred to these References. Following the International Telecommunications Union's (ITU) 64 kbits/s pulse code modulation (PCM) and 32 kbps adaptive PCM (ADPCM) G.721 standards, in 1986 the 13 kbits/s regular pulse excitation (RPE) [4,5] codec was selected for the Pan-European mobile system known as GSM, and more recently vector sum excited linear prediction (VSELP) [6,7] codecs operating at 8 and 6.7 kbits/s were favoured in the American IS-54 and the JDC wireless networks. These developments were followed by the 4.8 kbits/s American Department of Defence (DoD) codec [8]. The state-of-art was documented in a range of excellent monographs by O'Shaughnessy [9], Furui [10], Anderson and Mohan [11], Kondoz [12], Kleijn and Paliwal [13] and in a tutorial review by Gersho [14]. More recently the 5.6 kbits/s half-rate GSM Vector Sum Excited Linear Predictive (VSELP) speech codec standard developed by Gerson et al. [15] was approved, while in Japan the 3.45 kbits/s half-rate JDC speech codec invented by Ohya et al. [16] using the so-called pitch synchronous innovation (PSI) CELP principle was standardised. Other currently investigated schemes are the so-called prototype waveform interpolation (PWI) proposed by Kleijn [17], multi-band excitation (MBE) suggested by Griffin et al. [18] and interpolated zinc function prototype excitation (IZFPE) codecs advocated by Hiotakakos and Xydeas [19]. In the low-delay, but more error sensitive backward adaptive class the 16 kbps ITU G.728 codec [20] developed by Chen et al. from the AT&T speech team hallmarks a significant step. This was followed by the equally significant development of the more robust, forward-adaptive 10 ms delay G.728 ACELP arrangement proposed by the University of Sherbrook team [22,23], AT&T and NTT [24]. Lastly, the standardisation of the 2.4 kbps DoD codec led to intensive research in this very low-rate range and the mixed excitation linear predictive (MELP) codec by Texas Instrument was identified [25] in 1996 as the best overall candidate scheme. Following this brief speech coding review let us now concentrate on the specific 4.8 kbps codec proposed for our "IS-54-like" system.

20.2.2. The 4.8 kbit/s Speech Codec

In this brief exposure we follow the approach of Salami *et al.* [26,27], omitting the mathematical details for reasons of space economy. In code excited linear predictive (CELP) codecs a Gaussian process with slowly varying power spectrum is used to represent the residual signal after short-term and long-term prediction, and the speech waveform is generated by filtering Gaussian excitation vectors through the time-varying linear pitch and LPC synthesis filters. The Gaussian excitation vectors of dimension N are stored in a codebook or generated in real time without actually storing them and the optimum excitation sequence is determined by the exhaustive search of this virtual or pre-stored excitation codebook. The codebook entries $c_k(n)$, $k = 1, \ldots, L, n = 0, \ldots, N-1$, after scaling by a gain factor G_k, are filtered through the pitch synthesis filter $1/P(z)$ and the perceptual error weighting filter $W(z) = 1/A(z/\gamma)$ to produce the so-called weighted synthetic speech $\tilde{s}_w(n)$, which is compared to the weighted original speech $s_w(n)$ [26,27].

Let $x(n)$ be the weighted original speech after removing the memory contribution of the concatenated pitch synthesis and error weighting filters $W(z) \cdot 1/P(z)$ from previous frames and $h(n)$ be the impulse response of the weighting filter $W(z)$. Then the mean squared weighted error (mswe) between the original and synthesised speech is given by:

$$E = \sum_{n=0}^{N-1}[x(n) - G_k c_k(n) * h(n)]^2 \tag{20.1}$$

Setting $\partial E/\partial G_k = 0$ leads to the mean square weighted error expression [26,27]:

$$E_{min} = \sum_{n=0}^{N-1}x^2(n) - \frac{\left[\sum_{i=0}^{N-1}\psi(i)c_k(i)\right]^2}{\sum_{i=0}^{N-1}c_k^2(i)\,\phi(i,i) + 2\sum_{i=1}^{N-2}\sum_{j=i+1}^{N-1}c_k(i)c_k(j)\,\phi(i,j)} \tag{20.2}$$

where $\psi(i)$ represents the correlation between the weighting filter's impulse response $h(n)$ and the signal $x(n)$, given by $\psi(n) = x(n) * h(-n)$, while $\phi(i,j)$ represents the covariances of $h(n)$:

$$\phi(i,j) = \sum_{n=0}^{N-1}h(n-i)h(n-j) \tag{20.3}$$

The best innovation sequence is constituted by that codebook entry c_k with index k ($k = 1, \ldots, L$), which minimises the mean squared weighted error in equation 20.2.

Since $\psi(n)$ and $\phi(n)$ are computed outside the error minimisation loop, the computational complexity is predetermined by the number of operations needed to evaluate the second term of equation (20.2) for all the codebook entries. In our PCS transceiver we favoured a so-called transformed binary pulse excited (TBPE) speech codec proposed by Salami [26]. The great attraction of TBPE codecs, when compared to conventional CELP codecs, accrues from the fact that the excitation optimisation can be achieved in a direct computation step [26,27], as will be highlighted below. The sparse Gaussian excitation vector is assumed to take the form of

$$\mathbf{c} = \mathbf{Ab} \tag{20.4}$$

where the binary vector $\mathbf{b}$ has M elements of ± 1, while the $M \times M$ matrix $\mathbf{A}$ represents an orthogonal transformation. Due to the orthogonality of $\mathbf{A}$ the binary excitation pulse vector $\mathbf{b}$ is transformed into a vector $\mathbf{c}$, constituted by independent, unit variance Gaussian components. The set of 2^M binary excitation vectors gives rise to 2^M Gaussian vectors of the original CELP codec.

The block diagram of the TBPE codec is shown in Figure 20.1. As seen in the Figure, the weighted synthetic speech is generated for all $2^M = 1024$ codebook vectors and subtracted from the weighted input speech in order to find the one resulting in the best synthesized speech quality. The synthetic speech is generated at the output of the weighted synthesis filter, which is excited by the vectors given by the superposition of the adaptive codebook vector scaled by the long-term predictor gain (LTPG) and that of the orthogonally transformed binary vectors output by the binary pulse generator scaled by the excitation gain (EG).

The direct excitation computation of the TBPE codec accrues from the matrix representation of equation (20.2) using equation (20.4), which can be expressed as:

$$E = \mathbf{x}^T \mathbf{x} - \frac{(\boldsymbol{\Psi}^T \mathbf{A} \mathbf{b})^2}{\mathbf{b}^T \mathbf{A}^T \boldsymbol{\Phi} \mathbf{A} \mathbf{b}} \tag{20.5}$$

Salami showed that the denominator in equation (20.5) is nearly constant over the entire codebook and hence plays practically no role in the excitation optimisation. This is due to the fact that the autocorrelation matrix $\boldsymbol{\Phi}$ is strongly diagonal, since the impulse response $h(n)$ decays sharply. Due to the orthogonality of $\mathbf{A}$ we have $\mathbf{A}^T \mathbf{A} = \mathbf{I}$, where $\mathbf{I}$ is the identity matrix, rendering the denominator near-constant.

Closer scrutiny of equation (20.5) reveals that its second term reaches its maximum if the binary vector element is given by $b(i) = -1$, whenever the vector element $\boldsymbol{\Psi}^T \mathbf{A}$ is negative, and *vice versa*, i.e., $b(i) = +1$ if $\boldsymbol{\Psi}^T \mathbf{A}$ is positive. The numerator of equation (20.5) is then constituted by exclusively positive terms, i.e., it is maximum,

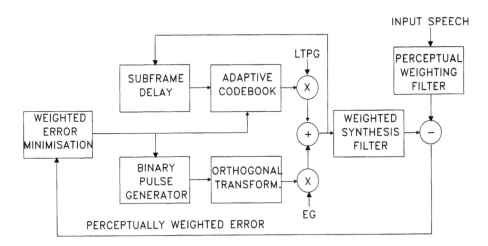

Figure 20.1 Block diagram of the 4.8 kbit/s TBPE codec (© Kluwer, 1995, Hanzo et al. [28]).

and the weighted mean squared error is minimum. The optimum Gaussian excitation is computed from the binary vector **b** using equation (20.4) in both the encoder and decoder. Only the M-bit index representing the optimum binary excitation vector **b** has to be transmitted. The evaluation of the vectors $\mathbf{\Psi}^T\mathbf{A}$ and $\mathbf{c} = \mathbf{Ab}$ requires $2M^2$ multiplications/additions, which gives typically five combined operations per output speech sample, a value 400 times lower than the complexity of the equivalent quality CELP codec.

The bit allocation of our TBPE codec is summarised in Table 20.1. The spectral envelope is represented by ten line spectrum frequencies (LSFs) which are scalar quantised using 36 bits. The 30 ms long speech frames having 240 samples are divided into four 7.5 ms subsegments having 60 samples. The subsegment excitation vectors **b** have 12 transformed duo-binary samples with a pulse-spacing of $D = 5$. The LTP delays (LTPD) are quantised with seven bits in odd and five bits in even indexed subsegments, while the LTP gain (LTPG) is quantised with three bits. The excitation gain (EG) factor is encoded with four bits, and the grid position (GP) of candidate excitation sequences by two bits. A total of 28 or 26 bits per subsegment is used for quantisation, which yields $36 + 2 \cdot 28 + 2 \cdot 26 = 144$ bits/30 ms, i.e. a bitrate of 4.8 kbit/s.

The TBPE codec was subjected to rigorous bit-sensitivity analysis [29] and the bits were assigned to three sensitivity classes, namely classes 1–3 (C1–C3) for embedded source-matched forward error correction to be detailed in section 20.5. Having described the proposed 4.8 kpbs TBPE speech codec we note that a variety of other similar-rate codecs can be invoked in our system, such as the 4.8 kbps DoD codec [8] or the 2.4 kbps DoD MELP codec [25], which would double the number of users accommodated.

20.3. VIDEO COMPRESSION ISSUES

20.3.1. Advances in Video Compression

The theory and practice of image compression has been consolidated in a number of established monographs by Netravali and Haskell [30], Jain [31], Jayant and Noll [32]. A plethora of video codecs have been proposed in the excellent special

Table 20.1 Bit allocation of the 4.8 kbps TBPE codec (© Kluwer, 1995, Hanzo *et al.* [28])

Parameter	Bit number
10 LSFs	36
LTPD	$2 \cdot 7 + 2 \cdot 5$
LTPG	$4 \cdot 3$
GP	$4 \cdot 2$
EG	$4 \cdot 4$
Excitation	$4 \cdot 12$
Total	144/20 ms

issues edited by Tzou et al. [33], by Hubing [34] and Girod et al. [35] for a range of bitrates and applications, but the individual contributions by a number of renowned authors are too numerous to review. Khansari et al. [36] as well as Pelz [37] reported promising results on adopting the International Telecommunications Union's (ITU) H.261 codec for wireless applications. They showed by invoking powerful signal processing and error-control techniques how to remedy the inherent run length coding induced error-sensitivity problems when its application domain is stretched from benign Gaussian wireline links to hostile wireless fading environments. Further important contributions in the field were due to Chen et al. [38], Illgner and Lappe [39], Zhang [40], Ibaraki et al. [41], Watanabe et al. [42] etc. and the MPEG4 consortium's endeavours [43]. The European Advanced Communication Technologies and Services (ACTS) programme has also fostered ambitious objectives in terms of creating mobile multimedia systems.

The ITU H.263 [44,45] scheme is in many respects similar to the H.261 codec, but it incorporates a number of recent advances in the field, such as using half pixel resolution in the motion compensation, or configuring the codec for a lower data rate or better error resilience. Furthermore, four so-called negotiable coding options were introduced. In references [46,47] Cherriman and Hanzo reported on the design of a low-rate video transceiver, where the H.263 codec was constrained to operate at a constant bitrate using an appropriate packetisation algorithm. This controlled the codec's quantisers such that it would output the required number of bits per frame for transmission by a multimode transceiver. As a design alternative, in References [3,48,49,50] Streit et al. proposed a range of fixed but arbitrarily programable-rate video codecs specially contrived for videotelephony over existing and future mobile radio speech systems. In the next section we follow the approach of Streit et al. [28,48].

20.3.2. The Fixed-rate Video Codec

The video codec's outline is depicted in Figure 20.2. It was designed to achieve a time-invariant compression ratio associated with an encoded video rate of 8.52 kbps.[1] The codec's operation is initialised in the intra-frame mode, but once it switches to the inter-frame mode, any further mode switches are optional and only required if a drastic scene change occurs.

In the intra-frame mode the encoder transmits the coarsely quantised block averages for the current frame, which provides a low-resolution initial frame required for the operation of the inter-frame codec at both the commencement and during later stages of communications in order to prevent encoder/decoder misalignment. For 176×144 pixel CCITT standard Quarter Common Intermediate Format (QCIF) images we limited the number of video encoding bits per frame to 852. In order to transmit all block averages in the initial intra-frame mode with a 4-bit resolution, while not exceeding the 852 bits/frame rate the intra-frame update block size is fixed to 11 × 11 pixels.

---

[1]The associated video quality at various bitrates can be viewed under the WWW address http://www-mobile.ecs.soton.ac.uk

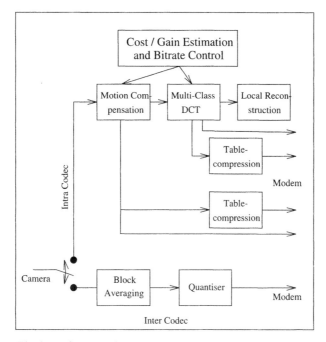

Figure 20.2 The intra-frame codec schematic (© Kluwer, 1995, Hanzo *et al.* [28]).

In the motion-compensation 8×8 blocks are used. At the commencement of the encoding procedure the motion compensation (MC) scheme determines a motion vector (MV) for each of the 8×8 blocks using full search [31]. The MC search window is fixed to 4×4 pels around the centre of each block and hence a total of 4 bits are required for the encoding of 16 possible positions for each MV. Before the actual motion compensation takes place, the codec tentatively determines the potential benefit of the compensation in terms of motion compensated error energy reduction. Then the codec selects those blocks as "motion-active" whose gain exceeds a certain threshold. This method of classifying the blocks as motion-active and motion-passive results in an active/passive table, which consists of a one bit flag for each block, marking it as passive or active. In the case of 8×8 blocks and 176×144 pel images this table consists of 396 entries which is compressed using the following technique.

First the whole table is grouped in 2×2 blocks and a 4-bit symbol is allocated to those blocks which contain at least one active flag. These symbols are then run length encoded and transmitted to the decoder. This concept requires a second active table containing $396/4 = 99$ flags in order to determine which of the two-by-two blocks contain active vectors. Three consecutive flags in this table are packetised to a symbol and then run length encoded. As a result, a typical 396-bit active/passive table containing about 30 active flags, corresponding to 30 motion-active blocks, can be compressed to less than 150 bits. Due to their low correlation the motion vectors themselves are not run length encoded. If at this stage of the encoding process the number of bits allocated to the compressed tables and active motion vectors exceeds half of the total number of available bits/frame, a number of blocks satisfying the motion-active criterion will be relegated to the motion-passive class. This process takes account of the subjective importance of various blocks and does not ignore motion-active blocks in the central eye and lip regions of the image, while relegating those which are closer to the fringes of the image.

Pursuing a similar approach, gain control is also applied to the discrete cosine transform (DCT) based compression [31]. Every block is DCT transformed and quantised. In order to take account of the non-stationary nature of the motion-compensated residual error (MCRE) and its time-variant frequency-domain distribution, four different sets of DCT quantisers were designed. The quantisation distortion associated with each of the four DCT quantisers is computed in order to be able to choose the best one. Ten bits are allocated for each quantiser, all of which are trained Max-Lloyd quantisers catering for a specific frequency-domain energy distribution class. All DCT blocks whose coding gain exceeds a certain threshold are marked as DCT-active resulting in a similar active/passive table as for the motion vectors. For this second table we apply the same run length compression technique, as above. Again, if the number of bits required for the encoding of the DCT-active blocks exceeds half of the total maximum allowable number for the video frame concerned, blocks around the fringes of the image are considered DCT-passive, rather than those in the central eye and lip sections. If, however, the active DCT coefficient and activity-table do not fill up the fixed-length transmission burst, the thresholds for active DCT blocks is lowered and all tables are recomputed.

The encoded parameters are transmitted to the decoder and also locally decoded in order to be used in future motion predictions. The video codec's PSNR versus frame index performance is shown in Figure 20.3, where an average PSNR of about 33.3 dB was achieved for the MA sequence.

20.4. MODULATION ISSUES

In conventional mobile systems, such as the Pan-European GSM system [51] or the Digital European Cordless Telecommunications (DECT) scheme [27] constant envelope partial response Gaussian minimum shift keying (GMSK) [27] is employed. Its main advantage is that it facilitates the utilisation of power-efficient non-linear class-C amplification. In third generation personal communication systems, however, benign pico- and micro-cells will be employed, where low transmitted power and low signal dispersion are characteristic. Hence the employment of more bandwidth efficient multilevel modulation schemes becomes realistic. In fact the American and Japanese second generation digital systems have already opted for 2 bits/symbol modulation.

Multi-level modulation schemes have been considered in depth in reference [1] and we have shown that the bandwidth efficiency and minimum required signal-to-noise ratio (SNR) and signal-to-interference ratio (SIR) of a modulation scheme in a given frequency re-use structure is dependent on the bit error ratio (BER) targeted. The required BER in turn is dependent on the robustness of the source codecs used. Furthermore, in indoors scenarios the partitioning walls and floors mitigate the co-

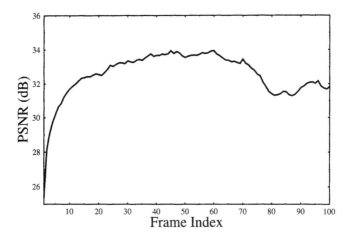

Figure 20.3 PSNR versus frame index performance of the 8.52 kbps video codec for the "Miss America" sequence (© Kluwer, 1995, Hanzo *et al.* [28]).

channel interference and this facilitates the employment of 16-level quadrature amplitude modulation (16QAM). In the proposed system we have opted for differentially detected Star 16-QAM [1], noting that the required channel SNR can be typically reduced by about 6 dB, when using somewhat more complex, coherently detected pilot-assisted 16QAM.

20.5. CHANNEL CODING

Both convolutional and block codes have been successfully used to combat the bursty channel errors [27]. In the proposed system we have opted for binary BCH codes, since they exhibit reliable error detection, which is useful in controlling handovers and error concealment. A set of appropriate FEC codes is constituted by the BCH5 = BCH(63,36,5), BCH2 = BCH(63,51,2) and the BCH1 = BCH(63,57,1) codes, correcting 5, 2 and 1 bits per 63-bit frame, respectively. Accordingly, the most sensitive so-called class 1 (C1) 36 speech bits are protected by the powerful BCH(63,36,5) code, while the less vulnerable 51 class 2 (C2) and 57 class 3 (C3) bits are encoded by the BCH(63,51,2) and BCH(63,57,1) codes, respectively. The total number of protected bits is 144. The packet header conveying control information is also BCH(63,36,5) coded, hence $4 \cdot 63 = 252$ bits per 30 ms are transmitted. After adding six ramp-symbols in order to assist the transceiver in its attempt to mitigate spurious adjacent channel emissions, the total bit rate becomes 8.6 kbit/s, yielding a signalling rate of 2.15 kBd.

The above FEC scheme has the advantage of curtailing BCH decoding error propagation across speech frame boundaries, although the speech codec's memory will inherently propagate errors. For a propagation frequency of 1.9 GHz, as in the future PCS, the wavelength is about 15 cm, and therefore interleaving over an interval of about 40 cm travelling distance ensures adequate error randomisation for the FEC scheme to operate efficiently. This corresponds to a transmission frame duration of 30 ms at a velocity of about 13 m/s or 30 m.p.h. However, for lower pedestrians speeds there is a danger of idling in deep fades, in which case the employment of a switch-diversity scheme is essential.

Let us now consider the FEC aspects of the video codec. The 852-bit video frame is encoded using 12 BCH(127,71,9) code words, yielding a total of 1524 bits. A pair of such code words form a video packet of 254 bits, which is expanded by four ramp symbols to deliver a 258-bit/30 ms video packet at a corresponding rate of 2.15 kBd. Six such video packets are needed to deliver the 1524-bit BCH-coded video frame, but during the 90 ms video frame repetition time there are only three 30 ms multiple access frames. This implies that two reserved time-slots per multiple access frame are required for video users. This is equivalent to a video signalling rate of $2 \times 2.15 = 4.3$ kBd. Having resolved the choice of FEC codecs let us now consider how the system components are amalgamated in the transceiver.

20.6. THE PROPOSED PCS TRANSCEIVER

The block diagram of the proposed PCS transceiver is shown in Figure 20.4. The TBPE encoder outputs a 4.8 kbit/s bit stream, which is mapped in three bit sensitivity classes, C1, C2 and C3. The bits belonging to these classes are FEC encoded along with the packet packet header conveying control information by the BCHE1 = BCH(63,36,5), BCHE2 = BCH(63,51,2) and BCHE1 = BCH(63,57,1) encoders, respectively. The 258 bits/30 ms = 8.6 kbit/s FEC-coded speech packets are then transmitted at 2.15 kBd using differentially encoded 16-QAM [1]. Packet reservation multiple access (PRMA) is used in the uplink, which is not detailed here due to lack of space. The interested reader is referred to Reference [1], for example. Suffice to say here that only active speech spurts are queued by the voice activity detector [27] (VAD) for transmission to the base station. If there is no other portable station (PS) attempting to acquire a slot reservation, the PS is allocated this particular time slot for its future communication. In case of collision further contention is enabled with a less than unity permission probability, until either a slot is reserved or the speech packet's life-span expires. The microcellular PSs receive the slot-status near-instantaneously, implying negligible propagation delays. If this cannot be ensured, adaptive time frame alignment must be used, as proposed for the Pan-European GSM system [51].

The transmission Baud rate of our transceiver was fixed to 20 kBd, in order for the PRMA signal to fit in a 30 kHz channel slot, as in the IS-54 system, when using a modem excess bandwidth of 50% [1]. Hence our transceiver can accommodate TRUNC(20 kBd/2.1 kBd) = 9 time slots, where TRUNC represents truncation to the nearest integer. Again, the slot duration was 30 ms/9 ≈ 3.33 ms and one of the PRMA users was transmitting speech signals recorded during a telephone conversation, while all the other users generated negative exponentially distributed speech spurts and speech gaps with mean durations of 1 and 1.35 s. The PRMA parameters

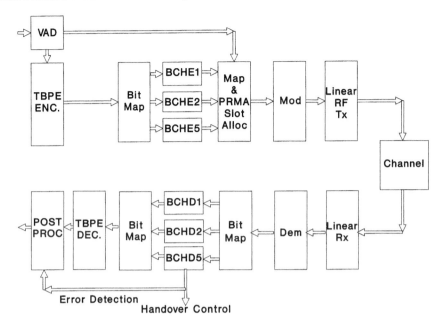

Figure 20.4 Transceiver schematic (© Kluwer, 1995, Hanzo *et al.* [28]).

used are summarised in Table 20.2. Observe that two time slots are reserved for a videophone user and 7 slots are dedicated to PRMA for the speech users. Let us now consider how the above system components are integrated in the proposed transceiver.

Again, the 852 bits per frame video encoded stream is BCH(127,71,9) coded to 1524 bits/frame and transmitted at 4.3 kBd, which is equivalent to the signalling rate of two speech users. Clearly, for video telephony two time-slots are required. The

Table 20.2 Summary of PRMA/TDMA parameters (© Kluwer, 1995, Hanzo *et al.* [28])

PRMA parameter	
Channel rate	20 kBd
Speech rate	2.15 kBd
Video rate	4.3 kBd
Frame duration	30 ms
Total no. of slots	9
No. of PRMA slots	7
No. of TDMA slots	2
Slot duration	3.33 ms
Header length	63 bits
Maximum speech delay	32 ms
Speech perm. prob.	0.6

video transceiver obeys the structure of Figure 20.4, simply the TBPE speech encoder must be replaced by the DCT video codec and no bit mapping is invoked, since a single-class BCH(127,71,9) codec is used.

The receiver seen in Figure 20.4 carries out the inverse functions of the transmitter. The error detection capability of the strongest BCH(63,36,5) decoder is exploited to initiate handovers and to invoke speech post-processing, if the FEC decoder happens to be overloaded due to interference or collision. The system elements of Figure 20.4 were simulated and the main transceiver parameters are summarised in Table 20.3, while the system performance will be characterised in the following section.

20.7. RESULTS AND DISCUSSION

The system performance was evaluated over narrowband Rayleigh-fading channels characterised by a propagation frequency of 1.9 GHz, vehicular speed of 30 m.p.h. and signalling rate of 20 kBd. The packet dropping probability versus number of users curve of the proposed system is portrayed in Figure 20.5. Observe that about 10–11 users can be supported by our 7-slot PRMA scheme with a packet dropping probability of $P_{drop} < 1\%$, a value inflicting almost negligible speech degradation. The overall objective SEGSNR degradation (SEGSNR-DEG) versus channel SNR performance of our diversity-assisted PCS transceiver is displayed in Figure 20.6 parameterised with the number of PRMA users supported. While for 7–10 users no speech degradation can be observed, if the channel SNR is in excess of about 24 dB, for 12 users the SEGSNR-DEG due to PRMA packet dropping becomes noticeable, although not subjectively objectionable. In case of 14 users, however, there is a consistent SEGSNR-DEG of about 1 dB due to the 4–5% packet dropping probability seen in Figure 20.5. Without diversity about 5 dB higher channel SNR is necessitated in order to achieve a similar performance to that of the diversity-assisted scheme. This is similar to the performance of a coherent-detection PSAM modem without diversity [1]. The PSNR versus channel SNR performance of the diversity-assisted video transceiver is portrayed in Figure 20.7, where in harmony with the voice transceiver, a channel SNR of about 22–25 dB is required for near-unimpaired video quality. Without diversity the video scheme

Table 20.3 Speech Transceiver Parameters (© Kluwer, 1995, Hanzo *et al.* [28])

1 Speech rate (kbps)	2 PRMA source rate (kBd)	3 TDMA user bandwidth (kHz)	4 No. of TDMA users/ carrier	5 No. of PRMA users/ carrier	6 No. of PRMA users/slot	7 PRMA user bandwidth (kHz)	8 Min SNR and SIR (dB)
4.8	2.15	4.3	7	10	1.43	3	24

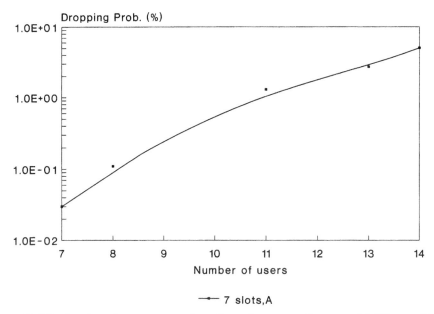

Figure 20.5 Packet dropping versus number of speech users performance (© Kluwer, 1995, Hanzo *et al.* [28]).

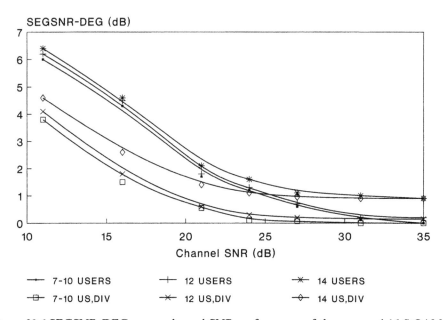

Figure 20.6 SEGSNR-DEG versus channel SNR performance of the proposed 16-StQAM transceiver at 30 m.p.h. with and without diversity parameterised with the number of PRMA users supported (© Kluwer, 1995, Hanzo *et al.* [28]).

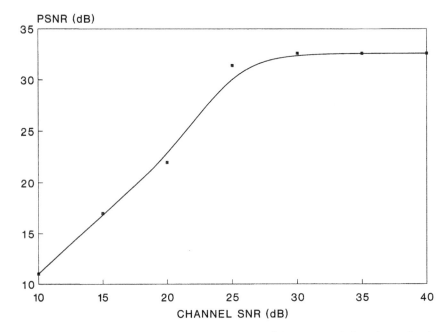

Figure 20.7 PSNR versus channel SNR performance of the proposed diversity-assisted videophone scheme (© Kluwer, 1995, Hanzo *et al.* [28]).

lacks robustness, since the corrupted run-length coded activity tables affect the whole of each video frame.

20.8. CONCLUSIONS

The potential of a bandwidth-efficient 2.15 kBd PRMA-assisted TBPE/BCH/16-QAM scheme has been investigated for employment in the future PCS under the assumption of benign channel conditions. Within the 30 kHz IS-54 bandwidth about 10 voice users plus a video telephone user can be supported, if channel SNR and SIR values in excess of about 24dB can be maintained. The main transceiver features are summarised in Table 20.3. The system performance can be further improved at the cost of higher implementational complexity, when using a more sophisticated pilot symbol assisted, block-coded coherent square 16-QAM modem. The system's bandwidth efficiency can be further improved for example by employing the 2.4 kbps DoD MELP codec [25], while its robustness can be enhanced by coherent detection. Future work will be targeted at improving the speech quality, implementational complexity, bit rate, bandwidth occupancy and error resilience trade-off achieved invoking the adaptive transceiver re-configuration algorithms we used in reference [1].

ACKNOWLEDGEMENT

The financial support of the following organisations is gratefully acknowledged: Motorola ECID, Swindon, UK; European Community, Brussels, Belgium; Engineering and Physical Sciences Research Council, Swindon, UK; Mobile Virtual Centre of Excellence, UK.

REFERENCES

[1] Webb, W. T. and Hanzo, L., *Modern Quadrature Amplitude Modulation, Principles and Applications for Fixed and Wireless Channels*, IEEE Press Pentech Press, London, 1994.

[2] Hanzo, L. and Woodard, J. P., *Modern Voice Communications: Principles and applications for fixed and wireless channels*, IEEE Press, in preparation. For detailed contents please refer to http://www-mobile.ecs.soton.ac.uk.

[3] Hanzo, L., Cherriman, P. and Streit, J., *Modern Voice Communications: Principles and applications for fixed and wireless channels*, IEEE Press, in preparation. For detailed contents please refer to http://www-mobile.ecs.soton.ac.uk.

[4] Kroon, P. and Deprettere, E. F., Regular pulse excitation – a novel approach to effective multipulse coding of speech. *IEEE Trans. on Acoustics, Speech and Signal Processing*, **34** (5), 1054–1063, 1986.

[5] Vary, P., Hellwig, K., Hofmann, R., Sluyter, R., Galland, C. and Rosso, M. Speech codec for the European mobile radio system. In *Proceedings of ICASSP*, 227–230, April 1988.

[6] Gerson, I. A. and Jasiuk, M. A., Vector sum excitation linear prediction (VSELP) speech coding at 8 kbps. In *Proceedings of ICASSP'90*, Albuquerque, New Mexico, 461–464, 3–6 April, 1990.

[7] Gerson, I. A. and Jasiuk, M. A., Vector sum excited linear prediction (VSELP). In Atal, B. S., Cuperman, V. and Gersho, A. (eds), *Advances in Speech Coding*, 69–80, Kluwer Academic Publishers, New York, 1991.

[8] Campbell, J., Welch, V. and Tremain, T., An expandable error-protected 4800 bps CELP coder (US Federal Standard 4800 bps Voice Coder). In *Proceedings of ICASSP*, 735–738, 1989.

[9] O'Shaughnessy, D., *Speech Communication, Human and Machine*, Addison-Wesley, Reading, MA, 1987.

[10] Furui, S., *Digital Speech Processing, Synthesis and Recognition*, Marcel Dekker, New York, 1989.

[11] Anderson, J. B. and Mohan, S. *Source and Channel Coding – An Algorithmic Approach*, Kluwer Academic Publishers, New York, 1991.

[12] Kondoz, A. M., *Digital Speech: Coding for Low Bit Rate Communications Systems*, Wiley, New York, 1994.

[13] Kleijn, W. B. and Paliwal, K. K., *Speech Coding and Synthesis*, Elsevier Science, Amsterdam, 1995.

[14] Gersho, A., Advances in speech and audio compression. In *Proceedings of the IEEE*, 900–918, June 1994.

[15] Gerson, I. A., Jasiuk, M. A., Muller, J. M., Nowack, J. M. and Winter, E. H., Speech and channel coding for the half-rate GSM channel. In *Proceedings ITG-Fachbericht 130, VDE-Verlag, Berlin*, 225–233, November 1994.

[16] Ohya, T., Suda, H. and Miki, T., 5.6 kbits/s PSI-CELP of the half-rate PDC speech coding standard. In *Proceeding of the IEEE Conference on Vehicular Technology*, 1680–1684, June 1994.

[17] Kleijn, W. B., Encoding speech using prototype waveforms. *IEEE Trans. on Acoustics, Speech and Signal Processing*, October, 386–399, 1993.

[18] Griffin, D. W. and Lim, J. S., Multiband excitation vocoder. *IEEE Trans. on Acoustics, Speech and Signal Processing*, August, 1223–1235, 1988.

[19] Hiotakakos, D. J. and Xydeas, C. S., Low bit rate coding using an interpolated zinc excitation model. *Proceedings of the IEEE Singapore International Conference on Communications Systems*, November, 865–869, 1994.

[20] Chen, J.-H., Cox, R. V., Lin, Y.-C., Jayant, N. and Melchner, M. J., A low-delay CELP coder for the CCITT 16 kb/s speech coding standard. *IEEE Journal on Selected Areas in Communications*, June, 830–849, 1991.

[21] Coding of Speech at 16 kbit/s Using Low-Delay Code Excited Linear Prediction. CCITT Recommendation G.728, 1992.

[22] Salami, R., Laflamme, C., Adoul, J.-P. and Massaloux, D., A toll quality 8 Kb/s speech codec for the personal communications system (PCS). *IEEE Transactions on Vehicular Technology*, August, 808–816, 1994.

[23] Coding of Speech at 8 kbit/s using conjugate-structure algebraic code-excited linear prediction (CS-ACELP), *ITU Draft Recommendation G.729*, February 1996.

[24] Kataoka, A., Adoul, J.-P., Combescure, P. and Kroon, P., ITU-T 8-kbits/s standard speech codec for personal communication services. In *Proceedings of the International Conference on Universal Personal Communications, Tokyo, Japan*, 818–822, November 1995.

[25] McCree, A. *et al.*, A 2.4 kbit/s MELP candidate for the new US federal standard. In *Proceedings of ICASSP'96, Atlanta, Georgia, US*, 200–203, 1996.

[26] Salami, R. A., "Binary pulse excitation: a novel approach to low complexity CELP coding. In *Advances in Speech Coding*, Kluwer Academic Publishers, New York, 145–156, 1991.

[27] Steele, R. (ed.), *Mobile radio communication*. IEEE Press Pentech Publishers, London, 1992.

[28] Hanzo, L. *et al.*, A low-rate multi-level voice/video transceiver for personal communications, *Wireless Personal Communications*, Kluwer Academic Publishers, **2** (3), 217–234, 1995.

[29] Hanzo, L., Salami, R., Steele, R. and Fortune, P. M., Transmission of digitally encoded speech at 1.2 KBd for PCN. *IEE Proc.-I*, **139** (1), 437–447, 1992.

[30] Netravali, A. and Haskell, B., *Digital Pictures: Representation and Compression*. Plenum Press, New York, 1988.

[31] Jain, A. K., *Fundamentals of Digital Image Processing*. Prentice-Hall, Englewood Cliffs, NJ, 1989.

[32] Jayant, N. and Noll, P., *Digital coding of waveforms, Principles and Applications to Speech and Video*. Prentice Hall, Englewood Cliffs, NJ, 1984.

[33] Tzou, K. H., Mussmann, H. G. and Aizawa, K. (eds), Special issue on very low bit rate video coding. *IEEE Transactions on Circuits and Systems for Video Technology*, **4**, 213–357, 1994.

[34] Hubing, N. (ed.), Speech and image coding. *Special Issue of the IEEE Journal on JSAC*, **10**, 793–976, 1992.

[35] Girod, B. *et al.* (ed.), Special issue on image sequence compression. *IEEE Transactions on Image Compression*, **3**, 465–716, 1994.

[36] Khansari, M., Jalali, A., Dubois, E. and Mermelstein, P., Robust low bit-rate video transmission over wireless access systems. In *Proceedings of International Communications Conference (ICC)*, 571–575, 1994.

[37] Pelz, R. M., An unequal error protected px8 kbit/s video transmission for dect. In *Proceedings of Vehicular Technology Conference*, 1020–1024, IEEE, 1994.

[38] Chen, T., A real-time software based end-to-end wireless visual communications simulation platform. In *Proceedings of SPIE Conference on Visual Communications and Image Processing*, 1068–1074, 1995.

[39] Illgner, K. and Lappe, D., Mobile multimedia communications in a universal telecommunications network. In *Proceedings of SPIE Conference on Visual Communications and Image Processing*, 1034–1043, 1995.

[40] Zhang, Y., Very low bit rate video coding standards. In *Proceedings of SPIE Conference on Visual Communications and Image Processing*, 1016–1023, 1995.

[41] H. Ibaraki *et al.*, Mobile video communication techniques and services. In *Proceedings of SPIE Conference on Visual Communications and Image Processing*, 1024–1033, 1995.

[42] K. Watanabe *et al.*, A study on transmission of low bit-rate coded video over radio links. In *Proceedings of SPIE Conference on Visual Communications and Image Processing*, 1025–1029, 1995.

[43] Sheldon, P., Cosmas, J. and Permain, A., Dynamically adaptive control system for mpeg-4. In *Proceedings of the 2nd International Workshop on Mobile Multimedia Communications*, 1995.

[44] ITU-T, Recommendation H.263: Video coding for low bitrate communication, 1996.

[45] Telenor Research and Development, P.O.Box 83, N-2007 Kjeller, Norway, H.263 Software Codec.

[46] Cherriman, P. and Hanzo, L., H261 and H263-based Programmable Video Transceivers. In *Proceedings of ICCS'96/ISPAC'96*, Singapore, Westin, 25–29, 1369–1373, 1996.

[47] Cherriman, P. and Hanzo, L., Programmable H.263-based wireless video transceivers for interference-limited environments, *IEEE Trans. on CSVT*, **8** (3), 275–286, 1998.

[48] Hanzo, L. and Streit, J., Adaptive low-rate wireless videophone schemes. *IEEE Transactions Video Technology*, **5**, 305–319, 1995.

[49] Streit, J. and Hanzo, L., Quad-tree based parametric wireless videophone systems. *IEEE Transactions on CSVT, Video Technology*, **6** (2), 225–237, 1996.

[50] Streit, J. and Hanzo, L., Vector-quantised low-rate cordless videophone systems. *IEEE Transactions on Vehicular Technology*, **47** (2), 340–357, 1997.

[51] Hanzo, L. and Stefanov, J., The Pan-European digital cellular mobile radio system – known as GSM. In Steele, R. (ed.), *Mobile Radio Communication*, IEEE Press Pentech Publishers, Chapter 8, 677–768, London, 1992.

21

Wavelet-based Video Compression at Low Bit Rates*

J. H. Wilkinson

21.1. INTRODUCTION

Although most current video coding standards such as MPEG-1, MPEG-2, H.261 and H.263 are based on a block-based discrete cosine transform (DCT), a significant amount of research work has demonstrated the benefits of alternative sub-band decompositions such as those based on the *wavelet transform*. Previous unpublished work by the author has shown that the use of wavelet coding combined with motion vector assisted logarithmic temporal decimation is capable of producing pictures which could be described as "entertainment" quality at 1.5 Mbps and "information" quality at 256 kbps. This earlier work has recently been extended to cover the lower bit rate of 64 kbps for motion video. This bit rate is low enough for videophone and

---

*This chapter is taken from "Motion video compression at very low bit rates", *Fifth International Conference on Processing and its Applications*, Edinburgh, July 1995. Published by the Institution of Electrical Engineers and reproduced with permission.

future wireless multimedia applications where there is a requirement for acceptable "communications" quality rather than broadcast reproduction fidelity.

This chapter describes a video coding method based on the use of "I" and "B" frames within a logarithmic decimation structure. Spatial decimation is performed by wavelet filtering and the quantised output is processed using an entropy coder similar to that used in JPEG.

The key areas developed in this chapter are:

- *Open-loop coding.* This can only be used with a temporal compression system based on "I" and "B" frames. The meaning of the terms "open-loop" and "closed-loop" will be described later.
- *Windowed motion vector compensation.* This produces considerable benefits at very low data rates where "blockiness" due to the macroblock structure of the motion compensation becomes evident even in a wavelet coded image sequence. This approach does not result in reduced reconstruction error; rather, the error is made less annoying to the viewer.
- *Temporal weighting.* With the "open-loop" method of coding, comes the possibility of weighting the reconstruction errors in a frame to ensure that low temporal frequency errors exceed the higher frequency errors. Unlike spatial coding, it is apparent that high frequency temporal errors are more visible than their low frequency counterparts. This effect was first observed in motion JPEG systems and can be exploited to improve perceived picture quality.
- *Noise coring.* This has been examined in the context of low bit-rate coding where the conventional method of quantisation provides too severe a reduction in quality. An improved technique is described in this chapter.

21.2. VIDEO CONVERSION AND DISPLAY

The source data used in this work originated at CCIR601 levels with the following key parameters:

- *luminance*: 576 × 720 pixels at 50 Hz interlaced field rate;
- *chrominance*: 576 × 360 pixels at 50 Hz interlaced field rate for each chrominance component.

The coding algorithms described, however, operate with QCIF source pictures at a frame rate of 12.5 Hz. The conversion from CCIR601 to QCIF and vice versa was achieved using half band filters. The simplest of these has the coefficient set: 1, 2, 1. However, whilst commonly used, such filters introduce a high level of pixel aliasing and hence higher order half-band filters have been applied in this study. Although more complex to implement, the results obtained with these are visibly superior. The 720 horizontal pixels of the CCIR601 source files were reduced to 704 by removing 8 pixels from the left and right hand sides of the picture. This facilitates conversion to CIF and QCIF via the application of 2:1 decimation filters.

The first conversion process, from 704 × 576, 4:2:2 format pictures to 352 × 288, 4:2:0 CIF pictures was achieved using 9 tap half-band decimation and reconstruction filters. Each chrominance component was subjected to an additional 2:1 vertical sub-sampling to make near-equal horizontal and vertical bandwidths (as for luminance). The conversion process from CIF to QCIF pictures used the same process but with a 5 tap filter for both decimation and reconstruction to minimise ringing artifacts.

The source rate of QCIF at 25 frames per second is too high for coding at 64 kbps. Rather than reduce the spatial resolution further, it is preferable to achieve source rate reduction by lowering the frame rate from 25 Hz to 12.5 Hz. Trials with simple sub-sampling for the decimation process, followed by frame replication for the interpolation process, showed clear temporal aliasing effects in the form of judder at an unacceptable level. The simple low pass temporal filter, 1-2-1, was found to provide adequate quality for both the decimation and interpolation stages and is simple to implement in both software and hardware.

The source rate before compression was: $(176 \times 144 \times 12.5) + 2 \times (88 \times 72 \times 12.5) = 475\,200$ bytes/s or 3.8016 Mbps. This is sufficiently low to make compression to 48 kbps for the video data alone (excluding motion vectors and audio) viable at about 0.15 bits/pixel.

For display of the results, the above algorithms were used to up-convert to CIF picture size at 25 Hz for display as a quarter area picture in the centre of a monitor. The resultant picture size is then closer to that likely to be used in real applications. A full sized interpolation to display on a regular sized monitor would not reflect likely viewing conditions and make picture assessment unnaturally critical.

21.3. SPATIAL CODING PARAMETERS

The spatial coding used consists of wavelet decimation and interpolation operating on individual frames (either "I", "P" or "B") passed from the temporal decimation process. The wavelet coded frames were quantised using an HVS quantiser developed from the work of Ngan et al. [1].

The wavelet coding uses bi-orthogonal odd-tap QMF filters reported by Wilkinson [2] with the coefficients shown in Table 21.1.

These filters are characterised by *integer values*, *perfect reconstruction*, *linear phase* and near-equal responses for each pair. The filters are used in a three-stage wavelet decimation in the following orders:

- luminance: $1 \rightarrow 2 \rightarrow 3$;
- chrominance: $2 \rightarrow 3 \rightarrow 3$.

Internal wordlength accuracy was set to 11 bits centred on a DC level of 2048 allowing 6 dB of overhead for filter ringing.

Table 21.1 Wavelet filter coefficients

Type	Decimation	Interpolation
1	38, 18, −4, −2, 1	1924, 992, −140, −139, 46, 13, −4, −2.
2	10, 4, −1	272, 135, −20, −16, 4, 1.
3	10, 4, −1	56, 25, −4, −1.

21.4. NOISE CORING

The quantiser employed was adjusted for HVS weighting, thus higher frequency bands were more severely quantised than low frequency bands. The simplest *noise coring* technique removes the values $+/-1$ after quantisation; (though not in the dc band of "I" frames). The effect of this is to remove low level detail (which is often noise) thereby enhancing the performance of the entropy coder. The visual effect is nearly always beneficial even if only slight because the removal of low level signals is counterbalanced by the improved quantisation accuracy associated with higher level signals.

The process of noise coring, applied after the quantisation process and before the unquantisation process, ensures that the errors introduced by the noise corer are HVS weighted (i.e. the effect is equally visible across all wavelet bands). The implementation is very simple and almost "cost" free. An example of the noise coring process is given below:

> *Input (after quantisation):*
> -4 -3 -2 -1 0 1 2 3
>
> *After noise coring:*
> -3 -2 -1 0 0 0 1 2
>
> *Expansion of noise corer:*
> -4 -3 -2 0 0 0 2 3

At the very high compression levels required for 64 kbps coding, it was found that such non-linear quantisers removed useful video content as well as noise. A modification to the basic process was introduced to improve picture quality. For each pixel having a value of $+/-1$ prior to noise coring, the surrounding pixels are tested. If these surrounding pixels are all zero, then the central pixel is zeroed, otherwise its value is left unchanged.

Two forms of this test have been used:

> *Test 1:*
> ```
>        *   *   *
>        *   X   *
>        *   *   *
> ```
>
> *Test 2:*
> ```
>            *
>        *   X   *
>            *
> ```

Test 1 allows removal of $+/-1$ values in the "X" position only if all 8 surrounding pixels are zero. Test 2 allows zeroing of the "X" value only if the surrounding 4 pixels are zero.

The effect of the second test is to prevent removal of horizontal and vertical lines whereas the first test prevents the removal of diagonal lines as well. The 4 element test (2) was found to be satisfactory and this has been used in all the results shown in this chapter.

21.5. MOTION VECTOR ESTIMATION

In order to minimise the data rate associated with the motion vectors the chrominance and luminance images were all up-converted to the luminance CIF size of 352×288 pixels. This gives half pixel resolution for the luminance vector prediction and quarter pixel resolution for each chrominance vector prediction. Both estimation and compensation use this resolution. Vector generation in the estimation process generates one vector for each macroblock from a combination of luminance and both chrominance vectors thereby improving efficiency by removing the need for separate Y and C vectors. The macroblock size is the same as MPEG (16×16 for the luminance component and 8×8 for each chrominance component) resulting in a macroblock array of 9 rows by 11 columns, leading to 99 vectors.

21.5.1. Using a Soft Edged Motion Vector Window

A "soft" window was applied to the macroblock structure, as originally proposed by Katto *et al.* [3]. Tests showed that in areas of rapid movement, the macroblock structure becomes visible and visually annoying. This was even more evident when reviewing sequences frame by frame and the effect was found to become progressively more noticeable as the data rate was reduced.

The pitch of a macroblock was retained at 16×16 for luminance and 8×8 for each chrominance component. However, an overall window area of 24×24 for luminance and 12×12 for chrominance was used for the vector compensated frame differences. In order for the overall compensated picture to have equal gain, the edges of the window were tapered as shown in Figure 21.1.

These windows form "tiles" which can be overlaid in such a way as to compensate for the tapered edges. A 2-D window is created simply from a convolution of identical horizontal and vertical 1-D windows. Since the 1-D window is scaled by a value of 16, then the 2-D window is scaled by a factor of 256.

This technique can be expensive to implement in high rate, high resolution systems and provides only marginal improvement. However, in low bit-rate applications the implementation is easier and provides clear visual quality improvements. The operation need only be used for motion vector compensation; no advantage was observed when using soft edged windows for motion vector estimation.

21.6. PREDICTIVE TEMPORAL CODING

The classic DPCM coding process requires a decoder section within the encoder forming a loop in order to prevent recursive errors. The same problem occurs with both "P" and "B" differential frame coders. The conventional method of overcoming this is by means of the loop shown in Figure 21.2.

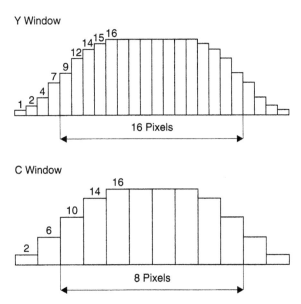

Figure 21.1 Y and C motion vector windows.

The "I" frame F_0 is processed first, resulting in a reproduced frame $F_0 + e_0$, the frame now having errors introduced by the quantisation process. The first frame is motion compensated prior to subtraction from the second frame F_1 to produce the difference frame P_1. This is then coded and produces its own error which is added to the original signal, $P_1 + e_1$. Since the subtraction and addition of the previous frame

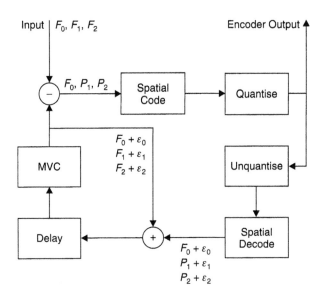

Figure 21.2 Closed loop predictive temporal coding.

is achieved with the same signal, the signal added to P_1 recreates the original F_1 with error e_1 added. This ensures no error extension.

Thus, although the encoder is more complex, the errors introduced by each frame are local to that frame. However, it should also be noted that the noise introduced by the prediction from the reconstructed previous frame, $F_0 + e_0$, introduces an imperfect prediction match that can significantly worsen the "P" frame prediction. Furthermore, visual results indicate that low level errors can still propagate. This happens when there is rapid movement of a foreground object in a plain background scene. The foreground movement leaves a trail of errors on the background. This effect is clearly illustrated in frame 22 of the "Susie" sequence as shown in Figure 21.3.

The effect shown is non-linear, and believed to be related to a form of *limit cycle* which occurs in the recursive loop. Rapid foreground movement leaves an error in the backgound at both the transmitter and receiver. However, the error is sufficiently small that it is quantised to zero in the encoder and therefore never removed until the next "I" frame. The only way this error can be reduced is by restricting movement, thereby raising the quantiser scale value which will reduce all errors in the picture. Alternatively, adaptive schemes may be possible which track this effect and implement special measures to reduce it. However, for the work reported here, no such methods were explored, as the logarithmic coder described below was found to eliminate such problems and produce significantly better results.

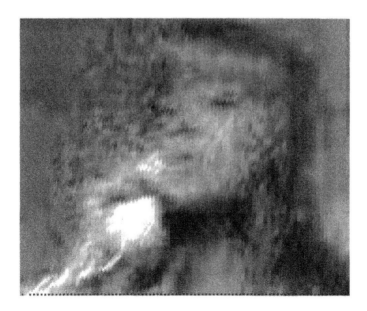

Figure 21.3 Predictive frame coding of "Susie".

21.7. OPEN LOOP CODING

Consider now, a system using only "I" and "B" frames. A logarithmic decimation can be achieved using a recursive coder as illustrated in Figure 21.4. Furthermore, the feedback element of Figure 21.2 can be eliminated, thus simplifying the coder in a way that is not practically possible in the conventional predictive case.

The use of motion compensated "B" frames allows better prediction compared to "P" frames and the coding of the inter-frames is much improved as a result. A slight limitation is that this system requires "I" frames every 2^n frames. The value of n is typically 2 or 3, giving a GOP sequence of 4 or 8 frames. The work described here uses a 4-frame GOP. Since the encoding operation is performed without the feedback loop, the codec would produce a sequence of frames as in Table 21.2.

Reconstructing the "I" frames: I_2', I_1' and I_3' from the "B" frames: B_2, B_1 and B_3 requires adding the average of adjacent "I" frames to invert the encoding operation. Since recreated adjacent "I" frames have their own errors, newly created "I" frames will contain their own errors and a contribution of errors from other frames as in Table 21.3.

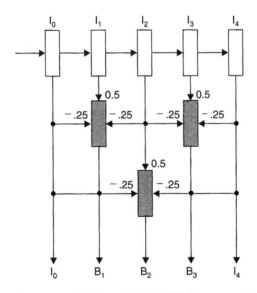

Figure 21.4 Logarithmic "B" frame encoding.

Table 21.2 Sequence of frames

Frame	Encoded frame	Decoded frame
I_0	I_0	$I_0 + \varepsilon_0$
I_4	I_4	$I_4 + \varepsilon_4$
I_2	$B_2 = I_2 - (I_0 + I_4)/2$	$B_2 + \varepsilon_2$
I_1	$B_1 = I_1 - (I_0 + I_2)/2$	$B_1 + \varepsilon_1$
I_3	$B_3 = I_3 - (I_2 + I_4)/2$	$B_3 + \varepsilon_3$

Table 21.3 Expanded error terms

$$I'_2 = B_2 + \varepsilon_2 + (I'_0 + I'_4)/2$$
$$E_2 = I_2 + \varepsilon_2 + \varepsilon_0/2 + \varepsilon_4/2$$
$$I'_1 = B_1 + \varepsilon_1 + (I'_0 + I'_2)/2$$
$$E_1 = I_1 + \varepsilon_1 + 3\varepsilon_0/4 + \varepsilon_2/2 + \varepsilon_4/4$$
$$I'_3 = B_3 + \varepsilon_3 + (I'_2 + I'_4)/2$$
$$E_{31} = I_3 + \varepsilon_3 + \varepsilon_0/4 + \varepsilon_2/2 + 3\varepsilon_4/4$$

The error terms introduced by the quantiser ε_n can be adjusted by changing the quantiser level to ensure that the decoded error terms E_n are equal. However, the distribution of errors from each frame is not equal but spread over several frames. The components due to errors in the "I" frames ε_0 and ε_4 have the slowest fading errors, lasting 7 frames each, whilst error ε_2 lasts for 3 frames and errors ε_1 and ε_3 last only 1 frame. The error contributions are shown in Figure 21.5.

Viewing tests conducted using both open and closed loop encoding methods indicated that the errors introduced by the open loop encoding process are less objectionable than those from the closed loop process. The reason appears to be related to the nature of the noise component. The closed loop encoder results in reconstruction errors which are largely independent and persist for 1 frame only. The open loop encoder produces errors in various components: errors in the I frames persist for seven frames, errors in the B frames persist for only one or three frames. The eye's temporal response is such that flickering types of distortions are more readily visible than slower changing errors. Anecdotal evidence demonstrates the effect as follows. When a picture is displayed as a still frame, some errors are not visible. When the sequence is played at normal speed these otherwise invisible *static* errors become visible. This effect has been observed on motion JPEG systems but as yet remains undocumented.

The results of coding the "Susie" sequence using the approach described are shown in Figure 21.6. The data rate is identical to that of the predictive coder results

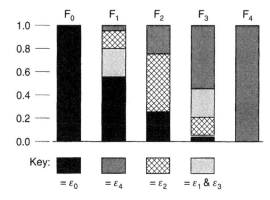

Figure 21.5 Error distribution of open loop coding.

Figure 21.6 Open-loop coding of "Susie".

in Figure 21.3, but the improvement is clear. The much improved result exhibited is supported by display of the motion sequences for each system.

The results for the open-loop codec are about 5 dB better than those for the predictive coder. The open-loop codec results at 48 kbps are shown in Table 21.4.

21.8. TEMPORAL WEIGHTING

Tests on various picture sequences showed that the errors introduced by the quantiser/unquantiser function did not exhibit a uniform distribution. In particular, "B" frames have significantly lower errors than "I" frames. The benefit is that, although the quantiser values for "B" frames have been raised for open-loop coding, they can be lowered again because of the lower error levels created. The cause is thought to be the error distribution though no justification is provided here.

Table 21.4 Results of 4-frame open loop coding

Seq.	Frames	Q value	Y S/N	C S/N
0	0..4	156	−36.02	−42.13
1	8..16	148	−35.37	−42.22
2	16..24	133	−35.92	−41.82
3	24..32	155	−37.70	−42.05
4	32..40	150	−36.10	−42.47
5	40..48	133	−35.61	−42.51

Experiments have shown that severe temporal weighting can be applied to maximise picture quality. Optimum weightings were found to be (for each frame):

For Y: 1.0, 0.4, 0.7, 0.4
For C: 1.0, 0.0, 0.25, 0.0

The weighting is applied as a multiplier to the quantisation value for its respective frame. Weightings of less than unity will cause larger quantisation errors.

Note that the weighting for the "C" component is so severe that no data is sent for frames 1 and 3, thereby implying a 6.25 Hz frame rate for "C". However, this is not simply a temporal sub-sampling of the "C" components since motion vector compensation is still applied.

21.9. CONCLUSIONS

This chapter has described new and established techniques developed for low bit-rate coding of motion images. The conclusions of the simulation studies are that the conventional technique of predictive coding has severe quality limitations at these bit rates and that wavelet open-loop coding produces much better picture quality. It was shown that the spatial quantiser profile and temporal weighting add considerable coding benefit. In addition, soft window MVC results in a distinct benefit to the picture quality and is, at the low source rates considered, practical to implement. An improved form of noise coring has been developed which reduces isolated noise elements rather than picture information. One area not explored here is the coding delay which occurs as a result of the improved coding efficiency. The total loop delay of approximately 1 s may be difficult to accept in practice. This is an area for future study.

REFERENCES

[1] Ngan, K. N., Leong, K. S. and Singh, H., Adaptive cosine transform coding of images in the perceptual domain. IEEE 0096-3518/89, 1100–1743, 1994.
[2] Wilkinson, J. H. Wavelet transforms in a digital video tape recorder. IEE Colloquium Digest 1993/009, 1993.
[3] Katto, J., Ohki, J., Nogaki, S. and Ohta, M., A wavelet codec with overlapped motion compensation for very low bit-rate environment. *IEEE Trans Circuits and Systems for Video Technology*, **4** (3), 328–338, 1994.

22

Morphological Methods for Image and Video Coding: An Overview

P. Czerepiński and D. R. Bull

22.1. INTRODUCTION

Mathematical morphology investigates the properties of size, shape and structure of objects known as *complete lattices*. Unlike linear signal processing methods, mathematical morphology is based on set-theoretic relations [1], which make it a powerful

Insights into Mobile Multimedia Communication
ISBN 0-12-140310-6

tool for pattern analysis [2,3,4] and image segmentation [5,6,7,8]. According to a widely-known conjecture, the edge-texture [9] paradigm leads to an efficient scene description, and hence to compression. This description can be provided by morphological methods, and auxiliary algorithms have been developed that efficiently encode the contour–texture information [10,11,12,13,14,15].

This chapter presents an overview of morphological algorithms for image and video compression. Section 22.2 provides an introduction to morphological operators on lattices. Then, three applications of morphological filters to image and video compression are presented:

1. *Region based image and video compression.* In section 22.3, an overview of region-based approaches to picture and video compression is provided. Such methods adopt the contour–texture approach to describe two- or three-dimensional objects. Although the aim is to segment the signal into regions corresponding to real-life objects, semantic object classification is not attempted, and simpler, statistical criteria are employed for segmentation. See reference [16] for a recent overview.

2. *Segmentation based coding of the DFD (displaced frame difference) signal.* As demonstrated by Girod [17] and Strobach [18], video coding techniques gain from employing motion compensation and the temporal DPCM loop, rather than from transform coding of the resulting DFD data. This led to the development of morphological segmentation-based approaches to DFD coding [19,20,21,22,23,24], as described in more detail in section 22.4.

3. *Morphological subband decomposition.* This is another area of active research [25,26,27,28]. Morphological decompositions have been found to provide a good rendition of smooth image areas and sharp edges, at the expense of a relatively poor performance in textured areas, compared to linear filters. In addition, such decompositions exhibit good error resilience properties. Subband techniques are discussed in section 22.5.

22.2. MORPHOLOGICAL OPERATORS

A number of books [29,30,31] and journal papers [32,33,34,35,36,37] provide an in-depth description of this subject. A brief introduction, aimed at familiarising the reader with mathematical morphology, is presented below.

22.2.1. Complete Lattices

Morphological operators are mappings on *complete lattices*. A lattice is a set with an associated ordering relation ($\leq$) [31, Chapter 2]. The lattice is complete if its every subset has an infimum and a supremum. In the context of image processing, two lattices are of practical importance: the lattice of planar sets, $P(\Re^2)$ (also referred to as the *power set* of $\Re^2$ – the set of all subsets of the Euclidean plane), where ordering corresponds to set inclusion ($\supset$); and the lattice of all functions on $\Re^2$, $\Re^{\Re^2}$, where for two functions f and g: $f \le g \Longleftrightarrow f(x) \le g(x)$, for all spatial locations x. The first lattice has found application in binary morphology, and the latter in greyscale morphology. We shall find it convenient to identify the power set with the lattice of binary functions, by noting that the lattice of sets, $P(\mathfrak{R}^2)$, is *isomorphic* to the lattice of binary functions, $\{0, 1\}^{\mathfrak{R}^2}$, with the rule that elementhood in $P(\mathfrak{R}^2)$ is mapped onto the value "1" in $\{0, 1\}^{\mathfrak{R}^2}$.

22.2.2. Basic Operators

From this point onwards, we focus our attention on digital signal representations, by considering operators on discrete lattices $\{0, 1\}^{Z^2}$ and Z^{Z^2}. Let f and B be functions on Z^2. Moreover, let B be a *flat structuring element*, i.e. such that $B(x) = 0$, for every x in the domain of B, D_B. Then *erosion*, $\varepsilon_B(f)$, and *dilation*, $\delta_B(f)$, of function f by the structuring element B are defined by:

$$\varepsilon_B(f)(x) = \min\{f(x+y), y \in D_B\}$$
$$\delta_B(f)(x) = \max\{f(x-y), y \in D_B\}$$

Morphological *opening*, α, and *closing*, β, are defined as compositions of erosion and dilation, namely:

$$\alpha_B = \delta_B \varepsilon_B$$
$$\beta_B = \varepsilon_B \delta_B$$

Morphological opening and closing are idempotent, i.e. they satisfy $\alpha_B^2 = \alpha_B$, $\beta_B^2 = \beta_B$. The *median* operator, μ_B:

$$\mu_B(f)(x) = \text{median}(f(x+y), y \in D_B)$$

is closely related to erosions and dilations, as it can be expressed as a maximum of a number of erosions, or a minimum of a number of dilations.

22.2.3. Geodesic Operators and Filters by Reconstruction

Geodesic operators and filters by reconstruction [38] are always defined in terms of two functions, f and m. Intuitively, m will correspond to the "marker" function. A section of the function f will be reconstructed as long as it has been "marked" in m. Let I be a structuring element of size one.[1] The *geodesic dilation* and *erosion*[2] of size one are then defined respectively by:

$$\delta^1(m, f) = \min(\delta_I(m), f)$$
$$\varepsilon^1(m, f) = [\delta^1(m^*, f^*)]^*$$

---

[1] The domain of such an element will depend on the type of connectivity assumed. In the context of 8-connectivity, I is a 3×3 window, centred on the origin. In the case of 4-connectivity, I is a "cross", i.e. its domain is the set $\{(0,1), (-1,0), (0,0), (1,0), (0,-1)\}$.

[2] The "*" symbol stands for the negation operator. Negation reverses lattice ordering, i.e. if $f \le g$, then $f^* \ge g^*$. To negate a function f, multiply $f(x)$ by (-1). To negate a set, use set complement.

The *geodesic reconstruction* by dilation or erosion is defined by:

$$\rho(m,f) = \delta^{\infty}(m,f)$$
$$\sigma(m,f) = \varepsilon^{\infty}(m,f)$$

where the superscript '∞' denotes iterating the corresponding operation of size one, until stability. The above notions lead to a class of operators introduced by Salembier *et al.* [39], called opening and closing by *partial reconstruction*:

$$\alpha_{A,B}(f) = \rho(\varepsilon_A(f), \alpha_B(f))$$
$$\beta_{A,B}(f) = \sigma(\delta_A(f), \beta_B(f))$$

where $D_A \geqslant D_B$. Note, that if $A = B$, the above filters become morphological opening and closing. If $D_B = \{(0,0)\}$, then $\alpha_B(f) = \beta_B(f) = f$, and they are referred to as opening and closing by (full) reconstruction. Morphological opening and opening by reconstruction operate by removing those high-valued parts of the signal whose spatial support is small compared to the structuring element. Dually, morphological closing and closing by reconstruction operate by removing those low-valued parts of the signal whose spatial support is small compared to the structuring element.

22.2.4. Watershed Segmentation

Watershed segmentation [5,6,7] has been the subject of significant research interest in recent years. Consider a greyscale image as a three dimensional landscape. Suppose some rain falls on the landscape. Then every drop will follow the path of the steepest descent, until it is caught by a local minimum. Thus, it is possible to segment the image in terms of zones of influence of local minima, or, in other words, according to watersheds. Such an approach is usually applied to the gradient rather than the original image. The result of a watershed segmentation is shown in Figure 22.1.

Figure 22.1 Result of watershed segmentation.

22.2.5. Morphological Filters and Operators

Along with morphological operators, the notion of morphological filters can also be found in the literature. Two different definitions exist: Serra [29] and Heijmans [31] define the morphological filter as any increasing mapping ψ that is idempotent, whereas Giardina and Dougherty [30] require ψ to be increasing and translation invariant. Here we do not attempt to make any distinction between "filters" and "operators". Instead, we rely on the fact that mathematical morphology studies the properties of size, shape and structure, and hence the output of morphological filters (or operators) depends on geometrical relations between the input signal and the structuring element.

22.2.6. Implementation

It should be noted that the above equations only serve the purpose of defining morphological operators. Indeed, the computational cost required by filters by reconstruction may seem prohibitive in the absence of parallel processing hardware. However, efficient implementation algorithms for morphological filters have been reported. Erosions, dilations and medians can be implemented recursively, avoiding overlapping local maximum, minimum or sorting operations, as described by Ko *et al.* [40]. Vincent [41,42] reports a fast implementation of a reconstruction filter, by employing sequential and region-growing algorithms, and Vincent and Soille report a fast implementation of the watershed segmentation [6]. Algorithms for erosion, dilation, opening, closing, morphological methods and watersheds are also described by Bleau *et al.* in [43,44].

22.3. REGION-BASED CODING OF IMAGES AND VIDEO

Region- and object-based compression techniques are an active area of research [45,46,47,48,49]. As opposed to statistical compression methods, based on rate-distortion theory, here the scene is modelled as a collection of two or three dimensional objects, uniform with respect to some criteria. The compression process can be split into two major stages: *segmentation*, and *shape and texture coding* of the objects. Eryurtlu and Kondoz [46] perform a two dimensional image decomposition using quadtrees, followed by motion estimation of control points. Texture is represented by mean luminance value. Rajala *et al.* [47] employ a region-growing algorithm to obtain a three dimensional segmentation, contour information is coded using a three dimensional arithmetic coder, and object textures are represented by mean intensity values. Morphological filters are sensitive to the geometric properties of signals, hence they are a convenient tool for performing the segmentation step. References [16,50] offer a recent review of region-oriented morphological methods for image compression.

22.3.1. Segmentation

The objective of segmentation is to partition an image into connected regions of uniform texture, colour or intensity. Ideally, the segmentation should semantically correspond to real image objects. Salembier [51] presents a hierarchical, top-down image compression method. Three stages of the segmentation process can be distinguished:

- *Simplification*. The purpose of this stage to "clean" the input. This includes removing random noise and small regions, by utilising some type of smoothing filter. At the same time, good edge preservation of the remaining regions is mandatory, which immediately rules out the application of linear low-pass filters. Comparisons of the simplification and edge preservation performances of several filters (linear, median, open_close, open_close by reconstruction) presented in [39,51] demonstrate superior performance of opening and closing by reconstruction. These filters belong to a class of connected operators [52,53]. Filters by partial reconstruction may also be used, these enable the control over the degree of edge simplification. Figure 22.2 shows an example of simplification.
- *Marker extraction*. At this stage, homogeneous regions are marked with unique labels. A number of labelling techniques must be in place, so that both uniform and high-contrast regions are labelled correctly. In addition, labelling must be consistent with the segmentation passed on from a higher level of the hierarchy.
- *Decision*. Marker extraction still fails to classify a small group of pixels, usually corresponding to region boundaries. At this stage, the watershed [5,6,7] algorithm is employed in order to locate edges and classify remaining pixels.

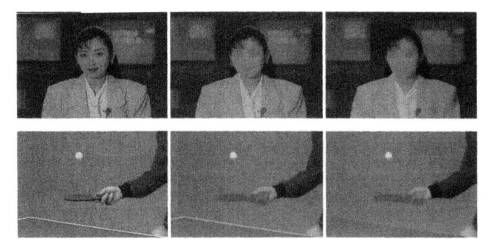

Figure 22.2 First column: original "Akiyo" and "Table Tennis", 352×240; second column: filters by "full" reconstruction; third column: filters by partial reconstruction.

A recent alternative segmentation method is the *region-growing* version of the *watershed* [16]. This algorithm is applied to the input image, starting from a marker, that identifies initial object interiors, and is propagated until all pixels are classified.

22.3.2. Region Coding

After segmentation has been performed, the resulting partitions must be encoded. This is accomplished by recording object shape and texture information. The following techniques have been developed for efficient shape representation:

- *Contour coding by chain codes* [54] The algorithm for 4-connected contours [10] operates by recording relative directions: left, straight and right, resulting in an average bit rate of $\log_2 3 \approx 1.58$ bit/pixel. If contextual information is also exploited, this can be reduced to 1.27 bit/pixel. Some authors have suggested lossy techniques: Gu and Kunt [14] have developed a contour simplification algorithm, that reduces the contour coding cost by 20% while keeping the contour location error within one pixel. Schuster and Katsaggelos [55] describe a method for finding the best boundary in a rate-distortion sense, given a bit budget.
- *Skeleton coding* There exists a morphological method of coding region shapes, namely, the skeleton transformation. It has been successfully applied to binary image coding [11,56]. Brigger *et al.* [12,13] show that the classical skeletons contain information that can be removed either without or with minor loss of information, and introduce the notion of a geodesic skeleton. It is concluded that geodesic skeleton is preferred to chain coding, whenever there are many isolated and short contour arcs to be coded.
- *Texture coding* Various methods have been proposed for texture coding. Pardàs and Salembier [57] employ the simplest representation using mean luminance value. Salembier [51] uses second order polynomials on smooth regions, and block truncation coding [58] on high-contrast ones. Recently, a novel technique referred to as "morphological interpolation" has been described [59].

22.3.3. Extensions to Video

Morphological methods have also been extended to video data [57,60,61]. Notably, a segmentation-based proposal has been submitted to MPEG-4 [62]. Motion estimation and compensation enable considerable bit-rate reductions in the context of video compression [63,64]. If object shapes in a frame n are known, it is possible to use parametric, object-based techniques to find motions between frames n and $n + 1$. An affine motion model has been employed in [57]. The codec described in [61] includes motion compensated contour and texture coding blocks, in order to differentially encode signal variations. More advanced techniques for texture coding have also been reported: for example using wavelets [62] and orthogonal transforms [61]. In addition to exploiting spatial signal characteristics, such as luminance value or texture, some authors [65,66] take the motion information into account. This leads to motion-based fusion of regions with differing colours, that semantically belong to the same objects.

22.4. MORPHOLOGICAL SEGMENTATION FOR LOW BIT RATE DFD CODING

To date, DCT (discrete cosine transform) based techniques have been adopted in both image [67] and video compression [68,69] standards. It is known that the DCT is a special case of the optimum Karhunen–Loève expansion for stationary and first order Markov processes, whose autocorrelation coefficient, ρ, approaches 1. Most still images possess good correlation properties, typically with $\rho > 0.9$. This justifies the application of the DCT for intra-frame coding, since it will provide near optimum decorrelation and energy compaction capabilities. The situation changes drastically in the case of video. Here, motion compensation, combined with the temporal DPCM loop, is employed in order to reduce the temporal correlation of the video signal. This pair can be viewed as a whitening filter, and the resulting DFD as the innovations signal. In reality, the value of the first order autocorrelation coefficient, ρ_1, falls between 0.3 and 0.5 [70,18,20]. Strobach [18] demonstrated that, for a considerable number of cases, the application of DCT to DFD subblocks resulted in data expansion rather than reduction! For the remaining blocks, the transform coding gain was minor compared to that achieved in the intra-frame mode. Similar results have been reported by others [70,17,71]. Moreover, quantisation of the coefficients corresponding to impulse-like DFD signal, associated with moving edges, leads to "ringing" artefacts, due to the Gibbs phenomenon.

The above argument leads to the conclusion that it is not always appropriate to apply DCT coding to the DFD signal, and that spatial approaches have the potential of matching the performance of transform techniques. Also, when high fidelity is not an important issue, as is the case in low bit rate coding applications, selected motion failure areas can be removed without causing objectionable artefacts. Segmentation methods, utilising morphological tools, rely on extraction of perceptually relevant information from the DFD signal, and focus on encoding these areas only. Such an approach can lead to edge preservation and a lack of ringing artefacts. This type of algorithm is described in this section.

22.4.1. From DFD Pixels to Regions

Various algorithms can be employed for motion estimation [63,64]. In the following, we ignore the motion compensation details, and focus on the morphological segmentation step that follows. A displaced frame difference signal can be loosely described in terms of three types of areas: (i) elongated, high amplitude regions along moving edges, due to slight inaccuracies of motion compensation; (ii) small amplitude variations, due to illumination changes or moving smooth objects; (iii) changes due to object deformation or appearance of new objects. The goal of the segmentation is to filter out types (i) and (ii), as these can be tolerated by a human observer, and encode type (iii) artefacts.

Typically, a binary mask is created by thresholding the DFD signal [20,21,23,24]. The mask is a collection of irregular objects, as well as a large number of isolated pixels. Some filtering is required so that the mask is transformed into a set of regions, with well defined boundaries. This is accomplished by close-open filters [72], or, more often, by median filtering, with structuring elements of size 3×3 [20,21], 5×5 [73] or 7×7 [23]. Now the task is to remove small and narrow regions, and retain large ones while keeping their shapes intact. This is comfortably accomplished by an opening by reconstruction: first an erosion is performed, and then those regions, whose interiors are marked, are brought back to original shape. DFD pixels, coinciding with regions preserved in the final mask, are selected for transmission. The segmentation process is illustrated in Figure 22.3.

Alternatively, equivalent filters can be applied to the greyscale DFD, thus postponing the thresholding to the last stage. An example greyscale segmentation mask is shown in Figure 22.4. Such methodology enables the control of the area preserved and thus the number of bits needed. In this manner rate can be controlled without having to re-calculate the binary mask for a different threshold value.

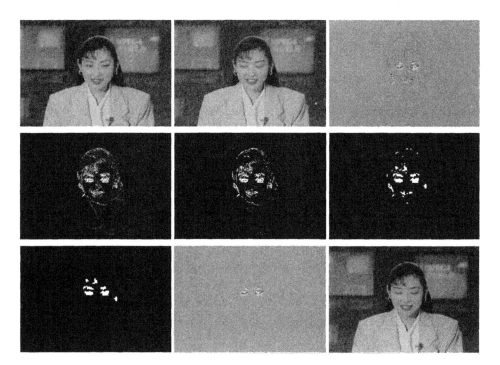

Figure 22.3 DFD segmentation example, "Akiyo", 352×240. First row: frame 153, frame 154, DFD signal (scaled by 4 and offset by 128); second row: DFD signal (absolute value, scaled by 20), mask after thresholding ($T=6$), mask after median filtering; third row: final mask (small regions removed), quantised update signal ($Q=9$) coded at 0.031 bpp, reconstructed frame 154.

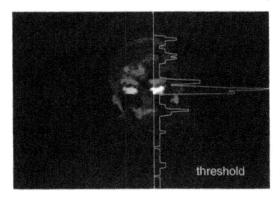

Figure 22.4 Final greyscale segmentation mask. The threshold value can be varied to match a given bit budget.

22.4.2. Region Coding

Chain- and skeleton-based coding schemes, as described in section 22.3, can readily be used to encode the shape of the segmentation mask. This may impose restrictions on the shape of contours, such as 4-connectivity or lack of interior subregions. Moreover, segment boundaries can be overly complex, resulting in high coding costs. Typically, these problems are overcome by close/open operations with a small kernel filter (2×2 or 3×3). Interior subregions can be merged in by a closing by reconstruction.

If the shape of the segmentation mask is recorded, the *update signal* can be coded spatially. Scalar quantisation has been employed [20,22,24], followed by arithmetic coding of the quantised values, using first [19,20] or second [21] order models. The quantiser and the threshold values determine the reconstruction quality and the compression ratio. The right choice of these parameters poses a problem. For example: (1) a small threshold with a coarse quantiser as well as (2) a large threshold with a fine quantiser may lead to an identical compression ratio. Subjective tests have been performed in order to establish the best quantisation strategy. The performances of both uniform and non-uniform quantisers have been investigated. The conclusion is that, given a threshold value, T, best subjective quality versus bit rate trade-off is achieved with uniform quantisation, and a quantiser value, Q, comparable to the threshold value.

The above approaches involve two stages: contour coding and interior coding. An alternative entropy coding algorithm, which obviates the need for contour coding, has been developed. This algorithm does not impose any restrictions on the shape or connectivity of the coded regions, and consists of the following stages:

- *Restructuring*. Pixels are arranged into 8×8 blocks. For every block, either a CODED or NOT_CODED symbol is encoded, indicating whether or not the block contains any active pixels.
- *Re-scanning*. If a block is marked CODED, pixels within the block are rearranged into a one dimensional array, as shown in Figure 22.5. This pattern ensures that the resulting structure retains any neighbour-to-neighbour predictability.

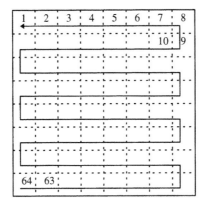

Figure 22.5 Pixel scanning pattern.

- *Conditional coding.* The 64-pixel structure, resulting from the new scan, is encoded using a conditional coder (arithmetic implementation). The coder state, when encoding a pixel at a location k, corresponds to the value of its preceding neighbour at a location $k-1$, $1 < k < 64$.

A comparison between coding performances of the above algorithm and the contour-interior coding approaches (first and second order coding of interior pixels) is shown in Figure 22.6. The data have been obtained using "Akiyo" (352×240) and "Silent Voice" (352×288) sequences.

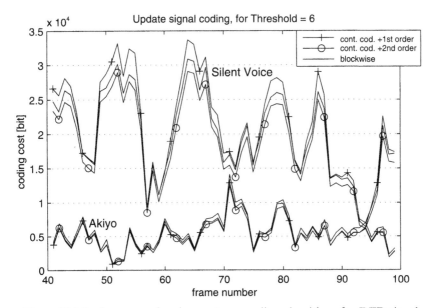

Figure 22.6 Performance of various entropy coding algorithms for DFD signal.

On the other hand, some authors use frequency domain techniques for encoding the *update signal*. This is justified, as most of the impulsive components have been removed from the signal, although no experimental proof of this fact has been reported in the literature. Subband coding has been employed by Qian [73], whereas Talluri [72] and Özcelik and Katsaggelos [23] apply the DCT.

22.4.3. Post Processing

A disadvantage of segmentation-based DFD coding lies in step-wise intensity changes, which may appear at the edges of transmitted regions, due to the thresholding and quantisation processes. Such distortion is particularly noticeable in areas of low intensity gradient. Non-linear, edge preserving filters have been designed [21,22,24] to overcome this problem.

22.5. MORPHOLOGICAL SUBBAND DECOMPOSITIONS FOR IMAGE CODING

Another active area of research is that of subband decomposition using morphological filters. Linear filterbanks [74] are a powerful method of decomposing signals into a set of *critically decimated* subbands. Bit allocation can then be varied across subbands, in order to exploit the characteristics of the human visual system. Besides good compression, such systems offer the capability of progressive transmission, useful for transmission over low bandwidth channels. Unfortunately, at high compression ratios linear filters produce "ringing" artefacts around sharp edges, due to the Gibbs phenomenon. Morphological filter banks have the capability to overcome this type of distortion.

22.5.1. Size Decompositions

Early morphological decompositions [75,76] attempted to split the signal into sub-images, containing objects of a specific size. These methods, however, did not achieve critical subsampling, and proved inferior to linear filterbanks. Recent methods [25,26,27,28] satisfy this requirement and provide excellent compression performance. The term "morphological" may be slightly misleading here, as such methods do not really adopt the shape-oriented processing. In fact, non-linear (primarily median) operators are in place in order to split the signal into different frequency bands.

22.5.2. Critically Decimated Decompositions

Redmill [77] and Florêncio and Schafer [25] implement non-expansive pyramidal methods. At every level of the pyramid, the input signal is decimated by a factor of 2 in each direction (the low-frequency band), and the remaining samples are interpolated using median filters. The interpolated version is subtracted from the original, and the resulting error signal encoded.

The above approach has been found to provide a very good rendition of smooth image areas and sharp edges, at the expense of a relatively poor performance in textured areas. Egger *et al.* adopt an adaptive approach: a texture detection algorithm operates on the decimated low-frequency subband. If an area is classified as "flat" or "edge", a median interpolation filter is used. If an area is classified as "texture", a linear prediction is used instead. Thus, the method combines best parts of both linear and median filters into a single system. There is no need for sending any overhead information – an identical texture detection filter is employed at the decoder. Note that this is possible only because the low-frequency subband is created by decimation only, without any anti-aliasing filter in place.

Lack of any low-pass analysis filter in systems described in [25,26,27] is restrictive, and can limit compression performance. Improved techniques, equivalent to the structure depicted in Figure 22.7, have been proposed by Florêncio and Schafer [78], Hampson and Presquet [79] and Redmill and Bull [28]. If the function $b(.) = 0$, low frequency signal is obtained by decimation only. An example pair of non-linear filters, designed for quincunx sampling (see Figure 22.8), reported in [80] is:

$$a(.) = -\text{median}(p_1 + p_2 + p_3 + p_4)$$
$$b(.) = \tfrac{1}{2}\text{median}(p_1 + p_2 + p_3 + p_4)$$

Figure 22.9 gives an example of the performance of these filters.

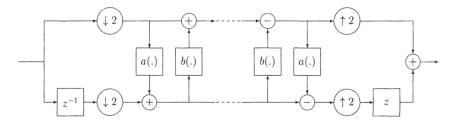

Figure 22.7 Two-stage non-linear filter bank structure.

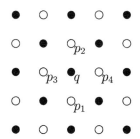

Figure 22.8 Quincunx sampling system.

Figure 22.9 Lena: original (256×256) and compressed using non-linear filters, 0.5 bpp.

When compared to other architectures, median filterbanks have the advantages of short wordlength arithmetic and multiplier-free implementation. Moreover, Redmill and Bull [80] demonstrate that such filterbanks can have excellent error resilient properties, mainly due to the "outlier rejection" properties of the median filter.

22.6. CONCLUSIONS

In this chapter, we have provided an overview of morphological methods for image and video compression. Morphological filters have found application in three areas: (i) region-based coding methods, where they provide scene simplification, while preserving edge locations, and are also used for shape and texture representation; (ii) displaced frame difference segmentation, where they enable removal of impulsive signals of small spatial support and (iii) non-linear subband decomposition, yielding "ringing-free" and error resilient compression systems. Much research is ongoing to further improve the properties and performance of this approach to image and video coding.

REFERENCES

[1]   Matheron, G., *Random Sets and Integral Geometry*, Wiley, New York, 1975.
[2]   Skolnick, M., Application of morphological transformations to the analysis of two-dimensional gels of biological electrophoretic materials. *Comput. Vision Graph. Image Process.*, **35**, 306–332, 1986.
[3]   Meyer, F., Automatic screening of cytological specimens. *Comput. Vision Graph. Image Process.*, **35**, 356–369, 1986.

[4] Bhagvati, C., Grivas, D. and Skolnick, M., Morphological analysis of pavement surface condition. In Dougherty, E. (ed.), *Mathematical Morphology in Image Processing*, Marcel-Dekker, New York, 121–150, 1993.

[5] Meyer, F. and Beucher, S., Morphological segmentation. *J. Visual Commun. Image Representation*, **1** (1), 21–46, 1990.

[6] Vincent, L. and Soille, P., Watersheds in digital spaces: an efficient algorithm based on immersion simulations. *IEEE Trans. Pattern Analysis and Machine Intelligence*, **13**, (6), 583–598, 1991.

[7] Beucher, S. and Meyer, F., The morphological approach to segmentation: the watershed transformation. In Dougherty, E. (ed.), *Mathematical Morphology in Image Processing*, Marcel-Dekker, New York, 433–481, 1993.

[8] Li, W., Hease-Coat, V. and Ronsi, J., Robust morphological features for texture classification. In *Proceedings of the IEEE International Conference on Image Processing*, **3**, 173–176, 1996.

[9] Kunt, M., Ikonomopoulos, A. and Kocher, M., Second generation image coding techniques. *IEEE Proceedings*, **73** (4), 549–574, 1985.

[10] Eden, M. and Kocher, M., On the performance of a contour coding algorithm in the context of image coding. Part I: Contour segment coding. *Signal Processing*, **8**, 381–386, 1985.

[11] Maragos, P. and Schafer, R., Morphological skeleton representation and coding of binary images. *IEEE Trans. Acoustics, Speech and Signal Processing*, **34** (5), 1228–1244, 1986.

[12] Brigger, P., Ayer, S. and Kunt, M., Morphological shape representation of segmented images based on temporally modelled motion vectors. In *Proceedings of the IEEE International Conference on Image Processing*, **3**, 756–760, 1994.

[13] Brigger, P. and Kunt, M., Morphological shape representation for very low bit rate video coding. *Signal Processing: Image Communication*, **7**, 297–311, 1995.

[14] Gu, C. and Kunt, M., Contour simplification and motion compensated coding. *Signal Processing: Image Communication*, **7**, 279–296, 1995.

[15] Czerepiński, P. and Bull, D., Morphological segmentation based video coding employing conditional smoothing. In *Proceedings of the IEEE International Symposium on Circuits and Systems*, Atlanta, **2**, 628–631, 1996.

[16] Salembier, P., Brigger, P., Casas, J. and Pardàs, M., Morphological operators for image and video compression. *IEEE Transactions on Image Processing*, **5**, (6), 881–898, 1996.

[17] Girod, B., The efficiency of motion-compensated prediction for hybrid coding of video sequences. *IEEE J. on Selected Areas in Communications*, **5**, 1140–1154, 1987.

[18] Strobach, P., Tree-structured scene adaptive coder. *IEEE Trans. on Communications*, **38** (4), 477–486, 1990.

[19] Wei, L. and Mateo, F., Segmentation based coding of motion compensated prediction error images. *Proceedings of the IEEE International Conference on Acoustics, Speech and Signal Processing*, **5**, 357–360, 1993.

[20] Wei, L. and Kunt, M., Morphological segmentation applied to displaced frame difference coding. *Signal Processing*, **38**, 45–56, 1994.

[21] Wei, L., Bhaskaran, V. and Kunt, M., Very low bit-rate video coding with DFD segmentation. *Signal Processing: Image Communications*, **7**, 419–434, 1995.

[22] Bhaskaran, V., Wei, L. and Kunt, M., A segmentation based scheme for very low bit-rate video coding. *Proceedings of the SPIE Visual Communications and Image Processing Conference*, **2419**, 81–89, 1995.

[23] Özcelik, T. and Katsaggelos, A., Detection and encoding of model failures in very low bit rate video coding. *Proceedings of the SPIE Visual Communications and Image Processing Conference*, **2727**, 820–831, 1996.

[24] Czerepiński, P. and Bull, D., Enhanced interframe coding based on morphological segmentation. *IEE Proceedings – Vision, Image and Signal Processing*, **144** (4), 220–226.

[25] Florêncio, D. and Schafer, R., A non-expansive pyramidal morphological image coder. *Proceedings of the IEEE International Conference on Image Processing*, **2**, 331–335, 1994.

[26]   Egger, O., Wei, L. and Kunt, M., High compression image coding using an adaptive morphological subband decomposition. *IEEE Proceedings*, **83** (2), 272–287, 1995.

[27]   Egger, O. and Wei, L., Very low bit rate image coding using morphological operators and adaptive decomposition. *Proceedings of the IEEE International Conference on Image Processing*, **2**, 331–335, 1994.

[28]   Redmill, D. and Bull, D., Non-linear perfect reconstruction filter banks for image coding. *Proceedings of the IEEE International Conference on Image Processing*, **1**, 593–596, 1996.

[29]   Serra, J., *Image Analysis and Mathematical Morphology*, Academic Press, New York, 1982.

[30]   Giardina, C. and Dougherty, E., *Morphological Methods in Image and Signal Processing*, Prentice Hall, Englewood Cliffs, NJ, 1988.

[31]   Heijmans, J., *Morphological Image Operators*, Academic Press, New York, 1994.

[32]   Serra, J. and Vincent, L., An overview of morphological filtering. *Circuits, Systems and Signal Processing*, **11** (1), 47–108, 1992.

[33]   Serra, J., Introduction to mathematical morphology. *Comput. Vision Graph. Image Process.*, **35**, 283–305, 1986.

[34]   Sternberg, R., Grayscale morphology. *Comput. Vision Graph. Image Process.*, **35**, 333–355, 1986.

[35]   Haralick, R., Sternberg, S. and Lee, J., Image analysis using mathematical morphology. *IEEE Transactions Pattern Analysis and Machine Intelligence*, **9** (4), 532–550, 1987.

[36]   Maragos, P. and Schafer, R., Morphological filters. Part I: Their set-theoretic analysis and relation to linear filters. *IEEE Transactions. Acoustics, Speech and Signal Processing*, **35**, 1153–1169, 1987.

[37]   Maragos, P. and Schafer, R., Morphological filters Part II: Their relation to median, order statistic and stack filters. *IEEE Transactions. Acoustics, Speech and Signal Processing*, **35**, 1170–1184, 1987.

[38]   Lantuéjoul, C. and Maisonneuve, F., Geodesic methods in quantitative image analysis. *Pattern Recognition*, **17**, 117–187, 1984.

[39]   Salembier, P., Serra, J. and Bangham, J., Edge versus contrast estimation of morphological filters. In *Proceedings IEEE International Conference on Acoustics, Speech and Signal Processing*, **5**, 45–48, 1993.

[40]   Ko, S., Morales, A. and Lee, K., Fast recursive algorithms for morphological operators based on the basis matrix representation. *IEEE Transactions on Image Processing*, **5** (6), 1073–1076, 1996.

[41]   Vincent, L., Morphological grayscale reconstruction in image analysis: applications and efficient algorithms. *IEEE Transactions on Image Processing*, **2** (2), 176–201, 1993.

[42]   Vincent, L., Morphological algorithms. In Dougherty, E. (ed.), *Mathematical Morphology in Image Processing*, Marcel Dekker, New York, 255–288, 1993.

[43]   Bleau, A., de Guise, J. and Leblanc, R., A new set of fast algorithms for mathematical morphology: idempotent geodesic transforms. *Computer Vision, Graphics and Image Processing: Image Understanding*, **56** (2), 178–209, 1992.

[44]   Bleau, A., de Guise, J. and Leblanc, R., A new set of fast algorithms for mathematical morphology: identification of typographic features on grayscale images. *Computer Vision, Graphics and Image Processing: Image Understanding*, **56** (2), 210–227, 1992.

[45]   Gerken, P., Object based analysis-synthesis coding of image sequences at very low bit-rates. *IEEE Transactions Circuits and Systems for Video Technology*, **4** (3), 228–236, 1994.

[46]   Eryurtlu, F. and Kondoz, A., Very low bit rate segmentation based video coding. *IEEE Proceedings – Vision, Image and Signal Processing*, **142** (5), 253–261, 1995.

[47]   Rajala, S., Civanlar, M. and Lee, W., Video data compression using three-dimensional segmentation based on HVS properties. *Proceedings IEEE International Conference on Acoustics, Speech and Signal Processing*, **2**, 1092–1095, 1985.

[48]   Musmann, H., Hotter, M. and Ostermann, J., Object oriented analysis-synthesis coding of moving images. *Signal Processing: Image Communications*, **1** (2), 117–138, 1989.

[49] Dieh, N., Object-oriented motion estimation and segmentation in image sequences. *Signal Processing: Image Communications*, **3**, 23–56, 1991.

[50] Salembier, P., Meyer, F., Brigger, P. and Bouchard, L., Morphological operators for very low bit rate video coding. In *Proceedings of the IEEE International Conference on Image Processing*, **3**, 659–662, 1996.

[51] Salembier, P., Morphological multiscale segmentation for image coding. *Signal Processing*, **38**, 359–386, 1994.

[52] Salembier, P. and Serra, J., Flat zones filtering, connected operators and filters by reconstruction. *IEEE Transactions on Image Processing*, **4** (8), 1153–1160, 1995.

[53] Oliveras, A. and Salembier, P., Generalised connected operators. *Proceedings SPIE Visual Communications and Image*, **2727**, 762–772, 1996.

[54] Freeman, H., On the encoding of geometric configurations. In Lipkin, B. S. and Rosenfeld, A. (eds), *Picture Processing and Psychopictorics*, Academic Press, New York, 241–263, 1970.

[55] Schuster, G. and Katsaggelos, A., An optimal segmentation encoding scheme in the rate distortion sense. *Proceedings IEEE International Symposium on Circuits and Systems*, Atlanta, **2**, 640–643, 1996.

[56] Reinhardt, J. and Higgins, W., Toward efficient morphological shape representation. *Proceedings IEEE International Conference on Acoustics, Speech and Signal Processing*, **5**, 125–128, 1993.

[57] Pardàs, M. and Salembier, P., '3D morphological segmentation and motion estimation for image sequences. *Signal Processing*, **38**, 31–43, 1994.

[58] Clarke, R., *Digital Compression of Still Images and Video*, Academic Press, New York, 175–181, 1995.

[59] Casas, J., Salembier, P. and Torres, L., Morphological interpolation for texture coding. *Proceedings IEEE International Conference on Image Processing*, **1**, 526–529, 1995.

[60] Salembier, P. and Pardàs, M., Hierarchical morphological segmentation for image sequence coding. *IEEE Transactions on Image Processing*, **3**, 639–651, 1994.

[61] Salembier, P., Torres, L., Meyer, F. and Gu, C., Region-based video coding using mathematical morphology. *Proceedings IEEE*, **83** (6), 843–857, 1995.

[62] Salembier, P., Marqués, F., Pardàs, M., Morros, J., Corset, I., Jeannin, S., Bouchard, L., Meyer, F. and Marcotegui, B., Segmentation-based video coding system allowing the manipulation of objects. *IEEE Transactions on Circuits and Systems for Video Technology*, **7** (1), 60–74, 1997.

[63] Musmann, H., Pirsch, P. and Grallert, H., Advances in picture coding. *Proceedings IEEE*, **73** (4), 523–548, 1985.

[64] Dufaux, F. and Moscheni, F., Motion estimation techniques for digital TV: a review and a new contribution. *Proceedings IEEE*, **83** (6), 858–876, 1995.

[65] Kruse, S. and Kauff, P., Fine segmentation of image objects by means of active contour models using information derived from morphological transformations. *Proceedings of the SPIE Visual Communications and Image Processing Conference*, **2727**, 1164–1172, 1996.

[66] Choi, J., Lee, S. and Kim, S., Spatio-temporal video segmentation using a joint similarity measure. *IEEE Transactions on Circuits and Systems for Video Technology*, **7** (2), 279–286, 1997.

[67] ISO 10918. JPEG digital compression and coding of continuous-tone still images, 1991.

[68] ISO/IEC CD 13818. Generic coding of moving pictures and associated audio (MPEG-2), 1995.

[69] ITU-T Draft Recommendation H.263. Video coding for low bitrate communication, May 1995.

[70] Kaneko, M., Hatori, Y. and Koike, A., Improvements of transform coding algorithm for motion-compensated interframe prediction errors – DCT/SQ coding. *IEEE Journal on Selected Areas in Communications*, **5**, 1068–1078, 1987.

[71] Pearlman, W. and Jakatdar, P., The effectiveness and efficiency of hybrid transform/DPCM interframe image coding. *IEEE Transactions on Communications*, **32**, 832–838, 1984.

[72]   Talluri, R., A hybrid object-based video compression technique. *Proceedings of the IEEE International Conference on Image Processing*, **3**, 387–390, 1996.

[73]   Qian, D., A motion compensated subband coder for very low bit-rates. *Signal Processing: Image Communications*, **7**, 397–418, 1995.

[74]   Vaidyanathan, P. P., *Multirate Systems and Filter Banks*, Prentice Hall, Englewood Cliffs, NJ, 1993.

[75]   Toet, A., A morphological pyramidal image decomposition. *Pattern Recognition Letters*, **9**, 255–261, 1989.

[76]   Zhou, Z. and Venetsanopoulos, A. N., Morphological methods in image coding. *Proceedings IEEE International Conference on Acoustics, Speech and Signal Processing*, **3**, 481–484, 1992.

[77]   Redmill, D., Image and video coding for noisy channels. PhD thesis, Cambridge University, 1994.

[78]   Florêncio, D. and Schafer, R., Perfect reconstruction non-linear filter banks. *Proceedings IEEE International Conference on Acoustics, Speech and Signal Processing*, **3**, 1815–1818, 1996.

[79]   Hampson, J. and Presquet, J., A nonlinear subband decomposition with perfect reconstruction. *Proceedings IEEE International Conference on Acoustics, Speech and Signal Processing*, **3**, 1523–1526, 1996.

[80]   Redmill, D. and Bull, D., Non-linear filter banks for error-resilient image compression. *Proceedings of the IEE International Conference on Image Processing and Applications*, 91–95, 1997.

23

Scalable Image and Video Coding Algorithms

S. Thillainathan, D. R. Bull and C. N. Canagarajah

23.1. INTRODUCTION

For many applications, especially those involving mobility, it is important to be able to support a range of terminal types and bit rates. This requires the image and video codecs to be capable of optimising their performance dependent on the available bandwidth and decoder capability. Thus, scalability has become an important requirement for current and future video coding applications. In the context of image and video coding, scalability is used to denote the property of the encoded bitstream that enables it to support various bandwidth constraints or display resolutions. In order to illustrate the importance or usefulness of scalability, consider a video-conferencing scenario involving a number of participants with end-user equipment of different hardware capabilities, requiring access to video information through different types of networks. In such a situation, it is important not to compromise the quality achievable with a high specification equipment via a high bandwidth network while supporting a user at the other end of the spectrum. On the other hand, the transmission of a high-rate bitstream to all receivers would result in the corruption of data received by decoders connected via low bandwidth networks due to buffer overflow problems.

Recent years have seen the boundaries between the different facets of multimedia applications – telecommunications, broadcasting, computing and various interactive multimedia applications – disappearing, leading to a myriad of potential digital video applications. Interactive features in broadcasting could allow the viewers to choose their favourite presenter and language, view a scene from different angles or click on an object in the scene to obtain more information about it. Such content-based interactivity is also at the heart of many other interactive services like content-based audio-visual database access or off-line editing in the bitstream domain. These emerging digital video applications have led to standardisation efforts in the form of MPEG-4 and MPEG-7 to support a content-based environment capable of addressing new services. In order to provide these content-based services over heterogeneous networks, content-based coding needs to incorporate the concept of scalability.

The different types of scalability can be summarised as follows:

1. *Spatial or temporal scalability:* (multiresolution scalability) for providing video sequences at different frame sizes and/or frame rates. This is useful in video database browsing, video-conferencing applications and for format conversions when supporting related standards.
2. *SNR scalability*: for providing different qualities of reconstruction while supporting different bandwidth constraints and also enabling prioritisation of data which is useful for certain error protection schemes or packet-based networks.
3. *Content-based scalability*: includes one or all of the scaling methods mentioned above in the context of objects within an image. This is useful for video database browsing and transmission over heterogeneous networks while supporting content-based interactivity.

23.2. SCALABILITY WITHIN MPEG-2

MPEG-1 and MPEG-2 [1–4] were designed primarily to provide highly efficient coding for storage and transmission of pixel-based video and audio. MPEG-2 is a generic video coding standard supporting a range of applications at bit-rates from about 2 to 30 Mbps and is based on a toolkit approach with different algorithmic tools or profiles developed for various applications. Provisions for scalability are included in the form of three scalable profiles: SNR, spatial and high profile [4–6]. These scalable profiles are extensions of the non-scalable main profile coder. The non-scalable hybrid coder consists of a motion prediction/compensation stage (MCP) for predicting the current frame. The displaced frame difference (DFD), the difference between the current and the predicted frame, is transformed using a discrete cosine transform (DCT) followed by quantisation of the DCT coefficients and entropy coding before transmission or storage.

The *SNR scalable profile* provides layers of different qualities of reconstruction at a fixed spatial and temporal resolution. The coder that supports SNR scalability in MPEG-2 is illustrated in Figure 23.1. The DCT coefficients corresponding to the DFD signal are coarsely quantised and transmitted as the base layer. The resulting error is quantised using a finer step size and transmitted as the next higher layer. The base layer decoder output will exhibit some prediction drift error as the prediction stage at the decoder, which uses only the coarsely quantised coefficients, is not identical to that at the encoding end where finely quantised coefficients are used. However, the prediction drift in the lower layer is barely perceptible and the resulting quality is acceptable for many applications [5]. SNR scalability is intended for tele-communications and video services requiring multiple qualities. Example applications include digital terrestrial television broadcasting and delivery over ATM.

The *spatial scalable profile* uses a lower layer coder that is identical to the main profile non-scalable coder. The upper layer coder uses a prediction that is based on a weighted combination of the decoded up-sampled image from the lower layer (spatial prediction) and a motion compensated prediction from the upper layer (temporal prediction). Such a spatial scalable scheme provides optimum performance at low

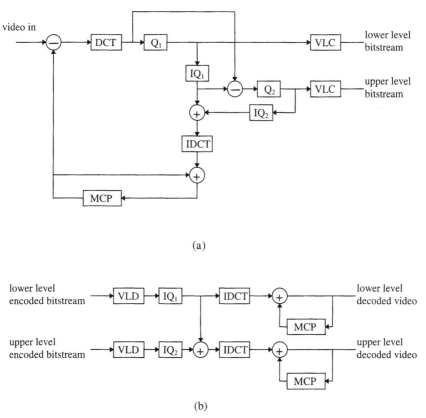

Figure 23.1 SNR scalable video codec; (a) a two level video encoder; (b) a two level video decoder.

resolution. However, the penalty paid for the spatial scalable property is in terms of coding performance of the highest layer which lies between that of simulcast and single resolution coding. Spatial scalability is useful for video standard conversion as in the provision of HDTV service with downward-compatibility to standard definition TV. It could also find applications in video database browsing systems. The *high scalable profile* is used for transmission or storage of studio quality (4:2:2) video signals as enhancements to the 4:2:0 signal and is based on the SNR and spatial scalable profiles.

23.3. WAVELET METHODS

One of the most efficient wavelet-based image coders for a noiseless channel is the zerotree algorithm developed by Shapiro [7] and improved further by Said and Pearlman [8,9]. The zerotree coding algorithm generates an embedded bitstream from which subsets can be extracted to support a range of bit-rates with reconstructions that are optimum in the mean squared error (MSE) sense. The highly scalable and high compression properties are achieved by using a variable-length coding method that depends on the previously coded data. The 2-D zerotree method has been extended to a 3-D technique for video coding by including the temporal domain [10,11]. The technique provides scalable SNR property with continuous scalability over the bit-rate range supported. In other words, the decoding can be stopped at any arbitrary point in the bitstream to yield a reconstruction that is optimum in the MSE sense.

In zerotree coding algorithm, the wavelet transformed coefficients are encoded by utilising the inherent hierarchical data structure. With the exception of a few coefficients in the lowest spatial frequency subband, every coefficient is related to a set of 2×2 coefficients at the next higher level thus giving rise to spatial orientation trees as shown in Figure 23.2. The coefficients in the subbands are scanned in a pre-determined fashion and their values are compared with an octavely decreasing threshold. The success of the zerotree coding technique mainly arises from the efficient coding

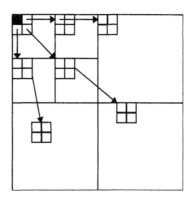

Figure 23.2 Relationship between coefficients in the spatial-orientation tree.

of the significance information. At low bit-rates where large values are used for the threshold, it is highly likely that, if the coefficient at the root of the spatial orientation tree is insignificant, then all the coefficients further down in the tree are also insignificant. In this case, the insignificance of the tree with respect to a particular threshold value can be conveyed by a single symbol.

The performance of the zerotree algorithm (SPIHT) is compared with the JPEG in Figure 23.3. It is clear from this figure that the performance of the SPIHT is superior to that obtained with JPEG algorithms. The main reason for this improvement is that SPIHT is based on a wavelet decomposition of the image which is better than the DCT scheme used in JPEG. Furthermore the JPEG header information needs to be coded, which increases the overhead in coding this image. This explains the horizontal shift of the JPEG graph. It is important to note that the scalable nature of zerotree allows the decoding of the image at any desired bitrate unlike the JPEG system which must be encoded for each required bitrate. Figure 23.4 shows the progressive increase in quality of the zerotree coding scheme at different target bitrates.

3-D subband coding methods are being considered as an alternative to the hybrid motion compensated/DCT approach used in almost all current standards. Podilchuk *et al.* [12] use a 3-D subband coder together with a number of quantisation methods to achieve fixed rate systems suitable for different bit-rate applications. For sequences with small amounts of motion, as is often encountered in videophone sequences, the amount of energy in the higher spatio-temporal subbands will be

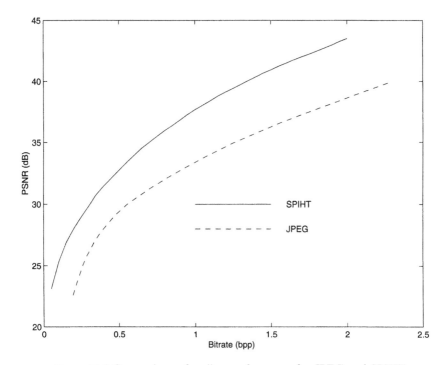

Figure 23.3 Comparison of coding performance for JPEG and SPIHT.

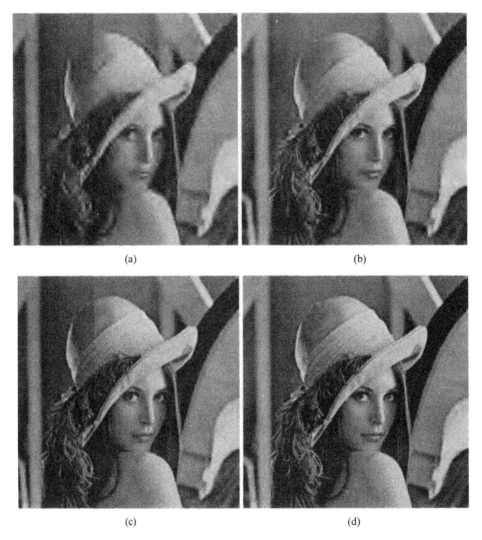

(a)   (b)

(c)   (d)

Figure 23.4 Decoded images using SPIHT at (a) 0.1 bpp, 25.32 dB, (b) 0.2 bpp, 28.00 dB, (c) 0.5 bpp, 32.74 dB, (d) 1.0 bpp, 37.67 dB.

negligible. For such cases, energy compaction using methods that quantise the coefficients corresponding to a few low frequency spatio-temporal subbands, produce acceptable performance without any motion adaptation stage.

However, for high motion sequences, large-valued coefficients are distributed across the high spatio-temporal subbands and hence efficient quantisation methods are required to encode the significant coefficients prior to the less significant ones. One such method is the zerotree coding technique which was developed for coding images [7–9] and subsequently extended for coding video sequences [10,11]. A number of 3-D video coding techniques incorporating motion estimation and compensation have also been proposed to overcome this problem. Ngan and Chooi [13] propose an algorithm where the subband analysis is performed in the temporal direction followed by motion estimation in the temporal frequency domain. Ohm [14] proposes a video coding algorithm where motion estimation and compensation are performed prior to subband analysis.

The highly scalable video compression method proposed by Taubman and Zakhor [15,16] achieves both mulitresolution and multi-rate scalability with fine gradation of bit-rates over the range of several tens of kilobits per second to several megabits per second. As noted earlier, 3-D subband technique without any motion adaptation is only suitable for video sequences that exhibit very small amounts of motion. In most sequences, a significant proportion of scene motion results from camera pan and hence the scalable method of Taubman and Zakhor [16,17] uses an invertible camera pan compensation to eliminate the effects of such motion before using separable 3-D subband analysis. The different spatio-temporal sub-bands are selected on the basis of their relative importance in the reconstruction. Thus, the 3-D subband/wavelet structure provides an inherent method for achieving multi-resolution scalability. The progressive quantisation and coding of each sub-band leads to multi-rate scalability. This work has been extended further to facilitate simple, generic scaling of the highly scalable compressed data using constant bit-rate and constant distortion scaling criteria [17].

23.4. SCALABILITY WITHIN MPEG-4

The objective of MPEG-4 is to formulate an audio-visual representation that can support a wide variety of multimedia applications. Some of the functionalities specified in the MPEG-4 specification include:

- content-based interactivity and bitstream manipulation;
- content-based scalability;
- robustness in error-prone environments;
- coding of both natural and synthetic scenes.

These are in addition to the requirement for improved coding efficiency especially at low bit-rates. The most important feature that differentiates MPEG-4 from all the other video coding standards is the data representation that supports access to content or objects in the video sequences over all types of networks, i.e. content-based scalability.

There is currently no standard system for achieving content based coding. The MPEG-4 Video Verification Model [18], which provides an important platform for collaborative development of the standard, introduces the concept of video objects to enable the content-based interactive features. Each video frame is segmented into regions known as video object planes (VOPs) that may or may not correspond to the actual or physical objects in the scene. Relevant information such as shape, motion and texture corresponding to the same regions in the successive frames (video objects) are coded into separate video object layers (VOL). Additional data conveying the relation of one video object layer to another, necessary to compose the scene at the receiver (compositor), is also included in the bitstream. Such a video representation allows separate decoding of each video object layer and hence, enables manipulation of video sequences at the bitstream level. This leads to a sequence that is identical to that at the encoding end or an entirely different sequence formed by including objects from a number of different sequences.

To achieve content-based scalability, i.e. to support spatial and temporal scalability at the VOP level, the shape, motion and texture information corresponding to each VOL needs to be coded in a scalable manner. As the MPEG-4 Video Verification Model gives some indication about the structure of the final MPEG-4 video coding standard, we shall first consider the techniques adopted in the Verification Model for coding these information.

The location of each arbitrarily shaped video object plane to be coded is defined with respect to a reference window of constant size. The shape information, referred to as alpha plane, is encoded using either a binary or grey-scale shape coding technique. A shape adaptive macroblock grid of dimensions 16×16 pixels is used for alpha plane coding, motion estimation and compensation as well as for block-based texture coding. Motion compensation is performed on a block (8×8 pixels) or macroblock basis. The motion-compensated prediction error is calculated by subtracting each pixel in a block or macroblock belonging to a VOP in the current frame from a motion-compensated VOP in the previous frame. The motion-compensated prediction error and the first VOP in the video object sequence (Intra-VOP) are texture coded using 8×8 DCT followed by quantisation of the DCT coefficients and subsequent run-length and entropy coding.

Spatial scalability is achieved by spatially downscaling the video object to the required resolution level and encoding the lowest resolution video object as the base layer. The low resolution image is upscaled and used as prediction for the next higher resolution video object image. The prediction difference is coded in the enhancement layer. Similarly, temporal scalability can be achieved by providing a temporal prediction for the enhancement layer based on coded video from lower layers.

A number of segmentation based methods have been proposed [19,20] to extract and encode separately the shape, motion and texture information of a video sequence. Salembier *et al.* [19] propose a video coding system that uses a spatio-temporal segmentation method to allow efficient manipulation of objects in a scene. Each video frame is partitioned into regions and partition and texture information corresponding to the intraframe necessary to initiate the coding process and to refresh the information at the receiver is periodically transmitted. The partition projection stage adapts the partition of the previous frame to the current frame before estimating motion for each region.

The hybrid motion-compensated DCT structure has also been proposed for object-based coding. Neff and Zakhor [21] and Banham and Brailean [22] propose a system in which a matching pursuits algorithm is used to code the motion residual (DFD) signal. The motion estimation stage uses a technique similar to that used in the H.263 standard with an overlapping motion model to improve performance at very low bit-rates. The matching pursuit technique developed by Mallat and Zhang [23] uses a dictionary of overcomplete basis functions to encode the DFD signal structures more accurately. The algorithm decomposes the motion residual signal iteratively using the dictionary elements (atoms) that provide the greatest reduction in the MSE between the true and the coded signal. This enables the visually prominent features to be encoded prior to the less significant ones. The dictionary comprises an overcomplete collection of 2-D separable Gabor functions. The algorithm can be easily extended to provide SNR scalability and also region-based scalability by selectively coding the residual structures from the DFD signal that correspond to the required objects.

23.5. ERROR-RESILIENT ZEROTREE CODING ALGORITHM

23.5.1. The Influence of Errors on the Bit-stream

In the zerotree coding algorithm, the wavelet transformed coefficients are encoded by utilising the inherent hierarchical data structure. The bits in a data stream generated by the Said and Pearlman coder [9] can be one of the following three types:

- *SIG bit* – conveys the decision of significance comparisions;
- *sign bit* – conveys the sign of significant coefficients;
- *refinement bit* – conveys the refinement information for coefficients that have been encoded.

If an error occurs in either the *sign bit* or the *refinement bit* of an EZW coded image, it does not propagate to the adjacent data. It affects the value of just one coefficient in the reconstructed wavelet transform and a region in the reconstructed image. This is dependent on (i) the number of levels of wavelet decomposition used, (ii) the subband level at which the coefficient error occurs and (iii) the length of the synthesis filter. An error in the *SIG* bit however causes the rest of the bits in the data stream to be decoded incorrectly as the decoder goes into an unknown state (i.e. out of step with the encoding decisions). Therefore, even a single bit error can cause the decoding algorithm to fail if it occurs in the *SIG bit*. Figure 23.5 illustrates the effect of bit errors. Figures 23.5(a) and 23.5(b) show the effect of single bit errors in *SIG bit* and

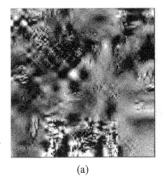  

(a)   (b)   (c)

Figure 23.5 Image Lena coded at 1 bpp using the original zerotree algorithm: (a) with a bit error in a SIG bit; (b) with a bit error in a sign bit; (c) with multiple bit-errors half-way down the bitstream.

sign bit near the start of the bitstream and Figure 23.5(c) shows the effects of bit errors in 3 bits (each belonging to the 3 different categories) half-way down the scalable bitstream. It can be noted that the data at the start of the zerotree encoded bitstream has a greater influence on the quality of the reconstruction compared to the rest of the bitstream.

23.5.2. Error-resilient Zero-tree Coding Algorithm

The error-resilient entropy coding (EREC) algorithm [24] (see Chapter 25) is a simple technique that improves the error-resilience of variable-length data produced by many block-based algorithms. An important requirement for the EREC algorithm is that the decoder should be capable of identifying the end of each block in the absence of channel errors without any reference to previous or subsequent data. As the zero-tree algorithm of Said and Pearlman is not a block-based technique, we have modified this algorithm to produce blocks of variable length data. The data corresponding to each group of 2×2 coefficients in the lowest spatial frequency subband and their related descendants for a particular threshold value are coded in one block. In the original technique, data corresponding to all groups of 2×2 coefficients in the lowest spatial frequency subband were coded prior to the data corresponding to their descendants. Thus, the original technique is slightly more efficient in the MSE sense compared to that of the modified technique. However, this is traded off against the improved error-resilience that can be achieved.

For each value of threshold, there are M variable length blocks of data. The value of M depends on the frame size and the number of levels of wavelet decomposition used. Groups of N blocks (where N is a factor of M) are considered at a time and, converted into EREC frames composed of N slots each of s_i bits in length by using a simple bit reorganisation algorithm. If T is the total number of bits in a group of N blocks to be coded, then it satisfies

$$T = \sum_{i=0}^{N-1} s_i$$

The value of the total number of bits, T, encoded in an EREC frame and the number of slots, N, per frame need to be known at the decoder. These parameters are transmitted as highly protected header information before the data corresponding to each EREC frame. The EREC algorithm is described more fully in Chapter 25.

The effect of errors in an EREC frame can be summarised as follows. Any errors in an EREC slot may mean that the corresponding block is decoded erroneously. This could result in the end of the block being identified incorrectly. A half-decoded block that searches a slot with errors may also be decoded wrongly. Thus, in the worst case, it is possible for all N blocks of an EREC frame to be wrongly decoded. Therefore, it is desirable to keep the EREC frame dimension, N, small. However, as each EREC frame has an associated header that needs to be transmitted, increasing the number of EREC frames leads to an increase in the total amount of transmitted data for a given quality of reconstruction.

With zerotree coded data, the decoding of an EREC frame is not independent of the related previous EREC frames. The correct decoding of a block in the threshold region $T_0/2^i$; for $i = j$ (T_0 = initial threshold value) depends on correctly decoding the related blocks in the previous threshold regions $T_0/2^i$; for $i < j$. Thus, erroneous decoding of a block in the previous frame may cause the corresponding block and possibly a number of other blocks within the current EREC frame to be decoded incorrectly. In summary, errors can only propagate to related EREC frames and also within an EREC frame.

The following results were obtained with grey-scale 8 bpp, 256×256 image Lena. A separable biorthogonal 9/7 wavelet filter pair [25] was used to obtain a 2-D wavelet transformed data structure with four levels of wavelet decomposition. The initial header information containing the values of parameters such as image dimensions, initial threshold value and the subsequent header information for each EREC frame were encoded using the (24,12) extended Golay code which is capable of correcting up to 3 errors per codeword block. Figure 23.6 shows that the proposed error-resilient algorithm confines the effects of the error depicted in Figure 23.5(a) and 23.5(c) to a smaller region in the reconstructed frame.

Using a large number of levels of wavelet decomposition results in highly efficient compression, particularly, at low bit-rates. However, the area in the reconstruction affected by an error in a *SIG* bit is comparatively large for higher levels of decomposition as shown in Figure 23.7. We have therefore used four levels of wavelet decomposition in the proposed coder and each EREC frame was composed of $N=8$ blocks of data.

Figure 23.8 compares the variation in the quality of reconstruction (in terms of PSNR values) with bit-rate in an error-free channel for the original zerotree algorithm and the proposed error-resilient algorithm. It can be noted that the proposed algorithm is only slightly less efficient than the original SPIHT algorithm [9] but this is more than compensated for by the improved resilience to errors in channels with BER values up to 0.01 as shown in Figure 23.9. The effect of channel noise at different error probabilities for the proposed algorithm

(a)   (b)   (c)

Figure 23.6 (a) Original image, reconstructed image using the error-resilient zerotree image coder in the presence of (b) an error in a SIG bit, (c) multiple errors half-way down the bitstream.

(a)   (b)

Figure 23.7 Image coded at 1 bpp using the proposed error-resilient zerotree algorithm with (a) 5 levels, (b) 4 levels of wavelet decomposition, both in the presence of a bit error in a SIG bit.

is shown in Figure 23.10. A graceful degradation in the quality of reconstruction can be observed as the channel BER is increased.

23.6. CONCLUSIONS

In this chapter, we have described the requirements of and options for implementing a scalable image and video codec. We have presented an error-resilient version of the zerotree algorithm for coding still images and video sequences. The proposed algorithm provides significant resilience to channel errors at the expense of a minimum amount of added redundancy. It was noted in Section 23.5.1 that the data at the start

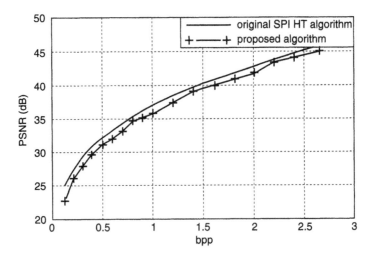

Figure 23.8 Variation of PSNR with bit-rate.

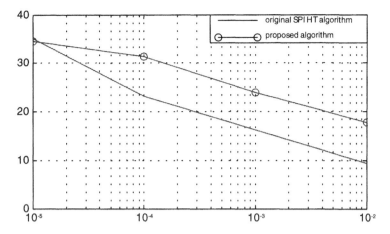

Figure 23.9 Variation of PSNR with bit error rate (BER) @ 1 bpp.

(a)　　　　　　　　　　(b)　　　　　　　　　　(c)

Figure 23.10 The effects of channel noise on the image Lena coded at 1 bpp for different values of channel BER: (a) 0.01%; (b) 0.1%; (c) 1%.

of the zerotree encoded bitstream has a greater influence on the quality of the reconstruction compared to the rest of the bitstream. This implies that it should be possible to further improve the performance of the error-resilient coder by using some prioritisation or selective error-correction scheme.

REFERENCES

[1]　MPEG Video Committee Draft, volume ISO-IEC/JTC1/SC2/WG8/MPEG, International Organisation for Standardisation, 18 December 1990.

[2]　Le Gall, D., MPEG: A video compression standard for multimedia applications. *Communications of the ACM*, **34** (9), 47–58, 1991.

[3]　Morris, J., MPEG 2: the main profile. In Toumazou, C. (ed.), *Circuits and Systems Tutorial, IEEE Symposium on Circuits and Systems*, 107–120, 1994.

[4]   Tudor, P. N., MPEG-2 video compression. *IEE Electronics & Communications Engineering Journal*, 257–264, December 1995.

[5]   Wells, N. D. and Tudor, P. N., Standardisation of scalable coding schemes. *Circuits and Systems Tutorials, IEEE Symposium on Circuits and Systems*, 121–130, 1994.

[6]   Chiang, T. and Anastassiou, D., Hierarchical coding of digital television. *IEEE Communications Magazine*, 38–45, May 1994.

[7]   Shapiro. J., Embedded image coding using zerotrees of wavelet coefficients. *IEEE Transactions on Signal Processing*, **41**, 3445–3462, 1993.

[8]   Said, A. and Pearlman, W. A., Image compression using the spatial-orientation tree. *IEEE International Symposium on Circuits & Systems*, Chicago, May, 279–282, 1993.

[9]   Said, A. and Pearlman, W. A., A new, fast and efficient image codec based on set partitioning in hierarchical trees. *IEEE Transactions on Circuits and Systems for Video Technology*, **6** (3), 243–249, 1996.

[10] Chen. Y. and Pearlman, W. A., Three-dimensional subband coding of video using the zero-tree method. *Proceedings of SPIE*, **2727**, 1302–1312, 1996.

[11] Kim, B.-J. and Pearlman, W. A., An embedded wavelet video coder using three dimensional set partitioning in hierarchical trees (SPIHT). In *Proceedings of the IEEE Data Compression Conference*, 252–260, March 1997.

[12] Podilchuk, C. I., Jayant, N. S. and Farvardin, N., Three-dimensional subband coding of video. *IEEE Transactions on Image Processing*, **4** (2), 125–139, 1995.

[13] Ngan, K. N. and Chooi, W. L., Very low bit rate video coding using 3-D subband approach. *IEEE Transactions in Circuits and Systems for Video Tech.*, **4** (3), 309–316, 1994.

[14] Ohm, J.-R., Three-dimensional subband coding with motion compensation. *IEEE Transactions on Image Processing*, **3** (5), 559–571, 1994.

[15] Taubman, D. and Zakhor, A., Highly scalable, low-delay video compression. *International Conference on Image Processing*, Austin, Texas, Vol I Ch. 198; pp. 740–744, 1994.

[16] Taubman, D. and Zakhor, A. Multirate 3D subband coding of video. *IEEE Transactions on Image Processing*, **3** (5), 572–589, 1994.

[17] Taubman, D. and Zakhor, A., A common framework for rate and distortion based scaling of highly scalable compressed video. *IEEE Transactions on Circuits and Systems for Video Technology*, **6** (4), 329–353, 1996.

[18] Sikora, T., The MPEG-4 video standard verification model. *IEEE Transactions on Circuits and Systems for Video Technology*, **7** (1), 19–31, 1997.

[19] Salembier, P., Marques, F. *et al.*, Segmentation-based video coding system allowing the manipulation of objects. *IEEE Transactions on Circuits and Systems for Video Technology*, **7** (1), 60–73, 1997.

[20] Talluri, R., Oehler, K. *et. al.*, A robust, scalable, object-based video compression technique for very low bit-rate coding. *IEEE Transactions on Circuits and Systems for Video Technology*, **7** (1), 221–233, 1997.

[21] Neff, R. and Zakhor, A., Very low bit rate video coding based on matching pursuits. *IEEE Transactions on Circuits and Systems for Video Technology*, **7** (1), 158–171, 1997.

[22] Banham, M. R. and Brailean, J. C., A selective update approach to matching pursuits video coding. *IEEE Transactions on Circuits and Systems for Video Technology*, **7** (1), 119–129, 1997.

[23] Mallat, S. and Zhang, Z., Matching pursuits with time-frequency dictionaries. *IEEE Transactions on Signal Processing*, **41**, 3397–3415, 1993.

[24] Redmill, D. W. and Kingsbury, N. G. The EREC: an error resilient technique for coding variable-length blocks of data. *IEEE Transactions on Image Processing*, **5** (4), 565–574, 1996.

[25] Villasenor, J. D., Belzer, B. and Liao, J., Wavelet filter evaluation for image compression. *IEEE Transactions on Image Processing*, **4** (8), 1053–1060, 1995.

24

Integrated Speech and Video Coding for Mobile Audiovisual Communications

F. Eryurtlu, A. H. Sadka and A .M. Kondoz

24.1. INTRODUCTION

In audiovisual communications, audio and video signals are usually compressed by different coders which work independently. In some cases where different transmission channels are allocated to audio and video signals, this strategy is appropriate. However, most of the time, both signals are transmitted through the same channel and therefore share the same bandwidth. When the available transmission bandwidth is very limited as in mobile communications, efficient bit rate allocation significantly improves the codec's performance. The work presented in this chapter is aimed at integrating speech and video codecs, addressing the advantages and disadvantages of this approach.

Insights into Mobile Multimedia Communication
ISBN 0-12-140310-6

Mobile audiovisual communications require very low bit rate coding of both speech and video signals. Numerous techniques for low bit rate speech or video coding have been proposed in the literature over the past 20 years [1,2]. Most of these try to optimise the trade-off between the output quality and the bit rate for each signal source independently. This may seem logical since the audiovisual information is received by the human eyes and ears separately. However, the ultimate receiver is the brain, hence the audio and video signals complement each other. In the following section, the advantages of variable rate coding of audiovisual signals are discussed. Then, the speech and video codecs chosen for this research are described. After briefly considering the requirements of mobile audiovisual communication applications, efficient bandwidth allocation techniques are presented.

24.2. VARIABLE RATE CODING

In most applications, the available transmission bandwidth is fixed, therefore fixed bit rate codecs are often preferred. It is rather easy to design efficient fixed rate speech codecs. However, for video signals, coding efficiency can be achieved only with variable rates and a buffer is required at the output of the encoder in order to smooth the bit rate fluctuations.

The bit rate should ideally be a function of the information content of the signal. As this changes with time in both speech and video signals, the bit rate also is expected to change. For instance, in telephone conversations, often only one of the parties is speaking at a time. Therefore, the codec switches to the silence mode for the listening party in which it just needs to send overhead information, which means that the bit rate should be very low. Similarly, in video coding, when the motion in the scene is very low, interframe coding is very effective, and therefore the coding rate is very low. In fixed bit rate audiovisual applications, instead of targeting fixed rates for both speech and video codecs, it is more efficient to allow variable bit rates at the coders, but with a buffer placed before the fixed bandwidth channel, smoothing the overall bit rate variations and also controlling the codec rates.

24.3. VIDEO CODING

Since a very low bit rate is the main requirement in mobile communications, a video codec working at bit rates lower than 64 kbits/s is desirable. Unfortunately, current video coding standards are not efficient for very low bit rate video coding. We present here a segmentation-based video codec which can operate at rates as low as 10 kbits/s.

In the algorithm, the image is first passed through a non-linear filter which suppresses impulsive noise while preserving the edges. This improves the segmentation performance and reduces the number of regions. The split-and-merge technique is preferred for segmentation. First, the frame is segmented to variable-sized squares by using a tree structure. In the merging stage, two regions are merged if the mean squared error of the resulting region is below a given threshold. The last two stages of segmentation reduce the number of regions by merging neighbouring regions with low contrast and eliminating small regions.

A segmented frame is represented by contour and texture information. Different techniques are used to encode the contour information of the first and subsequent frames. If the present picture is the first frame of a new scene, then the contours are represented by using 4-connected chains. The chain elements are grouped and Huffman coded. A code is used to indicate the 3- and 4-region points. This code reduces the number of bits required for encoding the positions of the contours. On the other hand, if the present frame is from the same scene as the previous one, then the contour information is encoded using the previous frame. To encode the contours of the present frame, segmentation information of the previous frame is exploited. The shapes of the regions are modified so that their boundaries are formed by several line segments. The connection points of these lines are referred to as control points (CP).

In motion estimation the previous frame is split into 16×16 non-overlapping square blocks, and the motion vector corresponding to the centre of each block is found by using the block-matching method. The number of motion vectors which have to be encoded does not change from one frame to another. However, since the motion vector differences are entropy coded, different numbers of bits are used for motion information in different frames. The motion parameters of the CPs are computed by interpolating the motion vectors of the neighbouring blocks. There are several ways to interpolate the motion vectors for the control points. The easiest method is to choose the motion vector of the block which contains the control point. A more sophisticated solution is to calculate the weighted sum of the neighbouring motion vectors. As the first condition, the response of the weighting function should be dependent only on the distance between the control point and the centre of the block. This means that the weighting function should be symmetric around the origin:

$$W(x, y) = W(-x, y) = W(x, -y) = W(-x, -y) \tag{24.1}$$

In addition, the sum of the weighting factors should be unity for each interpolated point. If only four closest block centroids are used for interpolation, then the following equation referring to Figure 24.1 can be satisfied:

$$W(x, y) + W(x - x_0, y) + W(x, y - y_0) + W(x - x_0, y - y_0) = 1 \tag{24.2}$$

$$\text{where } 0 < x < x_0 \text{ and } 0 < y < y_0$$

The two-dimensional weighting function used in this work is the product of two linear one-dimensional functions in the x and y directions from 1 to 0 in the corresponding direction:

$$W(x, y) = \frac{(x_0 - |x|)(y_0 - |y|)}{x_0 y_0} \text{ if } |x| < x_0 \text{ and } |y| < y_0$$

$$0 \text{ otherwise} \tag{24.3}$$

The motion vectors of the CPs are used to estimate their new positions in the current frame. Then, the CPs are joined by straight lines according to the connection information of the previous frame to approximate the region boundaries. This means that the motion parameters inherently possess the contour shape information as well as the contour positions.

Zero order polynomial approximation, which is based on the mean value of each region, is chosen for texture coding. The position of each region in the present frame is estimated in the previous frame using the motion parameters that are found in the motion estimation stage. Then, the mean value of the estimated region is calculated and the difference between the actual and estimated values is encoded by allocating less bits to the more probable values. Postprocessing is applied in the decoder side in order to enhance the output quality by smoothing the edge profiles of artificial contours and by rectifying the jagged edges that are caused by segmentation. These processes do not change the bit rate since the postprocessing parameters can be preset or adjusted by the decoder.

The proposed algorithm was simulated and tested on the "Miss America" and "Salesman" sequences at a frame rate of 10 frames per second. The PSNR values obtained for QCIF resolution are given in Figure 24.1 while the corresponding bit rates are plotted in Figure 24.2. The performance of the algorithm is clearly dependent on the motion estimation method employed. The block matching method with a block size of 16×16 does not perform well in certain image sequences since it can only estimate the translational motion. Another important point is the size of the search area. If the scene contains some objects moving at relatively high speeds, the range of the motion estimator has to be adjusted accordingly. The most problematic regions for encoding are the ones which do not exist in the previous frames. If the scene includes some CPs around those regions, the quality degradation is not significant. When the bit rate is not extremely low, the noisy background results in several regions which require many control points. Therefore, the control points around are transferred to the new regions. The quantisation step size for the region mean values can be used to control the output bit rate. With varying quantisation levels, the bit rate can be stabilised with some changes in the output quality. For

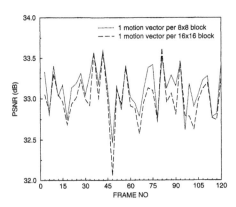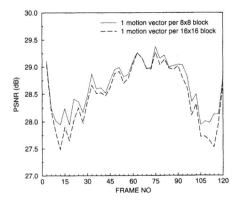

Figure 24.1 PSNR values for QCIF-size Miss-America (left) and Salesman (right) sequences encoded at 10 frames per second by using 3 different ways for motion vectors.

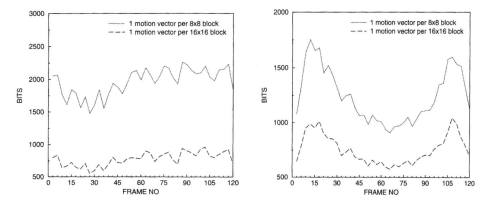

Figure 24.2 Number of bits per frame for QCIF-size Miss-America (left) and Salesman (right) sequences encoded at 10 frames per second by using three different ways for motion vectors.

applications which require a fixed output bit rate, the fluctuations in the rate can be smoothed out by the use of a buffer.

24.4. SPEECH CODING

A new LPC vocoder which splits the LPC excitation into two frequency bands using a variable cut-off frequency is presented [3]. The lower band is responsible for representing the voiced parts of speech, whilst the upper represents the unvoiced speech. In doing so, the coder's performance during both mixed voicing and speech containing acoustic noise is greatly improved, producing soft natural sounding speech. We also describe a new parameter determination and quantisation techniques vital to the operation of this coder at bit rates below 3 kbits/s. We also describe a new parameter determination and quantisation technique vital to the operation of this coder at bit rates below 3 kbit/s.

The split-band LPC vocoder uses vector quantisation techniques to efficiently encode the LPC parameters in the LSF domain, freeing bits which may then be used to encode additional information about the excitation in order to improve the speech quality. The excitation information is extracted by applying a similar technique to the IMBE [4] scheme to the LPC residual, which is then quantised using one of two schemes, resulting in an overall coded bit rate of 2.5 kbits/s or above.

24.4.1. Encoder

Figure 24.3 presents the schematic of the speech encoder. The higher frequencies of speech are pre-emphasised and are processed in 20 ms frames. LPC parameters are determined using a 10th order Durbin's algorithm which are then quantised in the LSF domain. The quantised LPC parameters are then used to find the LPC residual

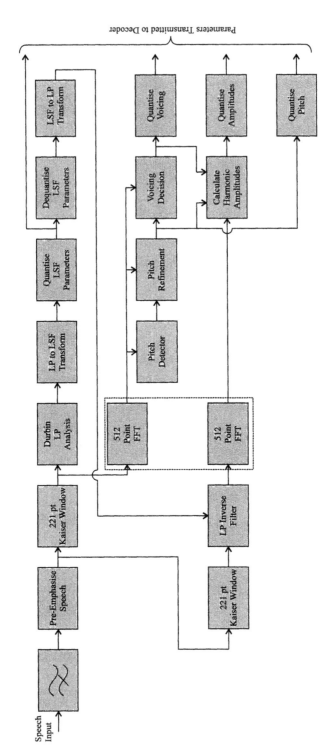

Figure 24.3 Encoder schematic.

required for determination of the excitation harmonic amplitudes. Both the speech signal and the LPC excitation are transformed into the frequency domain using a 512 point FFT (note that these two real FFTs may be calculated using one complex FFT in order to reduce complexity).

Pitch analysis is performed in the spectral domain using a modified version of the algorithm described by McAulay [5] which determines the pitch period to half sample accuracy. The pitch frequency is the value of ω_0 which maximises $\rho(\omega_0)$ in Equation (24.4), where $E(\omega)$ is the exponentially decaying envelope of the speech spectrum shown in Figure 24.4, A_l and ω_l are the magnitudes and frequencies of the local peaks in the speech spectrum. $D(\omega - k\omega_0)$ is given by Equation (24.5) which is non-zero only for the main lobe of the "sinc" function.

$$\rho(\omega_0) = \sum_{k=1}^{L(\omega_0)} E(k\omega_0)\{\max_l[A_l D(\omega_l - k\omega_0)] - \tfrac{1}{2}E(k)\} \tag{24.4}$$

$$D(\omega - k\omega_0) = \left[2\pi\left(\frac{\omega - \omega_0}{\omega_0}\right)\right] \tag{24.5}$$

This algorithm was further improved with regard to robustness to noise by incorporating a further energy based metric. All candidate pitch periods returned by equation (24.1) were re-examined using equation (24.6) and those whose value of exceeded treble the minimum value of $\phi(l)$ for all candidate values of l (the candidate pitch period), were rejected. $\phi(l)$ measures how much the RMS energy of the speech fluctuates as a function of the window length used in the RMS calculation. If the window length is equal to the pitch period, then the variation is small and $\phi(l)$ is also small. An example of $\phi(l)$ is plotted in Figure 24.5 for four consecutive speech frames. This combined scheme was found to be highly reliable on all of the speech material tested, which is fundamental to the overall coder performance.

$$\phi(l) = \left[\sum_{j=0}^{N} |d_l(j)|\right] / \left[\sum_{j=0}^{N} |e_l(j)|\right] \tag{24.6}$$

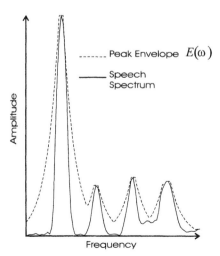

Figure 24.4 Example of spectral peak envelope.

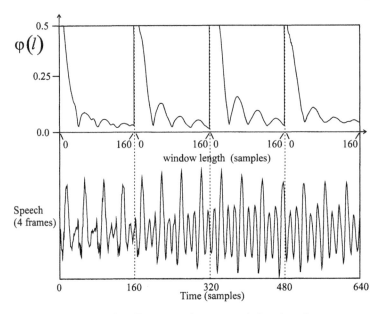

Figure 24.5 Energy variance vs. window length.

$$e_l(j) = \sum_{k=0}^{l}[s(j - \tfrac{1}{2} + k)]^2 \tag{24.7}$$

(where $s(n)$ represents the original speech samples)

$$d_l(j) = d_l(j - 1) + 0.95[e_l(j) - e_l(j - 1)] \tag{24.8}$$

Pitch refinement is performed on the speech spectrum using a method similar to that described by Griffin [4] and a binary voicing decision is performed for each pair of harmonics using a technique similar to the APCO scheme [6]. Finally the harmonic amplitudes are determined from the excitation spectrum, using weighted spectral matching, and the LSF, pitch, voicing and excitation parameters are finally quantised and transmitted to the decoder.

24.4.2. Decoder

Figure 24.6 shows the decoder schematic. Once the parameters have been decoded, the harmonic excitation amplitudes are modified to reduce noise in the LPC valleys, thereby perceptually improving the coder performance. This scheme maintains a flat excitation spectrum and is an alternative to using the more traditional post-filtering technique as proposed by Chen [7]. The excitation amplitudes are modified according to equation (24.9), where $H(i\omega_0)$ is the LPC spectrum sampled at the harmonic frequencies and $P(i\omega_0)$ is the peak interpolated LPC spectrum sampled at the harmonic frequencies, as shown in Figure 24.7. This also shows the suppressing effect on the valleys of the LPC spectrum.

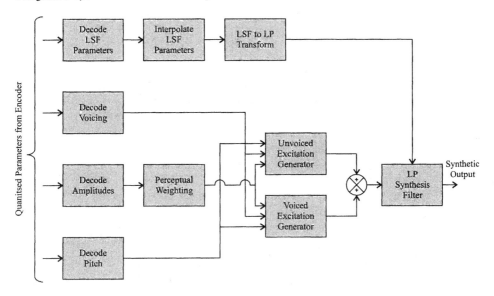

Figure 24.6 Decoder schematic.

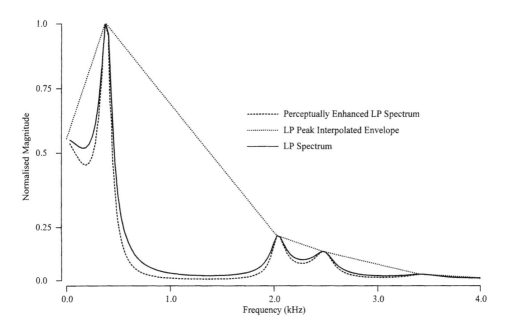

Figure 24.7 Effect of perceptual enhancing process on LPC spectrum.

$$a'(i) = a(i)\left[\frac{H(i\omega_0)}{P(i\omega_0)}\right]^{0.3} \tag{24.9}$$

The synthetic excitation is equal to the sum of the unvoiced and voiced generator outputs. The unvoiced generator is performed using FFT filtering, i.e. spectrally shaping a random noise source in the frequency domain according to the voicing and harmonic amplitude information, then transforming back to the time domain. The voiced excitation is generated by summing up each sinusoidal harmonic, scaled by the decoded harmonic amplitude as per equation 24.10. L is the number of harmonics in the 4 kHz band, which is dependent upon the pitch. $a'(i)$ is the ith perceptually modified amplitude. Note that equation (24.10) does not indicate the use of amplitude interpolation. However, this is performed but has been excluded from this discussion for simplicity. $\phi(i)$ is the phase of the fundamental given by equation (24.11), which is an integration of the fundamental frequency $\omega(i)$ (in radians) which is linearly interpolated between frames if the deviation is less than 20%.

$$e_v(i) = \sum_{j=1}^{j\leq L} a'(i)\cos[j\phi(i)] \tag{24.10}$$

$$\phi(i) = \phi(i-1) + \omega(i) \tag{24.11}$$

Finally the overall spectral shaping is added to the excitation using an LPC synthesis filter whose coefficients are linearly interpolated every 5 ms.

24.4.3. Quantisation

Although a voicing decision is performed for each pair of harmonics the limited number of bits available for encoding forces us to restrict the voicing to adhere to the following rule: all harmonics up to a certain frequency are declared voiced, and those above are declared unvoiced. This means that the voicing information is now represented by a single frequency value which can easily be quantised using three bits. Previously, Yeldener [8] suggested the use of a voicing probability to determine the voicing frequency, given by the ratio of the number of voiced harmonics to the total number of harmonics. This was found to give good performance in general, however in certain circumstances strongly voiced harmonics in the mid-range of the spectrum were incorrectly declared unvoiced, giving the synthetic speech a hoarse quality. A new single frequency voicing quantiser is defined by equation (24.12), which takes into account the original speech harmonic amplitudes $\alpha(j)$ and determines the voicing frequency using a soft-decision process. $v(j)$ represents the unquantised harmonic voicing decisions, where values $1 = $ voiced and $-1 = $ unvoiced. Additionally, another term was incorporated to give more weight to voiced harmonics, given by equation (24.14). This scheme greatly reduced the hoarseness in the synthetic speech.

$$q(i) = \sum_{j=1}^{L} \alpha(j)v(j)u_i(j)b(v(j)) \tag{24.12}$$

$$u_i(j) = \begin{cases} 1 & j \leq \frac{iL}{7} \\ -1 & j > \frac{iL}{7} \end{cases} \tag{24.13}$$

$$b(x) = \begin{cases} 1.5 & x \geq 0 \\ 1.0 & x < 0 \end{cases} \qquad (24.14)$$

LSF quantisation was performed using the linked split-vector approach described by Kim [9] which switches between quantiser tables based on already quantised LSF values. This gives good performance using 28 bits and requires only 448 words of storage. The LSF quantisation scheme is shown in Table 24.1.

The pitch and the RMS value of the harmonics amplitudes are quantised by logarithmic scalar quantisers, using 7 and 6 bits respectively. Finally, the harmonic amplitudes normalised to their RMS value are quantised. The first 8 values are vector quantised using 6 bits. The remainder are assumed to be equal to unity in the 2.5 kbit/s version. Alternatively, they are grouped into 8 bands and then vector quantised in the order of lower to higher frequencies according to the variable bit requirements. The overall bit allocation scheme is presented in Table 24.2.

24.5. INTEGRATION ASPECTS

Mobile audiovisual communications systems have to work at very low bit rates since the available channel bandwidth is very limited. Unlike the bit rate requirement, quality and complexity expectations are highly dependent on the user preferences. The message carried by the audiovisual signal is not evenly distributed to the speech

Table 24.1 LSF quantisation scheme

LSF group	Comment	Bits	Storage/Words
1,2	Switched 2×32	5	64
3,4	Single 128	7	128
5,6	Switched 2×64	6	128
7,8	Switched 2×32	5	64
9,10	Switched 2×32	5	64
	Total	28	448

Table 24.2 Bit allocation schemes

	Version	
Parameter	2.5kbits/s	variable
LSF	28	28
Voicing frequency	3	3
Pitch	7	7
Energy	6	6
1st 8 harmonics	6	6
Excitation Shaping	–	variable
Total	50	> 50

and video signals. In certain moments, the speech is more important, and the low quality video signal does not affect the communication. Obviously, the reverse case is also valid in some other situations. Since the objective is to convey the message, the speech and video codecs have to interact with each other, and a mechanism has to be designed to allocate the bandwidth resource to the speech and video coders.

In some cases, the bit rates of the speech and video codecs can be adjusted automatically. If the speech is in the silence mode, then the bandwidth can be allocated to the video coder. On the contrary, if there is no change in the scene, the previous frame can be used to approximate the current one and the available bandwidth can be shifted to the speech codec. This requires communications between the speech and video coders, and also some overhead information has to be transmitted so that the decoder can operate properly. In other cases, it is impossible to decide which signal is more important since it is highly dependent on our perception. Therefore, in a low bit rate audiovisual communication system, an external control has to be included in order to permit the users to emphasise one of the coders in real time.

Another trade-off is involved in the choice of the frame rate for the video codec. When the video codec encodes a document, it is better to use a low frame rate. On the other hand, in some applications, a higher frame rate with lower picture quality is preferred in order to avoid the jerkiness effect. An additional external switch has to be considered so that the user can choose between higher quality images with a lower frame rate and lower quality images with a higher frame rate.

Considering the bit rate allocation as a binary decision for either of the codecs reduces the efficiency of the system. In an efficient system, the speech or video codec should be able to work at any bit rate provided that the overall rate does not exceed the available bandwidth. However, the implementation of such a system requires several different codecs since different algorithms are efficient at different bit rates. The speech and video codecs described above have been integrated. Figure 24.8 depicts the integration of both speech and video in the same packets through a joint buffer and multiplexer. A voice activity detector (VAD) precedes the speech coder in order to detect the active talk-spurts and the idle silences of a speech signal. The video coder is notified of the result of this detection so that it adjusts its bit rate

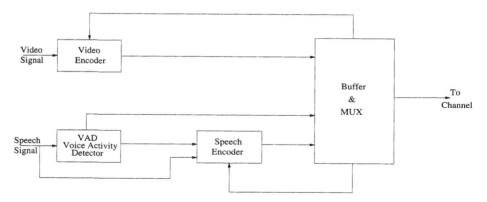

Figure 24.8 Integration of speech and video data through a joint buffer/MUX.

accordingly. The quality of audiovisual communications has improved both subjectively and objectively when the activity of the coders was taken into consideration to decide about the bandwidth allocation for each traffic source.

24.6. CONCLUSIONS

A segmentation based video coder and a high quality split-band LPC vocoder have been integrated for use in mobile audiovisual communications. The new LPC-based vocoder was shown to produce a higher quality speech at a bit rate of 2.5 kbit/s compared with both the IMBE coder operating at 4.5 kbit/s and the DoD 1016 CELP operating at 4.8 kbit/s. By splitting the speech into voiced and unvoiced bands, both mixed voicing speech and speech containing acoustic noise could be reliably coded without introducing excessive buzziness into the synthetic speech. The coder is made variable rate by grouping the harmonics, following the first eight, into groups of eight bands and then sending sets of eight vector quantised values depending on the desired output bit rate of the speech coder. A segmentation-based video coding algorithm, based on contour and texture prediction working at very low bit rates has also been presented. The region contours were approximated by line segments that were represented by some control points. Then, the motion parameters were estimated and entropy coded. At the decoder, the contour information was reconstructed by using the motion parameters and the connectivity information of the previous frame. The textures of each region was approximated by the mean values. However, the actual values were not encoded. Instead, the motion parameters were used to predict the region means from the previous frame, and the prediction residual was entropy coded. A joint output buffer shared by both of the coders was introduced and used to control the bit rates by adjusting the speech and video coding parameters. A better utilisation of the channel bandwidth was achieved through the negotiation made between the speech and video coders about the most efficient use of the capacity depending on their activity. Moreover, the user performance concerning the frame rate and the codec choice was integrated to the system and a higher coding efficiency was obtained with the integrated system.

REFERENCES

[1] Jayant, N. S. and Noll, P. *Digital Coding of Waveforms*, Prentice Hall, Englewood Cliffs, NJ, 1984.
[2] Kondoz, A. M., *Digital Speech Coding for Low Bit Rate Communications Systems*, Wiley, Chichester, 1994.
[3] Atkinson, I. A., Yeldener, S. and Kondoz, A. M., High quality split-band LPC vocoder operating at low bit rates. *Proceedings of ICASSP '97*, **II**, 1559, Munich, Germany.
[4] Griffin, D. W. and Lim, J. S., Multi-band excitation vocoder. *IEEE Transactions ASSP*, **36** (8), 1223–1235, 1988.
[5] McAulay, R. J. and Quateri, T. F., Pitch estimation and voicing decision based upon a sinusoidal speech model. *Proceedings of ICASSP*, **1**, 249–252, 1990.

[6] APCO, Vocoder description. Document No. IS102BABA, Association of Public-Safety Communication Officials, 1993.

[7] Chen, J. H. and Gersho, A., Real time vector APC speech coding at 4.8kb/s with adaptive post-filtering. *Proceedings of ICASSP*, **3**, 2185–2188, 1987.

[8] Yeldener, S., Sinusoidal model based low bit rate speech coding for communication systems. PhD Thesis, University of Surrey, April 1993.

[9] Kim, M. Y., Ha, N. K. and Kim, S. R., Linked split-vector quantiser of LPC parameters. *Proceedings of ICASSP*, **2**, 741–744, 1996.

Part 6

Error Resilient Coding for Multimedia Applications

Multimedia information will, in the future, be produced and consumed by a wide variety of systems and will be transmitted over a range of channels. Because wireless channels are notoriously variable in quality, one of the main functionalities required of image and video coding schemes is error resilience. In the past, most source coding schemes have been designed for clean channels and these can suffer considerable degradation if the data is corrupted during transmission. The major problems with most source coding schemes are associated with the use of variable length code words and predictive coding. These cause single errors to propagate both spatially and temporally. Conventional FEC schemes are inappropriate for low bit rate video coding because they must be designed for worst case channel conditions and necessarily reintroduce redundancy to the bitstream. Methods based on simple ARQ techniques are also generally unacceptable for real time delivery due to the delay introduced. A selection of alternative techniques are presented in this chapter which are more appropriate to video information.

An overview of error resilient source coding schemes is given in Chapter 25 by Redmill *et al*. This describes the causes of error propagation in a compressed video bitstream and outlines methods of restructuring the bitstream to mitigate the effect of these errors. An alternative approach is described by Ramstadt in Chapter 26 where the benefits of a joint modulation and coding regime are presented. Sadka *et al*. describe the benefits of increased intraframe frequency combined with reversible entropy code words to localise the impact of errors. Finally in Chapter 28, Girod *et al*. present a method based on the use of a feedback channel, proposed for incorporation within the H.263 standard. Here the propagation of errors is restricted through error tracking and selective replenishment using an ARQ approach.

25

Error Resilient Image and Video Coding for Wireless Communication Systems

D. W. Redmill, D. R. Bull, J. T. Chung-How and N. G. Kingsbury

25.1. INTRODUCTION

In recent years, the convergence and growth of both computing and communications has resulted in a requirement for multi-media communications using a variety of systems and over a range of channel conditions (e.g. wireless). Such channels usually have both a limited capacity (a result of total available spectrum), and a highly variable channel quality (resulting from interference, fading and limited transmission power). Thus, there is a need for low rate image and video coding systems which can be used over variable quality channels.

Most source coding techniques are designed for a clean channel and can suffer seriously if any of the compressed data is corrupted. Figure 25.1 shows an example of a JPEG [1] coded image transmitted over a channel with a random bit error rate (BER) of 10^{-4} (0.01%). Although this image contains only 5 bit errors it clearly shows that any error can result in large distortions in the decoded image. Increasing the error rate to about 10^{-3} (0.1%) usually results in an image which the decoder cannot decode (due to errors in important header information).

Channel control methods, such as channel coding and automatic repeat request (ARQ) protocols can be used to eliminate errors, but only at the cost of reduced channel capacity. Channel codes or forward error correcting codes (FEC) use additional parity data, which allows the decoder to detect and correct errors. The problem with this approach is that the channel code must be selected to cope with particular worst case channel condition. This worst case situation may imply the need for a very powerful code, which significantly reduces the data rate available for the compressed image or video, resulting in a loss of quality. Catastrophic failure can also result if the worst case design criterion is exceeded. Another problem with channel coding is the difficulty involved in designing codes for channels with bursty error characteristics. One method is to use interleaving; however, this is sub-optimal and introduces large delays.

ARQ protocols allow the decoder to request retransmission of any corrupted data. This method can be highly effective for coping with packet loss or large burst errors. However, the need to wait for data to be retransmitted can lead to significant delays, which are unacceptable for many real time applications (e.g.

Figure 25.1 JPEG with 10^{-4} BER (5 errors).

video-conferencing). ARQ is less suitable for dealing with random errors or short bursts, where the overhead needed for the protocol becomes excessive. ARQ also requires the use of a reverse channel and encoder interaction, which may not be possible for some broadcast or storage applications.

Error resilient methods can be used instead of or in conjunction with conventional channel coding and involve redesigning the compression system to be more resilient to channel errors. The aim is to reduce the propagation effects of channel errors so that each error causes a lower expected distortion. This offers graceful degradation with deteriorating channel conditions and should give improved performance for both good and poor quality channels A channel coded system however may provide superior performance for a channel with a constant intermediate quality. Figure 25.2 shows the expected performance trade-offs.

A final class of technique which is useful in the design of compression systems for noisy channels is error concealment. These systems use post-processing to detect and conceal errors. Channel errors can be detected either from explicitly included parity information, illegal or improbable data. Having detected an error, the decoder then attempts to conceal the corrupted data by predicting the probable content. Since error concealment is a post-processing process, it is well suited for use with existing standards. However, error detection and concealment strategies are often highly specific, and rely very heavily on the particular coding method used. For example a method which conceals square block artefacts associated with a DCT codec will not be successful with a wavelet-based codec.

This chapter provides a review of error resilient methods for coding both still images and video. Section 25.2 provides a brief analysis of the mechanisms by which individual channel errors can propagate within the decoded data. This is useful in understanding and designing error resilient systems which attempt to limit this propagation. Section 25.3 provides a review of the error resilient entropy code (EREC),

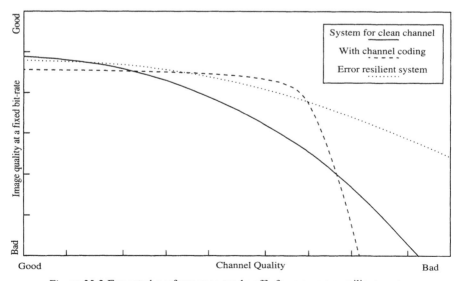

Figure 25.2 Expected performance trade-offs for an error resilient system.

a method which is compatible with existing coding systems. Section 25.4 presents a review of pyramid vector quantisation (PVQ), which is well suited for fixed rate coding systems. Section 25.5 reviews other alternative methods including predictive coding, layered coding and combined source channel coding. Section 25.6 provides a comparison of these techniques. Section 25.7 gives a brief summary, and goes on to propose areas for further consideration.

25.2. ERROR PROPAGATION MECHANISMS

In order to both design and use error resilient methods, it is necessary to understand how channel errors can propagate with compressed data. There are a variety of different types of propagation.

25.2.1. Catastrophic Error Propagation

Most compression systems contain a small amount of header information (e.g. image size and coding method) which is usually coded prior to each frame. Any error within this data often results in the decoder being unable to decode any meaningful information. In order to design an error resilient codec this information must be adequately protected by means of a suitably powerful channel code. Since the header is usually a small fraction of the total coded data, a powerful channel code can be used with only a small increase in total redundancy (or loss of quality).

25.2.2. Propagation in Predictive Coding

Predictive coding systems such as differential pulse code modulation (DPCM), are commonly used within a variety of systems (e.g. JPEG [1], H.263 [2], MPEG2 [3]) to code the low frequency data. Another form of predictive coding is temporal prediction in video systems (H.263 and MPEG). In predictive coding systems data is coded as the difference between the actual data value and a prediction derived from previously coded data. Any channel errors will affect not only the current value, but also subsequent predictions and thereby subsequent data. The lighter and darker regions of Figure 25.1 are an example of this form of error propagation.

Error propagation in a predictive coder can be limited either by incorporating a loss term in the prediction loop, or by periodically breaking the loop (e.g. the presence of an intra frame coded frame or region in video codec). The latter approach is usually preferable since it requires a lower loss of compression performance.

25.2.3. Loss of Synchronisation

Many compression systems employ some form of variable length entropy coding such as Huffman or arithmetic coding. With Huffman coding, an error in any codeword can cause the decoder to decode either a longer or shorter code-word, leaving it in the wrong position to decode the next code-word. This is referred to as a loss of synchronisation. In practice, most Huffman codes will resynchonise themselves on a statistical basis [4].With arithmetic coding the situation is usually much worse, since they rarely resynchronise themselves. Synchronisation can also be lost with run-length coding systems, where errors may cause the decoder to miscalculate the current position. Thus even when code-word synchronisation is regained, the meaning of code-words may still be mis-synchronised. The chequerboard blocks of Figure 25.1 are caused by this form of error propagation.

Loss of synchronisation can also occur within block-based systems (e.g. JPEG, H.263, MPEG). Loss of synchronisation within a block can cause the end-of-block (EOB) codes to be either missed or falsely detected. This causes the decoder to miscount the block number, which results in subsequent data being shifted (see Figure 25.1). Resynchronisation code-words can be used to periodically resyncronise the data. However, even sophisticated synchronisation code-words, (e.g. [5]) can incur a significant redundancy if used too often.

25.2.4. Spatial Propagation

Many coding systems achieve high compression by using a sub-band decomposition such as the DCT or wavelet transform. An inverse transformation is applied at the decoder. Any errors within the sub-band representation will propagate to a region of spatial pixels. This propagation can be viewed in terms of the impulse responses of the synthesis filters or basis functions. For example, the chequerboard blocks of Figure 25.1 correspond to DCT basis functions.

25.3. THE ERROR RESILIENT ENTROPY CODE (EREC)

25.3.1 Motivation and Basic Algorithm Description

One of the most serious forms of error propagation within a typical image or video codec (JPEG, H.263 or MPEG) is caused by the loss of block synchronisation, resulting in spatial shifting of decoded information. This form of propagation is a result of the sequential transmission of variable length coded blocks. The error resilient entropy code (EREC) [6,7] provides a very low redundancy alternative to sequential transmission, which offers significant improvements in error resilience.

The basic operation of the EREC is to rearrange the n variable length blocks of data (of length b_i) into a fixed length slotted structure (with slot lengths s_i), in such a way that the decoder can independently find the start of each block and start decoding it. The encoder first chooses a total data size T which is sufficient to code all the data (i.e. $T \geq \Sigma b_i$). The value T needs to be coded as a small amount of protected header information. Next the decoder splits the T bits into n slots of lengths s_i ($T = \Sigma s_i$). The lengths s_i are usually chosen to be approximately even. An n stage algorithm is used to place data from the variable length blocks into the fixed length slots. At each stage k, a block i with data left uncoded searches in slot $j = i + \phi_k \pmod{n}$, for space to code some or all of the remaining data from block i. ϕ is an offset sequence (usually pseudo-random). Figure 25.3 shows an example of the algorithm with 6 blocks (and the offset sequence 0,1,2,3,4,5,6). At stage 1 blocks 3,4 and 6 are completely coded in slots 3,4 and 6, with space left over. Blocks 1,2 and 5, however are only partially coded and have data left to be placed in space left in slots 3,4 and 6. At stage 2, remaining data from block 2 is coded in slot 3, and some data from block 5 is coded in slot 6. By the end of the nth stage all the data is coded. The decoder operates in a similar manner by decoding data for each block until the block is fully decoded (an end-of-block code). Thus, in the absence of channel errors the decoder will correctly decode all the data.

Channel errors however can cause the end-of-blocks to be missed or falsely detected. This then means that the decoder will incorrectly decode following information which was placed at later stages of the algorithm. This implies that data placed in later stages (that from towards the ends of long blocks) is more susceptible to error propagation than data placed in earlier stages. For typical image compression systems, these propagation effects occur in high frequency data from active regions of the image. Such data is subjectively less important than the low frequency data (see results below).

25.3.2. EREC for Image Compression

Figure 25.4 shows an example of a block-based image compression system both with and without the EREC. The system is based on JPEG and employs the DCT, zigzag scanning, run-length and Huffman coding. However, it differs from JPEG, in that the header information is reduced and coded with a high redundancy channel code. Also the predictive coding of DC coefficients has not been used in order to avoid the resulting predictive error propagation effects. These two modifications introduce a

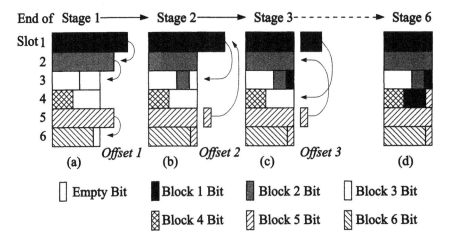

Figure 25.3 Example of the EREC algorithm.

Figure 25.4 Effect of errors: Without the EREC: (a) 10^{-3} BER; (b) 10^{-2} BER. With EREC (c) 10^{-3} BER; (d) 10^{-2} BER. (256×256 Lena Coded at 0.5 bpp using DCT and Huffman coding).

slight redundancy. However, the use of the EREC means that we will no longer require the relatively costly resynchronisation code-words. Figure 25.4 shows that the EREC system eliminates the spatial shifting of data caused by a loss of block synchronisation.

In [8], it was demonstrated that a system employing arithmetic coding is less resilient than one employing Huffman coding. However, when the EREC was used with arithmetic coding it provided similar objective results coupled with a significant improvement in subjective performance.

Many of the most annoying errors in this system are sharp blocking artefacts resulting from the use of the DCT. An alternative method which avoids the sharp blocks is the discrete wavelet transform. The choice of wavelet transform is reasonably important, since the impulse response of the synthesis filters controls the spatial error propagation properties. For error resilience it is important to choose filters with a short smooth impulse response. A good example are the 3,5 short symmetric filters of [9]. Images coded with these filters in conjunction with the EREC are shown in [7,10]. Note that the errors are much smoother in nature, but slightly more spread out.

Another alternative to the DCT or wavelet transform is to use a non-linear filter bank of the form proposed in [11,12,13]. These make use of median operators which have good outlier rejection properties. The result [13] is that impulsive errors in the sub-band domain are unlikely to propagate and result in impulsive errors in the spatial domain. These can be easily removed using a median post-filter.

25.3.3. Bursty Channels

One of the interesting properties of the EREC [7] is its resilience to bursty errors. The results of Figure 25.5 (using the wavelet transform) show that the EREC copes well when errors are grouped in bursts rather than randomly distributed. The reason for this is that successive errors in a burst are more likely to affect the same information and can often be considered as only a single error event. Burst errors have been simulated using a simple two state burst error model with 50% probability of error in the burst state.

Figure 25.5 Resilience to burst errors with an average of 10^{-2} BER: (a) random errors. Average burst length of (b) 5 and (c) 20 bits.

25.3.4. EREC for Video Compression

The EREC can also be used for video compression. With a video system, we have the added problem of temporal propagation of errors to subsequent frames (or previous B frames). This propagation needs to be limited by means of intra frame coded regions or frames (I frames). However, the intra-frame coding gives less compression and has been found to be more sensitive to channel errors [7,14,15] (this is a result of the fact that intra-frame coding contains the majority of the signal energy). Thus the frequency of replenishment by intra-frame coding should be a compromise between the need to limit temporal error propagation, maintain high compression, and mini- mise the increased errors resulting from intra-frame coding.

Figure 25.6 shows an example of a low bit rate video codec operating at 20 kbit/s. Although the EREC offers some improvement, for higher rates, the image is still dominated by low frequency errors. These errors actually occur in intra-frame coded portions of the image, and then propagate to subsequent frames.

25.3.5. Using the EREC in Conjunction with a Layered System and Channel Coding

The EREC can also be used in conjunction with other techniques such as layered coding and channel coding. The most noticeable errors in Figure 25.6 were a result of errors in the low frequency coefficients in intra-frame coded regions of the image. In order to overcome these difficulties it is sensible to use some channel coding for these

Figure 25.6 Video coding at 20 kbit/s. Without the EREC: (a) 3×10^{-4} BER; (b) 3×10^{-3} BER. With EREC: (c) 3×10^{-4} BER; (d) 3×10^{-3} BER.

coefficients, to correct the majority of the most annoying errors. To achieve this we can use a layered system (section 5.2) which codes the data in two distinct layers, which can then be protected using different channel codes. Figure 25.7 shows results for a layered system. The system codes each block with a embedded zero-tree codec. This allows us easily to separate the information into different priority bit-planes. Two EREC structures are used for the two different layers. In this way we ensure that all the data is efficiently and resiliently packed into the two layers. In this example 30% of the data is coded in the first layer and protected with a 23,12 Golay code, while the remainder of the data is packed into the second layer and coded with a 31,26 Hamming code.

Figure 25.7 shows how the layered system with channel coding offers superior performance in the presence of channel errors compared to the standard system (Figure 25.7). Note that the use of channel coding does lead to a slight loss of quality in the clean channel performance. The EREC can be seen to offer further performance improvements.

25.3.6. Compatibility With Standards

Since the EREC is only a component of a video codec rather than a complete system, it can be used in conjunction with a variety of standard block-based coding systems such as JPEG, H263 and MPEG. It is therefore possible to build a transcoder which recodes a standard bitstream into an error resilient form.

In [14,15] a transcoder is described that converts any standard MPEG-2 bitstream into an error-resilient bitstream *at the same bit rate* before transmission over an error-prone channel, and then converts it back at the receiver into a standard bitstream for MPEG-2 decoding. In the absence of errors the transcoding is lossless. In addition to the basic EREC scheme, non-linear interpolative coding was used to replace the differential coding of DC coefficients and motion vectors in MPEG, without increasing the bit rate. Statistically-based error concealment was used to minimise the visibility of large high frequency errors. For a typical test sequence (mobile and calendar), the error resilient bitstream requires a random bit error rate of at least 2×10^{-3} before the PSNR drops to 25 dB, whereas the original bitstream requires less than 2×10^{-5} for the same drop in PSNR.

The application of these techniques to H.261/263 bitstreams has also been investigated and present no significant additional problems. The techniques have also been considered for MPEG-4.

Figure 25.7 Layered video coding at 20 kbit/s. Without the EREC: (a) 3×10^{-4} BER; (b) 3×10^{-3} BER. With EREC: (c) 3×10^{-4} BER; (d) 3×10^{-3} BER.

25.4. PYRAMID VECTOR QUANTISATION (PVQ)

Pyramid vector quantisation (PVQ) [16] is a form of lattice vector quantisation (VQ), derived specifically for use with a memoryless Laplacian source. PVQ is thus well suited to coding DCT [17,18] and/or subband [19,20] and video [18] data which can be modelled as a Laplacian source. Like other forms of VQ, PVQ provides fixed length code-words which and thus a fixed length coding system. Fixed length code-words are useful for error resilient coding applications since channel errors cannot propagate between code-words.

25.4.1. Vector Quantisation (VQ)

Vector quantisation (VQ) [21], is a multi-dimensional extension of simple 1D scalar quantisation. A VQ system takes an L dimensional vector v and codes it as a quantised vector $\hat{v}$, taken from a codebook of possible vectors. The quantised vector can then be coded as the corresponding codebook index, and reconstructed by the decoder providing the decoder has the same codebook. Typically vectors are formed by taking a combination of adjacent correlated values. The encoder then needs to search the codebook for the closest match, minimising a distortion measure $D(v - \hat{v})$.

The achievable compression and reconstruction error depends on a number of factors, including vector dimension L, vector distribution, and codebook size. In theory, VQ can achieve a performance close to the rate-distortion bound as the vector dimension L increases to infinity. However, increasing the vector dimension and the number of vectors in the codebook results in an exponential increase in encoding complexity (due to the codebook search). Therefore, large vector dimensions are practically impossible in conventional VQ.

This has led to the development of lattice-based VQ where code-vectors are arranged at regular, predefined uniform points in the input vector space. The closest code-vector can be found analytically without the need to search the entire codebook. This enables the use of larger codebooks and larger vector dimensions. However, lattice VQs are only optimal if the data to be coded fits some regular distribution assumed by the lattice.

25.4.2. PVQ Theory

The development of the pyramid vector quantiser [16] is based on a source producing random variables having independent, identical Laplacian distributions. A zero mean Laplacian distribution with a standard deviation σ is given by:

$$P(x) = \frac{1}{\sqrt{2}\sigma} e^{-\frac{\sqrt{2}}{\sigma}|x|}$$

If these variables are grouped into vectors, X, of dimension L, the resulting probability density of the vector is given by:

$$f(X) = \prod_{i=1}^{L} P(x_i) = \left(\frac{1}{\sqrt{2}\sigma}\right)^L e^{\left(-\frac{\sqrt{2}}{\sigma}\sum_{i=1}^{L}|x_i|\right)}$$

where x_i is the ith component of the vector X. The equi-probability contours for this distribution are found to be L-dimensional hyper-pyramids:

$$r = \sum_{i=1}^{L} |x_i| = \| X \|_1 = \text{constant}$$

where $\| X \|_1$ is the l_1 norm of vector X, and r is known as the pyramid radius. The probability density function for r is given by [16]

$$P_r(r) = \frac{2\left(\frac{k}{2}\right) r^{L-1} e^{-\left(\frac{\sqrt{2}r}{\sigma}\right)}}{(L-1)!}$$

It can be shown that for sufficiently large vector dimensions L, vector X becomes highly localised around a particular hyper-pyramid with radius $r = L\sqrt{2}/\sigma$. Since all points lying on the pyramid surface have equal probability, a pyramid vector quantiser (PVQ) can be constructed which consists of points that lie both on a cubic lattice and on the hyper-pyramid surface.

The codebook points lying on the hyper-pyramid surface for the case when $L = 3$ and $r = 4$ are illustrated in Figure 25.8.

However, for moderate sizes of vector dimension which are normally used in practice, significant distortion can be introduced by quantising on a single hyper-pyramid. So instead of using only one pyramid, quantisation can be done on a number of concentric pyramids. This variation of PVQ is known as product code PVQ. The use of a number of concentric pyramids means that all codewords are no longer equiprobable. However, product code PVQ still gives good performance because of the possibility of using large vector dimensions compared to conventional VQ.

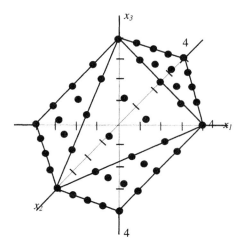

Figure 25.8 Codebook points on hyper-pyramid surface with $L = 3$ and $r = 4$.

The PVQ codebook vectors will then consist of points on the surface of a number of concentric pyramids, with an equal number of points on each pyramid. These points are chosen geometrically so that their relative positions are the same on each pyramid. The design bit-rate determines the number of points on each pyramid surface. The codebook points are illustrated in Figure 25.9 for $L=2$.

Thus, a PVQ codeword consists of the pyramid radius r, which indicates on which concentric pyramid the vector lies, followed by the position of the vector on the pyramid. The pyramid radius r can be coded using an optimum scalar quantiser [23] for the distribution. In order to code the position of the vector on the pyramid, a unique fixed length binary index is assigned to every possible position using an indexing algorithm. The indexing algorithm which minimises the effects of channel errors was presented in [19], referred to as product enumeration. This results in a fixed-length codeword for each codebook vector, making PVQ robust to channel errors.

25.4.3. Image Coding using PVQ

It follows directly from its derivation that PVQ is particularly suited for coding sources having a Laplacian distribution. It is well known that the distribution of the coefficients resulting from transform and subband/wavelet coding of images can be modelled by a Laplacian distribution. Thus a PVQ coding system for still images can be designed by first transforming the image using a block DCT or a wavelet transform, and then using PVQ to code each subband. In order to ensure that the vector components are uncorrelated, vectors are formed from non-adjacent components scattered over the subband. Figure 25.10 shows decoded images at error rates of 10^{-3} and 10^{-2}.

25.5. ALTERNATIVE TECHNIQUES

As well as the two methods discussed in the previous sections, there are also numerous other methods which can be used either together with or instead of these

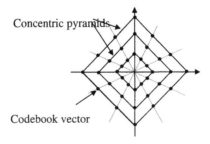

Figure 25.9 Concentric pyramids and codebook points for $L = 2$.

Figure 25.10 Images coding using PVQ at 0.5 bpp. Using DCT with (a) 10^{-3} BER; (b) 10^{-2} BER. Using a wavelet transform with (a) 10^{-3} BER; (b) 10^{-2} BER.

methods to provide an error resilient coding system. This section briefly reviews a few of the more common ones.

25.5.1. Robust Predictive Coding

Predictive coding such as DPCM is often used in simpler less complex codecs, as well as for the low pass subband within a subband or DCT-based codec. Predictive coding is also used to reduce temporal redundancies in video coding. However, errors can propagate within the prediction loop (see section 2.2). These propagation effects can be reduced by using more robust predictors, such as median, or FIR median hybrid predictors [23,24,7]. These predictors make use of the outlier rejection properties of the median operators. Median predictors [7,14,15] can also be used within a hierarchical interpolation system [25], which is suitable for coding low frequency subbands and motion vectors. This method can also be thought of as a simplified form of non-linear filter bank [11,12,13].

25.5.2. Layered Coding

Multi-layer coding provides another popular technique for error resilient image coding (e.g. [26,27]). These techniques code the image into two or more layers. The first layer is used to code more important information, while subsequent layers are used to code less important refinement information. The different layers are then given varying amounts of protection (e.g. channel coding) according to their priority. Alternatively, the different layers may be transmitted over different subchannels e.g. [28]. Layered coding systems are also well suited to applications requiring scalability of bitrate, since higher layers can be ignored.

25.5.3. Robust Vector Quantisation

Vector quantisation (VQ) is another common technique for use within error resilient coding systems [29,30,31,32]. The advantages stem from the use of fixed length code-words which limit error propagation to isolated code-words. Since the information in an image is non-uniformly distributed, the density of vectors over the image needs to be varied. PVQ (discussed in section 4) overcomes this problem by assuming that the vector components are randomly distributed over the image. Another solution is the error resilient positional code (ERPC) [29,32] which was a forerunner to the EREC (discussed in section 3).

25.6. COMPARISON OF TECHNIQUES

This chapter has presented a variety of techniques for error resilient coding of images and video signals. This section provides a brief comparison of their relative performance and suitability for different systems.

Two very different strategies, the EREC and PVQ were discussed in sections 3 and 4. Figure 25.11 shows a comparison of their performance for still image coding in the presence of random errors. From this figure we see that the EREC system offers superior performance for clean channels while the PVQ system offers greater resilience for high error rates. This difference is due to the fact that the EREC system is utilising a variety of effective techniques for exploiting correlations in the transformed image (including run-length and entropy coding). Meanwhile the fixed length codewords of the PVQ system leads to superior error resilience. Thus the choice of which system to use depends significantly on the expected characteristics of the channel available.

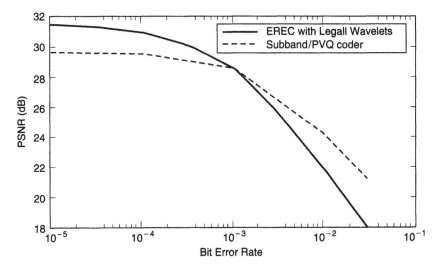

Figure 25.11 Comparison of PSNR for PVQ and EREC (for the 256×256 Lena image coded at 0.5 bpp).

25.6.1. Fixed or Variable Rate Coding

There are two main rate control strategies employed within video compression systems. The first is fixed rate coding which is well suited to fixed rate channels, but can yield to a variation in image quality. The second strategy is to use variable rate coding, which is well suited to either variable rate channels (e.g. ATM) or systems where delay is acceptable (allowing the buffering of data). The choice of which strategy is most appropriate is highly dependent on the application requirements.

However, this choice can significantly affect the choice of coding strategy to be used. The PVQ system of section 4, is inherently a fixed rate system and is less suited to variable rate applications. However, the EREC was designed to be used with variable rate systems. Note that if a suitable entropy coding strategy is employed (e.g. embedded zero-tree coding), then the EREC can also be used to provide a fixed rate system.

25.7. CONCLUSIONS

This chapter has reviewed several methods available for error resilient image and video coding. The methods have various different advantages and disadvantages, depending on the particular application requirements. For example, if a variable rate channel is available, then we may be best to use a variable rate system such as the EREC. However, for fixed rate systems the choices are more open. Another important factor is the expected channel characteristics. For a fixed quality channel, it is probably best to use a channel code specifically designed for the channel, while for variable quality systems we require an error resilient system. Channels which exhibit long bursts or fades may benefit from an ARQ protocol, providing the delay is acceptable. Other factors which affect the choice of strategy are the tolerable system delay, and the presence of and speed of a reverse channel for ARQ protocol.

REFERENCES

[1]   ISO 10918-1. Digital compression and coding of continuous-tone still images, 1994.
[2]   ITU-T. Draft recommendation H.263, video coding for low bit rate signals, 1995.
[3]   ISO/IEC 13818-2. Generic coding of moving pictures and associated audio, 1994.
[4]   Ferguson, T. J. and Rabinowitz, J. H., Self synchronizing Huffman codes. *IEEE transactions on information theory*, **30**, 687–693, 1984.
[5]   Lam, W. M. and Reibman, A., Self-synchronization variable length codes for image transmission. In *Proceedings of IEEE International Conference on Acoustics, Speech and Signal Processing*, **III**, 477–480, 1992.
[6]   Redmill, D. W. and Kingsbury, N. G., The EREC: an error-resilient technique for coding variable length blocks of data. *IEEE transactions on image processing*, **5**, 565–574, 1996.
[7]   Redmill, D. W., Image and video coding for noisy channels. PhD Thesis, Cambridge University, 1994.

[8]   Redmill, D. W. and Bull, D. R., Error resilient arithmetic coding of still images. In *Proceedings of IEEE International Conference on Image Processing*, Lausanne, **II**, 109–112, 1996.

[9]   Le Gall, D. and Tabatabai, A., Sub-band coding of images using symmetric short kernel filters and arithmetic coding techniques. In *Proceedings of IEEE International Conference on Acoustics, Speech and Signal Processing*, 761–764, 1998.

[10] Redmill, D. W. and Kingsbury, N. G., Still image coding for noisy channels. In *Proceedings of IEEE International Conference on Image Processing*, **1**, 95–99, 1994.

[11] Redmill, D. W. and Bull, D. R., Non-linear perfect reconstruction critically decimated filter banks. *Electronics Letters*, **32**, 310–311, 1996.

[12] Redmill, D. W. and Bull, D. R., Non-linear perfect reconstruction filter banks for image coding. In *Proceedings of IEEE International Conference on Image Processing*, **1**, 593–596.

[13] Redmill, D. W. and Bull, D. R., Non-linear filter banks for error resilient image compression. In *Proceedings of IEEE International Conference on Image Processing and its Applications*, 91–95, 1997.

[14] Swann, R. E. and Kingsbury, N. G., Transcoding of MPEG-II for enhanced resilience to transmission errors. In *Proceedings of IEEE International Conference on Image Processing*, **II**, 813–816, 1996.

[15] Swann, R. E., MPEG-II video coding for noisy channels. PhD Thesis, Cambridge University, 1997.

[16] Fischer, T. R., A pyramid vector quantizer. *IEEE Trans. Information Theory*, July, 568–583, 1986.

[17] Filip, P. and Ruf, M., A fixed-rate product pyramid vector quantization using a bayesian model. *Globecom*, 1992.

[18] D'Allesandro, P. and Lancini, R., Video coding scheme using DCT-pyramid vector quantization. *IEEE Transactions on Image Processing*, March, 309–318, 1995.

[19] Hung, A. C. and Meng, T. H.-Y., Error resilient pyramid vector quantization for image compression. In *Proceedings of the IEEE International Conference on Image Processing*, 583–587, 1994.

[20] Tsern, E. K. and Meng, T. H. Y., Image coding using pyramid vector quantization of subband coefficients. In *Proceedings of the International Conference on Acoustics Speech and Signal Processing*, 601–604, 1994.

[21] Gersho, A. and Gray, R., *Vector Quantisation and Signal Compression*, Kluwer, New York, 1992.

[22] Max, J., Quantizing for minimum distortion. *IRE Transaction Information Theory*, March, 7–12, 1960.

[23] Mickos, R., Okten, L., Cambell, T. G., Sun, T. and Neuvo, Y. 3-D median based prediction for image sequence coding. In *IEEE International Symposium on Circuits and Systems*, 1656–1659, 1992.

[24] Viero, T. and Neuvo, Y., 3-D median structures for image sequence filtering and coding. In Sezan, M. I. and Lagendijk, R. (eds), *Motion Analysis and Image Sequence Processing*, **4**, Chapter 14, Kluwer Academic Press, New York, 411–445, 1993.

[25] Arnold, L., Interpolative coding of images with temporally increasing resolution. *Signal Processing*, **17**, 151–160.

[26] Ghanbari, M., Two layered coding of video signals for VBR networks. *IEEE Journal on Selected Areas in Communications*, **7**, 771–781, 1989.

[27] Chen, Y. C., Sayood, K. and Nelson, D. I., A robust coding scheme for packet video. *IEEE Transactions on Communications*, **40**, 1491–1501, 1992.

[28] Steedman, R., Gharavi, H., Hanzo, L. and Steel, R., Transmission of subband coded images via mobile channels. *IEEE Transactions on Circuits and Systems for Video Technology*, **3**, 15–26, 1993.

[29] Cheng, N. T., Error resilient video coding for noisy channels. PhD Thesis, Cambridge University Engineering Department, 1991.

[30] Farvardin, A., A study of vector quantization for noisy channels. *IEEE Transactions on Information Theory*, **36**, 799–809, 1990.

[31] Modestino, J. W. and Daut, D. G., Combined source-channel coding of images. *IEEE Transactions on Communications*, **27**, 1644–1659, 1979.

[32] Cheng, N. T. and Kingsbury, N. G., The ERPC: an error resilient technique for encoding positional information of sparse data. *IEEE Transactions on Communications*, **40**, 140–148, 1992.

26

Combined Source Coding and Modulation for Mobile Multimedia Communication

Tor A. Ramstad

26.1. INTRODUCTION

In multimedia services there are two distinctly different information sources:

- analogue waveform sources, such as speech, music, scanned images and video;
- numeric or digital sources.

In order to integrate the transmission of these two basic signal types for transmission and storage, it seems practical that both have the same type of representation. The standard solution is to resort to a digital representation for both types of data. Digitisation will usually result in bandwidth expansion, and compression is the

Insights into Mobile Multimedia Communication
ISBN 0-12-140310-6

natural vehicle for again reducing the bandwidth requirement. This is well adapted to certain storage media (such as hard disks, RAMs, etc.) where a numerical representation is the common format. On the other hand, most physical transmission media (e.g. wireless) are analogue.

The question to ask is: are we sure that the prevailing strategy of digitising analogue waveforms followed by digital modulation onto an analogue channel is optimal? We will advocate the use of time-discrete and continuous amplitude channel representations for the analogue signals. Numeric source signals would still have to be digitally modulated onto the same physical channel for multimedia applications, but this combination can be easily handled. In this chapter we propose a novel technique where analogue source representation vectors are approximated and mapped to analogue channel vectors with reduced dimensionality. Sample rate reduction implies compression at a system level. To evaluate the performance of such a system we have defined compression as the ratio between the sample rate of a reference system without compression and the sample rate of the proposed system when both offer equal received signal quality. The robustness exhibited by the proposed method, which is especially valuable for wireless systems where the channel conditions vary considerably, is obtained when neighbouring channel vectors are mapped to neighbouring source vectors.

The proposed method performs better than any other known method and is close to the optimal performance theoretically attainable (OPTA) curves for the synthetic sources tested. For practical communication applications it renders far greater robustness against channel variations than earlier proposed methods. Due to space limitations only one image communication example will be presented. Several other examples can be found in the references.

26.2. SYSTEM MODEL

The following modelling assumptions are made:

- The signal can be ideally represented by uniform sampling at the rate f_s, where $f_s = 2B$, and B is the signal bandwidth.
- The channel is an ideal Nyquist channel that can transmit at a rate f_c over a channel with bandwidth $f_c/2$.

This means that a sampled signal of bandwidth B can be transmitted over a channel of bandwidth B. Consequently, the time discrete signal requires the same bandwidth as the original analogue signal. We shall use this later in the definition of system *compression* or *efficiency*.

We can, in practice, relax these idealised assumptions by incorporating some oversampling of the signal, and introduce a roll-off factor for the channel filters to make them implementable.

A block diagram of the idealised image communication system is shown in Figure 26.1.

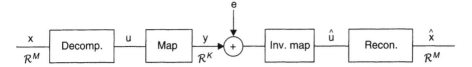

Figure 26.1 The idealised image communication system including signal decomposition, mapping from signal parameters to the channel representation, channel with additive noise, inverse mapping, and reconstruction.

The encoder section comprises two basic parts:

- a signal decomposition unit;
- a mapping device.

As will become clear below, the role of the decomposition unit is to cluster signal components into blocks with similar statistics. Compression is obtained through the mapping device, which takes a block of M source samples (collected into the vector **u**) and maps it to the vector **y** consisting of K channel samples, where $K < M$. This can be obtained in two ways:

- discarding samples;
- combining samples.

Sample skipping can be done if insignificant samples are clustered, and will imply signal distortion if some component is different from 0. Sample combinations will, for all practical purposes, generate *approximation* noise. These distortions plus additive noise on the channel are the main contributors to the resulting non-ideal performance.

The system objective is to minimise the overall noise

$$E[||\mathbf{x} - \widehat{\mathbf{x}}||^2], \tag{26.1}$$

subject to a channel constraint, as for example:

$$E[||\mathbf{y}||^2] \leq KS_{max}, \tag{26.2}$$

where S_{max} is the maximum average *power* or *energy per channel sample*. We have chosen here to limit the channel power. Other possible limitations could be maximum amplitude or the combination of the two.

To reduce the impact of the channel, it seems intuitive to make small channel errors result in small errors in the reconstructed signal. This is equivalent to saying that neighbouring symbols in the channel signal configuration must correspond to neighbouring symbols in the source signal representation. It transpires that there is a sharp distinction between the case when the channel signal dimension is at most equal to the dimension of the signal and the case with dimension expansion. As will become apparent later, for the compression case it is easy to make the system robust towards channel disturbances, while for the signal expansion case this is more difficult.

26.3. DEFINITION OF COMPRESSION

To evaluate the performance of a complete system, it is advantageous to define compression in terms of the advantage of the proposed system over some reference system without compression. A natural choice for reference system is an idealised PAM system with sample rate f_s and average channel power S_{max}. This system is without any compression, but is otherwise bandwidth efficient. If we assume that the rate after compression is $f_c = f_s K/M$, still with channel power S_{max}, and each received signal component has unaltered quality, we can define the compression factor as $\alpha = M/K$. This means that we can transmit M/K source channels on the same physical medium as one source channel without compression.

How is this possible? The compression process generates quantisation noise. This is compensated for in the following way: the channel noise per sample is the same with and without compression. However, when expanding the signal in the decoding unit, the average number of samples is increased. This means that the channel noise is spread over more samples. At the compression ratio where the extra approximation noise is equal to the reduction in channel noise due to this expansion, the reference system and the compression system render equal signal qualities. The corresponding rate ratio defines the compression factor.

26.4. OPTIMAL PERFORMANCE THEORETICALLY ATTAINABLE (OPTA)

For certain idealised sources and channels it is possible to find the best possible performance of the transmission system in terms of OPTA-curves. This is based on the *rate-distortion function* [2] of a source combined with the *channel's information capacity* [22].

The rate-distortion function for a Gaussian source of independent and identically distributed samples is given by

$$R = \frac{1}{2}\log_2\left\{\frac{\sigma_u^2}{\sigma_d^2}\right\} \text{ bits per source sample,} \tag{26.3}$$

where σ_u^2 is the signal power and σ_d^2 is the distortion.

Let us consider an AWGN memoryless channel with noise variance σ_n^2. Then the channel's information capacity is given by

$$C = \frac{1}{2}\log_2\left\{1 + \frac{\sigma_c^2}{\sigma_n^2}\right\} \text{ bits per channel sample.} \tag{26.4}$$

where σ_c^2 is the channel signal variance. The information capacity is the maximum amount of information that can be transmitted per sample over a channel without errors.

The OPTA curve can now be found as follows: assume we transmit M/K source samples per channel sample. If we can transmit at the information capacity the following relation holds:

$$C = R\frac{M}{K} \qquad (26.5)$$

Inserting equations (26.4) and (26.3) into this equation, and solving for the signal-to-noise ratio, we get

$$\frac{\sigma_u^2}{\sigma_d^2} = \left(1 + \frac{\sigma_c^2}{\sigma_n^2}\right)^{K/M} \qquad (26.6)$$

Plots on a decibel–decibel scale are shown in Figure 26.2.

The OPTA curves give the best possible performance as a function of the channel signal-to-noise ratio (CSNR) for different sample compression factors. It should be noted that the above example is applicable for Gaussian signal and noise, but the principle will be the same for other kinds of statistics.

26.5. SOURCE-TO-CHANNEL MAPPINGS

The mapping operations are illustrated in detail in Figure 26.3.

In the figure $q(\cdot)$ represents the approximation where any point in $\mathcal{R}^M$ is mapped to a point in a subset of the same space. If this were standard *vector quantisation*

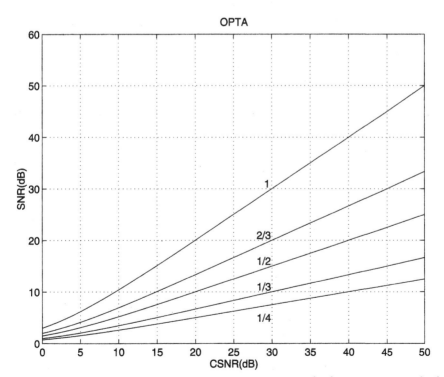

Figure 26.2 OPTA curves with k/M as parameter (SNR $= \sigma_u^2/\sigma_d^2$ and CSNR $= \sigma_c^2/\sigma_n^2$).

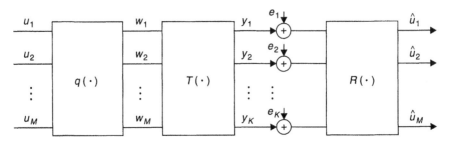

Figure 26.3 Mapping from higher to lower dimension (when $K < M$) as a two-stage process. $q(\cdot)$ indicates approximation while $T(\cdot)$ represents the mapping between dimensions. Included are also the additive chanel noise sources and the dimension expanding mapping $R(\cdot)$ in the receiver.

(VQ), the subset would be distinct points in $\mathcal{R}^M$ (called representation points or codebook vectors). For the general case the subset will be represented by super-surfaces. $T(\cdot)$ is an invertible transform that maps all points of this super-surface to a space with dimension equal to the dimension of the super-surface. An example is a subset of points on a surface in a three-dimensional space. This subset can be transformed to a two-dimensional space. The corresponding operations in VQ transmission systems are index assignment and mapping of the indices onto the channel.

The channel adds noise to the signal characterised by the channel signal-to-noise ratio (CSNR). If the channel were not power or amplitude constrained, the channel signal could be made large enough to overcome the channel noise problem, and also the approximation noise.

The receiver part of the system consists of $R(\cdot)$, which is the mapping from dimension K back to dimension M. At high CSNR, $R(\cdot) \approx T^{-1}(\cdot)$. Often this approximation is also used at low CSNR, but this is not optimal. The received signal block of M samples can be expressed as

$$\widehat{\mathbf{u}} = R(\mathbf{e} + T(q(\mathbf{u}))) \tag{26.7}$$

As an example, consider a mapping from two to one dimensions. The approximation must map all points in the two-dimensional space to a one-dimensional curve. We use Figure 26.4(c) to illustrate some important facts. We look at the continuous curve formed by the link between the circles in that figure. Each point in the plane is approximated by some close point on the given curve. This is very much like *vector quantisation* (VQ) where each point is approximated to the nearest of a finite set of given points.

The second stage is the mapping of the curve to the channel. As an example of a mapping, we choose for the channel representation *pulse amplitude modulation* (PAM). The PAM amplitude can be chosen proportional to the distance between the origin and the approximation point along the representation curve in the two-dimensional space. If no transmission errors affect the signal, any amplitude can be put back on the proper place on this curve by the receiver. There exists a unique nonlinear mapping that minimises the overall noise, but the improvements over the suggested method seems to be minor and the optimisation problem increases sub-

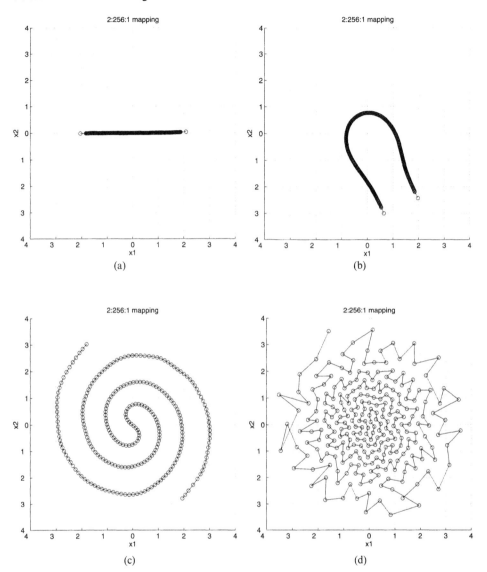

Figure 26.4 Four characteristic approximations for mapping from two to one dimensions. The channel signal-to-noise ratios are 0 dB, 7 dB, 25 dB, and 53 dB [1,11].

stantially. With transmission errors, we receive a changed amplitude, which mapped back to the two-dimensional space will represent a signal error. Because the typical errors are assumed to be small, there are no disasters: there will be minor changes in one or both of the source signals.

Another important aspect of the double spiral approximation shown in Figure 26.4c is that it runs through the origin and covers the plane in a symmetric manner for the negative and positive channel amplitudes. This implies that the channel representation will be symmetric (provided that the source samples are rotation symmetrically distributed), and the transmitted power will be low because the probability density function is peaky at the origin and will be represented by small channel amplitudes.

26.6. MAPPING OPTIMISATIONS FOR THE GAUSSIAN CASE [11]

Above, we introduced the OPTA curve for the case of Gaussian signal and channel noise. Let us now optimise mappings for this case and compare their performance to OPTA.

Mappings for different CSNRs have been optimised for the 2:1 case based on the minimisation of the noise expressed in equation (26.1) inserting $\hat{u}$ from equation (26.7) with $R(\cdot) = T^{-1}(\cdot)$ and $q(\cdot)$ representing linear stretching, under the constraint in equation (26.2) [1]. Four characteristic results are shown in Figure 26.4.

The mappings are designed as vector quantisers with size 256 codebooks, rather than a continuous representation. The circles show each vector and straight lines are drawn between them to indicate the continuous approximation case.

It is interesting to note that, for very low CSNR, the approximation is given by a straight line in one of the dimensions, and therefore one sample is simply discarded through the approximation. (As long as the two-dimensional probability density function is rotational symmetric, the straight line could go through the origin at any angle implying that we can share distortion among the two components.) When increasing the CSNR, the next phase is a horse-shoe approximation, which does not go through the origin. In the next figure, we see a double spiral which runs through the origin. And finally at very high CSNR, the smooth spiral breaks up into something which may look like a fractal behaviour (if we extend to an infinite number of points and infinite CSNR).

The system performance is given in Figure 26.5 together with OPTA and performance for smaller codebooks. Although the optimised systems do not perform as well as OPTA, the results are, to our knowledge, the best currently available.

26.7. IMAGE DECOMPOSITION

The objective of the decomposition block in Figure 26.1 is to cluster and classify parameters with different statistics, as e.g. variances. Each class will usually require different compression in order to *allocate* an optimal amount of resources for overall system performance, as described below.

Sample combinations are also easier in a decomposed domain. Usually we resort to different mappings for different components and different spatial locations, meaning different ratios M/K and possibly different energy per channel sample to obtain optimal performance.

The decomposition can be performed by linear or nonlinear transforms. Most common are the block transforms, for which case the Karhunen–Loève transform is

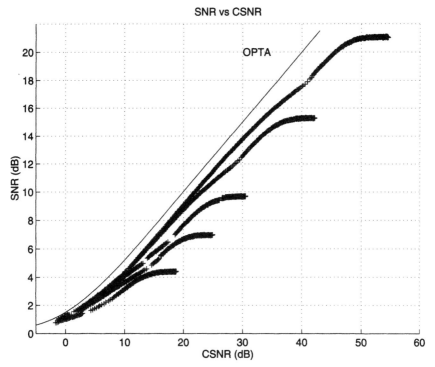

Figure 26.5 Performance of different size codebooks as a function of channel noise. The curves are (from top to bottom): OPTA, optimised coders for codebook sizes: 256, 64, 16, 8, and 4.

optimal in terms of *coding gain* [2], and subband filter banks [3,4]. As the filter banks have a region of support which is larger than the block transforms, they have potentially better performance.

26.8. POWER AND BANDWIDTH ALLOCATION [5]

Assume that we have obtained clustering of signals with similar statistics into easily definable blocks of data through the signal decomposition. Assume further that we have L different mapping devices available which use $r_i = (K/M)_i$, $i = 1, 2, \ldots, L$ channel symbols per source symbol. If device k is used for a portion P_k of the samples, the average channel symbol rate per source sample is given by

$$r = \sum_{i=1}^{L} P_i r_i \qquad (26.8)$$

If a certain maximum bandwidth is available, r must be chosen so that the compressed signal will occupy at most that bandwidth. Assume further that we want to restrict the energy per channel symbol to S, then

$$\sum_{i=1}^{L} P_i r_i S_i \leq rS \tag{26.9}$$

must be satisfied when S_i is the power used by mapping i. Both r_i and S_i must be nonnegative real numbers for physical reasons.

The optimal power and bandwidth allocation problem can now be stated as

$$\min_{\{r_i, S_i\}} \sum_{i=1}^{L} P_i D_i(S_i, r_i), \tag{26.10}$$

where $D_i(S_i, r_i)$ is the total distortion incorporating both approximation (mapping) noise and channel noise for class i.

If we go back to the special case of Gaussian signal and channel, and apply the OPTA curve given in equation (26.6), we can formulate an explicit minimisation problem:

$$\min_{\{r_i, S_i\}} \sum_{i=1}^{L} P_i (1 + S_i/\sigma_n^2)^{-r_i} \sigma_i^2 \tag{26.11}$$

The solution of this problem is that each non-zero channel symbol is allocated an equal amount of power and the rate for each source symbol is given by [6]

$$r_i = \begin{cases} 0, & \text{if } \sigma_i^2 \leq \sigma_L^2 \\ \frac{1}{2C} \log_2 \frac{\sigma_i^2}{\sigma_L^2} & \text{if } \sigma_i^2 > \sigma_L^2 \end{cases} \tag{26.12}$$

σ_L^2 is a parameter which must be chosen to balance the equation $\sum_{i=1}^{L} P_i r_i = r$, and the channel information capacity is given by $C = 1/2 \log_2(1 + S/\sigma_n^2)$.

In practice, we need to limit the number of mappings to a small number. The mappings also need to be simple in the sense that K and M must be small numbers. Selection of representative mappings will therefore be necessary.

26.9. EXAMPLE: STILL IMAGE CODING COMBINED WITH 81 PAM [5]

The following still image communication system is somewhat simpler than prescribed by the above theory. The channel representations will basically be discrete (except for a one-to-one dimensional mapping), the approximations will all be performed by scalar quantisers, and the resources distributed through three-level symbols. We hope through the example to clarify the principles and demonstrate some of the potential advantages of the proposed principles.

The example combines subband coding, allocation of scalar quantisers, and mappings from the quantised symbols of different resolutions to an 81 pulse amplitude modulation scheme. The signal is transmitted over an additive, white, Gaussian noise channel.

The complete block diagram of the system is shown in Figure 26.6. The system contains decomposition by an 8×8 filter bank followed by a selection of 4 different quantisers of 3, 9, 27, and 81 levels. The "ters (ternary symbol) allocation" block

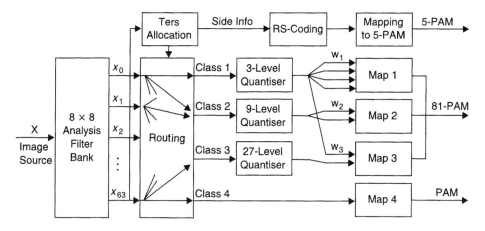

Figure 26.6 Subband coder combined with 81 PAM.

picks the right quantiser for each block of samples based on their power levels. This is very much the same as bit allocation, the only deviation is that we use three instead of two levels. The most interesting part of the system is the bank of mapping devices. This combines the multilevel symbols to 81 level symbols. The combinations used are

- mapping 0: Discard samples
- mapping 1: $3 \times 3 \times 3 \times 3$
- mapping 2: 9×9
- mapping 3: 3×27
- mapping 4: Continuous.

One of the mappings is illustrated in Figure 26.7. Figure 26.7 also shows the approximations used. Because the quantisers are scalar, the representation values are located on irregular rectangular grids. It should be noted that VQ, and especially continuous VQ, would perform better.

The mapping allocation has to be transmitted as side information. Errors in the side information would be disastrous, as it would cause loss of synchronism between transmitter and receiver. To avoid side information errors, RS codes in combination with 5 PAM are used for this part of the code.

Simulation results using the model on the test image "Lena" are shown in Figure 26.8 for 0.12 symbols/pixel together with coding results for a standard system where the mappings to the channel are done randomly, corresponding to what you might expect for a standard communication system. The visual quality for the two systems is demonstrated in Figure 26.9. For random mapping (left) the channel has 35 dB CSNR. For the proposed method (right) the signal is transmitted over a channel with 20 dB CSNR.

In this image transmission system the approximation noise and the channel noise are both additive. That is, both types of noise will generate visually equivalent distortions. If the approximation stage is designed to give pleasing artefacts, so will the channel distortion. This is very different from most signal transmission systems.

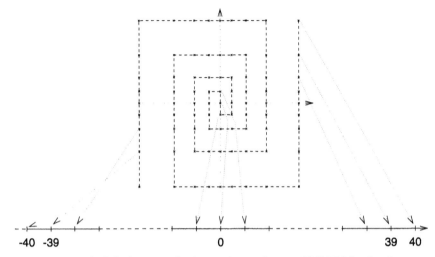

Figure 26.7 Scalar quantisation and mappings to 81 PAM for 9 × 9.

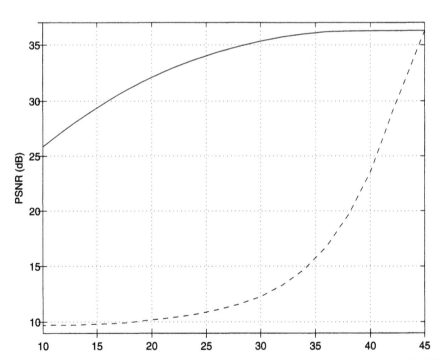

Figure 26.8 Simulation result for the "Lena" image. Solid line: proposed system. Dashed line: random mappings.

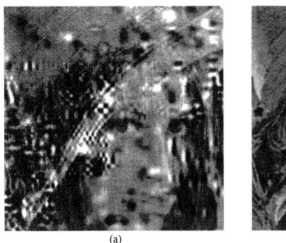

(a)   (b)

Figure 26.9 Decoded images. Left: random mappings at CSNR = 35 dB, right: proposed system at CSNR = 20 dB.

26.10. CONCLUSIONS

The main message of this chapter is that the potential gain by performing joint source coding and modulation is significant. Robustness against channel variations is an added benefit. This is extremely important for many wireless applications. Such a system can avoid the sudden breakdown experienced by traditional digital systems when the noise level surpasses a value corresponding to the capacity of the error protection codes used.

One aspect of the theory discussed in this chapter is that the relationship between neighbouring channel symbols and neighbouring source symbols can only be guaranteed when the encoder mappings reduce the signal dimensionality. In the case of dimension expansion, which is necessary either when the CSNR is very low, as for example in many satellite channels, or when the signal parameter dynamics are large, such as with audio signals, the problem is more difficult. Several authors have addressed this problem for the case of vector quantisation combined with a binary channel [7,8,9,10]. We believe that this work can be extended to embrace broader issues as discussed in this chapter to provide flexible implementations of future wireless communication systems.

ACKNOWLEDGEMENTS

This chapter is based on the Dr.Ing. theses by John Markus Lervik [6] and Arild Fuldseth [11]. Most of the material can also be found in references [12,11,14,1,15,16,17,18,19,20,21,5].

REFERENCES

[1] Fuldseth, A. and Ramstad, T. A., Bandwidth compression for continuous amplitude channels based on vector approximation to a continuous subset of the source signal space. In *Proceedings of ICASSP-97*, **IV**, 3093–3096, 1997.

[2] Jayant, N. S. and Noll, P., *Digital Coding of Waveforms, Principals and Applications to Speech and Video*. Prentice Hall, Englewood Cliffs, NJ, 1984.

[3] Vaidyanathan, P. P., *Multirate Systems and Filter Banks*. Prentice Hall, Englewoood Cliffs, NJ, 1993.

[4] Ramstad, T. A., Aase, S. O. and Husoy, J. H., *Subband compression of images – Principles and Examples*. Elsevier, North-Holland, Amsterdam, 1995.

[5] Lervik, J. M., and Fischer, T. R., Robust subband image coding for waveform channels with optimum power and bandwidth allocation. Submitted to ICASSP-97.

[6] Lervik, J. M., Subband image communication over digital transparent and analog waveform channels. Ph.D. thesis, Norwegian University of Science and Technology, December 1996.

[7] Farvadin, N. and Vaishamapayan, V., Optimal quantizer design for noisy channels: An approach to combined source-channel coding. *IEEE Transactions on Information Theory*, **33**, 827–838, 1987.

[8] Knagenhjelm, P., A recursive design method for robust vector quantization. In *Proceedings of the International Conference on Signal Processing Applications and Technology*, Boston, MA, 948–954, November 1992.

[9] Modestino, J. W. and Daut, D. G., Combined source-channel coding of images. *IEEE Transactions on Communications*, **27**, 1644–1659, 1979.

[10] Modestino, J. W., *et al.*, Combined source-channel coding of images using the block cosine transform. *IEEE Transactions on Communications*, **29**, 1262–1274, 1981.

[11] Fuldseth, A., Robust subband video compression for noisy channels with multilevel signaling. Ph.D. thesis, Norwegian University of Science and Technology, 1997.

[12] Fuldseth, A., Combined celp speech coding for quadrature amplitude modulation. In *Proceedings of EUSIPCO-94*, 928–931, 1994.

[13] Fuldseth, A. and Ramstad, T. A., Combined video coding and multilevel modulation. In *Proceedings ICIP-96 Lausanne, Switzerland*, **I**, 941–944, September 1996.

[14] Fuldseth, A. and Ramstad, T. A., Robust subband video coding with leaky prediction. In *Seventh IEEE Digital Signal Processing Workshop, Loen, Norway*, 57–60, September 1996.

[15] Lervik, J. M., Integrated system design in digital video broadcasting. *Piksel'n*, **10**(4), 12–22, 1993.

[16] Lervik, J. M. and Ramstad, T. A. An analog interpretation of compression for digital communication systems. In *Proceedings ICASSP-94*, Adelaide, South Australia, **V**, 28–284, April 1994.

[17] Fuldseth, A. and Levik, J. M., Combined source channel coding for channels with a power constraint and multilevel signaling. In *Proceedings of Nordic Signal Processing Symposium (NORSIG-94), Alesund, Norway*, 38–42, May 1994.

[18] Lervik, J. M., Joint optimization of digital communication systems: Principles and practice. In *Proceedings of the ITG Conference, Munich*, 115–122, October 1994.

[19] Fuldseth, A. and Levik, J. M., Combined source channel coding for channels with a power constraint and multilevel signaling. In *Proceedings of ITG Conference, Munich*, 429–436, October 1994.

[20] Lervik, J. M., Grovlen, A. and Ramstad, T. A., Robust digital signal compression and modulation exploiting the advantages of analog communication. In *Proceedings IEEE GLOBECOM, Singapore*, 1044–1048, November 1995.

[21] Lervik, J. M. and Ramstad, T. A., Robust image communication using subband coding and multilevel modulation. In *Proceedings SPIE VCIP-96, Orlando, Florida*, **2727**, part 2, 524–536, October 1996.

[22] Shannon, C. E., A mathematical theory of communication. *Bell Systems Technical Journal*, **27**, 379–423, 623–656, 1948.

27

Aspects of Error Resilience for Block-based Video Coders in Multimedia Communications

A. H. Sadka, F. Eryurtlu and A. M. Kondoz

27.1. INTRODUCTION

In recent years, there has been an increasing need for using video signals in multimedia services. As we move towards a third-generation universal telecommunication environment, the necessity to carry video information on mobile networks increases. Since mobile environments are highly hostile, error resilience techniques [1] must be developed to render their bit streams more immune to channel conditions.

Unlike speech and data signals, video information contains redundancies in both time and space. The role of a video coder is to exploit this characteristic in the best possible way to optimise the coding efficiency of the compression algorithm. However, the suppression of this redundant information has a very negative impact on the quality of service provided to the decoder when the compressed video signal is transmitted over erroneous channels. Therefore, special precautions must be taken

Insights into Mobile Multimedia Communication
ISBN 0-12-140310-6

during the video coding process to enhance the robustness of the coder and render the compressed bit stream more resilient to channel errors.

The effect of a mobile channel can be best simulated by error bursts affecting portions of the video bit stream at a certain frequency. However, using a good interleaving processor, the burst of errors can be converted into a random occurrence of errors affecting various small portions of the bit stream, thus smoothing out the effect of damage. In some cases, however, a single bit flip in the video stream plunges the decoder in a state of desynchronisation, forcing it to skip all the forthcoming bits until synchronisation is retrieved by decoding an error-free synch bit-pattern. In this case, a single bit error effectively leads to a burst error and drastic measures need to be taken to reduce the propagation of errors.

27.2. BIT RATE VARIABILITY IN BLOCK-BASED VIDEO CODERS

To ensure a constant perceptual quality of a video signal at the decoder, it is necessary to maintain a fixed quantisation step-size parameter at the encoder. Alternatively, varying the quantiser value within the encoding process can be employed to target a fixed bit rate at the expense of a variable subjective quality which is perceptually undesirable. Moreover, the variable bit rate aspect of a video traffic can be exploited to dynamically accommodate other types of traffic (e.g. speech, data) according to the activity of the video scene. Further details of this can be found in Chapter 24 of this book.

There are two main reasons why a block-based video coder exhibits this variable bit rate characteristic:

1. The temporal and spatial redundancies that are inherently available in a video sequence leads the video coder to determine the part (could be non-intersecting slices) of a video frame which needs to be compressed and transmitted. The remaining part is redundant and can be reconstructed by predicting its contents from transmitted information. In block-based video coders, like H.263 [2], the amount of redundancy in the temporal domain (INTER mode) can be estimated and controlled by the motion estimation threshold which determines how different a specific macroblock (MB) is from its best match in the previous frame. Obviously, this sort of redundancy controls the number of coded MBs within a particular frame. Non-coded MBs are represented by a one-bit flag (COD) to notify the decoder of the presence of a redundant MB. In the spatial domain, the amount of redundancy is best represented by the level of similarity in the pixel values of the same MB. Passing these pixel luminance and chrominance values through the DCT and zigzag-pattern transforms yield longer or shorter runs which can be eventually represented by a run-length encoder with a higher or lower number of bits respectively. Consequently, the compression efficiency of a video encoder is determined by the amount of redundancy that is detected and suppressed from a video signal in both the temporal and spatial domains. The temporal and spatial dependencies of a video signal are illustrated in Figure 27.1.

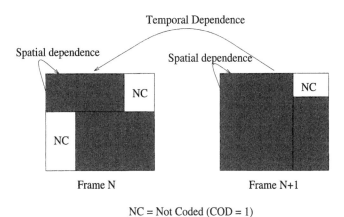

Figure 27.1 Effect of temporal and spatial redundancies on the output bit rate variability of a video coder.

2. Another major factor which leads to variable bit rate in a block-based video coder such as H.263 is the variable length representation of entropy coding. To optimise the compression efficiency, H.263 employs Huffman coding to represent the output parameters based on their frequency of occurrence within the bit stream. The Huffman encoder generates a stream of variable length code words.

27.3. EFFECTS OF CHANNEL ERRORS ON H.263

Errors in the video stream can significantly affect the motion compensation process defined in the recommendation of this video coding algorithm. When the best-match MB in the previous frame is identified and a motion vector (MV) for the current MB is calculated in both full and half pixel accuracy, the encoder applies differential coding to MVs instead of their initial x and y components. The motion compensation mechanism does not only affect the synchronisation of the decoder once an MV is corrupted, but it helps in speeding up the propagation of errors in the sequence due to error accumulation, and in widening the damaged area due to the temporal dependency between consecutive video frames. In the H.263 motion compensation process, an MV is sent as the difference between the calculated MV components and those of the median of three candidate MV predictors belonging to MBs to the top and left of the current MB. This is best illustrated in Figure 27.2. Consequently, if an error occurs at a particular MB, the decoder is unable to correctly reconstruct the MB whose MV depends on that of the affected MB as a candidate predictor. Similarly, the failure in the correct reconstruction of the current MB prevents the decoder from correctly recovering forthcoming MBs which depend on the current MB in the motion compensation mechanism. The accumulative damage due to this temporal dependency might occur because of a single bit error regardless of the

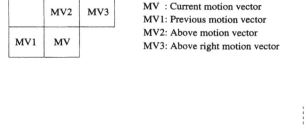

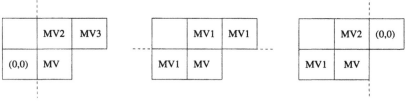

- - - - - - - - - - -   : Picture or GOB header

Figure 27.2 Encoding of motion vectors in H.263 video coding algorithm.

correctness of subsequent information which depends on the erroneous MB in the motion prediction process.

On the other hand, the occurrence of a bit error can be categorised into three different classes:

1. The bit flip that corrupts one of the parameters does not have any influence on segments of data other than the affected parameter. The error is limited in this case to a single MB that does not take part in any further reconstruction process. One example of this category can be a bit error in an INTRADC coefficient of an MB which is not used in the motion prediction. Since the affected MB is not used in a subsequent prediction process, the damage will be localised and confined only to the affected MB. Moreover, the decoder is not bound to enter any state of desynchronisation since it has skipped the correct number of bits, flushing its buffer before moving to the next parameter in the bit stream. This kind of error is the least destructive to the quality of service amongst the three categories of errors.

2. The second type of error is more problematic because it incurs an accumulative damage in both time and space due to prediction. When the prediction residual of motion vectors is sent, bit errors in motion code words propagate until the end of the frame. Moreover, the error propagates to subsequent INTER coded frames due to temporal dependency induced by the motion compensation process. This category of errors is obviously more detrimental to the quality of the decoded signal than the first one; yet, it does not cause any state of desynchronisation either, since the decoder flushes the correct number of bits of the erroneous motion code words.

3. The worst effect of bit errors occurs when synchronisation is lost, and the decoder is no longer able to determine which part of a frame the received information belongs to. In this situation, when the decoder detects an error in

a variable length code word (VLC), it skips all the forthcoming bits, regardless of their integrity, in the search for the first error-free synch pattern. Therefore, a single bit corruption is transformed into a burst error. The occurrence of a bit error in this case manifests in two different ways: The first problem arises when the corrupted VLC word results in a new bit pattern that is a valid entry in the Huffman tables of that specific parameter. In this case, the error cannot be detected. However, the resulting VLC word might be of a different length, causing the decoder to skip the wrong number of bits before moving forward to the next piece of information in the bit stream, hence creating a state of desynchronisation. This situation remains until an invalid code word is received, causing the decoder to jump to the first error-free synch word. The second problem appears when the corrupted VLC word (possibly in conjunction with subsequent bits) creates a bit pattern that is not approved nor readable by the Huffman decoder. In other words, the decoder fails to detect any readable VLC word for a particular parameter within a portion of the bit stream that corresponds to the maximum length code of the parameter. In this case, the decoder signals the occurrence of an error, skips the forthcoming bits and resumes decoding at the next intact synch word. Figure 27.3 illustrates these two cases.

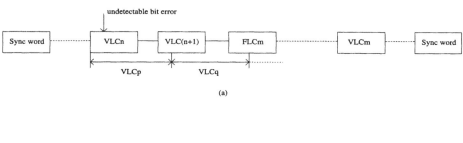

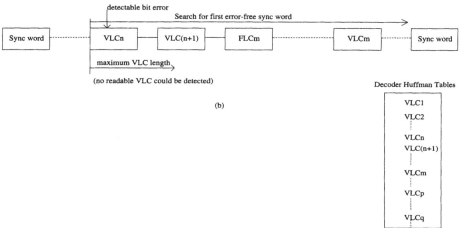

Figure 27.3 Bit error occurrence in a VLC: (a) undetectable error; (b) detectable error.

27.4. ERROR RESILIENT SCHEMES

To enhance the robustness of H.263 efficient error resilience techniques have to be employed. In this section, we present two algorithms developed to improve the robustness of a block-transform video coder to channel errors.

27.4.1. Robust I-Frame Algorithm

Because of the accumulative effect of errors, the distortion in the video sequence increases with time, and the quality experiences a gradual degradation on both subjective and objective scales. The number of frames (P-frames or INTER frames), in a video bit stream affects the quality of the video sequence when the latter is subject to channel errors. The higher the number of P-Frames, the less robust the video bit stream is. Therefore, one way to minimise the propagation of error is to refresh the scene by the inclusion of an I-Frame (INTRA Frame) at a predetermined frequency. This will of course reduce the compression efficiency. Several techniques have been applied to enhance the resilience of H.263 to channel errors:

1. *Change of the transmission order and error protection of INTRADC coefficients.*
   Increasing the frequency of I-frames fails when channel errors corrupt a VLC code word of an I-frame itself. The worst effect of this error occurs when the damaged VLC word is at the beginning of the I-frame. In this case, if one synch pattern is used per frame, the decoder has to skip all the frame bits searching for the recovery of synchronisation. Consequently, the I-frame will be completely unreconstructable, and the effect of such an error propagates for all the following P-frames until the next I-frame appears in the bit stream. This bursty effect of errors arises at the decoder due to a single bit flip that plunges the decoder into a state of desynchronisation until the next error-free synch word is received, and implies that the inclusion of an I-frame is a redundant overhead. To prevent such a situation from taking place, a method to protect I-frames must be devised and implemented in a way that stops channel errors from destabilising the decoder and skipping (possibly error-free) information due to loss of synchronisation. Since INTRADC coefficients are defined by the syntax of the H.263 video coding algorithm as fixed length code words, their fixed length can be exploited in changing the order of their transmission in an I-frame. For a QCIF-size (176×144), 594 INTRADC coefficients, 6 for each MB, are transmitted and each of them is represented by an 8-bit word. However, in the ordinary H.263, INTRADC coefficients are transmitted in the MB layer and thus followed by variable length parameters belonging to administrative data such as CBPY and MCBPC or to run-length coded AC coefficients. If a single bit error corrupts these VLC words, the decoder might lose synchronisation and skip all the INTRADC coefficients of the following MBs. This loophole in the resilience of ordinary H.263 can be tackled by transmitting the 594 DC coefficients of an I-Frame as a continuous stream of 4752 (8×594) bits before any of the frame VLC words appears in the bit stream. This will ensure that the decoder will complete the decoding of the INTRADC coefficients without

the possibility of losing synchronisation when a bit error corrupts a VLC word. To further enhance the resilience of the coder, a ½ rate convolutional coder is employed to protect the INTRADC coefficients which carry a high amount of energy of the video frame. Each 8-bit INTRADC coefficient is channel coded using a ½ rate convolutional coder with a constraint length of 7, hence generating an output of 16 bits onto the communication medium.

2. *Frame mode indicator coding.* Another drawback in the error resilience aspects of ordinary H.263 is attributed to the syntax of the picture coding indicator. This flag is allocated a single bit in the header of every video frame (both INTRA and INTER). However, due to the increase in the frequency of I-frames, there is a larger need to notify the decoder of the coding type for every received frame. If this bit is toggled on the channel, the decoder enters a state of desynchronisation. For this reason, this picture mode indicator is given 3 bits with a Hamming distance of 3 between its 2 possible values (000 for INTRA, 111 for INTER). Thus one bit error is allowed for the decoder to correctly identify the coding mode of the relative frame.

3. *Robust synch words at integer multiples of L.* When the decoder detects an error in a VLC code word it skips the bits following the error until the start code of the next frame to regain synchronisation. If the intermediate data bits are corrupted in a way that makes the decoder fall on a false synch word within the data stream, the start of a frame will be incorrectly identified and the decoder will be unable to notice the loss of synchronisation until it detects a fault in reading an unacceptable VLC code word. This state causes catastrophic damage to the quality of service at the decoder end. To reduce the possibility of occurrence of a false synch word, frame start codes are sent at the beginning of fixed-length intervals of L bits each (we have used $L = 32$ as a trade-off between bit overhead and efficiency of the error resilience algorithm). Consequently, synch words are made more robust to false decoding by being inserted at fixed intervals within the bitstream. This makes them less likely to be falsely decoded within the data segment of the video stream. Figure 27.4 shows the insertion of synch words at integer multiples of L within the bitstream. When the decoder is in the course of seeking synchronisation, it locates the start of these intervals for possible synch words and skips 32 bits instead of skipping over the stream bit by bit. A Hamming distance of 2 is given between the real synch word, defined by the syntax of H.263, and that of an equal-length word at the beginning of each of those fixed-length intervals. Figure 27.5 depicts the new layering structure of H.263 after introducing these three methodologies to enhance its error resilience.

In the experiments, the H.263 software from Telenor R&D [3] was used and many ITU test sequences were tried. An I-Frame was coded every 20 frames in video processed at 25 fps. A random value generator was used to simulate the effect of channel errors. A variety of error patterns have been tried on many test sequences, and average results were collected and used to assess the effectiveness of the error resilience algorithm. Figures 27.6 and 27.7 show the subjective and objective improvements achieved by applying the proposed error resilience algorithms on an erroneous "Grandmother" sequence. When an error is detected in a VLC word of an

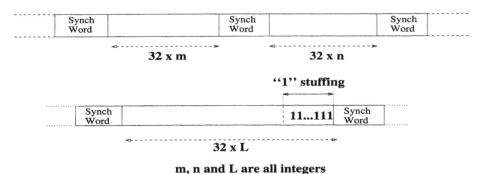

m, n and L are all integers

Figure 27.4 Robust synch words inserted at integer multiples of L.

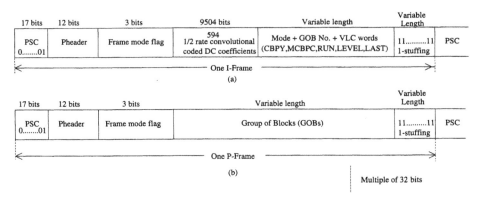

Figure 27.5 Modified picture layering structure: (a) I-frame; (b) P-frame.

Figure 27.6 First frame of grandmother sequence encoded at 52 kb/s, frequency of I-frames = 1.2 frames/s, BER = 0.0001: (a) ordinary H.263; (b) error resilient H.263.

I-Frame after all INTRADC coefficients are received, the decoder sets the values of skipped AC coefficients to zeros, searches for the start code of the next frame and resumes decoding. The algorithm proves that without any increase in the output bit rate (coarser quantiser), H.263 provides better objective and subjective results when video bit streams are sent over error-prone environments such as mobile radio links.

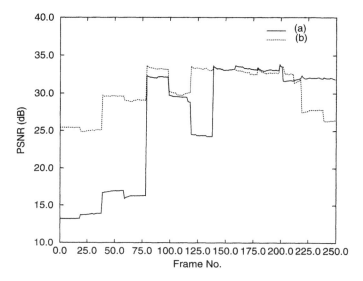

Figure 27.7 PSNR graph for 250 frames of Grandmother sequence encoded at 52 kbit/s, frequency of I-frames = 1.2 frames/s, BER = 0.000: (a) ordinary H.263, (b) error resilient H.263.

27.4.2. Two-way decoding and RVLC Algorithm

Two-way decoding makes use of the reversible VLC (RVLC) code words. These are decodable in two directions and thus help to locate and confine the corrupted area of a bit stream. If an error is detected in the forward direction, the decoding process is immediately interrupted and the next sync code is searched for. Once the sync word is located, the decoding process is resumed in the reverse direction. This scheme enables the decoder to start decoding at two different locations of the bit stream without the need to send additional information. The combination of reversible VLCs and the two-way decoding are very effective at reducing the effects of both random and burst channel errors.

To generate an RVLC, one of many techniques can be used [1]. One approach is to adopt a predetermined number of 1s and use it in any particular code word. For example, if three 1s are included in one code, a possible RVLC would be: (111, 1011, 1101, 11001, 10101,). Another methodology to generate a RVLC assumes the same number of 0s and 1s in each code word. In this case, a possible RVLC would be: (01, 10, 0011, 1100, 001011, 000111, 110100, ...).

In the simulations, an RVLC table whose words contain a fixed number of the first symbol (either 0 or1) was used. An error-free first INTRA frame is assumed and a random error generator is used to simulate the effects of errors. Although mobile channels introduce burst errors, the effect of an error burst on a portion of a video bit stream is quite similar to that of a random error. When a segment of a bit stream is affected by errors in one-way decoding, resulting in a state of desynchronisation, the whole segment of data is skipped until the decoder locates the first upcoming error-free synch word. This will simulate the effect of loss due to a burst of errors corrupting all the skipped bits regardless of their integrity. This is shown in Figure 27.8.

Therefore, a single bit error in a video bit stream usually has bursty error effects. To enable the fast recovery of synchronisation, a GOB start code was used at the beginning of each GOB. However, in two-way decoding, the occurrence of an error does not indicate the loss of the relative GOB since backward decoding can identify the correctly received bits which were skipped due to loss of synchronisation (Figure 27.9).

Figures 27.10 and 27.11 show the subjective and objective improvements achieved on the Carphone video sequence encoded at 28 kbit/s by applying the two-way decoding algorithm on video streams with RVLC code words.

Since a single bit error in a video bit stream usually implies bursty error effects, the effective bit error rate (BER) may be much higher than the channel BER. Assuming that there are L bits in a GOB (between two consecutive correctly detected synch words) and the channel BER is e_{ch}, then the effective BER e_{eff} can be derived as follows [4]:

$$e_{eff} = \sum_{m=0}^{L-1} p_m \frac{L-m}{L} \qquad (27.1)$$

where p_m is the probability of having the first corrupted bit at the bit position m as illustrated in Figure 27.8, and it is a function of e_{ch} and m:

$$p_m = (1 - e_{ch})^m e_{ch} \qquad (27.2)$$

Then equation (27.1) can be rewritten as:

$$e_{eff} = \frac{e_{ch}}{L} \sum_{m=0}^{L-1} (1 - e_{ch})^m (L - m) \qquad (27.3)$$

Equation (27.3) demonstrates that the effective error rate is actually much larger than the channel error rate in one-way decoding.

The block length L is an important parameter which affects the relation between effective and channel BERs. Due to the variable rate nature of a video signal, the block length varies depending on the encoder parameters such as the quantisation parameter. However, the average block length can be estimated by using the overall bit rate r, frame rate f, number of synch words per frame s and the length of the synch word l_s.

In the case of two-way decoding, when an error is detected in forward decoding, the decoder can find the next synch word and start decoding in the reverse direction.

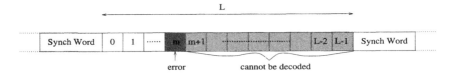

Figure 27.8 One-way decoding of variable-length codes.

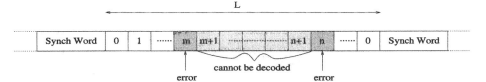

Figure 27.9 Two-way decoding of variable-length codes.

Figure 27.10 Carphone sequence encoded at 28 kbit/s and subject to one bit error for: (a) one-way decoding: (b) two-way decoding with reversible code words.

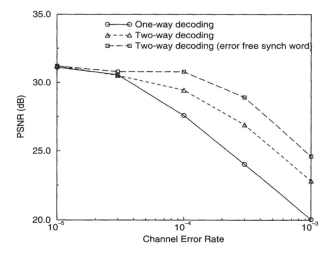

Figure 27.11 PSNR values for Carphone sequence encoded at 28 kbit/s using original and two-way decodeable codes.

In this way, more information can be correctly decoded. With reference to Figure 27.9, the effective BER for two-way decoding can be calculated as follows:

$$e_{eff} = \sum_{m=0}^{L-1}(1 - e_{ch})^m e_{ch}\left((1 - e_{ch})^{L-m-1}\frac{1}{L} + \sum_{n=0}^{L-m-2}(1 - e_{ch})^n e_{ch}\frac{L-m-n}{L}\right) \quad (27.5)$$

$$e_{eff} = (1 - e_{ch})^{L-1}e_{ch} + \frac{e_{ch}^2}{L}\sum_{m=0}^{L-1}\sum_{n=0}^{L-m-2}(1 - e_{ch})^{m+n}(L - (m + n)) \quad (27.6)$$

Figure 27.12 shows that the effective BER for two-way decoding converges to the channel rate at low BER values while there is always an off-set between them in the case of one-way decoding.

27.5. CONCLUSIONS

In order to stop the accumulation of errors in a video sequence, INTRA frames are sent at a higher rate in order to refresh the error-affected scene. Due to their sensitivity, I-frames need to be protected against errors in order to meet the requirements of their transmission. The robust I-frame error resilience algorithm used with the H.263 video coder yields better robustness to errors than conventional codec. We have also showed that two-way decoding algorithm is an efficient error resilience algorithm when used with reversible code words. This helps to locate the erroneous bits and the affected area to minimise the effect of error on the overall QoS.

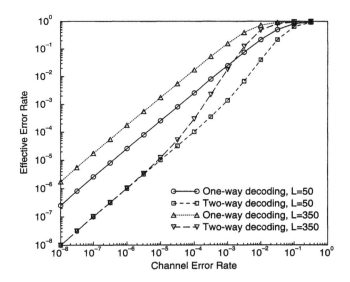

Figure 27.12 Effective error rate in one and two-way decoding as a function of channel error rate and block length.

Synchronisation of the video decoder can be recovered with a negligible increase in the overall bitrate.

REFERENCES

[1] International organisation for standardisation, ISO/IEC/JTC1/SC29/WG11. Description of error resilient core experiments, Document No. N1327, July 1996.
[2] ITU Telecommunication standardisation sector, ITU-T recommendation H.263 Video coding for low bitrate communication, November 1995.
[3] Telenor R&D H.263 video codec test model, November 1995.
[4] Eryurtlu, F., Sadka, A. H., Kondoz, A. M. and Evans, B. G., Error robustness improvement of video codecs with two-way decodable codes, *IEE Electronic Letters*, **33** (1), 41–43, 1997.

28

Error-resilient Coding for H.263

Bernd Girod, Niko Färber and Eckehard Steinbach

28.1. INTRODUCTION

The ITU-T Recommendation H.324 describes terminals for low bit rate multimedia communication, that may support real-time voice, data, and video, or any combination, including videotelephony. Because the transmission is based on V.34 modems operating over the widely available public switched telephone network (PSTN), H.324 terminals are likely to play a major role in future multimedia applications. In fact, an increasing number of H.324 terminals is already being implemented and purchased by various companies and vendors. One important reason for this success was the availability of the H.263 video coding standard [1], which achieves acceptable image quality at less than 32 kbps. Other recommendations in the H.324 series include the H.223 multiplex, H.245 control protocol, and G.723 audio codec.

As mobile communication becomes a more important part of daily life, the next step is to support mobile multimedia communication. Recognizing this development, the ITU-T started a new "Mobile" Ad Hoc Group (AHG) in 1994 to investigate the use of H.324 in mobile environments. In the following paragraphs we will try to give a short summary of the main issues discussed in the Mobile AHG, till 1997, that are related to robust video transmission.

During the work of the Mobile AHG, it turned out that one major requirement for a mobile H.324 terminal is its ability to interwork with terminals connected to the PSTN at a reasonable complexity and low delay. This resulted in the decision to use the H.263 video codec and G.713 audio codec unchanged, because transcoding was considered to be too complex. With this decision, many promising proposals became obsolete. For example, the reordering of the H.263 bit stream into classes of different sensitivity has been proposed to enable the use of unequal error protection [2,3]. Though this approach is known to be effective [4–6], the parsing and re-assembling of the bit stream with added error protection cannot be implemented in a low complexity, low delay interworking unit.

Two other useful proposals, which did not have the drawback of high complexity, could not be adopted by the ITU-T, because they would have required minor changes in the H.263 bit stream syntax. Because H.263 was already frozen in January 1996, neither the SUB-VIDEO approach [7] nor the NEWPRED approach [8] has been included in H.263 for increased error robustness. However, both approaches are included in a slightly modified and extended version in the new H.263 + video coding standard (Annex K and Annex N respectively).

Given the above restriction on changes in H.263, a standard compatible extension for robust video transmission, as presented here, was a valuable enhancement of H.324 for the use in mobile environments and could therefore be quickly included into the standard. Because of its compatibility, no technical changes were needed in H.263. However, an informative appendix (Appendix II) was added to explain the basic concept. In addition, minor extensions of the H.245 control standard were necessary to include the additional control message "videoNotDecodedMBs". Because our approach was developed in close relationship to the Mobile AHG, we will base our description on H.263. However, it should be noted that the approach could also be used with other video coding standards like H.261 or MPEG-2. Previous publications of our work on this approach include [9,10].

The rest of this chapter is organized as follows. In section 28.2 we will discuss why residual errors are difficult to avoid in mobile environments for real-time transmission and why encoded video is particularly sensitive to residual errors. In section 28.3 we describe the error concealment technique employed in this paper and investigate the typical distortion caused by error propagation in H.263. The new approach that uses feedback information to effectively mitigate the effects of errors is described in detail in section 28.4. In section 28.5, a low complexity algorithm for real-time reconstruction of spatio-temporal error propagation is described that can be used for more effective evaluation of feedback messages. Experimental results demonstrate the performance of our method in section 28.6.

28.2. MOBILE VIDEO TRANSMISSION

Many existing mobile networks cannot provide a guaranteed quality of service, because temporally high bit error rates cannot be avoided during fading periods.

Transmission errors of a mobile communication channel may range from single bit errors up to burst errors or even a temporal loss of signal. Those varying error conditions limit the effective use of forward error correction (FEC), since a worst case design leads to a prohibitive amount of overhead. This is particularly true if we have to cope with limited bandwidth requirements. Also note that the use of interleaving is constrained by low delay requirements. Closed-loop error control techniques like automatic repeat on request (ARQ) have been shown to be more effective than FEC [11–13]. However, retransmission of corrupted data frames introduces additional delay, which is critical for real-time conversational services. Given an upper bound on the acceptable maximum delay, the number of retransmissions is mainly determined by the round-trip delay of data frames. For networks like Digital European Cordless Telephony (DECT), where data frames are sent and received every 10 ms, several retransmissions may be feasible. On the other hand, retransmission of video packets over a satellite link would introduce a prohibitive delay. Therefore, residual errors are typical for real-time transmission in mobile environments even when FEC and ARQ are used in an optimum combination.

In the presence of residual errors, additional robustness is required because the compressed video signal is extremely vulnerable against transmission errors. Low bit rate video coding schemes rely on INTER coding for high coding efficiency, i.e. they use the previous encoded and reconstructed video frame to predict the current video frame. Due to the nature of predictive coding, the loss of information in one frame has considerable impact on the quality of the following frames.

The severity of residual errors can be reduced if error concealment techniques are employed to hide visible distortion as well as possible [14–18]. However, even a sophisticated concealment strategy cannot totally avoid image degradation and the accumulation of several small errors will finally result in a poor image quality. Only if the affected part of the picture is encoded in INTRA mode, i.e. without reference to a previous frame, the error propagation is terminated.

Summarizing the above paragraphs, we must accept that residual errors cannot be totally avoided for real time transmission in mobile environments, and that error propagation is the main problem in mobile video transmission. To illustrate the second statement, we investigate the effects of error propagation in the framework of the H.263 video coding standard in the following section.

28.3. ERROR PROPAGATION AND CONCEALMENT IN H.263

The basic concept of H.263 is a hybrid interframe prediction exploiting temporal redundancy and transform coding of the residual prediction error exploiting spatial redundancy and adaptively reducing spatial resolution. Temporal prediction is based on block-based motion estimation and compensation, while a discrete cosine transform (DCT) is used for spatial redundancy reduction.

The bit stream syntax of H.263 is based on the hierarchical video multiplex. For QCIF resolution (176×144 pel), pictures are subdivided into nine groups of blocks (GOB, 176×16 pel), each consisting of 11 macroblocks (MB, 16×16 pel). One MB consists of 6 blocks (8×8 pel), 4 for luminance and 2 for chrominance, respectively.

Two basic modes of operation can be used for each MB. In the INTER mode, motion compensated prediction is utilized and the residual prediction error is DCT coded. In the INTRA mode, no temporal dependency from previous frames is introduced and the picture is directly DCT coded. The choice between INTRA and INTER mode is not a subject of the recommendation and may be selected as part of the coding control strategy.

After this short review of H.263 terminology, we will define the error concealment technique that is employed in the following simulations. H.263 supports fast resynchronization after transmission errors by optionally inserting start codes for each GOB. Based on this option the decoder discards corrupted GOBs and replaces the corresponding image content with data from the previously decoded frame. This simple approach has been shown to yield good results for sequences with little motion [14,19]. However, severe distortions are introduced for image regions containing heavy motion. Note that an erroneous GOB can be detected by the use of a cyclic redundancy check (CRC), or by discovering syntax violations (e.g. invalid VLC codes, more than 64 DCT coefficients, wrong number of MBs per row, etc.). More sophisticated but more complex error concealment techniques such as described in [14–18] could be used instead.

Figure 28.1 illustrates an example for spatio-temporal error propagation due to motion compensated prediction for the case that one GOB is corrupted. With increasing time, the error inside a particular GOB propagates across the border in spatial direction and may, in a worst case, spread over the entire frame.

Figure 28.2 shows the loss of picture quality ($\Delta PSNR$) after concealment of three successive GOBs. The correctly decoded sequence serves as a baseline, i.e. for temporally corresponding frames $\Delta PSNR = PSNR_{lost} - PSNR_{correct}$ is computed. The QCIF sequence *Foreman* is coded at 64 kbps and 8.33 fps, resulting in a $PSNR_{correct}$ of about 34 dB. Twenty-five simulations are conducted with different temporal and spatial locations of the lost GOBs (dotted lines). The solid line represents the averaged result, indicating that a residual loss of approximately 3 dB still remains in the sequence after 3 seconds due to temporal error propagation. Please note that the visible distortion is concentrated in certain image regions (e.g. inside the lost GOB) and the PSNR is measured over the entire frame. In other words, even a small decrease in PSNR may represent very annoying artifacts in a small area.

Periodic INTRA refresh of image regions would accelerate the image quality recovery. However, the overall image quality decreases due to the lower efficiency of INTRA coding in comparison to INTER coding. This is particularly true if the

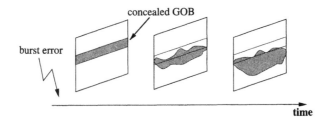

Figure 28.1 Spatio-temporal error propagation.

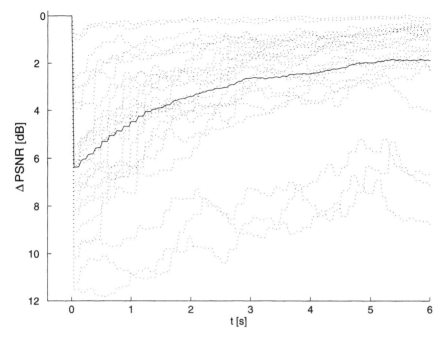

Figure 28.2 Decrease in PSNR for error concealment of three successive GOBs.

channel conditions are good most of the time. In this case, additional performance gain due to INTER coding should not be wasted, especially because the existing mobile networks exhibit severe bandwidth limitations. Therefore no periodic INTRA refresh is applied in the simulations in order to achieve highest coding performance for the error-free case, which serves as a baseline.

28.4. ERROR COMPENSATION BASED ON A FEEDBACK CHANNEL

As shown in the previous section, the distortion caused by the concealment of corrupted GOBs may remain visible in the image sequence for several seconds. Our approach utilizes the INTRA mode to stop temporal error propagation but limits its use to severely affected image regions only. During error-free transmission, the more effective INTER mode is utilized and the system therefore adapts effectively to varying channel conditions. Note that this approach requires that the encoder has knowledge of the location and extent of erroneous image regions at the decoder. As will be shown below, this can be achieved by utilizing a feedback channel.

The feedback channel is used to send negative acknowledgements (NAKs) back to the encoder. NAKs report the temporal and spatial location of GOBs that could not be decoded successfully and had to be concealed. For H.263, the temporal and spatial location of a GOB can be encoded by the time reference (TR, 8 bit) and group number (GN, 5 bit). The resulting rate on the feedback channel is then mainly determined by the GOB error rate, but will in general be very small.

Based on the information of a NAK, the encoder can reconstruct the resulting error distribution in the current frame. To do so, the encoder could store its own bit stream and simulate the loss of the reported GOB while decoding it again. Note that the encoder will have to use the same concealment strategy as the decoder when simulating the loss. While this approach is not feasible for a real time implementation, it shows that the encoder itself "knows" all the information necessary for the reconstruction of spatio-temporal error propagation. For a practical system, the error distribution has to be estimated with a low complexity algorithm, which will be described in section 28.5. For the rest of this section, however, we assume that the encoder gains complete knowledge of the true error distribution. Then, the coding control of the encoder can be modified to effectively stop error propagation by selecting the INTRA mode whenever an MB is severely distorted. On the other hand, if error concealment was successful and the error of a certain MB is only small, the encoder may decide that INTRA coding is not necessary.

Figure 28.3 shows averaged simulation results for the test sequence *Foreman*. Identical simulation conditions as in Figure 28.2 are assumed and the round-trip delay equals 700 ms. Compared to the case without error compensation, the picture quality recovers rapidly as soon as INTRA coded MBs are received. Note that the round trip delay is much greater than 250 ms, which is a common requirement for conversational services. It is also important to note that our approach does not add any additional delay. A longer delay just results in a later start for the error recovery.

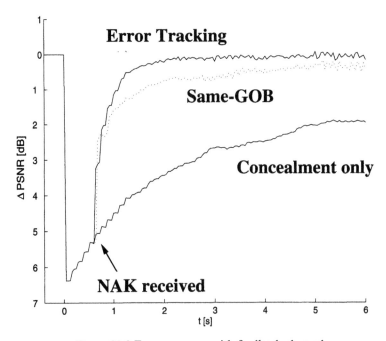

Figure 28.3 Error recovery with feedback channel.

Considering the slow recovery for "concealment only", NAKs may still be useful after several seconds.

Error Tracking represents the approach described above including the reconstruction of spatio-temporal error propagation. *Same-GOB* is a simplified version, where the spatial error propagation is not taken into account and the entire reported GOB is INTRA refreshed. The same-GOB strategy does not require the reconstruction of spatial error propagation and therefore exhibits lower complexity. However, it does not consider the propagation of errors across the borders of the GOB. Though the difference between the two strategies seems to be small in the average case (0.5 dB), annoying artifacts may be avoided by error tracking for particular unfavourable cases.

A perfect reconstruction of the error propagation as part of the coder control, while theoretically possible, would be computationally extremely demanding. In the next section we present a low complexity algorithm that estimates the spatio-temporal error propagation with a sufficient accuracy. This scheme is of particular interest to real-time implementations since its usage of memory and the processing load are very modest.

28.5. LOW COMPLEXITY ESTIMATION OF ERROR PROPAGATION

The spatio-temporal dependencies of MBs in successive frames arise from motion compensated prediction when coding MBs in INTER mode. These dependencies, together with an error severity measure, have to be stored at the encoder in order to provide enough information for rapid reconstruction of spatio-temporal error propagation. This can be achieved by the following algorithm.

Assume N macroblocks within each frame enumerated $mb = 1, \ldots, N$ in transmission order from top-left to bottom-right. Let $\{n_{err}, mb_{first}, mb_{last}\}$ be the content of an NAK sent to the encoder, where $mb_{first} \leq mb \leq mb_{last}$ indicates a set of erroneous macroblocks in frame n_{err}. For the case that complete GOBs are discarded and concealed, as described in section 28.3, mb_{first} and mb_{last} would correspond to the first and last MB of lost GOBs.

To evaluate the NAK, the encoder must continuously record information during the encoding of each frame. First, the initial error "energy" $E_0(mb, n)$ that would be introduced by the loss of macroblock mb in frame n needs to stored. For the error concealment strategy described in section 28.3, $E_0(mb, n)$ may be computed as the summed absolute difference (SAD) of macroblock mb in frame n and $n - 1$. Second, the number of pixels transferred from macroblock mb_{source} in frame $n - 1$ to macroblock mb_{dest} in frame n is stored in dependencies $d(mb_{source}, mb_{dest}, n)$. These dependencies can be derived from the motion vectors. Figure 28.4 shows an example for the dependencies introduced by the motion vector $(-4, -11)$ in MB 30. The prediction is formed from MB 18, 19, 29 and 30 using 44, 132, 20 and 60 pixels respectively. According to the above notation, $d(18, 30, n)$ would then for example be set to 44.

In practice, error tracking information will be stored in a cyclic data structure covering the last M frames. The value of M has to be chosen depending on the maximum round trip delay and the encoded frame rate. Because the typical frame rate for low bit rate video is rather low (e.g. 10 fps), and the round trip delay for

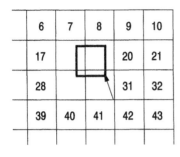

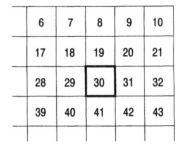

Frame *n* - 1   Frame *n*

Figure 28.4 Motion compensation of MB 30 with motion vector $(-4, -11)$.

conversational services should be in the order of 250 ms, typical values for M will be less than 10. Note that the area used for motion compensation of a particular macroblock can only overlap with a maximum of 4 macroblocks. Therefore, only 4 dependencies will have to be stored for each macroblock, resulting in small storage requirements.

The evaluation of the stored error tracking information is as follows. Assume that an NAK arrives before frame n_{next} is encoded, such that $n_{next} > n_{err}$. Then, the estimated error $E(mb, n_{err})$ in macroblock mb and frame n_{err} is initialized as

$$E(mb, n_{err}) = \begin{cases} E_0(mb, n_{err}) & \text{for} \quad mb_{first} \leq mb \leq mb_{last} \\ 0 & \text{else} \end{cases} \tag{28.1}$$

For subsequent frames n, with $n_{err} < n < n_{next}$, the error may be estimated recursively as

$$E(mb, n) = \sum_{i=1}^{N} \frac{E(i, n-1)d(i, mb, n)}{256} \tag{28.2}$$

where a uniformly distributed error in each macroblock is assumed after each iteration. Note that the evaluation of (28.2) does not actually require N multiplications for each macroblock but a maximum of 4 (see above). Therefore, only $4N$ multiplications are required for the whole frame, which results in low computational complexity for the algorithm.

After the last iteration, the error distribution $E(mb, n_{next} - 1)$ is available to the encoder and may be incorporated into the mode decision of the next frame as described in section 28.3. For example, macroblock mb is encoded in INTRA mode, if $E(mb, n_{next} - 1)$ exceeds a threshold.

28.6. EXPERIMENTAL RESULTS

In this section we try to investigate the performance of the proposed error compensation technique in a mobile environment. Our simulations are based on two bit error sequences named DECT1 and DECT2. As the name implies, those bit error sequences were generated by simulating a Digital European Cordless Telephony (DECT) channel. For both test channels, Rayleigh fading and a velocity of 1.4 km/h are assumed. Furthermore, the assumed signal to noise ratio (SNR) is 20 and 30 dB respectively, which results in a bit error rate (BER) of 2.1×10^{-2} for DECT1 and 2.1×10^{-3} for DECT2. Both channels exhibit severe burst errors and provide a total bit rate of 32 kbps.

Because the proposed error compensation techniques can be used together with FEC and ARQ, we will present simulation results for different combinations. In particular, the following simulation scenarios are compared:

- no protection of the video bit stream (NP);
- forward error correction (FEC);
- hybrid ARQ type I (ARQ-FEC);
- same-GOB strategy + FEC (SG-FEC);
- error tracking strategy + FEC (ET-FEC);
- same-GOB strategy + ARQ + FEC (SG-ARQ-FEC);
- error tracking + ARQ + FEC (ET-ARQ-FEC).

FEC and ARQ are based on a packet size of $n = 255$ bit. Because of the real time constraint, only one retransmission of corrupted packets is allowed. For FEC, the remaining code parameters have been selected from a BCH code table [20] to optimise the throughput for a given BER under a worst case assumption. For more detail on the reasoning behind this selection see [10]. The resulting code rate for the DECT1 channel is $R = 0.875$, which allows to correct $t = 4$ bit errors per packet. For DECT2, the selected code allows to correct $t = 2$ bit errors at a code rate of $R = 0.937$.

Because ARQ and FEC both require a certain percentage of the total bit rate, the net bit rate (NBR) for video has to be reduced. For example, the redundancy added by FEC on the DECT1 channels corresponds to a data rate of 4 kbps. In addition, retransmitted packets can reduce the NBR significantly at high packet loss rates. For the selected codes, the average bit rate spent for ARQ is 11.4 kbps and 4.2 kbps for DECT1 and DECT2, respectively. Note that these values strongly depend on the characteristic of the channel. Because the remaining bit rate for video can be as low as 18 kbps, the initial picture quality will already be significantly lower than in the error free case, where the total bit rate of 32 kbps is used for video. Table 28.1 summarizes the used NBR in the above scenarios.

To complete the description of simulation conditions, we note that the feedback channel is assumed to be error free and that the round trip delay is 300 ms. Furthermore, sequence numbering and CRC calculation for ARQ was not actually implemented. Therefore, the ARQ protocol cannot be disturbed by transmission errors and no additional overhead is considered.

Figures 28.5 and 28.6 show simulation results for the sequence *Mother and Daughter*. The first 400 frames are coded at a frame rate of 6.25 fps. Twenty-one simulations are averaged for different offsets in the bit error sequence, such that each simulation starts 10^5 bits after the preceding one in order to distribute the burst errors at varying temporal and spatial locations in the sequence.

Table 28.1 Net bit rate for video (kbps)

Scenario	DECT1	DECT2
ET-ARQ-FEC	18	26
SG-ARQ-FEC	18	26
ET-FEC	28	30
SG-FEC	28	30
ARQ-FEC	18	26
FEC	28	30
NP	32	32

The PSNR decreases dramatically if the video bit stream is not protected (NP). FEC alone improves the result somewhat, but is still far from satisfactory. ARQ combined with FEC gives a dramatic improvement of the resulting image quality. Because of error propagation, however, even a reduced number of residual errors does affect the image quality severely. The combination of FEC, ARQ and one of the two feedback channel strategies (error tracking or same-GOB) leads to the best

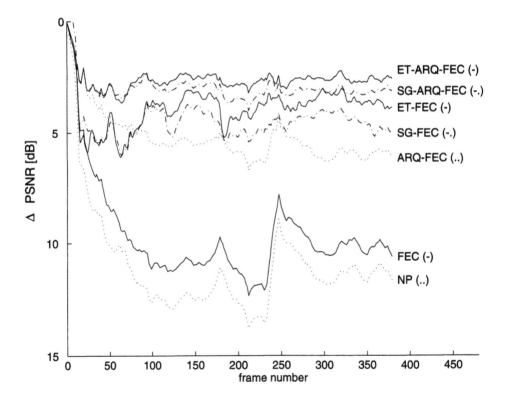

Figure 28.5 Loss of picture quality compared to error free case (DECT1, BER = 0.021, *Mother and daughter*).

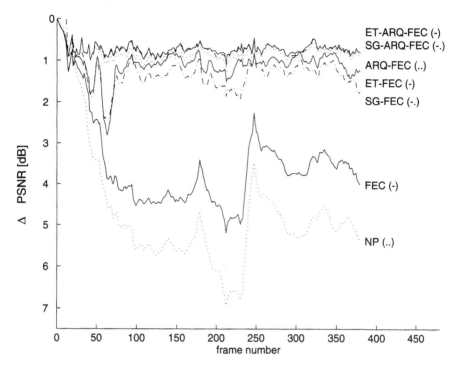

Figure 28.6 Loss of picture quality compared to error free case (DECT2, BER = 0.0021, *Mother and daughter*).

result. Especially for high BERs, the gain due to adaptive error compensation is significant (see DECT1, scenarios ARQ-FEC vs. ET-ARQ-FEC). For lower BERs, the gain decreases. If ARQ is not feasible due to a large round-trip delay, the error tracking and same-GOB strategies in combination with FEC achieve about the same level as ARQ-FEC. In terms of average PSNR, error tracking outperforms same-GOB by approximately 0.5 dB. Please note that the added amount of redundancy for FEC already reduces the average PSNR in the error free case by 0.49 dB and 0.23 dB for DECT1 and DECT2, respectively. These values are computed by subtracting the average PSNR of the sequence coded at 32 kbps from the average PSNR at $R \times 32$ kbps, with R being the code rate.

Simulations with identical conditions are performed for 200 frames of the sequence *Carphone* and 400 frames of the sequence *Foreman*. In order to provide an overview of all results, the obtained PSNR values are averaged in time and presented in Figure 28.7 for the DECT1 channel. Note that the decrease in average PSNR for the error free case due to redundancy added for FEC equals 0.46 dB and 0.44 dB for *Carphone* and *Foreman*, respectively.

Similar observations as for the *Mother and Daughter* sequence can also be made for the other two sequences. Adaptive error compensation using a feedback channel gives a significant gain of up to 4 dB (see scenarios ARQ-FEC vs. ET-ARQ-FEC or FEC vs. ET-FEC). With one exception (*Carphone*, SG-FEC vs. ET-FEC), the error

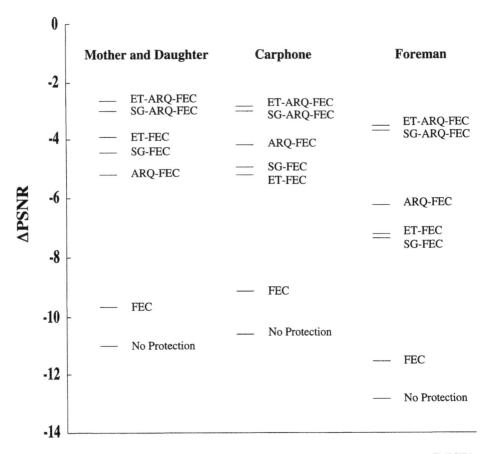

Figure 28.7 Average loss of picture quality comparred to error-free case (DECT1, BER = 0.021).

tracking strategy always outperforms the same-GOB strategy. However, the PSNR gain is only marginal in the average case. A significant advantage can be observed only if the subjective quality of individual error events is considered. To illustrate this statement, example frames from the *Foreman* sequence after the loss of 2 GOBs are presented in Figure 28.8.

Because the error in frame 75 has propagated from its original location, the same-GOB strategy cannot remove it successfully in frame 90. Only when error tracking is employed can the propagation be stopped entirely. Because these error events are relatively seldom and the resulting errors are concentrated in a small part of the image, they do not significantly affect average PSNR values. The subjective image quality, however, can be severely affected. Error tracking provides increased robustness for particular unfavourable cases, which may otherwise cause annoying artifacts.

Figure 28.8 (a) Frame 90 of sequence *Foreman* after two GOBs were lost and concealed in frame 75; (b) Same-GOB strategy; (c) error tracking (low complexity algorithm); (d) frame 90 without GOB loss in frame 75.

28.7. CONCLUSIONS

In this contribution we have presented a new method for error-resilient transmission of video in mobile environments. The investigated feedback channel approach leads to rapid image quality recovery after transmission errors since the spatio-temporal error propagation is reconstructed at the encoder and severely affected image regions are INTRA refreshed. No additional delay is introduced in the system, since the round-trip delay only determines the time until error recovery. We compared different combinations of FEC, ARQ, and two strategies for rapid error recovery, namely the error tracking strategy and the same-GOB strategy. It has been shown that a combination of the error tracking or same-GOB strategies with ARQ and FEC leads to robust video transmission even for bit error rates of 0.021 (DECT1). ARQ with a limited number of retransmissions is very helpful to combat burst errors and reduces the frequency of residual errors which need to be compensated by error tracking. The scheme adapts automatically to varying channel conditions since NAKs are only used in fading situations. The high coding efficiency of H.263 is preserved for good channel conditions, since no periodic INTRA refresh is used, and the redundancy added for FEC remains small. The low complexity algorithm proposed for reconstruction of spatio-temporal error propagation approximates the actual error propagation and is of particular interest to real-time implementations. It is important to note that the scheme presented in this paper does not require a modification of H.263. It is standard compatible, supporting interoperability of mobile and PSTN terminals.

The usage of a feedback channel for error robust video transmission is recommended by the ITU-T Study Group XV as an integral part of the mobile extensions of H.324. The low complexity algorithm for reconstruction of spatio-temporal error propagation described in this contribution forms an informative appendix in H.263 (Appendix II). For convenient use in a H.324 terminal, the H.245 control standard has been extended to include the feedback message "videoNotDecodedMBs", that can be used to indicate the loss of macroblocks.

ACKNOWLEDGEMENTS

This work was supported in part by the Deutsche Forschungsgemeinschaft (SFB 182) and Samsung Electronics Company (SEC). The authors are grateful for the helpful discussions with John Villasenor and Thomas Wiegand.

REFERENCES

[1] ITU-T recommendation H.263. Video coding for low bitrate communication (draft) December 1995.

[2] ITU-T/SG15/LBC-95-194. Suggestions for extension of recommendation H.263 towards mobile applications. Robert Bosch GmbH, June 1995.

[3] Fischer, R., Mangold, P., Pelz, R. M. and Nitsche, G., Combined source and channel coding for very low bitrate mobile visual communication systems. In *Proceedings of the International Picture Coding Symposium*, Melbourne, Australia, 231–236, 1996.

[4] Illgner, K. and Lappe, D., Mobile multimedia communications in a universal telecommunications network. In *Proceedings of SPIE Conference on Visual Communications and Image Processing*, 1034–1043, 1995.

[5] Horn, U., Girod, B. and Belzer, B., Scalable video coding with multiscale motion compensation and unequal error protection. In *Proceedings of the International Symposium on Multimedia Communications and Video Coding*, New York, October 1995.

[6] Pelz, R. M., An unequal error protected px8 kbit/s video transmission for dect. In *Proceedings of the Vehicular Technology Conference*, 1020–1024, 1994.

[7] ITU-T/SG15/LBC-95-309. Sub-videos with retransmission and intra-refreshing in mobile/wireless environments. National Semiconductor Corporation, Darmstadt, October 1995.

[8] ITU-T/SG15/LBC-96-033. An error resilience method based on back channel signalling and FEC, Telenor R&D, San Jose, January 1996.

[9] Färber, N., Steinbach, E. and Girod, B., Robust H.263 compatible video transmission over wireless channels. In *Proceedings of the International Picture Coding Symposium (PCS)*, 575–578, Melbourne, March 1996.

[10] Steinbach, E., Färber, N. and Girod, B., Standard compatible extension of H.263 for robust video transmission in mobile environments. *IEEE Transactions on Circuits and Systems for Video Technology*, **7** (6), 872–881, 1997.

[11] Lin, S., Costello, D. J. and Miller, M. J., Automatic repeat error control schemes. *IEEE Communications Magazine*, **22**, 5–17, 1984.

[12] Heron, A., MacDonald, N., Video transmission over a radio link using H.261 and DECT. *IEE conference publications*, **354**, 621–624, 1992.

[13] Khansari, M., Jalali, A., Dubois, E. and Mermelstein, P., Low bit-rate video transmission over fading channels for wireless microcellular systems. *IEEE Transactions on Circuits and Systems for Video Technology*, **6** (1), 1–11, 1996.

[14] Chen, C. T., Error detection and concealment with an unsupervised MPEG2 video decoder. *Journal of Visual Communication and Image Representation*, **6** (3), 265–278, 1995.

[15] Haskell, P. and Messerschmitt, D., Resynchronization of motion compensated video affected by ATM cell loss. In *Proceedings ICASSP*, **3**, 545–548, March 1992.

[16] Wada, M., Selective recovery of video packet loss using error concealment. *IEEE Journal on Selected Areas in Communications*, **7**, 807–814, 1989.

[17] Tzou, K., Post filtering for cell loss concealment in packet video. *SPIE Visual Communications and Image Processing IV*, **1199**, 1620–1628, 1989.

[18] Lam, W.-M., Reibman, A. R. and Lin, B., Recovery of lost or erroneously received motion vectors. In *Proceedings ICASSP*, **5**, 417–420, April 1993.

[19] ITU-T/SG15/LBC-95-186. Definition of an error concealment model (TCON). Telenor Research, Boston (20–23) June 1995.

[20] Clark, G. C., Jr. and Cain, J. B., *Error-Correction Coding for Digital Communications*, Plenum Press, New York, 1988.

Part 7

Advanced Radio Techniques

At present, the majority of wireless communication networks support voice and simple low bit-rate data services. However, with the introduction of the Universal Mobile Telecommunications System (UMTS), third generation mobile telecommunications are scheduled to start in Europe between 2000 and 2005. These systems will provide a new range of telecommunications services to mobile and stationary users in a variety of application areas and operating environments. The concept of UMTS is expected to extend far outside of Europe, to provide global roaming for UMTS users and to provide roaming with other networks implementing the recommendations of the Future Public Land Mobile Telecommunications Systems (FPLMTS). These new networks will offer a range of highly flexible multimedia services, in addition to basic voice telephony. To support these challenges, new advanced multi-format, multi-standard radio architectures will be required. In this section, issues ranging from the efficient design of DS-CDMA and adaptive equalisation architectures to the application of adaptive antennas and software radio concepts are introduced.

Understanding the radio propagation medium is vital for any wireless communications system. In Chapter 29 the authors describe the specific propagation challenges for direct sequence (DS) and frequency hopping (FH) systems. Issues such as handover and mobility are discussed, and the chapter concludes with a comparison of FH and DS capacity estimates.

Chapter 30 covers advanced reception techniques for multipath fading channels, such as the use of sectorised antennas, interference cancellation (using trellis-coded modulation) and adaptive equalisation. The use of adaptive antennas for personal communication systems is reviewed in Chapter 31. The concept of adaptive antennas is explained and the use of hierarchical cell structures introduced. The benefits of adaptive antennas are explained for both small and large cell applications. The chapter ends with a summary of the results obtained from the RACE TSUMANI DECT field trial.

One feature of future mobile radio systems will be the increased flexibility of the handsets. Chapter 32 explores the development of flexible adaptive equaliser algorithms for use in a universal mobile radio system. Such technology could allow a single solution to the problem of equalisation in radio standards such as GSM, DECT, HIPERLAN and D-AMPS.

The subject of orthogonal multirate direct sequence CDMA is explored in Chapter 33. The authors describe the radio link design, the code properties, and the proposed use of coherent rake combining. The performance advantages and capacity of such a network are derived. Based on this technology, the authors are currently developing a mobile multimedia testbed.

Chapter 34 explores the capacity of a combined voice and data CDMA a system when variable bit rate source coding is employed. For a DS-CDMA system, the impact of data on the voice services is explained and mathematical models are developed. Numerical outage probability results are given for the case of an integrated voice/fax CDMA wireless system.

Finally, returning to the theme of flexibility, a key element in future mobile radio terminals is expected to be the concept of the software radio. Chapter 35 explains the structure required and discusses the basic specifications for a third generation hand-portable terminal. Key areas such as A/D and D/A converters, anti-aliasing filters, DSP requirements and RF power amplifiers are specified.

29

Propagation Aspects of Mobile Spread Spectrum Networks

Mark A. Beach, Mike P. Fitton and Chris M. Simmonds

29.1. INTRODUCTION TO SPREAD SPECTRUM TECHNIQUES

The use of wideband air interface techniques is regarded by many as a suitable means for obtaining high capacity and flexibility of service provision for future generation wireless systems. This is certainly true in Europe where there are numerous research initiatives currently underway addressing the selection of the primary air interface technique for the *Universal Mobile Telecommunication System* (UMTS). Relevant examples include the RACEII ATDMA [1] and CODIT [2] projects, UK DTI/ SERC LINK CDMA [3] programme, as well as the ACTS FRAMES project [4].

Networks employing *spread spectrum* [5] access methods are extremely good examples of wideband radio access. Here a narrowband message signal is spread to a much wider bandwidth than is strictly necessary for RF transmission, then at the receiver the wideband representation is despread to yield an estimate of the original narrowband message signal. Ratios of 10 kHz message bandwidth to several MHz RF channel bandwidth are commonplace in such systems.

There are principally two techniques employed to spread the spectrum, direct sequence (DS) and frequency hopping (FH). Given the considerable interest and investment by companies such as Qualcomm [6], InterDigital [7], Geotek [8] and others in this technology, the propagation aspects of both DS and FH spread spectrum networks are now discussed with respect to the challenges encountered in the mobile radio channel.

29.2. DIRECT SEQUENCE SPREAD SPECTRUM

Direct sequence (DS) spread spectrum systems take an already modulated narrowband message signal and apply secondary modulation in the form of *pseudo random noise* (PN), thereby spreading the spectrum. This secondary modulation process is usually implemented in the form of phase shift keying (PSK), and the PN sequence is known as the *spreading waveform* or *sequence*, or *code*. At the receiver, the incoming spread spectrum waveform is multiplied with an identical synchronised spreading sequence thus resulting in the recovery of the information or message signal. By associating a unique spreading code with each user in the network in order to spread the spectrum, multiple users can be simultaneously overlaid in both the time and frequency domains, and the concept of *code division multiple access* (CDMA) is realised.

29.2.1. Impact of the Mobile Channel

The long term fading statistics, path and shadow loss, of both narrowband and wideband signal propagation in the mobile radio channel have been shown to exhibit very similar characteristics according to theory and also by measurement [9]. However, the short-term statistics or the multipath fading characteristics of these channels differ significantly, since spread spectrum systems tend to operate using channel bandwidths far in excess of the coherence bandwidth of the mobile channel.

The impact of the mobile radio channel upon the DS-SS waveform is most readily understood by considering the form of the correlation function [5] after transmission through an environment similar to that shown in Figure 29.1(a). The correlator output (see Figure 29.1(b)) shows three peaks corresponding to the time synchronisation between the local spreading code in the mobile receiver and the multipath components *a*, *b* and *c*. The relative delay between these peaks corresponds to the different path delays in the channel, and thus if the total multipath delay (T_m) is known, then the maximum number of multipath components (L_m) which can be

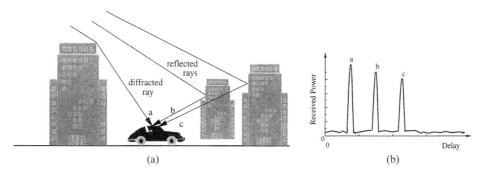

Figure 29.1 Resolution of multipath components using DS-CDMA.

resolved by the receiver is given by [10], where T_c is the chip[1] duration of the spreading code:

$$L_m \le \frac{T_m}{T_c} + 1 \tag{29.1}$$

Thus, it can be seen that the degree of multipath resolution is directly related to the chipping rate of the CDMA waveform. With reference to Figure 29.1, resolution of individual multipath rays would be possible if an infinite chipping rate[2] was employed. However, it is necessary to bandlimit such systems, and thus the multipath activity observed is clustered together in discrete time intervals, or *bins*, separated by T_c. Hence, as the mobile moves within the environment the signal energy within each bin of the correlator output will fade with the vector addition of these components.

The severity of fading will depend upon the amount of multipath activity present in each bin, and this is related to both the physical nature of the scattering volume as well as the time resolution of the system. Since each bin of the correlator contains energy contributions from different multipath components, these will tend to fade independently and also have different instantaneous Doppler components. Thus as shown in section 29.2.2 the bin level fading statistics in an urban cellular environment vary from "Rayleigh-like" characteristics through to "Rician" distributions with relatively high K-factors as the chipping rate is increased.

29.2.2. Channel Measurements and Models

The use of wideband correlation channel sounding techniques allows the extraction of the impulse response of the multipath channel. By employing continuous sampling techniques, the statistics directly related to the operation of DS-SS receivers [11] can be obtained. From soundings taken in Bristol [9], the multipath

---

[1]Each component of the spreading code is termed a chip in order to distinguish it from the message data bits or symbols.
[2]This is defined as the clocking rate of the spreading code.

resolution and bin statistics were obtained for an urban operational link. Figure 29.2 illustrates the multipath activity for a system operating at 1.25 Mchip/s, whereas Figure 29.3 gives the channel characteristics for the same environment for a chipping rate of 10 Mchip/s.

The signal strength variability of these results can also be considered in terms of the coefficient of variation [12] and the normalised standard deviation. The former is defined as the ratio of the standard deviation relative to the mean, whereas the latter is given by first normalising the received powers measured in dBs, and then calculating the standard deviation relative to the mean. These results are illustrated in Figure 29.4, where it can be seen that as the chip duration is increased[3] the coefficient of variation of the first diversity branch tends to 1, and also the value of the normalized standard deviation approaches 5.57 dB. Consequently, as the chip duration is increased the fading in each bin becomes less deterministic, and the distribution tends towards Rayleigh.

29.2.3. DS-CDMA Performance

The inherent ability of the DS-CDMA technique to produce multiple replicas of the same information bearing signal leads to the important concept of *path diversity* reception. Here multiple despreading circuits can be used to decode the wanted signal contributions contained within the active bins of the correlator, and then combining these using the familiar techniques of *switched*, *equal gain* or *maximal ratio combining* in order to obtain a better estimate of the message signal. This is also

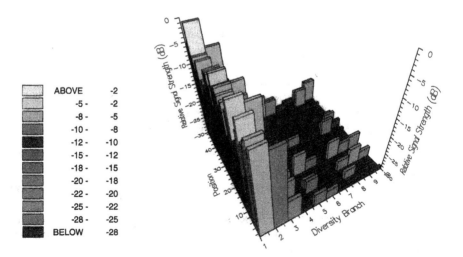

Figure 29.2 Urban DS-CDMA Multipath Activity at 1.25 Mchip/s (with permission from IEE [9]).

---

[3]Increasing the chip duration corresponds to a decrease in the chipping rate or bandwidth of the DS-CDMA transmission.

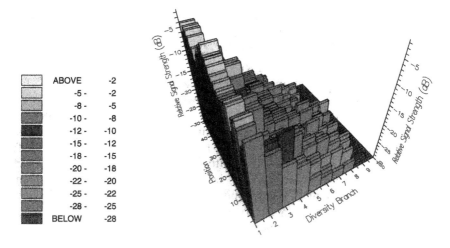

Figure 29.3 Urban DS-CDMA Multipath Activity at 10 Mchip/s (with permission from IEE [9]).

known as *rake* reception based on the analogy with a garden rake being used to rake the impulse response of the channel, with each prong, branch or finger corresponding to a time bin in the correlator.

In [9] a capacity or bandwidth efficiency model, where the sensitivity of DS-CDMA to channel parameters (bin fading statistics) and receiver architecture

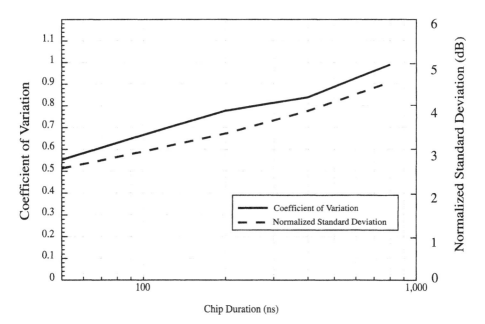

Figure 29.4 Coefficient of variation and normalised standard deviation versus chip duration (with permission from IEE [9]).

(diversity combining method), is reported. A summary of this analysis is given in Figure 29.5, where the bandwidth efficiency of a DS-CDMA network is given in terms of the diversity order extracted from the channel and the fading statistics of these samples. For the Rician case, a Rician K-factor of 3 dB has been assumed, and a log normal standard deviation of 8 dB (Rayleigh channel) and 2.77 dB (Rician channel) in order to represent long term shadowing effects.

From the results presented above it can be seen that the capacity of a DS-CDMA network can be traded against the complexity of the diversity signal processing architecture employed within the receiver. Furthermore, the gain available from path diversity processing is extremely sensitive to the short-term propagation characteristics of the channel, as indicated by the difference in performance between the Rayleigh and Rician channels. The chipping rate of these systems must be sufficient to ensure that multipath energy can be extracted from the channel for use in the diversity process, with the RF bandwidth being directly related to this value.

It has also been shown that the channel statistics of the mobile radio channel vary when observed through a DS-CDMA receiver as the chipping rate is varied, and this has a significant impact on the overall system performance. For low chipping rates ($\leqslant 1.25$ Mchip/s) the measured channel displayed "Rayleigh-like" statistics. Whereas, when the chipping rate was increased to 10 Mchip/s, strong Rician statistics were observed.

29.2.4. Mobility and Handover

Early mobile communications systems provided users with coverage in only one of two ways: either point-to-point communications or connection to the hard-wired public telephone system. Both of these restricted not only a user's movements to a localized area but also the total number of users that could be supported.

The first *cellular* systems were developed only relatively recently by Bell Laboratories in the 1970s [13]. These divided the total coverage area required for a single mobile phone system into a series of smaller, lower power coverage areas or

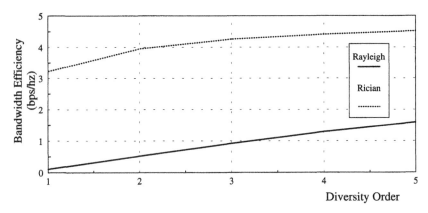

Figure 29.5 Bandwidth efficiency for an MRC rate architecture operating in both log-normal Rayleigh and Rician channels (with permission from IEE [9]).

cells. The main advantage behind this, other than reducing the power requirements of the individual base stations, was to allow the re-allocation or *re-use* of the finite number of channels allocated for that system. The re-use of these channels in cells far enough apart to not cause co-channel interference, allows the operator to effectively increase the number of channels available and hence increase the system capacity [14].

However, to allow mobile users full mobility, communications links must be maintained not only as the users *roam* within their current cell, but also allow roaming between cells. Therefore, the user's communications link to a current cell's base station must undergo a *handover* to the base station in the adjoining cell, into which the user has moved and which can offer a better coverage potential to main-tain the link.[4] Ideally, this handover process maintains call integrity and quality, irrespective of user location, services required from the link or communications device used, in other words the process remains *transparent* to the user.

The *traditional* handover technique adopted by most cellular systems to date is that of *hard handover*. This process is known as a "break-before-make" handover, because once the handover process starts, the old link with the mobile station is terminated before the new link to mobile and an adjacent cell's base station is established. Initiation of the process is usually done on a hysteresis window compar-ison of received signal levels from adjacent base stations to the current base station signal level. However, this process leaves the user open to losing the link at the point of handover if any stage of the new link set-up fails, e.g. through a lost or corrupted handover message.

More evolutionary methods now exist, such as *seamless handover* and *soft hand-over*. Seamless handover (used by DECT), allows a new signal path to be established between the new base station and mobile, whilst maintaining the old link. At the point of handover the paths are merely switched.

Soft handover is a form of macroscopic diversity [16]. This process allows the mobile to establish a link with a new base station[5] *before* terminating the old con-nection. As a result this process has numerous benefits over the other techniques [17]. For example, during handover, signals received at the mobile can then be combined whilst those received at the base stations can be switched, hence optimal combining of signal levels can be achieved via the two links with the user at the mobile. This potential gain in received signal level allows the user to receive lower signal levels from each link, to achieve the same received signal as from a single, higher power base station. In an interference limited system, such as DS, this results in a reduction of signal level required at cell boundaries to maintain a communications link and hence a reduction in interference levels between cells, providing a potential capacity increase to the system.

---

[4]Although, handover is also to be considered from a network point of view. There is little point in handing a user over to another base station if the improved signal quality, provided from that link, will only be transient. One example of this would be the unnecessary handing over of a user to an out-of-sight base station at a street crossing [15].
[5]The soft handover principle even allows the link to the mobile user to be with more than one other base station.

If carefully implemented, according to environmental conditions and handover threshold levels used, the soft handover technique could effectively allow more users/cell and increase the capacity of the system over the traditional hard handover technique. Figure 29.6 shows a simple simulated study of a cell capacity comparison for both hard and soft handover techniques under different propagation path loss exponents in a DS-CDMA system.

29.3. FREQUENCY HOPPING SPREAD SPECTRUM

In *frequency hopping spread spectrum* (FH-SS) systems the narrowband message signal is modulated with a carrier frequency that *hops* at regular intervals, in a pattern determined by a *spreading code*. This spreading code is also available at the receiver, and enables it to "re-tune" to the correct channel at the end of each hop. The majority of commercial systems currently available employ *slow frequency hopping* (SFH), where a number of symbols are transmitted on each hop frequency. The alternative is *fast frequency hopping* (FFH), with each symbol transmitted on several hop frequencies, resulting in a very robust system that requires a highly complex hopping synthesizer.

Multiple users can be accommodated by assigning each user a unique spreading code, resulting in *code division multiple access*. The characteristics of an FH spread spectrum network in terms of sensitivity to the mobile environment is very different to that of DS, and is described in more detail here.

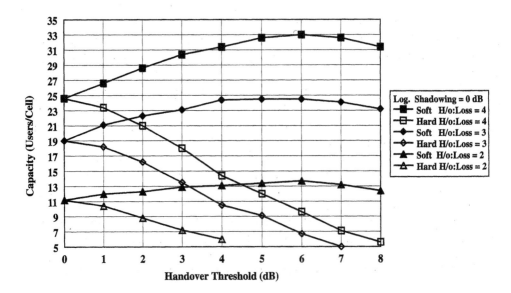

Figure 29.6 Simulated cell capacity vs. hard and soft handover threshold.

29.3.1. FH Design Issues

This section describes certain aspects of the design of a frequency hopping architecture, some of which are common to communication systems in general, and others that are unique to FH-SS.

Forward Error Correction and Interleaving

The performance of a slow FH system with forward error correction (FEC) coding can be maximised by interleaving the data, and transmitting each symbol of the codeword on a different hop frequency, affording inherent *codeword diversity*. The interleaving/de-interleaving operation randomises the position of the errors, and improves the likelihood of the coding scheme correcting the errors that are present.

Automatic Repeat Request Protocols

In automatic repeat request (ARQ) schemes, when a particular frame is detected as corrupted, then that frame is retransmitted. The nature of the frequency-hopped channel is highly suited to the application of an ARQ scheme [18]. Provided the hopping statistics are appropriate, the system will exhibit uncorrelated fading between hop frames, rendering the channel effectively *memoryless* over hop boundaries. Consequently, it is likely that any packet retransmission will experience uncorrelated channel statistics.

Hopping Parameters

A number of the characteristics of FH-SS are peculiar to a frequency-hopped system, and will be clarified here. These parameters are of particular importance in the classification of any frequency-hopped system, as they fundamentally govern overall performance.

- The number of *hop frequencies* (or *hop bins*) available for use is determined by the overall system bandwidth, and the separation between bins.
- The *hop rate* is used to characterise the number of transitions between frequencies that occur each second. This parameter impacts on overall performance in two main criteria. First, the hop rate will influence the instantaneous channel parameters, such as mean fade duration and level crossing rate [19], as shown in section 29.3.2. Furthermore, the hop rate affects the trade-off between interleaving depth and throughput delay in a hopped system. If the modem is not hopping at a fast enough rate, the maximum interleaving depth that is allowed for intelligible voice transmission will not allow for sufficient uncorrelated symbols in the de-interleaved code word.
- The *hop bin separation* defines the frequency separation between adjacent frequencies in an FH system, and is particularly important in determining the degree of correlation between hop frames in a particular hop pattern. Ideally, all symbols

should be transmitted on different frequencies, separated by greater than the coherence bandwidth [20].

29.3.2. Propagation Aspects of Frequency-Hopped Spread Spectrum

In this section, the important time domain statistics of a frequency-hopped mobile channel are analysed employing a combination of analysis and propagation measurements. Propagation analysis [21] indicates that frequency hopping does not alter the long-term statistics of the channel. A frequency hopping system does not actually change the channel probability statistics, it merely *randomizes* them. This is clearly shown by the cumulative characteristics included in Figure 29.7 for theoretical Rayleigh fading, and a hopped signal with 800 kHz spacing at 500 hops per second. This demonstrates the necessity for efficient coding and interleaving to take advantage of the randomizing effects of a frequency-hopped system.

Channel-Coherence Time

The *coherence time* is an important statistic in terms of predicting the performance of a mobile system, as it characterises the mean duration of any error bursts. The coherence time parameter indicates the time span over which the channel can be assumed to be constant. This is defined as the time offset required for the correlation coefficient to fall below a certain threshold (usually 0.5 or 0.9).

If the hop frame duration exceeds the coherence time of the channel, the overall coherence time perceived for the frequency-hopped channel will be largely determined by the system hop rate. Provided that adjacent hop frequencies are separated by greater than the coherence bandwidth, fading at the different frequencies will be

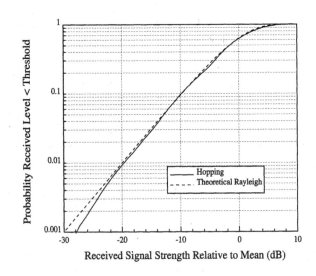

Figure 29.7 Cumulative probability distribution. (With permission from IEE [21].)

uncorrelated. Consequently, the hopped channel is unlikely to exhibit correlated fading over a hop boundary. Therefore, the coherence time is fixed at a level less than the hop frame duration.

Figure 29.8 demonstrates the coherence time of a hopped channel, incorporating the effects of multipath fading and frequency hopping. In this model, a Rayleigh channel was employed, and the velocity and hop rates were altered. Examining the results contained in Figure 29.8, it can be seen that frequency hopping provides a significant improvement in the channel coherence time. At Doppler frequencies exceeding a certain level (say, 100 Hz) hopping provides a small improvement. However, at lower Doppler frequencies, the effects of frequency hopping dominate the multipath fading. Consequently, the hopped systems provide a coherence time that is largely independent of Doppler frequency. For example, a system operating at 500 hops per second exhibits a maximum coherence time of 0.34 ms at low values of Doppler frequency. For a non-hopped system to demonstrate comparable performance, a Doppler frequency of approximately 211 Hz would be required at all times (from Figure 29.8). At a typical cellular radio frequency of 1.8 GHz, this corresponds to a velocity exceeding 148 km/h at all times.

This is an important result in the provision of high quality service, since outage duration will be determined by hop rate, rather than the Doppler frequency. It is demonstrated above that a non-hopped system can experience considerable periods of outage in a slowly changing channel. Conversely, a hopped system provides a coherence time that will not exceed a certain level, irrespective of the Doppler frequency.

Level Crossing Rate in a Hopped Channel

The level crossing rate [19] is defined as the number of positive going transitions of the signal magnitude with respect to a certain threshold. A high level crossing rate implies a rapidly changing channel, with problems arising from static nulls becoming

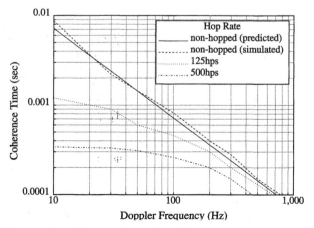

Figure 29.8 Coherence time for non-hopped and hopped systems. (With permission from IEE [21].)

statistically unlikely. Figure 29.9(a) shows the improvements in level crossing rate resulting from increasing the hop rate. Performance is improved with increased hop rate, even at a relatively low hop bin separation of 100 kHz. However, the initial performance improvement associated with hopping is not reflected by a continued increase in the hop rate. The trade-off between hardware complexity and improved channel characteristics becomes unfavourable as hop rate is increased too far.

Mean Fade Duration in a Hopped Channel

Mean fade duration [19] is defined as the time spent with a magnitude less than a certain threshold in a particular fade. It is possible to relate fade duration to an indication of system quality, such as outage duration. The results included in Figure 29.9(b) indicate the improvement in mean fade duration resulting from the randomising effects of frequency hopping.

As would be expected, the mean fade duration becomes shorter as the hop bin separation is increased, and the correlation of adjacent hop periods is reduced. Interestingly, as the hop bin separation exceeds the coherence bandwidth, there are diminishing returns in increasing it further. That is to say, there is little point in separating adjacent hop frequencies beyond the minimum required for uncorrelated fading. Although the mean fade duration at low hop bin separation will be influenced by the Doppler frequency, at higher separations the mean fade duration becomes largely independent of velocity, provided there are enough frequency bins in the hop pattern.

29.3.3. FH-CDMA Performance

Frequency hopping code division multiple access can provide an inherently robust air interface technique. Users co-exist in the same time and frequency, and are delineated by the user's unique code that governs the pattern of hopping. This

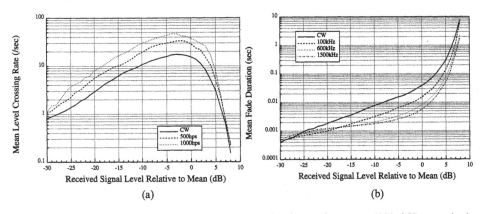

Figure 29.9 (a) Short term channel statistics: (a) level crossing rate (100 kHz spacing); (b) mean fade duration (500 hops/second).

approach results in a number of advantages. Primarily, the hopping nature of the system negates the need for rigorous frequency planning. This is in direct contrast to a direct sequence CDMA system, which has complex frequency planning, arising from the system's susceptibility to the *near–far effect* [5]. Furthermore, synchronizing all users in a given cell with a unique offset in the same code theoretically creates no intra-cell interference. In this way, all cells employ the same portion of the RF spectrum, each cell with a unique code. By judicious choice of these codes to have minimal cross-correlation properties, the inter-cell interference that exists can be minimised.

A frequency-hopped mobile link experiences a sampled version of both the channel's fading characteristic and any interference that co-exists in the spreading bandwidth. A given FH link will only suffer the effects of a significant interferer once in a particular hop sequence. In this way frequency hopping spread spectrum provides inherent resilience to the effects of interference, referred to as *interference diversity*. The entire system will experience composite interference, which is the mean of the interference experienced in the interleaved frame, referred to as *interference averaging*. This effect is illustrated in Figure 29.10, which indicates the most significant interferers in a particular hop frame. At each hop the number and magnitude of the interferers is randomised.

Due to interference diversity, the symbols within an interleaved codeword will experience uncorrelated fading, resulting in the efficiency of any coding scheme being maximised. Alternatively, the throughput of an automatic repeat request (ARQ) scheme can be improved. The enhancement in performance results from the coherence time, or channel memory, becoming fixed in the frequency-hopped channel [21]. Interference diversity in an FH-ARQ system provides uncorrelated interference for each packet retransmission.

A conventional non-hopped system can produce error bursts of significant duration, due to wideband fading or co-channel interference. The randomising effect of

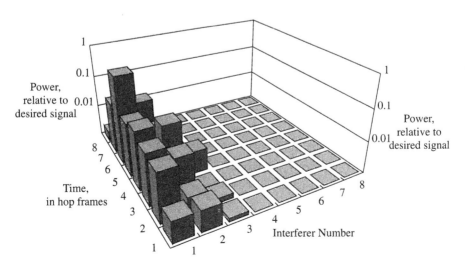

Figure 29.10 Interference diversity in a frequency hopped system.

frequency hopping improves the short-term channel statistics, and therefore limits the impact of error bursts. Furthermore, it is possible to implement interleaving or ARQ to exploit the inherent advantages of the frequency-hopped channel. Improvement in quality-of-service is more significant at low Doppler frequencies, when significant bursts of errors occur in a non-hopped system. Consequently, FH is particularly appropriate when the channel is slowly changing or stationary, which often occurs in wireless LAN and wireless local loop techniques.

29.4. COMPARISON OF DS AND FH

To provide a comparison between the spread spectrum techniques described in this chapter, this section describes a multi-user voice channel analysis of frequency hopping and direct sequence CDMA techniques, employing a Monte Carlo approach. The capacity performance of DS and FH-CDMA are analysed in a network supporting voice links only. In this case, the network is interference limited, and time dispersion is negligible.

29.4.1. Link Configuration

The network topology comprises a central base station, surrounded by three rings of interfering base stations, arranged in a hexagonal omni-cell pattern. Mobiles are positioned randomly within the network coverage area, and are connected to the base station with the highest link power (incorporating both attenuation and shadowing). The up-link cellular environment parameters are characterised in Table 29.1. The cellular capacity model incorporates the effects of cell loading, voice activity, and propagation characteristics.

Table 29.1 Up-link cellular scenario.

Parameter	FH-CDMA Specification	DS-CDMA Specification
Path loss coefficient	4	
Shadowing standard deviation	8 dB	
Handover margin	0 dB	
User bit rate	8 kbps	
Acceptable BER	10^{-3}	
Outage probability	1%	
Modulation	QPSK	BPSK
Detection	Coherent	Differential
Coding	½ rate convolutional, constraint length 7	
Interleaving delay	40 ms	
Antenna diversity	Dual, with maximal ratio combining	
Chipping rate	–	1.25 Mbps
Processing gain	–	21 dB [22]
C/I threshold for BER $< 10^{-3}$	6 dB [23]	−14 dB [22]

29.4.2. Comparison of DS and FH-CDMA Capacity

The FH-CDMA system architecture employed in this chapter is relatively simple. It is proposed that acceptable quality-of-service can be achieved without the complexities of equalisation, or a DS-CDMA approach. A traditional DS-CDMA architecture is employed [22]. The frequency hopping CDMA architecture employs orthogonal hopping within a cell, resulting in no intra-cell interference. Furthermore, adjacent cells utilise unique hopping patterns, resulting in randomisation of observed interference. An interference diversity factor of 8 is utilised. Employing the C/I thresholds shown in Table 29.1 it is possible to calculate the probability of the system being in outage (shown in Figure 29.11).

Employing an outage probability of 1%, the DS and FH system provide reasonably similar capacity, with a loading level of 19 and 25 users/MHz/cell respectively (from Figure 29.11). This performance is comparable to the results obtained by other authors [22]. Both CDMA systems exhibit a *soft capacity* limitation, where more users can be added to the network, at the expense of overall quality of service. However, in the case of FH, the degradation in quality is more gradual, and consequently more users can be added for a comparable decrease in performance. For example, at a link availability of 5%, DS will support 22 users/MHz/cell, compared to 44 for an FH system.

29.5. CONCLUSIONS

In this chapter, the propagation aspects of frequency hopping and direct sequence spread spectrum are characterised in terms of providing a high quality, high capacity wireless communication network. In particular, the impact of the mobile channel on service provision is examined. It is indicated that both DS and FH-SS can be

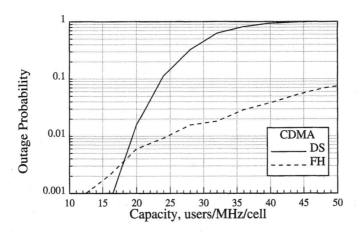

Figure 29.11 Outage probability for DS and FH CDMA.

employed to mitigate the deleterious effects of the mobile environment, resulting in improved quality-of-service for a wireless multi-media network.

Comparison of direct sequence and frequency hopping CDMA indicates similar voice channel capacity performance. In addition, FH-CDMA displays a greater soft capacity facility than DS-CDMA, which is important in the provision of a flexible cellular system. Furthermore, the nature of FH-CDMA eases the frequency planning requirements in a cellular network, since the interference pattern is randomised at each hop, and consequently no one interferer can dominate. In contrast, DS-CDMA requires complex closed-loop power control to mitigate the effects of the near–far effect. However, it is easier to exploit soft handover in a DS-CDMA network to provide an enhancement in the quality of service, as described in Section 29.2.4.

A number of multimedia applications will require high data rate capabilities. This is problematic in a highly time-dispersive environment, since errors arise due to intersymbol interference (ISI). For example, a typical urban macrocell may exhibit an RMS delay spread in the order of 1 μs, resulting in significant ISI in systems with a symbol rate exceeding 100 kbps. The frequency diversity of FH-SS will randomise any error bursts which occur due to ISI, thus improving the efficiency of FEC and ARQ. However, as the data rate is increased beyond a certain level, frequency hopping is unable to support adequate quality of service without additional techniques, such as equalisation. Consequently, the inherent *path diversity* of a DS-CDMA approach provides a significant advantage in the provision of high data rate services. The application of FH-CDMA is limited to medium data rate services in high time dispersion, such as urban macrocellular environment environments.

ACKNOWLEDGEMENTS

The authors wish to express their sincere thanks to colleagues at the Centre for Communication Research (University of Bristol), in particular to Prof. Joe McGeehan, Dr Simon Swales, Dr Steve Allpress, Dr Andy Nix and Dr Joseph Cheung for their invaluable help and support of this work. Furthermore, the authors are grateful to the Engineering and Physical Sciences Research Council for funding a variety of aspects of this work.

REFERENCES

[1] Urie, A., Streeton, M. and Mourot, C., An advanced TDMA mobile access system for UMTS. *IEEE Personal Communications Magazine*, **2** (1), 28–47, 1995.
[2] Andero, P. and Ewerbring, L., A CDMA-based radio access design for UMTS. *IEEE Personal Communications Magazine*, **2**(1), 48–53, 1995.
[3] Beach, M., A CDMA air interface for mobile access. In Damosso, E. (ed.), *Digital Mobile Radio*, Springer, Berlin, Chapter 7.2, ISBN 3-540-76038-5, 1997.
[4] Mäkeläinen, T., *et al.*, *FRAMES Demonstration Implementation Roadmap, ACTS Mobile Telecommunication Summit*, Granada, Spain, 27–29 November 1996, **1**, 104–108.
[5] Dixon, R., *Spread Spectrum Systems*, 2nd edn, Wiley, New York, 1984.

[6] Gilhousen, K., *et al.*, On the capacity of a cellular CDMA system. *IEEE Transactions on Vehicular Technology*, **40**, 303–312, 1991.

[7] Schilling, D., Broadband CDMA overlay. *Proceedings of IEEE 43rd VTC*, New Jersey, 452–455, May 1993.

[8] Livneh, N., *et al.*, Frequency hopping CDMA for future cellular radio. *Proceedings of IEEE 42nd VTC*, Denver, 400–404, May 1992.

[9] Beach, M. and Allpress, S., Propagation aspects of mobile spread spectrum networks. In *Radiowave Propagation*, IEE (Peter Peregrinus), Chapter 12, 235–256, 1996.

[10] Simon, M., *et al.*, *Spread Spectrum Communications*, Vols I–III, Computer Science Press, Rockville, 1985.

[11] Allpress, S., Optimising signalling rate and internal diversity order for the mobile cellular DS-CDMA systems. Ph.D. Thesis, University of Bristol, Bristol, December 1993.

[12] Miller, I. and Freud, J., *Probability and Statistics for Engineers*, 3rd edn, Prentice-Hall, Englewood Cliffs, NJ, 1985.

[13] Padgett, J., Gunther, C. and Hattori, T., Overview of wireless personal communications, *IEEE Communications Magazine*, **33**, 28–41, 1995.

[14] Johannsson, P., Evolution of wireless technologies. *Global Communications Congress and Exhibition*, Vancouver, B.C., Canada, 456–458, INTER COMM 95, February 1995.

[15] Ostling, P., Implications of cell-planning on handoff performance in Manhattan environments. In *Proceedings of the 5th International Symposium on Personal, Indoor and Mobile Communications*, The Hague, The Netherlands, IEEE, September 1994.

[16] Berhardt, R., Macroscopic diversity in frequency reuse radio systems. *IEEE Journal on Selected Areas in Communications*, **SAC-5** (5), 862–870, 1987.

[17] Simmonds, C. M., Soft handover parameter optimisation for DS-CDMA downlink design. Ph.D. Thesis, University of Bristol, September 1995.

[18] Guo, N. and Morgera, S., Frequency-hopped ARQ for wireless network data services, *IEEE Journal on Selected Areas in Communications*, **12**, 1324–1337, 1994.

[19] Jakes, W., *et al.*, *Microwave Mobile Communications*, Wiley, New York, 1974.

[20] Parsons, J. and Gardiner, J., *Mobile Communications Systems*, Blackie, Oxford, 1988.

[21] Fitton, M., Nix, A. and Beach, M., Propagation aspects of frequency hopping spread spectrum. In *IEE Proceedings on Microwaves, Antennas and Propagation*, **145** (3), June 1998.

[22] Skold, J., *et al.*, Analysis of a CDMA system, *RACE/RMTP/CM/J115*, Issue 1.4, CEC Deliverable: 43/ERN/CM10/DS/A/071/a1, May 1991.

[23] Fitton, M., The application of frequency hopping CDMA in flexible future wireless networks, Ph.D. Thesis, University of Bristol, Bristol, December 1996.

30

Advanced Reception Techniques for a Multipath Fading Channel

Susumu Yoshida and Hidekazu Murata

30.1. INTRODUCTION

Wireless personal communication channels are characterized by multipath fading and co-channel interference. It is well-known that the bit error rate (BER) performance of such a channel suffers from deterioration due to fading, intersymbol interference and co-channel interference. The degradation is especially severe when the multipath delay spread is large and co-channel interference is non-negligible.

To overcome this degradation, we have been studying a sector antenna diversity combined with a maximum likelihood sequence estimation (MLSE) equalizer and a trellis-coded co-channel interference canceller. Our long experience of multipath propagation study lead us to the use of a directive antenna or sector-antenna diversity. This has been combined with an MLSE equalizer to further enhance the capability to overcome the effect of long-delayed multipath propagation.

Insights into Mobile Multimedia Communication
ISBN 0-12-140310-6

However, even if we use a directive antenna, we cannot necessarily separate the desirable signal from co-channel interference. This led us to the idea of the interference canceller with enhanced cancelling capability by trellis-coded modulation (TCM).

This paper introduces some of the result we obtained regarding a sector-antenna MLSE receiver and a trellis-coded co-channel interference canceller (TCC).

30.2. EFFECT OF DIRECTIVE ANTENNA AND SECTOR ANTENNA DIVERSITY

A four-sector antenna diversity reception, or a four-direction antenna pattern diversity reception, was proposed as a method to reduce long-delayed multipath waves which often arrive from nearly opposite directions, as well as fading [1].

Extensive field tests were performed to measure the BER of the anti-multipath modulation scheme DSK [2] and found that a directive antenna becomes a powerful means to mitigate the effect of long-delayed multipath propagation. It should be noted that even BPSK showed a marked BER improvement if received by a directional antenna.

In the paper [3], a wireless LAN with a sectored antenna is proposed. It uses a six-sectored antenna and proved that a sector antenna is useful even in an indoor environment.

30.3. SECTOR-ANTENNA DIVERSITY COMBINED WITH MLSE EQUALIZER

Sector-antenna diversity substantially reduces multipath delay spread. Nevertheless, it is not enough especially for high bit-rate transmission. Therefore, an adaptive receiving system consisting of an MLSE, or a Viterbi equalizer, and sector-antenna diversity has been proposed [4][5]. In this system, by selecting appropriate sectors out of an M-sectored antenna adaptively, the receiver accepts multipath signals which can be equalized by MLSE, rejects multipath signals with longer delay time, i.e. out of equalizable range, and also rejects co-channel interference by distinguishing them from desirable signals.

Figure 30.1 is a block diagram of the adaptive receiving system. Each diversity branch has a delay profile estimator based on the recursive least squares (RLS) algorithm [6] and a branch metric calculator, which is a part of MLSE implemented with the Viterbi algorithm.

The time division multiple access (TDMA) burst consists of a 200-symbol sequence preceded and followed by 10-symbol training sequences. Accordingly, the delay profile is measured twice and a linear interpolation is made to obtain an accurate delay profile at any symbol location within the burst [7]. We assume $f_D T_s$, the maximum Doppler frequency f_D normalized by symbol rate $1/T_s$, is 1/2560. The number of states of the Viterbi algorithm is fixed to be 16.

Two types of simplified sector antenna patterns are assumed, i.e., $g_1(\theta)$ and $g_2(\theta)$.

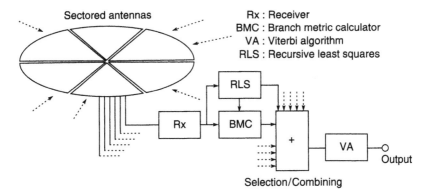

Figure 30.1 Proposed sector-antenna diversity receiver.

$$g_1(\theta) = \begin{cases} 1, & -\dfrac{\pi}{M} \le \theta \le \dfrac{\pi}{M} \\ A_{FB}, & \text{otherwise.} \end{cases} \tag{30.1}$$

$$g_2(\theta) = \max(G \cos^n \theta, G \cdot A_{FB}) \tag{30.2}$$

$$n = \frac{\log 0.5}{\log(\cos(\pi/M))} \tag{30.3}$$

where $g_1(\theta)$, $g_2(\theta)$ are the normalized power gains of antenna 1 and antenna 2, respectively, θ is the azimuthal angle. We assumed 20 dB front-to-back ratio, i.e., $A_{FB} = 0.01$, and directional gain $G = 1.0$ (0 dB gain) to make a clear comparison between antenna 1 and antenna 2.

The propagation model is selected to be as general as possible and yet bad enough to demonstrate the BER improvements clearly with the proposed receiving system. Three desirable waves and three undesirable waves which consist of both a long-delayed wave and two co-channel interference waves are assumed. The first wave is referred to as a desired (D) wave and delayed waves, delayed by τ_n second, as undesired (U_n) waves. Also, co-channel interference waves are denoted by I_n. All of the incoming signals received by the directional antennas are supposed to be subject to mutually independent Rayleigh fading.

To evaluate the performance of the proposed system by computer simulations, $\pi/4$-shift QPSK modulation is assumed as an example. Two simple arrival direction models are assumed. One is that the incident angles of incoming waves are fixed so that each antenna branch receives one incident wave (referred to as fixed incident angle model). The other is that the arrival angle of each ray is assumed to be random and uniformly distributed (referred to as random incident angle model).

As a criterion to distinguish both multipath signals with longer delay time and co-channel interference from desirable signals, the squared error between a regenerated ideal received signal and an actual received signal is employed. This squared error method is the best one we have so far tried. The branch selection strategy can be summarized as follows:

- Accept all branches at which the squared error $\Omega <$ *threshold*, assumed to be 1.0, holds.
- If no branch is accepted, accept a branch which has the minimum Ω.

Figure 30.2 shows the BER performance of the proposed receiver in a six-ray fading channel compared against three other receivers.

The performance of a receiving system which equalizes a single antenna branch with the maximum received power is indicated by "Select max. power". The BER performance attained by equalizing an antenna branch with minimum Ω is indicated by "Select min. Ω". Also, a conventional receiving system using an omni-directional antenna and 16-state MLSE is indicated by "OMNI".

The BER of "OMNI" is inferior because τ_3 is chosen out of the equalizable range of 16-state MLSE and there exists co-channel interference. It is clear that the BER of the proposed receiver is superior to those of "OMNI" and "Select max. power". This is because the proposed receiver can reject U_3, I_1 and I_2 signals. It should be noted that the proposed receiver achieves better performance than that of "Select min. Ω" by selecting more than one sector according to the branch selection strategy. It is seen that the error floor encountered with "OMNI" and "Select max. power" has been overcome.

In Figure 30.3, the BER performance of $M = 6$, for instance, is observed to be degraded if compared with that in Figure 30.2. This is mainly due to the fact that there happens to be a case where every antenna branch receiving a desirable signal is, at the same time, receiving an undesirable signal. The performance dependency on

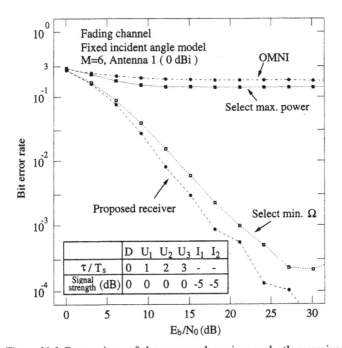

Figure 30.2 Comparison of the proposed receiver and other receivers.

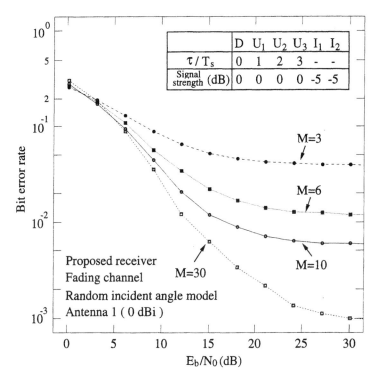

Figure 30.3 BER dependency on the number of sectors M.

the number of sectors M is also shown. The performance is confirmed to be improved by increasing M.

Figure 30.4 shows the cumulative probability distribution of BER at $E_b/N_0 = 18$ dB. This shows how the number of sectors M improves BERs more clearly than Figure 30.3. It is seen that the percentage of bursts at which BER is less than 10^{-2} is about 62% ($M=3$, antenna 1), 75% ($M=6$, antenna 2), 81% ($M=6$, antenna 1) and 90% ($M=10$, antenna 1).

Thus, it is shown that the proposed receiving system can achieve good BER performance in spite of the existence of U_3, I_1 and I_2. However, it is clear that the BER performance depends on the probability that at least one antenna branch receives desired signals without any influence of undesired signals.

30.4. TRELLIS-CODED CO-CHANNEL INTERFERENCE CANCELLER (TCC)

In a cellular system, a frequency reuse distance is determined by the required carrier-to-interference ratio of the receiver employed in the system. Therefore, a receiving system resistant to co-channel interference is the key to achieving high spectral efficiency.

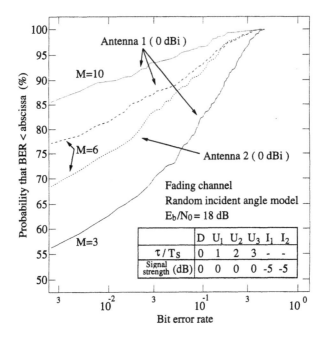

Figure 30.4 Cumulative probability distribution of BER.

A non-linear co-channel interference canceller is a promising candidate for achieving high spectral efficiency. Yoshino and Suzuki [8] have proposed an interference cancelling equalizer (ICE) based on MLSE and confirmed that ICE can achieve high cancelling performance over a Rayleigh fading channel by computer simulations. A blind algorithm has been incorporated into ICE so that ICE can operate without prior knowledge of training signals [9]. However, it has been shown that the cancelling performance of ICE is degraded over a static channel [8]. Since Rician fading is between static and Rayleigh fading, the performance of ICE over a Rician fading channel is poor.

It has been reported in [10] that the Rician factor k is greater than 2 in an urban microcell indicating that the fading is never as extreme as Rayleigh fading. This means that the performance over a Rician fading channel is important for a data-transmission system in a microcellular system.

In this paper, a new co-channel interference canceller suitable for microcellular radio is discussed. The proposed canceller, trellis-coded co-channel interference canceller (TCC) [11], utilizes trellis-coded modulation (TCM) [12] to improve the cancelling performance over a Rician fading channel. The approach developed here is to combine equalization, TCM decoding and co-channel interference cancelling. In addition, an adaptive algorithm is treated jointly.

30.5. SYSTEM DESCRIPTION

30.5.1. System Model

A digital communication system with single co-channel interference is shown in Figure 30.5. Symbol sequences $\{s_j^{(1)}\}$ and $\{s_j^{(2)}\}$ represent sequences of messages of desired signal and co-channel interference to be transmitted over the channel, respectively. The complex baseband signal sequences $\{x_j^{(1)}\}$ and $\{x_j^{(2)}\}$ are obtained from the transmitters. These are transformed by multipath fading channel into channel output sequences $\{y_j^{(1)}\}$ and $\{y_j^{(2)}\}$, respectively. The received signal $\{r_j\}$ consists of the desired signal, thermal noise and interference and, therefore, can be expressed as

$$\{r_j\} = \{y_j^{(1)} + y_j^{(2)} + w_j\} \tag{30.4}$$

where $\{w_j\}$ is the Gaussian noise sequence. The proposed canceller determines the sequence $\{y_j^{(1)} + y_j^{(2)}\}$ which is closest in Euclidean distance to the sequence $\{r_j\}$.

The reason that the bit error rate (BER) performance of the non-linear co-channel interference canceller ICE is degraded in a static channel can be explained as follows. Figure 30.6 shows examples of a constellation diagram of phaser points to be processed by ICE. There are at most sixteen points, when two QPSK modulated signals are received, i.e., the desired signal (carrier) and co-channel interference. In Figure 30.6(a), there are sixteen points on the constellation. On the other hand, there are only nine points in Figure 30.6(b). This means that the receiver cannot distinguish between two phasers with the same point. The lack of free distance between signal points causes BER performance degradation, if no measure to cope with this situation is taken.

To expand the free distance, the proposed data-transmission system introduces trellis-coded 8PSK into both transmitters of desired signal and interference, and employs super-trellis [13] in the receiver to decode these signals. The transmitter of the proposed system consists of a rate-2/3 convolutional encoder and signal mapper.

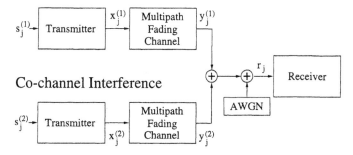

Figure 30.5 Data-transmission system with co-channel interference.

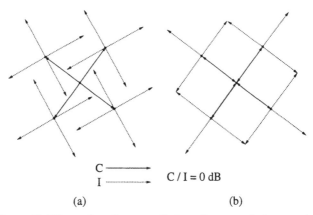

$$\begin{array}{l} \text{C} \longrightarrow \\ \text{I} \cdots\cdots\cdots\cdots\triangleright \end{array} \quad C/I = 0 \text{ dB}$$

(a)   (b)

Figure 30.6 Examples of a constellation diagram of phaser points.

30.6. SIMULATION RESULTS

The BER performance of TCC has been studied by computer simulation. The modulation schemes for ICE is assumed as QPSK. The spectrum efficiency of TCC is the same as that of ICE, since the trellis-coded 8PSK of coded rate 2/3 is used for TCC. The square-root Nyquist filters with roll-off factor 0.5 are used for both transmitter and receiver filters. For the sake of simplicity, a single co-channel interference source is assumed. Perfect sample timing is also assumed. The number of states of the Viterbi algorithm is reduced from 65 536-state to 256-state by using the DDFSE [14].

30.6.1. Performance over Quasi-Static Channel

Figure 30.7 shows the variation of BER versus E_b/N_0 at $C/I = 0$ dB. Observe in the figure that while TCC increases the BER for very low E_b/N_0 values, TCC ensures a BER of about 10^{-4} at an E_b/N_0 of 15 dB. For E_b/N_0 in excess of around 10 dB, the performance of TCC is dramatically superior to that of ICE and at values of E_b/N_0 in excess of 15 dB it is virtually error free, while the BER of ICE at $E_b/N_0 = 30$ dB is around 10^{-2}.

30.6.2. Performance over Rician Fading Channel

The BER performances of TCC and ICE using two-branch spatial diversity over a Rician fading channel are investigated. Independent fading between antennas is assumed, and $f_D T_s$ is also assumed to be 1/500. Figure 30.8 shows the BER versus Rician factor k for TCC and ICE at $E_b/N_0 = 15$ dB and $C/I = 0$ dB. As can be seen in the figure, TCC proved to be effective, particularly with Rician factor k in excess of about 3. The BER performance is improved and the channel is rendered suitable for data-transmission.

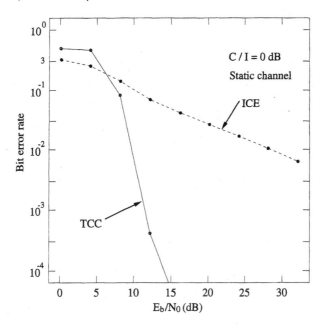

Figure 30.7 BER versus E_b/N_0 for ICE and TCC over a quasi-static channel at $C/I = 0$ dB.

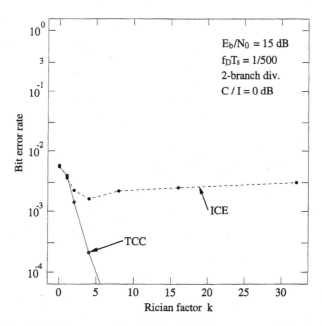

Figure 30.8 BER versus k for ICE and TCC at average $C/I = 0$ dB.

30.7. CONCLUDING REMARKS

Two advanced adaptive reception techniques have been discussed in this chapter.

Based on the research works performed during the last ten years at Kyoto University, the directional antenna or sector-antenna diversity was shown to be very effective to mitigate the BER degradation caused by large multipath delay spread. Thus, sector-antenna diversity is a very powerful means to overcome the effect of multipath propagation.

By combining the sector-antenna diversity with a Viterbi equalizer, we have confirmed further BER improvement is possible at the cost of increased computation.

Also discussed is a novel interference canceller called TCC which dramatically improved the BER performance over a Rician fading channel with co-channel interference.

It was found that both TCC and ICE achieve acceptable BER performance over a Rayleigh fading channel, whereas TCC significantly outperforms ICE over a Rician fading channel. The proposed TCC would be an effective solution to the co-channel interference encountered in microcellular systems.

Future works include the research on the optimum coding for TCC and the more accurate channel impulse response (CIR) estimation techniques.

REFERENCES

[1] Takeuchi, T., Ikegami, F. and Yoshida, S., Transmission bandwidth improvement by directive antenna in urban mobile radio communication. In *Proceeding of IEEE VTC'86*, Dallas, 57–61, May 1986.

[2] Yoshida, S., Ikegami, F., Ariyavisitakul, S. and Takeuchi, T., Delay-spread-resistant modulation techniques for mobile radio. In *Proceeding of IEEE GLOBECOM'87*, Tokyo, 811–817, November 1987.

[3] Buchholz, D., Odlyzko, P., Taylor, M. and White, R., Wireless in-building network architecture and protocols. *IEEE Network Magazine*, 31–38, November 1991.

[4] Murata, H. and Yoshida, S., Intelligent receiver consisting of Viterbi equalizer and sector-antenna diversity for cellular radio. In *Proceeding of International Conference on Communication Systems (ICCS'92)*, Singapore, 919–923, November 1992.

[5] Murata, H., Yoshida, S. and Takeuchi, T., Adaptive receiver consisting of MLSE and sector-antenna diversity for mobile radio communications. *IEICE Transactions on Communications*, **E77-B** (5), 573–579, 1994.

[6] Haykin, S., *Adaptive Filter Theory*, Prentice-Hall, Englewood Cliffs, NJ, 1986.

[7] Okada, M. and Sampei, S., Performance of DDFSE equalizer with interpolation type channel estimator under land mobile frequency selective fading channels. *Transactions on IEICE Japan*, **J73-B-II** (11), 727–735, 1990 (in Japanese).

[8] Yoshino, H. and Suzuki, H., Adaptive interference canceler extended from RLS-MLSE. In *Proceedings of Radio Communication Systems Group of IEICE of Japan*, RCS92-120, 25–30, January 1993.

[9] Fukawa, K. and Suzuki, H., Blind interference cancelling equalizer for mobile radio communications. *Transactions on IEICE*, **E77-B** (5), 580–588, 1994.

[10] Green, E., Radio link design for microcellular systems. *British Telecom Technology Journal*, **8** (1), 85–96, 1990.

[11] Murata, H., Yoshida, S. and Takeuchi, T., Trellis-coded co-channel interference canceller for mobile communication. In *Proceedings of Radio Communication Systems Group of IEICE of Japan*, **RCS93-75**, 39–46, 1993.

[12] Ungerboeck, G., Channel coding with multilevel/phase signals. *IEEE Transactions in Information Theory*, **IT-28** (1), 55–67, 1982.

[13] Eyuboglu, M. V. and Qureshi, S. U. H., Reduced-state sequence estimation for coded modulation on intersymbol interference channels. *IEEE Journal of Selected Areas in Communications*, **7** (6), 989–995, 1989.

[14] Duel, A. and Heegard, C., Delayed decision feedback sequence estimation. *IEEE Transactions in Communication*, **COM-37** (5), 428–436, May 1989.

31

Adaptive Antennas for Personal Communication Systems

George V. Tsoulos, Joe P. McGeehan and Mark A. Beach

31.1. INTRODUCTION

Over the last few years the demand for service provision via the wireless communication bearer has risen beyond all expectations. At the beginning of the last decade some analysts had predicted that few people would use cellular radio services by the year 2000 [1]; however, today in excess of 40 million utilise this technology, just in Europe. At present the number of cellular users grows annually by approximately 50% in North America, 60% in western Europe, 70% in Australia and Asia and more than 200% in South America's largest markets. The extraordinary fact that some half a billion subscribers to mobile networks are predicted by the year 2000 worldwide introduces the most demanding technological challenge: the need to increase the spectrum efficiency of wireless networks.

The two systems that have been proposed to take wireless communications into the next century are the International Mobile Telecommunications 2000 (IMT-2000) and the European Universal Mobile Telecommunications System (UMTS) [2]. The core objectives of both of these systems is to take the "personal communications

Insights into Mobile Multimedia Communication
ISBN 0-12-140310-6

user" into a new information society where mass market low-cost telecommunications services will be provided. In order to be universally accepted, these new networks will have to offer mobile access to voice, data and multimedia facilities in an extensive range of operational environments, as well as economically supporting service provision in environments conventially served by other wired systems. It is against these forecasts that existing mobile communication systems will require radical reformation in order to meet the UMTS/FPLMTS goals. Currently favoured proposals include improved air interface and modulation schemes, deployment of smaller radio cells with combination of different cell types in hierarchical architectures, and advanced signal processing. However, none of these schemes fully exploits the multiplicity of spatial channels that arises because each mobile user occupies a unique spatial location.

Filtering in the space domain can separate spectrally and temporally overlapping signals from multiple mobile units, and hence the spatial dimension can be exploited as a hybrid multiple access technique complementing FDMA, TDMA and CDMA.

This approach is usually referred to either as Space Division Multiple Access (SDMA) [3], which enables multiple users within the same radio cell to be accommodated on the same frequency and time slot (Figure 31.1), or Spatial Filtering for Interference Reduction (SFIR) [3], which reduces the overall interference and allows smaller cell repeat patterns (Figure 31.2). Realisation of these filtering techniques is accomplished using an adaptive antenna array which is effectively an antenna system capable of modifying its time, frequency and spatial response by means of amplitude and phase weighting and internal feedback control, as shown in Figure 31.3.

Numerous approaches using adaptive antennas have been considered in order to exploit the spatial domain, for example null steering to isolate co-channel users [4], optimum combining to reduce multipath fading and suppress interference [5–10],

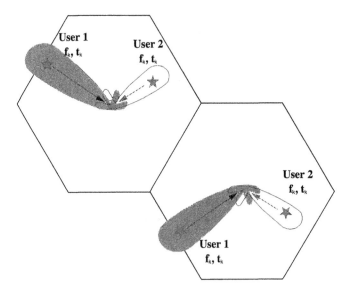

Figure 31.1 SDMA concept.

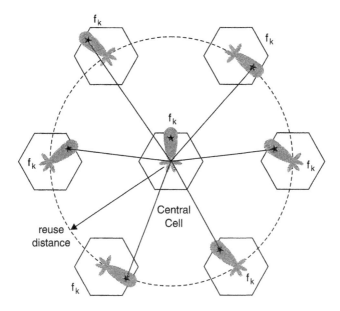

Figure 31.2 **SFIR** concept.

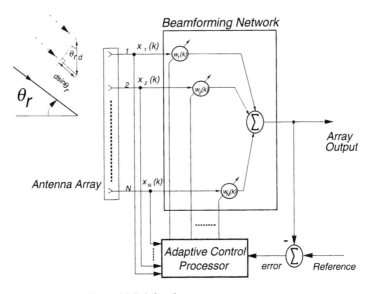

Figure 31.3 Adaptive antenna concept.

and beam steering to focus energy towards desired users [11–13]. By exploiting the spatial domain via an adaptive antenna, the operational benefits to the network operator can be summarised as follows [14]:

- capacity enhancement;
- coverage extension (smart link budget balancing, power reduction, longer battery life);
- ability to support value added services (e.g. high data rates, user location, etc.);
- increased immunity to the "near–far" and handover problems;
- ability to support hierarchical cell structures.

In the following sections the application of adaptive antennas is considered for different operational environments and for different air interface techniques, in an attempt to demonstrate the potentials of this technology.

31.1.1. Hierarchical Cell Structures

Recognising that the requirements of UMTS cannot be fulfilled with the known cellular architectures (macro, micro, pico cells) led to the conception of the idea of a mixed or hierarchical cell structure. The key issue is to apply multiple cell layers to each service area, with the size of each layered cell tailored to match the required traffic demand and environmental constraints (Figure 31.4). In essence, microcells will provide the basic radio coverage but they will be overlaid with umbrella cells to maintain the ubiquitous and continuous coverage required. This mixed cell tech-

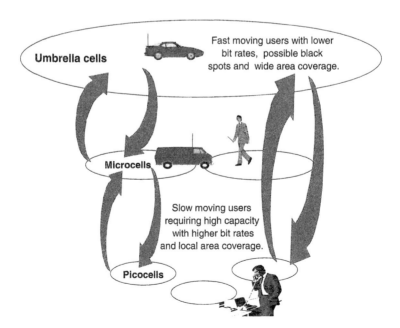

Figure 31.4 Hierarchical cell structure concept.

nique gives answers to situations where a possible performance degradation may occur, e.g. fast moving users requiring handover, or black spots in coverage.

Handover between different cell types introduces a few problems. First, the mobile user must adjust its transmit power to meet the link budget requirements of the new cell. Also the propagation conditions in the new cell may differ considerably and so it would be desirable to either upgrade or downgrade the call, e.g. downgrade a video call to voice only as the user moves from a microcell to a larger macrocell.

Furthermore, the power control requirement becomes increasingly difficult to implement in any form of practical system employing multiple uncoordinated operators within the same geographical area, such as with the case of a mixed cell structure. For the DS-CDMA system which uses the power control technique, macro diversity involving both macro and micro cells within the same frequency band is very difficult due to the closed loop power control algorithm which would reject transmission at the increased macrocell power level. In order to cope with the above problems the simplest solution proposed is to have different cell types on different carrier frequencies. The disadvantages of this technique include:

- Increased hardware complexity added from the extra receiver needed to monitor different frequencies for a mobile assisted handover (MAHO).
- Reduced spreading gain. It has to be emphasised that a bandwidth restricted DS-CDMA air interface is unlikely to meet the goals of UMTS because greatest benefits from CDMA are obtained when a large number of users operate within the same DS-CDMA channel and also the propagation characteristics of wideband CDMA are found to be less hostile to mobile data communications.
- Reduction in the overall system's spectrum efficiency.

By exploiting the spatial filtering that an adaptive antenna offers, it is possible to confine the radio energy associated with a given mobile to a small addressed volume, thus reducing interference experienced from and to co-channel users and multipath from distant reflectors. These properties can be exploited in order to accomplish several goals, as it was mentioned above. Of particular interest for a DS-CDMA system in a mixed cell structure is that in order to support the quality enhancement technique of handover, the different cell types should operate on the same frequency band. A possible way to provide seamless handover, achieve the necessary RF power balancing within each area and simultaneously avoiding the near–far effect, is by exploiting the spatial filtering properties offered by an adaptive antenna [15–16].

By employing adaptive antennas in umbrella cells, the interference from the microcells remains the same, while the interference from the umbrella cells is decreased by a factor equal to the directivity of the radiation pattern of the adaptive antenna (D_{AA}) (here the simplified assumption, of equal achieved performance for each umbrella cell with an adaptive antenna, is made). Hence, with adaptive antennas the total interference seen at the microcellular base stations is:

$$(I_{total})_{AA} = I_{micro} + I_{umbrella}/D_{AA} \qquad (31.1)$$

In the simulations performed here, the simple single slope path loss model with exponent 4 was employed with a value of 8 dB for the log-normal shadowing standard deviation and it was assumed that one umbrella cell overlays three tiers of microcells and that there are three tiers of umbrella cells. Figure 31.5 shows that when an adaptive antenna which can achieve 20 dB directivity is employed at the umbrella cell base station, the system can exploit the directivity to reduce the high transmitted power by the umbrella cell mobile users and hence mitigate the capacity reduction which is caused with an omnidirectional antenna. It can be seen from the figure that there is only a small capacity reduction when the umbrella cell loading is up to 15 users and the directivity achieved by the adaptive antenna is 20 dB. Simulations were also performed for different directivity values and it was noticed that reductions in the final capacity start to emerge whenever the achieved directivity is less than 15 dB, for the case of 15 umbrella cell users.

If the spectral efficiency of this system is compared with that of a system where the different cell types use different frequency bands to avoid interference, it can be seen that the system with an adaptive antenna increases the spectrum efficiency, as shown in Table 31.1 where the results from Figure 31.5, were considered along with the following assumptions:

- One microcell covers an area of 1 km².
- One umbrella cell overlays 37 microcells, i.e. covers an area of 37 km².
- 1 MHz bandwidth is used for the case of the mixed system with adaptive antennas.
- 1 MHz bandwidth is used for each layer for the case of different frequency bands for different cell types.

31.2. ADAPTIVE ANTENNAS FOR SMALL CELLS

The simulation model for this scenario can be separated into three basic elements [15, 10]:

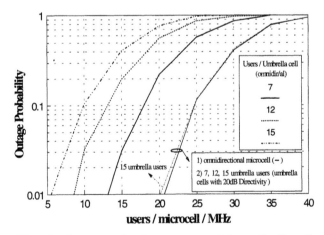

Figure 31.5 Omnidirectional versus adaptive antennas at the umbrella cell of a mixed cell structure.

Table 31.1 Spectrum efficiencies of different approaches for a mixed cell structure [15]

	Adaptive antennas	Different frequency bands
Microcell efficiency	20 users/MHz/km^2	20 users/MHz/km^2
Umbrella cell efficiency	15 users/MHz/37 km^2	20 users/MHz/37 km^2
Total efficiency	20.4 users/MHz/km^2	10.3 users/MHz/km^2

(a) *Multipath channel model.* Impulse responses from the environment under investigation are generated using a ray-tracing simulation tool [17]. The input parameters to this tool include a geographical data base of the service area, the number of reflections, transmissions and diffractions, the transmitted power, antenna radiation patterns, etc. The output file includes the electric field, the time delay and the angle of arrival of the four most dominant received rays from each user.

(b) *Adaptive Antenna Array.* The output from the adaptive array at instant $t = kT$, where T is the sampling interval and k is the number of the sample, is [15]:

$$y(k) = \sum_{n=1}^{N} w_n(k) \left\{ \sum_{m=1}^{M} \sum_{r=1}^{R} h_{mr} e^{jk_u d(n-1)\sin(\vartheta_r)} r_m(k - t_r) + N(k) \right\} \qquad (31.2)$$

where N is the total number of antenna elements, h_{mr} and $r_m(k)$ are the elements of the vectors of the impulse response and the DS-CDMA signal from the mth user respectively, $r_m(k) = d_m(k) \cdot PN_m(k) \cdot e^{j\xi_m}$, with $d_m(k)$ representing the binary data and ξ_m the carrier phase of user m. M is the total number of users, R is the total number of rays, d is the inter-element distance, k_u is the electromagnetic wave number, ϑ_r and t_r are the angle of arrival and the delay of each ray and $N(k)$ represents the additive random Gaussian noise.

This model for the adaptive antenna offers the capability of selecting one of several adaptive processing algorithms, such as the LMS, normalised LMS, RLS and square root RLS. Here only results from the RLS algorithm are presented. Parameters used in the analysis include: 8 element antenna array with half wavelength spacing, 255-chip M-sequence, averaging over 5 different runs and a value of 1 for the *forgetting factor* of the RLS algorithm. In addition, each user in the simulation is stationary and ideally power controlled by the base station.

Typical outputs from this model are illustrated in Figure 31.6. These also serve to demonstrate the concept of a "smart antenna" for any given spatial distribution of interfering and desired signals, since the array always attempts to generate the optimum radiation pattern in order to optimise the output signal to interference plus noise ratio. For example, in Figure 31.6(a), the main beam has been steered towards the second strongest multipath with the strongest multipath received by a secondary beam, whilst multipath contributions from interfering sources are minimised. When compared with Figure 31.6(b) where the number of interferers has been increased, it can be seen that the main lobe has switched to the strongest desired multipath and no other beams are formed.

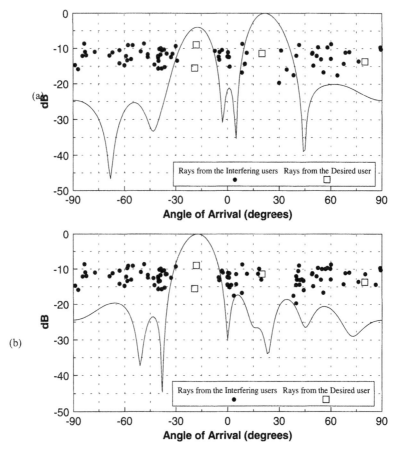

Figure 31.6 Radiation patterns produced from the adaptive antenna after convergence has
  been achieved (25 samples): (a) 20 users; (b) 25 users [15].

For different simulation scenarios and for different number of users, the above
simulations were repeated and from the produced radiation patterns, values for the
directivities ranging between 6 dB and 9 dB were calculated.

(c) *DS-CDMA capacity analysis.* The capacity of the DS-CDMA system is defined
here as the number of users/cell/MHz that can be supported with an outage probability less than 1%, where the outage probability is the probability that the signal to
interference level is less than a threshold value which is required for a minimum
transmission quality, i.e. BER $\leq 10^{-3}$. The total interference seen by the central base
station due to both the intracell and the inter-cell interferers, is [15]:

$$I_{tot} = \int_0^{(2Tiers+1)R_{mic}} \int_0^{2\pi} \frac{P(l,\varphi)}{\pi R_{mic}^2} D(\varphi) l \, dl \, d\varphi = \frac{I}{D} \tag{31.3}$$

where *Tiers* is the number of tiers of cells considered in the simulation, R_{mic} is the
microcell radius, $P(l,\varphi)$ is the power transmitted by a user at distance l from the
central base station and angle φ, I is a constant which represents the value of the total

interference seen by the base station before the spatial analysis is considered and D is the directivity of the base-station antenna which is assumed omni-directional in the vertical plane. The key parameters considered for the DS-CDMA system are:

- total number of base stations: 37, (3 tiers);
- path loss exponent: 4;
- log-normal shadowing std. dev.: 8 dB;
- power control: shadowing and path loss;
- voice activity: 0.5;
- data rate: 8 kbps;
- E_b/N_0 for BER $\leq 10^{-3}$: 7 dB;
- total spreading bandwidth: 1 MHz.

These parameters were chosen for the purpose of initial simulations to enable a comparison to be made between an omnidirectional and an adaptive antenna, hence the exact values have no effect on the calculated capacity gain with an adaptive antenna.

In order to find an approximate lower bound for the predicted improvement, we considered the worst case of the simulation scenarios, i.e. for the scenario that the adaptive antenna responds with the lowest directivity radiation pattern, which was almost 6 dB. Furthermore, the BER for the DS-CDMA system is calculated based on the simplified formula [18]:

$$P_b = Q(\sqrt{3SF \times CIR}) \tag{31.4}$$

with $Q(z)$ the standard Q-function and SF the spreading factor. This equation is accurate for perfect power control and for large number of users.

It can be seen from Figure 31.7 that the capacity has increased from 24 users/cell/MHz to 108 users/cell/MHz (BER $= 10^{-3}$), i.e. almost a fivefold increase has been achieved.

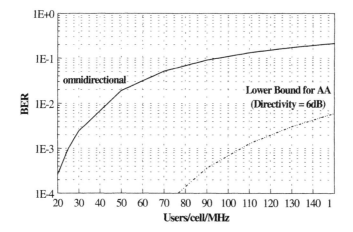

Figure 31.7 BER improvement for the DS-CDMA system with adaptive antennas [15].

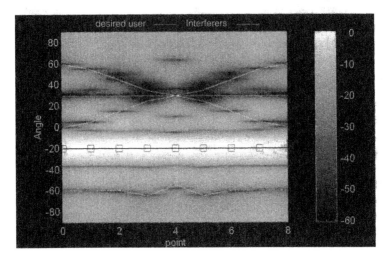

Figure 31.8 Ground plan of the 3D radiation pattern for a moving users scenario [15].

All the results presented up to here have considered static users. In order to see how the adaptive antenna performs when the users are moving, a simple scenario with one desired user and four interferers was considered. For simplification reasons, only the strongest ray of the impulse response of each user is considered and the users are supposed to move slowly (less than 20 m.p.h.). The route that the users follow is shown in Figure 31.8. The desired user is moving straight at −20°, two interferers are moving almost straight at −60° and 30° and two interferers start from 0° and 60° and finish at 60° and 0° degrees respectively. It can be seen from Figure 31.8 that the system is always able to both support the desired user and produce deep nulls for the interferers (better than −40 dB). The situation considered here is a simple one and as such a reduction in the performance of the algorithm is expected when the users travel with speeds higher than 20 m.p.h. and the LMS algorithms are used, or more than 60 m.p.h. and the RLS algorithms are used. However, as shown in [6], for the RLS algorithms and for speeds up to 60 m.p.h., the performance reduction is within 0.2 dB from the performance with the static case.

31.3. A TDMA SYSTEM WITH ADAPTIVE MULTI-BEAM ANTENNAS

A large cell operational environment which employs a predetermined reuse pattern is considered here. A Monte Carlo type set of tools similar to the one employed for the DS-CDMA analysis was developed in [15] for a TDMA system like GSM. For the case of the uplink, the simulation technique is illustrated in Figure 31.9. The simulation generates a random deployment of uniformly distributed users and then steers the main beam towards the desired user. For each new user the CIR is calculated and compared with the threshold value. If it is exceeded then the user is handed over to another channel. It can be seen that whenever one interfering user is within the main

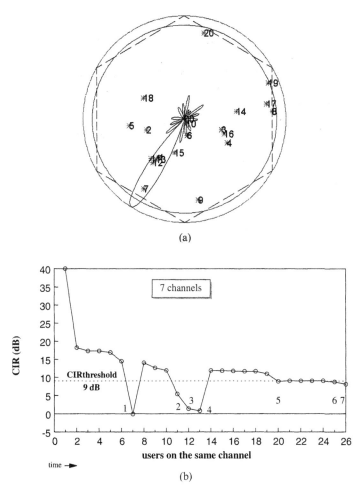

(a)

(b)

Figure 31.9 Adaptive antennas with TDMA simulation concept: (a) one snapshot of the user distribution (the first 25 users; (b) CIR history (the numbers in the figure denote the number of handovers required) [15].

beam (e.g. users 7, 11, 12, 13 in Figure 31.9(b)), the CIR value falls below the threshold and the user must be handed over to another channel. As the number of users in the same channel increases, the overall CIR value decreases, which effectively means that while at the beginning only the users within the main beam should be handed over to other channels, as time processes, users on primary sidelobes must also be handed over and only users falling on secondary sidelobes of the radiation pattern, can be accepted in SDMA mode. The above calculations are repeated for 10^4 times and then the probability density function and the outage probability of the number of co-channel users are calculated. For the simulations a radiation pattern with 22° 3 dB beamwidth with the first sidelobe at approximately -13 dB, and -60 dB maximum null depth is considered.

The effect that the propagation characteristics can have on the achieved capacity when one tier of interferers is considered, is shown in Figure 31.10. It can be seen that the capacity prediction for the case of one central cell is optimistic. The capacity is decreased when one tier of interferers is considered, with a path loss exponent of 4 and no shadowing. This again produces rather optimistic results, and is further decreased when more realistic propagation characteristics are included. Here, a dual slope model for the path loss with exponent 3 up to the cell radius and 4 beyond, and 5 dB standard deviation for the shadowing, has been included. The adaptive multi-beam system can now support approximately four users with 1% outage probability.

Figure 31.11 illustrates the effect that parameters such as maximum null depth and mispointing of the main beam have on capacity. In a real SDMA system the digital beam former calculates sets of weights which are used to produce the different beams corresponding to the various users. By using an orthogonal transformation the DBF can then produce a radiation pattern which supports a desired user and nulls the other users out. It is the effect of these nulls that is effectively studied here. From Figure 31.11(a) can be seen that when the ideal radiation pattern with approximately 22° of beamwidth and −60 dB maximum null depth is replaced with one with −30 dB maximum null depth, there is almost 40% reduction in capacity for 1% outage probability. Note that this result is correct when point sources are considered but if additional scattering around the mobile user is considered, then the signal "seen" by the base station will have a small angular spreading, and hence wider nulls will in some cases improve the capacity.

For the case of main beam pointing errors, the simulation generates a beam which is steered towards the correct angle plus an offset angle uniformly distributed throughout the 10 000 iterations for either ±3° or ±6°. The reduction in the capacity at 1% outage probability is about 12% and 28% for the cases of mispointing within 6° and 12° respectively, Figure 31.11(b). The above have a direct impact on practical implementation issues related with adaptive antennas, such as the accuracy of the calibration process employed by the adaptive antenna system.

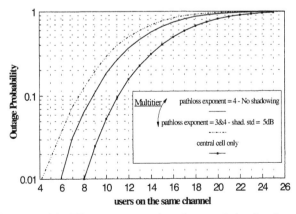

Figure 31.10 Capacity with different propagation characteristics for the system with one tier of interfering cells [15].

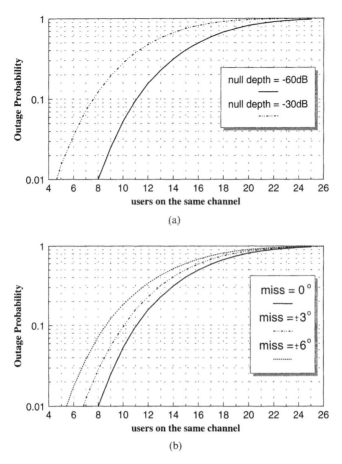

Figure 31.11 The effect of the radiation pattern on the achieved performance: (a) null depth; (b) mispointing [15].

Although current systems usually exploit a frequency reuse factor of 12 or 7 in the field, all the results presented here have assumed a frequency reuse factor of 3. The results given in Figure 31.12 indicate that it may be desirable not to reduce the reuse distance but to increase the capacity of the existing system with current reuse distances, through SDMA. A frequency reuse factor of 7 can almost double the capacity of the system, compared with a frequency reuse factor of 3, for 1% outage probability. Clearly, there is a trade-off between capacity enhancement due to the better performance of the adaptive antenna with higher frequency reuse distances, and the lower spectral efficiency that these higher frequency reuse distances imply. In Table 31.2 the percentage increase of the capacity due to the larger cell clusters and the corresponding percentage decrease in spectrum efficiency, have been calculated. It can be seen that the best improvement is achieved when the cell cluster is expanded from 3 to 7.

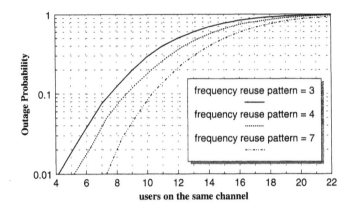

Figure 31.12 Capacity for different reuse patterns for the system with one tier of interfering cells [15].

Table 31.2 Relative spectrum efficiency decrease and capacity increase for different cell cluster sizes [15]

Cluster size	Efficiency E_i/E_j (%)	Capacity C_i/C_j (%)
3 → 4	−25	+25
4 → 7	−43	+40
3 → 7	−57	+75

31.4. FIELD TRIAL DEMONSTRATION OF SDMA

The results shown in this section have been produced with the testbed that was built from the consortium of the RACE TSUNAMI project. The DECT radio standard was selected as the operational wireless bearer since it could be readily integrated with the adaptive antenna platform and furthermore DECT can operate in an isolated radio cell mode, thus allowing networking aspects (e.g. handover) to be addressed at a later phase. The testbed hardware consisted of an 8 channel system employing a patch antenna array which could be deployed in various configurations and 8 independent linear up and down conversion chains which transform the signals to quadrature baseband. In order to compensate for mismatches and component drift within the multiple analogue signal paths a calibration system was employed. Calibration is required for DBF systems where each channel contains a complete transceiver, each of which must be amplitude and phase matched. Any differences will preclude precise pattern control and the ultimate performance achieved with this technology. Two SDMA channels were supported on two independent DSPs, thus enabling the trial system to establish two links within the same timeslot and frequency channel of the DECT air interface through the spatial domain. For the SDMA tests the MUSIC [19] algorithm was used as the basic spatial reference algorithm. Each iteration of the adaptive algorithm consisted of three steps: 1. Estimate the number of signals and directions of arrival using the MUSIC algorithm, 2. Apply the DOAs to the tracking algorithm [20] and 3. Synthesise beams for the users.

The results presented here are from the trials performed in Bristol. More details for the field trials and more results can be found in [15] and [21,22]. The area is a typical outdoor urban environment adjacent to the Engineering Faculty building. The base station array was a $\lambda/2$ linear deployment at a height of -30 m above ground level. During the test the users initially move at walking pace towards each other up to a point where their angular separation is almost $15°$ as seen from the base station; here they stop and return to their original positions. In Figure 31.13 the tracked Directions Of Arrival (DOAs) are given, illustrating the ability of the spatial reference algorithm to track the users. It is known that the user resolution that can be achieved using SDMA is much better than the main beam width of the antenna array, but as discussed in [15,21,22], practical limitations usually lead to a deviation from the optimum performance.

Clearly in Figure 31.13 can be seen that the adaptive antenna can successfully track the two users and that it can also provide user resolution approaching the 3 dB beamwidth of the antenna array. Figures 31.14(a–d) contain the measured bit error rate of the four independent channels (uplink–downlink for both SDMA users). Most of the time[1] the BER is better than 10^{-3}, which is an impressive demonstration of the SDMA capability of the adaptive antenna system.

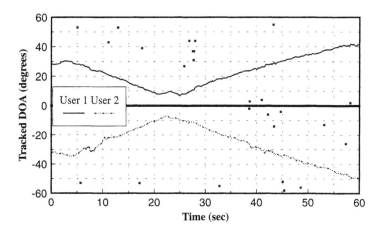

Figure 31.13 Tracked DOAs [15].

---

[1]It should be noted that some of the spurious peaks shown in the BER figures are because there are some bad slots during the whole time of the experiment, which are due to a problem of interference caused by other sources within the prototype base station, e.g. transmitter gating.

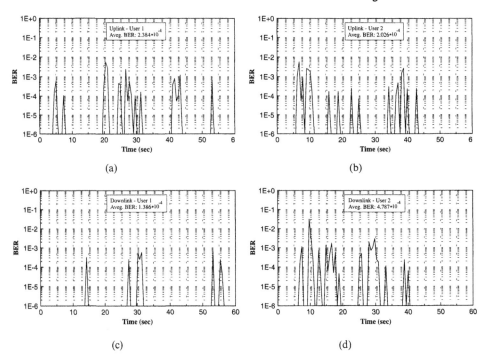

Figure 31.14 Uplink BER: (a) user 1; (b) user 2. Downlink BER: (c) user 1; (d) user 2 [15].

31.5. CONCLUSIONS

Future commercial wireless systems are expected to provide communication tailored to the specific requirements of each individual user, and this inevitably means value added services and more capacity problems. It is against these forecasts that existing mobile communication systems will require radical reformation in order to meet the goals of third generation systems. In this scenario, new technologies like the adaptive antennas, the potentials of which were briefly discussed here, will almost certainly play a key role.

REFERENCES

[1] Zysman, G. I., Wireless networks, *Scientific American*, special issue on key technologies for the 21st century, 52–55, September 1995.
[2] Rapeli, J., UMTS: targets, system concept, and standardisation in a global framework. *IEEE Personal Communications*, February 1995, **2** (1), 20–28.
[3] Tsoulos, G. V., Beach, M. and McGeehan, J., Wireless personal communications for the 21st century: European technological advances in adaptive antennas. *IEEE Communications Magazine*, September, 102–109, 1997.

[4] Marcus, M. J. and Das, S., The potential use of adaptive antennas to increase land mobile frequency reuse. In *Proceedings of the IEE 2nd International Conference on Radio Spectrum Conservation Techniques*, 113–117, Birmingham, UK, September 1983.

[5] Winters, J. S., Optimum combining in digital mobile radio with co-channel interference. *IEEE Transactions on Vehicular Technology*, **VT-33** (3), 144–155, 1984.

[6] Winters, J. S., Signal acquisition and tracking with adaptive arrays in digital mobile radio system IS-54 with flat fading. *IEEE Transactions on Vehicular Technology*, **VT-42** (4), 377–384, 1993.

[7] Andersson, S., Milnert, M., Viberg, M. and Wahlberg, B., An adaptive array for mobile communications systems. *IEEE Transactions on Vehicular Technology*, **VT-40** (1), 230–236, 1991.

[8] Naguib, A. F. and Paulraj, A., Performance of CDMA cellular networks with base station antenna arrays. In *Lecture Notes in Computer Science*, no. 783, pp. 87–100, Springer-Verlag, Berlin, 1994.

[9] Kohno, R., Imai, H., Hatori, M. and Pasupathy, S., Combination of an adaptive array antenna and a canceler of interference for direct sequence spread spectrum multiple access. *IEEE Journal on Selected Areas in Communications*, **8** (4), 675–681, 1990.

[10] Tsoulos, G. V., Beach, M. A. and Swales, S. C., DS-CDMA capacity enhancement with adaptive antennas. *IEE Electronics Letters*, **31** (16), 1319–1320, 1995.

[11] Swales, S. C., Beach, M. A., Edwards, D. J. and McGeehan, J. P., The performance enhancement of multi-beam adaptive base station antennas for cellular land mobile radio systems. *IEEE Transactions on Vehicular Technology*, **VT-39** (1), 56–67, 1990.

[12] Kennedy, J. and Sullivan, M., Direction finding and "smart antennas" using software radio architectures. *IEEE Communications Magazine*, **33** (5), 62–68, 1995.

[13] Tsoulos, G. V. and Beach, M. A., Sensitivity study for the capacity enhancement of DCS1800 with adaptive multibeam antennas. *IEE Electronics Letters*, **32** (19), 1745–1746, 1996.

[14] Tsoulos, G. V., Adaptive antenna technology, Tutorial, *3rd residential course on digital techniques in radio systems*, University of Bristol, Bristol, UK, 14–18 September 1997.

[15] Tsoulos, G. V., Smart antennas for third generation wireless personal communications. Ph.D. thesis, University of Bristol, December 1996.

[16] Tsoulos, G. V., Athanasiadou, G. E., Beach, M. and Swales, S. C., Adaptive antennas for microcellular and mixed cell environments with DS-CDMA, to be published in the *Wireless Personal Communications Journal*, issue 7, 2–3, 147–169, June 1998, Kluwer Academic Publishers, New York.

[17] Athanasiadou, G. E., Nix, A. R. and McGeehan, J. P., A ray tracing algorithm for microcellular wideband modelling. In *Proceedings of the 45th Vehicular Technology Conference*, Chicago, Illinois, USA, 261–265, July 25–28, 1995.

[18] Pursley, M. B., Performance evaluation for phase-coded spread spectrum multiple access communications with random signature sequences – Part I: system analysis. *IEEE Transactions in Communications*, **COM-25**, 795–799, August 1977.

[19] Schmidt, R. O., A signal subspace approach to emitter location and spectral estimation. *IEEE Transactions on Antennas and Propagation*, **AP-34** (3), 276–280, 1986.

[20] Riba, J., Goldberg, J., Vazquez, G. and Lagunas, M. A., Signal selective DOA tracking for multiple moving targets. *Proceedings IEEE ICASSP 96*, Atlanta, GA, 2559–2562, 1996.

[21] Tsoulos, G. V., McGeehan, J. and Beach, M. Space division multiple access (SDMA) field trials – Part I: tracking and BER performance. *IEE Proceedings on Radar, Sonar and Navigation*, special issue on Antenna Array Processing Techniques, 73–78, February 1998.

[22] Tsoulos, G. V., McGeehan, J. and Beach, M., Space division multiple access (SDMA) field trials – Part II: calibration and linearity issues. *IEE Proceedings on Radar, Sonar and Navigation*, special issue on Antenna Array Processing Techniques, 79–84, February 1998.

32

Adaptive Equalisation Methods for Next Generation Radio Systems

R. Perry, A. R. Nix and D. R. Bull

32.1. INTRODUCTION

To date there has been relatively little widespread deployment of wireless radio local area networks (LANs) and the systems available have been relatively expensive. Only proprietary solutions are currently available, mostly operating in the licence-free 900 MHz or 2.4 GHz frequency bands. This situation is now likely to change, following the ratification of several new standards and licensing of new frequency spectrum, both in the US [1] and Europe [2,3]. Such standardisation activities enable multi-vendor interoperability and thus avoid limitations of mobility due to technical boundaries.

In Europe the High Performance Radio LAN (HIPERLAN) standard was developed primarily to support *ad hoc* networking. The data rate over the air is ~23.5 Mb/s, which if used in typical office and business premises, will result in severe intersymbol interference (ISI). Due to this high data rate, the only available frequency spectrum for operation of HIPERLAN terminals lies between 5 and 6 GHz, which will impact on the overall cost of the radio. The 802.11 standard supports a data rate of 1 Mb/s, and optionally 2 Mb/s, which allows reception without significant ISI. At this lower rate, operation in the 2.4 GHz ISM band is possible.

The allocation of frequency spectrum in the 5 GHz band in the US also allows the deployment of HIPERLAN terminals there. However, the regulations do not specify an implementation and so other types of system can be considered. The rules, however, favour the deployment of broadband systems with transmission bandwidths exceeding the 20 MHz necessary, if use is to be made of the allowable peak power transmission [3].

In order to counteract the ISI introduced during transmission, equalisation is often used at the receiver. This form of ISI countermeasure has been expressly catered for in the HIPERLAN standard by the provision of a 450-bit training sequence at the start of each transmission. The requirement for channel equalisation does not arise in the 802.11 modems, due to the lower symbol rates.

The problem of ISI is not limited to wireless systems. The planned use of the potential transmission capacity of older twisted pair telephone wires for transmission of new digital services to the residential sector [4] requires echo cancellation or decision feedback equalisation [5]. In HDSL digital subscriber lines, equalisation has been investigated to counteract echoes in the channel [5].

There are alternative approaches to combat the effects of ISI other than equalisation. For example, orthogonal frequency division multiple access (OFDM) has been proposed for use in wireless LANs [6]. In this approach, a high rate serial transmission is converted to multiple parallel streams, which are transmitted at a reduced transmission rate. The multi-carrier signal can be efficiently generated using a Fast Fourier transform [7]. At the receiver, no equalisation is required, provided that the bandwidth, required for each parallel stream, is less than the coherence bandwidth of the radio channel. By coding across each block of data, the effects of a narrowband signal fade in any sub-channel can be corrected.

A difficulty of implementing this type of system is the large dynamic range of the transmitted signal. This occurs because the parallel streams of data can add either constructively or destructively, depending on the particular combination of symbols being transmitted. To transmit the signal, therefore, requires the very inefficient linear amplification. Recently, the use of alternative coding schemes has been developed to reduce the dynamic range of the transmitted signal as well as providing error correction capability [8]. This increases transmitter efficiency and provides improved performance in noise.

32.2. EQUALISATION METHODS

There is a vast number of equalisation methods that have been proposed over the last thirty years [9]. There are several ways in which these methods can be categorised. The approach taken here is to group the methods according to the way in which the equalisation is performed. On this basis, three main approaches have been identified which are listed in Figure 32.1.

It is possible to introduce other forms of classification depending on whether the method is inherently linear or non-linear, and whether a training sequence is provided for initialisation of the equaliser or whether the equaliser is trained blindly [10].

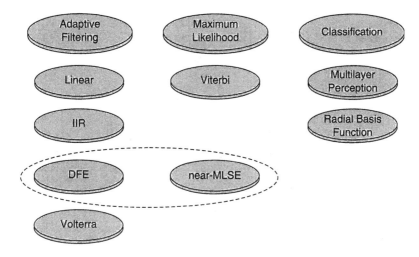

Figure 32.1 Classification of equalisation methods.

Blind equalisation algorithms are of considerable interest since no training data has to be transmitted, but initialisation of the equaliser is more difficult to achieve.

A linear transversal filter, with coefficients suitably selected, minimises the ISI, whilst avoiding excessive noise enhancement. At high signal-to-noise ratios, the filter impulse response approximates the inverse of the channel impulse response. However, when the channel frequency response possesses deep nulls, such an equaliser results in excessive noise enhancement. This problem has been discussed extensively in the literature [9]. As this occurs frequently in wireless channels, linear equaliser structures are not generally applicable and more complex non-linear techniques must be used.

Probably the simplest of these non-linear equaliser methods is decision feedback equalisation. Decision feedback equalisation has been a major subject of interest in its own right. The canonical form for the DFE is shown in Figure 32.2.

The performance of a DFE is superior to a linear equaliser, because the feedforward filter (FFF) is no longer required to realise the inverse of the channel. Instead the FFF, in cascade with the channel, results in a combined impulse response with only post-cursor ISI and a small residual pre-cursor ISI component. The FBF cancels the post-cursor distortion, without any additional noise enhancement. However, when an incorrect decision is made, this will increase the level of post-cursor ISI, potentially generating more errors and causing catastrophic failure.

In wireless applications, the channel conditions during each transmitted frame are unknown and often time varying. The coefficients of the DFE must, therefore, be obtained adaptively using a training algorithm. An integral part of the DFE is therefore the selection of a suitable training algorithm. A large number of adaptive algorithms exist, for both transversal and other filter structures. For transversal filters, the conventional recursive least square (RLS), fast Kalman [11] and fast transversal filter algorithms [12] may be considered to achieve fast convergence. In [13], the square-root Kalman algorithm was used for equaliser training, and has

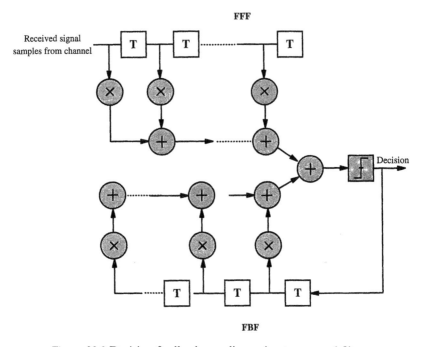

Figure 32.2 Decision feedback equaliser using transversal filters.

superior numerical stability to the conventional RLS algorithm. The least mean square (LMS) algorithm can also be used for training provided a slow speed of convergence is acceptable. There is a vast number of variants of the LMS algorithm, several of which are described in [14]. Related to the fast RLS algorithms are the least squares lattice algorithms. Using a multi-channel lattice filter, a lattice-filter-based DFE was developed [15]. Adaptive filters implemented as systolic arrays have also been developed, the recursive modified Gram–Schmidt (RMGS) algorithm being suitable for use as a decision feedback equaliser [16].

A large amount of effort has been directed towards the simplification of the adaptive algorithms to reduce the total operation count. The lattice and fast RLS algorithms were obtained by exploiting the structure of the conventional RLS algorithm to reduce the number of computations from $O(N^2)$ to $O(N)$, although the saving is only apparent for long filters. Despite this reduction, the stability of the FTF and FK algorithms is a significant disadvantage to their use [17]. The lattice and systolic algorithms have been reported to be more stable under finite precision arithmetic than the transversal-based algorithms, but require greater numbers of division and square root operations [16].

Non-linear filters based on Volterra series have also been investigated [18]. Such filters are able to synthesise and correct for non-linear distortion in the channel. In many mobile applications employing equalisation, the receiver is designed to be as linear as possible and, therefore, the requirement for non-linear filtering is not apparent. Such non-linear filters can, however, be constructed and trained using conventional linear filtering methods. Another alternative structure, similar to the

DFE, is the linear feedback equaliser [19]. Instead of quantising the equaliser output, the output is passed directly to the FBF. The structure is however potentially unstable, which has precluded its use in practice.

The use of maximum likelihood sequence estimation (MLSE) has been extensively studied for use in the GSM system and is known to be an optimal detector. This method is most efficiently implemented using the Viterbi algorithm [20], which is similar to dynamic programming optimisation methodologies. This method predicts the most likely sequence of transmitted data from samples of the received signal and an estimate of the channel impulse response. Since cancellation of the previously detected symbols is unnecessary, this method does not suffer from error propagation. The complexity of the algorithm is, however, exponentially dependent on the number of significant interfering terms. For the time dispersion predicted for high data rate wireless LANs, the complexity of MLSE has discouraged its use.

As a means of reducing complexity, whilst retaining the desirable performance characteristics of MLSE, *near*-MLSE approaches have been described [21]. An example of this approach is to use an MLSE in combination with a feedback filter to cancel postcursor ISI and thereby reduce the number of possible states used in the maximum likelihood sequence estimator. This method uses the initial tentative decisions from the MLSE section in the feedback filter and is therefore prone to error propagation. Bit error rate performance tests show that near-MLSE provides performance between that of a DFE and an MLSE [21].

As an alternative to the more traditional approaches to equalisation described above, the use of neural networks has also been investigated [22,23]. The problem is encapsulated as a classification process. This involves classifying input vectors, made up from the most recent received signal samples and past decisions, according to the decision that should be obtained. The classification of these vectors will, in general, result in highly non-linear decision regions and, therefore, non-linear neural networks have been used. An approach to learning that has been investigated is the use of *clustering*, which exploits the fact that the channel outputs will cluster around *centres* in the decision space. During training the neural network learns the position of these centres and the correct labelling for each centre. During data reception, decisions are made on the basis of the label applied to the centre, which is closest to the current input vector. Currently, the major disadvantages of these approaches for mobile applications has been the relatively high complexity and the numbers of training samples required to teach the network.

Although the equaliser is able to minimise the effects of ISI, if the signal strength is poor due to signal fading or shadowing, then a packet failure will be likely, irrespective of the method of equalisation used. For these reasons, diversity reception has been extensively studied for use in conjunction with equalisation. Several example combining strategies are proposed in [24,25] and several other approaches were discussed in [26]. There are three main classes of diversity combining that can be distinguished, these are (i) *switch selection* according to some measurable parameter, usually received signal strength, (ii) *independent and parallel equalisation* of each diversity branch and, (iii) *combined diversity and equalisation* in which all the diversity signals are processed by a single equaliser. Using a DFE for method (iii), this can be achieved by doubling the length of the FFF so that an equal number of samples from each diversity branch can be processed, while a single shared FBF is used.

Method (i) does not require complete demodulation of all diversity branches, and is thus significantly cheaper to implement than both methods (ii) and (iii).

The conclusions from these studies strongly suggest that an order of magnitude performance improvement is possible using dual antenna diversity compared to single branch diversity. However, the level of improvement is dependent on the exact form the diversity combining takes and there is considerable performance variation. Diversity selection on the basis of the received signal strength alone is generally a poor indicator of eventual performance, because ISI and co-channel interference will add to the total received signal strength measure. The most significant performance improvement is obtained when all diversity branches are demodulated and more sophisticated signal processing can be applied in the digital section of the receiver.

32.3. LOW COMPLEXITY AND HIGH-THROUGHPUT DFE ARCHITECTURES

In this section, DFE architectures are focused on because of their applicability in the new HIPERLAN standard. Some studies on the use of decision feedback equalisation for wireless LANs have already been reported [24,25]. These have concentrated on the use of the relatively simple LMS training algorithm as the preferred training algorithm which is shown below,

$$\mathbf{W}(n) = \mathbf{W}(n-1) + \beta e^*(n)\mathbf{U}(n) \tag{32.1}$$

$$e(n) = d(n) - \mathbf{W}^H(n-1)\mathbf{U}(n) \tag{32.2}$$

where $e(n)$ is the estimation error, $\mathbf{W}(n)$ is the vector of FFF and FBF coefficients, $\mathbf{U}(n)$ is the vector of FFF and FBF inputs and β is the step-size parameter used to control the convergence speed. Despite its slow rate of convergence, the high data rate and large payloads of HIPERLAN frames reduces the bandwidth efficiency loss incurred from using a long training sequence. Although simple, the implementation of this algorithm, at the high data rates required for HIPERLAN, remains a significant challenge. In this section, several techniques which aim to facilitate implementation are described and their relative suitability for wireless LANs assessed.

In the implementation of a fixed filter there are several approaches that the designer may use to enhance throughput at the expense of silicon area. These are parallelism, pipelining and look-ahead [27]. These methodologies cannot be applied directly to adaptive systems, because of the recursive nature of the computations, namely all the coefficients must be updated before the next input sample can be processed. Several ways to overcome this bottleneck have been reported. An essential feature of these approaches is that they are inherently approximate and the algorithm designer must, therefore, consider the performance impact of the chosen method. In a large number of cases, the performance impact is minimal or can be reasonably compensated.

Using look-ahead, an algorithm is iterated as many times as necessary to create the required level of concurrency. For example, the coefficient update for the LMS algorithm, given by equation (32.1), can be expressed in the alternative form

$$\mathbf{W}(n) = \mathbf{W}(n - D) + \sum_{i=0}^{D} e^*(n - i)\mathbf{U}(n - i) \tag{32.3}$$

It is clear that this transformation results in considerably more computation. However, by approximating the update as

$$\mathbf{W}(n) = \mathbf{W}(n - D) + e^*(n)\mathbf{U}(n) \tag{32.4}$$

the number of computations is reduced to the same number as the conventional LMS algorithm. There remain an additional D delays in the coefficient update which can now be used to pipeline the coefficient update. In [28] an alternative realisation of the update in (32.3) was obtained. An exact update was obtained, with reduced computation, by exploiting some redundancy in the computations.

Block-based algorithms were originally suggested as a means of achieving throughput increases [29]. In these algorithms, the update for the coefficients is performed once in every N clock cycles. This reduces the number of updates, but since the coefficients are adapted at a rate $1/N$, the training period is correspondingly increased by a factor of N. This reduces the bandwidth efficiency of the LAN.

An alternative parallel approach was suggested in [30]. In this method short known patterns of data samples are placed between data blocks, whilst a longer training sequence is appended to the start of a data packet which is used to initialise the coefficients of a single *master* DFE. Training of the master DFE is performed first, at a lower rate than the system sampling rate and, therefore, buffering of the input data is necessary. At the end of training, multiple *slave* DFEs are assigned different data blocks to process. The known patterns of symbols, included between the data blocks, are used to correctly initialise the FBF input of the slave DFEs. The coefficients of the slave DFEs are all initialised to the current values of the master DFE coefficients. Each slave DFE then performs equalisation of its respective data blocks. Now that multiple DFEs are used, the receiver is able to process the stored backlog of received data. The major disadvantages of this method are that the initial coefficients of the slave DFEs operating on the later data blocks will be out of date if the channel changes appreciably, and that there is a large area increase associated with the use of multiple DFEs. Furthermore, the storage of the input data, requires the use of buffering control and introduces additional receiver latency.

An improvement to the above was suggested in [30], referred to as the extended LMS (ELMS) algorithm. The ELMS algorithm attempts to improve the DFE tracking behaviour and avoid the occurrence of "glitches" in the DFE coefficients that can occur due to changes in the channel. This algorithm was based on the use of parallel DFEs as before, but jumps in the coefficient values, as a result of processing disconnected data blocks, were compensated for by using the coefficient updates from all the parallel DFEs within each individual DFE update. This incurs a considerable computational overhead since the modified coefficient update now involves matrix vector operations.

The use of a number of delays in the coefficient update has been used extensively [31,32]. Originally the delay was incurred due to a decoding delay in the feedback loop of the adaptive algorithm [31]. However, it has been recognised that the delay may be used to pipeline the filter [32]. This gives rise to the delayed LMS (DLMS) algorithm given by

$$\mathbf{W}(n) = \mathbf{W}(n-1) + e^*(n-D)\mathbf{U}(n-D) \tag{32.5}$$

$$e(n) = d(n) - \mathbf{W}^H(n-1)\mathbf{U}(n) \tag{32.6}$$

Highly modular and efficient filters for a DFE have been derived using the DLMS algorithm, by redistributing the delays in the coefficient update. Throughput improvements of greater than a factor of two are potentially possible [33] by this restructuring. The DLMS algorithm is more economic in silicon area, since the speed up is improved at the cost of only pipelining registers. In addition for DFE filter lengths required in wireless applications, the convergence speed is comparable to the conventional LMS algorithm.

Despite these increases in the throughput rate, there is an inherent upper limit on the throughput rate associated with the feedback of the decisions to the feedback filter, i.e. during decision directed mode, the data inputs to the FBF are generated sequentially and, therefore, the iteration period of the DFE and training algorithm will always be lower bounded by the time to perform the multiplication between the last DFE decision and the first FBF coefficient.

The convergence speed of the DFE using both the LMS and DLMS algorithms is illustrated in Figure 32.3 for a HIPERLAN type receiver and a (6,6) DFE. The results are shown assuming a stationary channel profile with multipath components (0.4, 1.0, 0.3, 0.2, 0.1 0.05). The convergence curves are averaged over 500 independent realisations.

Using a step size of 0.01, the speed of convergence of the DLMS algorithm is marginally faster than the LMS. However, for a step-size value of 0.02 the conver-

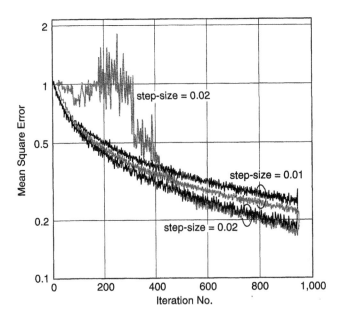

Figure 32.3 Performance of a DFE using the LMS and DLMS training algorithms.

gence of the LMS algorithm is faster and more stable. On average the convergence of the LMS algorithm is faster than the DLMS algorithm.

In addition to satisfying the throughput requirements, it is desirable to minimise area and power. Optimisation of these parameters can be performed at several levels in the design cycle, but significant reductions of both area and power can be achieved at an early stage, through the design of efficient algorithms and by simplifying the type of arithmetic operations that are required [34]. A frequently used example of this latter approach is the sign-LMS algorithm, which quantises the estimation error to a single sign bit [35]. This eliminates a multiplication operation from each coefficient update.

Recently, work has been carried out to eliminate multiplication operations by non-uniformly quantising the input data to a transversal filter. Originally applied in a spread spectrum system to suppress a narrowband interferer [36], the method has also been used in decision feedback equalisation [33,37]. In this approach the input data is approximated by the closest sum of N signed power-of-two terms as

$$x(n) = \sum_{m=1}^{N} s(m)2^{g(m)} \quad (m) = -1, 0, 1 \tag{32.7}$$

Using this approximation, any multiplication involving the input data sample and a twos complement operand can be performed using shifters and adders. In the LMS algorithm, the input data is used in both the DFE and the coefficient updates enabling a multiplierless realisation. Although this introduces additional quantisation noise which is dependent on the number of power-of-two terms used, rather than the wordlength [38], the impact on performance is negligible [33,37]. The number of power-of-two terms that are used will be influenced by the number of signal constellation points used in the signalling alphabet, for example, in a system using high level QAM, a larger number of power-of-two terms is likely to be necessary than in a system using QPSK. In Figure 32.4, the simulated performance of a DFE for a HIPERLAN receiver is shown when using input data that is both quantised uniformly and when quantised to the closest 2-SPT space.

An exponential power delay profile was assumed for the channel model. Two different values of normalised root mean square delay spread, 1.5 and 1.0, were considered. The data was quantised using both 6-bits and 8-bits. The performance curves when using the uniformly quantised data are shown as solid lines; for the 2-SPT input data, only points are plotted.

It is clear from Figure 32.4 that the performance when using 8-bit 2-SPT data is virtually identical to the case of uniformly quantised data. There is a small loss in performance when using 6-bit input data.

Reductions in the number of multiplications in the DFE have also been obtained by reformulating each complex multiplication so as to use three real multiplications, with five add/subtract operations [39]. An example of this type of transformation is

$$(x + \mathrm{j}y)(g + \mathrm{j}h) = ((x + y)g + (g + h)(-y)) + ((g - h)(-x) + (x + y)g)\mathrm{j} \tag{32.8}$$

Although extra additions are necessary, these consume significantly less area than is saved by eliminating a multiplier. This transformation can be extended to a complex filtering operation by generating three independent data streams and moving the

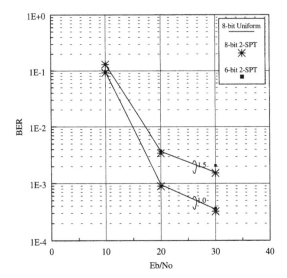

Figure 32.4 Decision feedback equaliser using 2-SPT quantised input data and uniformly quantised input data.

final addition operations outside of the filter. Although this eliminates a complete filter, the throughput rate of an adaptive filter is reduced, because modified sets of coefficients must be obtained after each update. This loss can be eliminated, by introducing a delay in the coefficient updates as in the DLMS algorithm. The coefficient update is then performed in parallel with the computation of the DFE output.

32.4. CONCLUSIONS

In this chapter, the current status of wireless LAN standards has been described. In order to provide a competitive alternative to wired networks, high data rate transmission is required. At the proposed rates, reception is very likely to be impaired by ISI as a result of multipath propagation. The successful deployment of these systems will, therefore, require the development of sophisticated receivers, to counteract the ISI. These must be both area and power efficient to support portable use. The use of equalisation has been focused on in this chapter, which is currently of interest for HIPERLAN type receivers. For HIPERLAN type-II systems, currently under development, multi-carrier transmission is the favoured approach.

By careful design of the DFE and training algorithm, the potential for considerable savings in area and power exists. However, achieving high throughput is more difficult to accomplish because of the inherent throughput bottleneck associated with the generation and feedback of an error signal. Several approaches were described to overcome this by the introduction of approximations to the algorithm. These do not appreciably alter the behaviour of the algorithm, but allow pipelining to be carried out, which would not otherwise be possible. Two methods of reducing silicon area were also described, (i) by reformulating complex multiplications to use just three real multiplications, and (ii) non-uniform quantisation of input data to allow multiplier implementation using only a limited number of shifters and adders.

REFERENCES

[1] Radio Equipment and Systems. HIgh PErformance Radio Local Area Network (HIPERLAN) Type 1; Functional Specification.
[2] IEEE 802.11. Wireless LAN medium access control (MAC) and physical layer (PHY) specifications. Draft Standard IEEE 802.11, P802.11/D5.0; July 1996.
[3] Summary of the Report and Order on Unlicensed NII Devices in the 5GHz Band. ET Docket No. 96–102, January 1997.
[4] Hawley, G. T., System considerations for the use of xDSL technology for data access. *IEEE Communications Magazine*, **35** (3), 56–60, 1997.
[5] Samueli, H., Daneshrad, B., Joshi, R. B., Wong, B. C. and Nicholas, H. T., A 64-tap CMOS echo canceller/decision feedback equalizer for 2B1Q HDSL transceivers. *IEEE Journal Selected Areas of Communications*, **9** (6), 839–847, 1991.
[6] Aldinger, M., Multicarrier COFDM scheme in high bit rate radio local area networks. *Proceedings of the IEEE Conference On Wireless Computer Networks*, September 1994.
[7] Bingham, J. A. C., Multicarrier modulation for data transmission: an idea whose time has come. *IEEE Communications Magazine*, May, pp. 5–14, 1990.
[8] Jones, A. E., Wilkinson, T. A. and Barton, S. K., Block coding scheme for reduction of peak to mean envelope power ratio of multicarrier transmission schemes. *Electronics Letters*, **30** (25), 2098–2099, 1994.
[9] Qureshi, S., Adaptive equalisation. *Proceedings of IEEE*, **73**, 1349–1384, 1983.
[10] Benveniste, A. and Goursat M., Blind equalizers. *IEEE Transactions in Communications*, **32**, 871–883, 1984.
[11] Falconer, D. D. and Ljung, L., Application of fast Kalman estimation to adaptive equalisation. *IEEE Transactions in Communications*, **26** (10), 1439–1446, 1978.
[12] Carayannis, G., Manolakis, D. G. and Kalouptsidis, N., A fast sequential algorithm for least-squares filtering and prediction. *IEEE Transactions in Acoustics, Speech and Signal Processing*, **31**, 1394–1402.
[13] Nix, A., Marvill, M. J., Wilkinson, T., Johnson, I. and Barton, S., Modulation and equalisation considerations for high performance radio LANs (HIPERLAN). In *Proceedings of the IEEE Personal, Indoor and Mobile Radio Communications Conference*, **III**, 964–969, 1994.
[14] Clarkson, P. M., *Optimal and Adaptive Signal Processing*, CRC Press, Boca Raton, FL, 1993.
[15] Ling, F. and Proakis, J. G., A generalised multichannel least squares lattice algorithm based on sequential processing stages. *IEEE Transactions in Acoustics, Speech and Signal Processing*, **32**, 381–389, 1984.
[16] Ling, F., Manolakis, D. and Proakis, J. G., A recursive modified Gram-Schmidt algorithm for least squares estimation. *IEEE Transactions in Acoustics, Speech and Signal Processing*, **34**, 829–835, 1986.
[17] Lin, D. W., On the digital implementation of the fast Kalman algorithm. *IEEE Transactions in Acoustics, Speech and Signal Processing*, **32** (5), 998–1005, 1984.
[18] Benedetto, G. and Biglieri, G., Non-linear equalisation of digital satellite channels. *Journal Selected Area of Communications*, **1**, 57–62, 1983.
[19] Proakis, J. G., *Digital Communications*, 2nd edn, McGraw-Hill, New York, 575–576.
[20] Forney G. D., Maximum-likelihood sequence estimation of digital sequences in the presence of intersymbol interference. *IEEE Transactions in Information Theory*, **18**, 363–378, 1972.

[21] Lee, W. U., A maximum-likelihood sequence estimator with decision feedback equalisation. *IEEE Transactions Communications*, **25** (9), 971–979, 1977.

[22] Chen, S., Mulgrew, B. and Mclaughlin, S., Adaptive Bayesian decision feedback equaliser based on a radial basis function network. In *Proceedings of IEEE International Conference on Communications*, 1267–1271, 1992.

[23] Chen, S., Gibson, G. J. and Cowan, C. F. N., Adaptive channel equalisation using a polynomial perceptron structure. *Proceedings of IEE*, **137** (5), 257–264, 1990.

[24] Sun, Y., Nix, A. and McGeehan, J. P., HIPERLAN performance analysis with dual antenna diversity and decision feedback equalisation. In *Proceedings of IEEE Vehicular Technology Conference*, 3, 1549–1553, 1996.

[25] Tellado-Mourelo, J., Wesel, E. K. and Cioffi, J. M., Adaptive DFE for GMSK in indoor radio channels. *IEEE Journal Selected Areas of Communications*, **14** (3), 492–501, 1996.

[26] Liu, Q., Scott, K. E., Wan, Y. and Sendyk, A. M., Performance of decision feedback equalisers with dual antenna diversity. *Proceedings of IEEE Vehicular Technology Conference*, 637–640, 1993.

[27] Parhi, K. K., Algorithm transformation techniques for concurrent processors. *Proceedings of IEEE*, **77** (12), 1879–1895, 1989.

[28] Benesty, J. and Duhamel, P., A fast exact least mean square adaptive algorithm. *IEEE Transactions in Signal Processing*, **40** (12), 2904–2920, 1992.

[29] Clark, G. A., Mitra, S. K. and Parker, S. R., Block implementation of adaptive digital filters. *IEEE Transactions in Circuits and Systems*, **28** (6), 584–592, 1983.

[30] Gatherer, A. and Meng, T. H. Y., A robust adaptive parallel DFE using extended LMS. *IEEE Transactions in Signal Processing*, **41**, 1000–1005, 1991.

[31] Long, G., Ling, F. and Proakis, J. G., The LMS algorithm with delayed coefficient adaptation. *IEEE Transactions in Acoustics, Speech and Signal Processing*, **37** (9), 1397–1405, 1989.

[32] Meyer, M. D. and Agrawal, D. P., A high sampling rate delayed LMS filter architecture. *IEEE Transactions in Signal Processing*, **40**, 727–729, 1993.

[33] Perry, R., Bull, D. R. and Nix, A. An adaptive DFE for high data rate applications. *Proceedings of IEEE Vehicular Technology Conference*, 686–690, 1996.

[34] Chandrakasan, A. P., Potkonjak, M., Mehra, R. and Brodersen, R. W., Optimizing power using transformations. *IEEE Transactions in Computer-Aided Design of Integrated Circuits and Systems*, **14** (1), 12–31, 1995.

[35] Eweda, E., Analysis and design of a signed regressor LMS algorithm for stationary and non-stationary adaptive filtering with correlated Gaussian data. *IEEE Transactions in Circuits and Systems*, **37**, 1367–1374, 1990.

[36] Li, D. and Lim, Y. C., Multiplierless realization of adaptive filters by nonuniform quantization of input signal. *Proceedings of IEEE International Symposium on Circuits and Systems*, 457–460, 1994.

[37] Chen, C., Khoo, K. and Willson, A. N., Jr., A simplified signed powers-of-two conversion for multiplierless adaptive filters. In *Proceedings of IEEE International Symposium Circuits and Systems*, 364–367, 1996.

[38] Perry, R., Bull, D. and Nix, A., Mean quantisation noise power for an N-SPT analogue-to-digital converter. *IEE Electronics Letters*, **32** (4), 295–296, 1996.

[39] Bull, D. R., Efficient IQ filter structure for use in adaptive equalisation. *IEE Electronics Letters*, **30** (24), 2018–2019, 1994.

33

Orthogonal Multi-rate DS-CDMA for Multimedia Mobile/Personal Radio

F. Adachi, K. Ohno, M. Sawahashi and A. Higashi

33.1. INTRODUCTION

The major services provided by current cellular mobile radio systems are limited to voice, facsimile, and low bit rate data services. The next generation systems must be flexible enough to support a variety of data services. Much higher-rate data services (e.g. 128 kbps and more) will become increasingly important to provide the simultaneous transmission of voice and image or computer data and also to allow a personal terminal to efficiently access the Internet. Requirements for the next generation of personal/mobile cellular systems can be summarized as follows: voice services of the same quality as the fixed network, multi-rate data services, flexible system deployment and minimum transmit power as well as increased link capacity.

The cellular direct sequence code division multiple access (DS-CDMA) technology [1]–[4], is the most promising candidate for next generation cellular personal/mobile radio systems and is now being extensively investigated [5]–[7]. Cellular DS-CDMA has a number of advantages: it allows universal single-frequency reuse and thus, diversity (or soft) handover can be relatively easily implemented compared with cellular time division MA (TDMA) systems (in cellular TDMA systems, surrounding cell sites use different carrier frequencies); each user can transmit low-to-high-rate data in a flexible manner; and multipath fading can be effectively exploited through rake combining to enhance transmission performance.

This chapter presents an overview of an orthogonal multi-rate coherent DS-CDMA system [8]–[10] as designed for multiple access in multimedia mobile/personal communications. Radio link design is described in section 33.2. In section 33.3 the average bit error rate (BER) performance of the power controlled reverse link is evaluated by computer simulation. Using the computer simulation results of BER performance and TPC error, the cell capacity under multi-user and multi-cell environments is theoretically evaluated.

33.2. ORTHOGONAL MULTI-RATE DS-CDMA

33.2.1. Overview

Radio link parameters of the orthogonal multi-rate DS-CDMA system are listed in Table 33.1. The orthogonal multi-rate DS-CDMA system has the following features:

1. two-layer spreading code assignment for asynchronous cell site operation;
2. orthogonal multirate forward link;
3. pilot-symbol aided coherent forward/reverse links,
4. fast transmit power control (TPC) based on signal-to-interference plus noise (SIR) measurement;

Table 33.1 Radio link parameters

Access	Coherent DS-CDMA
Carrier spacing	1.25/5/10/20 MHz
Chip rate	0.96/3.84/7.68/15.36 Mcps
Code-channel	Variable spreading factor
Spreading codes	Orthogonal multi-rate codes
Short	Pseudo random codes
Long	
	Data: QPSK
Modulation	Spreading: BPSK (forward)
	QPSK (reverse)
	Convolutional coding
Channel coding	($R = 1/3$, $K = 7$)
Diversity	Antenna + rake
Power control	SIR-based closed loop + Open loop

5. variable rate transmission within a single code-channel using blind rate detection;

6. two-stage concatenated channel coding.

To flexibly utilize limited radio frequency resources, multiple bandwidths can be used. Four different bandwidths (or spreading code chip rates) are currently being considered: 1.25, 5, 10, and 20 MHz (the corresponding chip rates are 0.96, 3.84, 7.68, and 15.36 Mcps). The bandwidth is selected according to the peak data rate requested by each user. For voice and voice-band data communications using a portable terminal, battery life is a primary concern and thus, it is advantageous to design portable terminals that can use a narrow bandwidth, e.g. 1.25 or 5 MHz, in order to reduce the receiver front-end signal processing complexity and power consumption. The entire 20 MHz radio spectrum may be shared by all systems using different bandwidths (of course, collaborative, precise transmit power control among all systems is required [11]).

Flexible system deployment is achieved by the use of two-layer code assignment (see Figure 33.1) and SIR-based TPC. A near-infinite number of spreading codes for the forward link is generated using a combination of short and long codes. Long codes are unique to each cell site and all the cell sites are assigned the same set of short orthogonal multi-rate codes; traffic channels in a cell are distinguished by different orthogonal short codes with different code lengths (in this paper, we call these codes the orthogonal multi-rate codes and their generation method is described later). On the other hand, the personal terminals use their own unique long codes to spread the transmitting signals.

SIR-based TPC [12] obviates the careful arrangement of cell site positions and transmit powers since the transmit power of each personal terminal is adaptively controlled to the minimum value necessary to maintain the targeted SIR at the best cell site which is selected by diversity (or soft) handover. Furthermore, new cell sites

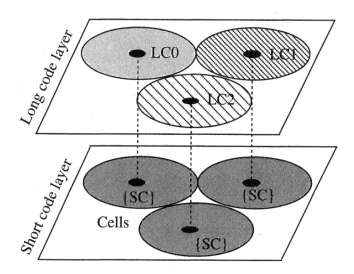

Figure 33.1 Spreading code assignment.

can be freely added as required, for example, to increase system capacity or to cover some dead spot area. When a particular cell site becomes overloaded, the transmit power of that cell site can be temporally reduced to distribute the traffic to surrounding cells.

33.2.2. Radio Link Design

The forward and reverse links are symmetrical except for spreading modulation. A transmission block diagram is shown in Figure 33.2. Transmitting user data sequence is divided into a sequence of 10 ms data blocks. To each 10 ms data we add the signalling data as well as the frame overhead (i.e., the frame number indicator, cyclic redundancy code (CRC) for error detection and the convolutional code tail bits) to form the frame data. Frame data are encoded by a rate 1/3, constraint length $K = 7$ convolutional code, block interleaved, and mapped over 20 slots, each 0.5 ms long. Data modulation is quaternary phase shift keying (QPSK). As shown in Figure 33.3, each slot contains the 2-symbol pilot needed for channel estimation and SIR measurement at the receiver. A TPC command (equivalent to one QPSK symbol) is sent at every slot in the forward and reverse link. Since the repetition interval for a 2-symbol pilot and the TPC command is 0.5 ms, both coherent rake combining and TPC can cope with fading up to a maximum Doppler frequency of 240 Hz with only a slight performance degradation.

The spreading modulation uses BPSK for the forward link (cell site-to-mobile) and offset QPSK in the reverse link (mobile-to-cell site). The forward radio link can use orthogonal spreading because all code-channels are synchronous. One way to flexibly provide data services of different rates is to use multiple orthogonal code-channels simultaneously in the forward link according to the data rate requested. However, this requires multiple Rake combiners at the mobile receivers, each combiner belonging to a different code channel. This increases the complexity of the mobile receiver. The spreading short codes used for the forward link are orthogonal

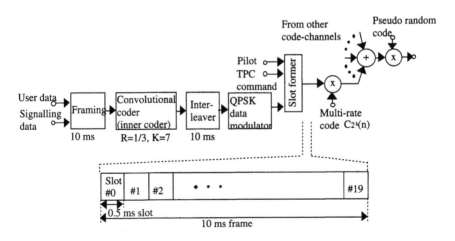

Figure 33.2 Forward link transmission structure.

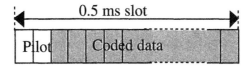

Figure 33.3 Slot structure.

multi-rate codes with different processing gains (or code lengths) and are generated by the recursive generation method described later [13]. After combining all code channels, the pseudo random code unique to the cell site is multiplied so that the interference given to other cells is made noise-like. A single code-channel can be used to transmit data at various rates. In the reverse radio link, on the other hand, all users are asynchronous so only random long spreading codes unique to each user are used (no short codes are necessary). If the spreading code with processing gain of 128 chips/symbol is used, the QPSK symbol rate is 60 ksps when the chip rate is 7.68 Mcps. This code-channel can transmit user data at rates of up to 32 kbps in addition to the overheads totalling 4 kbps.

Because of different requirements on channel quality for speech and data transmission, concatenated channel coding [14] is applied (see Figure 33.4). Outer coding is a burst error correction code using a high-rate Reed–Solomon code. Inner coding (1/3-rate convolutional code) is shared by speech and data.

33.2.3. Tree-structured Orthogonal Multi-rate Codes

Let $\mathbf{C}_{2^k}$ denote the set of 2^k binary spreading codes, $\{C_{2^k}(n)\}_{n=1}^{2^k}$, where $C_{2^k}(n)$ is the row vector of 2^k elements (k is a positive integer); it is generated from $\mathbf{C}_{2^{(k-1)}}$ as

$$
\mathbf{C}_{2^k} =
\begin{bmatrix}
C_{2^k}(1) \\
C_{2^k}(2) \\
C_{2^k}(3) \\
C_{2^k}(4) \\
\vdots \\
C_{2^k}(N-1) \\
C_{2^k}(N)
\end{bmatrix}
=
\begin{bmatrix}
C_{2^{k-1}}(1) & C_{2^{k-1}}(1) \\
C_{2^{k-1}}(1) & \bar{C}_{2^{k-1}}(1) \\
C_{2^{k-1}}(2) & C_{2^{k-1}}(2) \\
C_{2^{k-1}}(2) & \bar{C}_{2^{k-1}}(2) \\
\vdots \\
C_{2^{k-1}}(N/2) & C_{2^{k-1}}(N/2) \\
C_{2^{k-1}}(N/2) & \bar{C}_{2^{k-1}}(N/2)
\end{bmatrix}
\tag{33.1}
$$

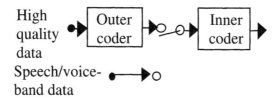

Figure 33.4 Concatenated channel coding for high quality data transmission.

where $\bar{C}_{2^{k-1}}(n)$ is the binary complement of $C_{2^{k-1}}(n)$ and is the row vector of 2^{k-1} elements. As a result, tree-structured spreading codes are generated recursively as shown in Figure 33.5. Starting from $C_1(1) = 1$, a set of 2^k spreading codes are generated at the k-th layer from the top. The code length of the k-th layer is 2^k chips and can be used for the code channels transmitting the user data at the rate of 32 kbps when $k = 7$ for 7.68 Mcps.

Code properties

1. Generated codes of the same layer constitute a set of Walsh functions and they are orthogonal. (It should be noted that, of course, code generation can start from any k-th layer using a set of 2^k orthogonal codes other than Walsh functions, e.g. orthogonal Gold codes.)
2. Any two codes of different layers are also orthogonal except where one code is the mother code of the other.

For example, all of $C_{64}(2)$, $C_{32}(1)$, $C_{16}(1)$, $C_8(1)$, $C_4(1)$, and $C_2(1)$ are mother codes of $C_{128}(3)$, and so are not orthogonal to $C_{128}(3)$. From this observation, we can easily find that if $C_{16}(1)$ is assigned to a user requesting an 8 times higher data rate service, all 14 codes $\{C_{32}(1), C_{32}(2), C_{64}(1), \ldots, C_{64}(4), C_{128}(1), \ldots, C_{128}(8)\}$ generated from this code cannot be assigned to other users requesting lower rates; in addition, mother codes $\{C_8(1), C_4(1), C_2(1)\}$ of $C_{16}(1)$ cannot be assigned to users requesting higher rates (of course, the use of codes of excessively short code lengths may also be impractical). This is a restriction imposed by the proposed tree-structured code assignment in order to maintain orthogonality; however, the use of $C_{16}(1)$ code is equivalent to the simultaneous use of 8 consecutive codes $\{C_{128}(1), \ldots, C_{128}(8)\}$ in the case of orthogonal multicode assignment.

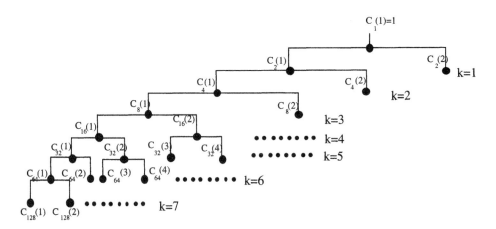

Figure 33.5 Generated code tree.

33.2.4. Coherent Rake Combining

A block diagram of a pilot symbol-aided coherent rake receiver is shown in Figure 33.6. D-branch antenna diversity is assumed. The multipath signal received on the d-th diversity antenna is despread ($d = 0, 1, \ldots, D - 1$) and resolved into L path components by a matched filter (or a bank of synchronous correlators). A channel estimator estimates the fading complex envelope of the l-th path at time $t = ((P - 1)/2 + kN)T$ using the received signal pilot samples (pilot symbols are assumed to be all 1s) associated with the k-th slot ($l = 0, 1, \ldots, L - 1$; $k = 0, 1, \ldots, 19$), where T is the coded symbol duration, P is the number of pilot symbols per slot, and N is the slot length (in this paper, $P = 2$ and $N = 30$ for 7.68 Mcps and processing gain of 128). The channel estimate is denoted by $\hat{\xi}_{d,l,k}$, which is given by

$$\hat{\xi}_{d,l,k} = \frac{1}{P} \sum_{m=0}^{P-1} r_{d,l}(m + kN) \qquad (33.2)$$

where $r_{d,l}(m + kN)$ is the l-th path component received on the d-th antenna and sampled at time $t = (m + kN)T$. The fading complex envelope at the coded data symbol position $t = (m + kN)T$, $m = P, P + 1, \ldots, N-1$, is estimated by a first order interpolation [15]:

$$\hat{\xi}_{d,l,k}(m) = \left(1 - \frac{m - (P - 1)/2}{N}\right)\hat{\xi}_{d,l,k} + \frac{m - (P - 1)/2}{N}\hat{\xi}_{d,l,k+1}. \qquad (33.3)$$

The received signal samples of resolved paths are then weighted by the complex conjugate of the channel estimates to remove random phase variations due to fading and summed coherently. The rake and diversity combiner output sample at time $t = (m + kN)T$ is given by

$$\hat{r}_k(m) = \sum_{d=0}^{D-1} \sum_{l=0}^{L-1} r_{d,l}(m + kN)\hat{\xi}_{d,l,k}^*(m), \quad m = P, P + 1, \ldots, N-1 \qquad (33.4)$$

where $*$ denotes the complex conjugate. Each rake combiner output sample is decomposed into the real component and imaginary component which are the soft decision samples associated with first and second bits of m-th QPSK symbol in the k-th slot, respectively, and then de-interleaved for subsequent soft decision Viterbi decoding. Note that for the forward link, decision on the power control command should be done immediately upon reception of the first data symbol of each slot.

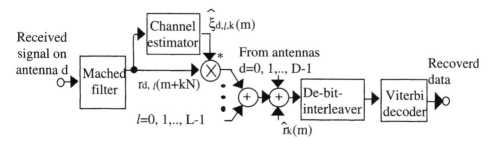

Figure 33.6 Coherent rake receiver.

33.2.5. Variable Transmission within a Single Code-channel

The processing gain of the code is changed according to the peak transmission rate requested. Once the processing gain is determined, the same orthogonal short code is used (note that for the reverse link, only the long code is used). During communication, the data rate may vary from frame to frame. For variable rate transmission within a single code-channel, discontinuous transmission is adopted to reduce the interference to other users. This is important for speech transmission. The blind rate detection using CRC decoding [16] requires no rate information but the receiver must know the set of possible rates. While Viterbi decoding of 1/3-rate convolutionally encoded data is performed at the receiver, the surviving trellis path ending at the "zero-state" at each possible end bit position is traced back to recover the data sequence and is error detected by a CRC code. If no error is detected, that recovered data sequence is declared to be a correct data sequence.

33.3. PERFORMANCE EVALUATION

The receiver can be a totally coherent receiver. Coherent chip synchronization as well as coherent rake combining is achieved based on pilot symbol interpolation. It was shown [17] that the average bit error rate (BER) performance using coherent chip synchronization is about 1 dB superior to that using noncoherent chip synchronization. In this simulation, we assume perfect chip synchronization. Since the link capacity is in general limited by the reverse link, we consider here only the reverse link. By computer simulation, we first evaluate BER performance of the power controlled reverse link and then, using the simuation results, the relationship between the cell capacity and mobile transmit power is evaluated under multi-user and multi-cell environments.

33.3.1. BER Performance

The computer simulation results for average BER performance are shown in Figure 33.7 as a function of the median signal energy per bit-to-interference plus noise spectrum density ratio $\bar{E}_b/I_0$ per branch for $f_D = 120$ Hz. Two-branch antenna diversity reception ($D = 2$) can reduce the required $\bar{E}_b/I_0$ by about 3 dB at the average BER=0.1 % required for high quality speech. As the number of paths increases, the required $\bar{E}_b/I_0$ increases due to the estimation error resulting from the reduced power per path. However, this performance degradation can be almost compensated by the decrease in other-cell interference resulting from the reduction in transmit powers when more than two paths exist.

High-quality data transmission may require BER better than 10^{-6}. To achieve this quality, concatenated channel coding using RS(40, 34) as an outer code is applied [18]. The simulated BER performance is shown in Figure 33.8 for two-branch diversity reception ($D = 2$) and two-path rake ($L = 2$). The required value

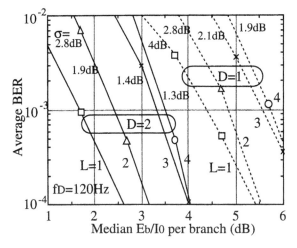

Figure 33.7 BER performance.

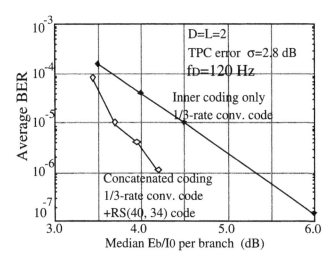

Figure 33.8 Concatenated channel coding for high-quality data transmission.

of $\bar{E}_b/I_0$ is only about 1.5 dB larger than that for speech transmission, which requires BER $= 0.1\%$.

33.3.2. Average Mobile Transmit Power

For the DS-CDMA reverse link, precise power control is essential to avoid the near/far problem. Using the SIR ratio as the indicator of the received signal quality instead of the received power, the transmit power of a personal terminal can be adaptively controlled according to the traffic variations in the cell. The computer simulation results for the excess transmit power over the single user case are plotted in Figure 33.9 as a function of the traffic load defined as the number of users per cell normalized by processing gain, C/pg, with the number of propagation paths as a parameter (pg = spreading factor × the inverse of channel coding rate).

It can be clearly seen that the transmit power of the personal terminals decreases with decreased traffic load. Unnecessarily large transmit power radiation is avoided. Also seen is the effect of multipaths. Increasing the number of resolvable paths provides higher order path diversity and thereby reduces the transmit power due to the smaller fluctuation in received power at the cell site. The power reduction is quite dramatic with two paths, and with more than three paths the reduction is much less. Accordingly, the use of spreading bandwidth sufficiently wide to resolve at least two paths greatly contributes to the reduction in transmit power, thereby increasing the capacity due to the resulting decrease in other-cell interference.

33.3.3. Approximate Cell Capacity

Here, we obtain the cell capacity for 7.68 Mcps and processing gain of 128. An approximate formula for the Erlang capacity of the reverse link is given in [19] assuming that the effects of multipath fading are completely removed. Although the use of antenna diversity reception can increase the equivalent number of resolvable paths, multipath fading effects still remain. Taking this effect into account, the expression for cell capacity C/pg can be refined as [10]

$$\frac{C}{pg} \approx \frac{1 - \eta^{-1}}{\bar{E}_b/I_0} \frac{1}{\exp\{(\sigma\beta)^2/2\} + f \times g} \frac{\alpha(N)}{\rho} \tag{33.5}$$

where $\eta = I_0/N_0$ represents the allowable maximum interference power density relative to the background noise power density, σ is the TPC error in dB ($\beta = \ln 10/10$),

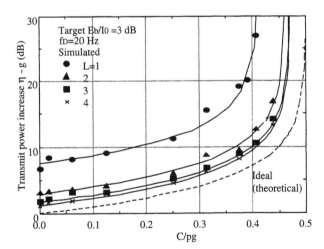

Figure 33.9 Transmit power increase.

g is the transmit power increase from the single path, no fading case, ρ is the proportion of transmission time, $\alpha(N)$ is the cell sectorization factor (N is the number of sectors per cell site), and f is the intercell-to-intracell interference power ratio when multipath fading does not exist. The first term of equation (33.5) is the capacity of an isolated cell, the second is the reduction factors due to TPC imperfection and other-user interference, and the third factor is the improvement factor owing to the variable rate (discontinuous) transmission and cell sectorization. For ideal sector antennas, $\alpha(N) = N$. However, for practical antennas, radiation patterns of adjacent sectors may overlap. This produces intersector interference within the same cell site so $\alpha(N) = 0.75N$ [20]. Since $E_b/N_0 = \eta(E_b/I_0)$, we notice that η is equivalent to the allowable increase in the transmit power from the single user case. When TPC is used, the average transmit power increases and accordingly, the average interference power from other cells increases. However, the value of f given in [21] takes into account the effects of distance dependent path loss and shadow fading, while neglecting the effects of multipath fading. The effect of multipath fading is included in g and thus, the average interference power ratio is given by $f \cdot g$.

Based on the BER performance of Figure 33.7, the transmit power is computed from equation (33.5) for a cell of 2 km radius taking into account the distance dependence path loss (dB) $= 6.2 + 38 \log_{10} r$, where r is the distance between the cell site and a mobile terminal; log-normal shadowing is assumed. The values of f and shadowing margin $d_{Q=5\%}$ at 5% outage were obtained by a Monte Carlo simulation assuming 19 hexagonal cells and that mobile users are distributed uniformly. In the simulation, we generated Gaussian random variable x_j representing the log-normal shadowing and the path loss between the cell site j and a mobile terminal, and obtained the cumulative distribution of

$$\delta = 10^{x/10} = \min_{\{j\}} 10^{x_j/10} \tag{33.6}$$

The results are $d_{Q=5\%} = 2$ dB and $f = 0.71$ for shadowing with standard deviation of 8 dB. The value of g was obtained theoretically from $g = DL/(DL - 1)$ [20]. It was found that the theoretical value of g is very close to the computer simulation results except for $L = 1$ [10].

The results are plotted in Figure 33.10 for $d_{Q=5\%} = 2$ dB, $f = 0.71$, the receiver noise figure NF $= 5$ dB, total transmit/receive antenna gain $G = 6$ dB including cable loss, $\rho = 0.5$ (thus, the average data rate $= 16$ kbps) and $N = 1$ (omni cell). For obtaining the average transmit power, we used the median-to-average conversion factor $\Delta = [DL/(DL - 1)] \exp\{(\sigma\beta)^2/2\}$ [20], which is listed in Table 33.2. From Table 33.2, we find that the rake gain is 2.25 dB for $L = 2$ and 3.17 dB for $L = 4$ when $D = 2$; these small rake gains in terms of transmit power are attributed to the fact that TPC can reduce the channel variations due to fading. It is understood from Figure 33.10 that narrowband DS-CDMA significantly increases other-cell interference if antenna diversity reception is not used ($D = 1$) or significantly decreases cell capacity if the transmit power is limited. This is because quite large transmit power is sometimes required to compensate the deep fades experienced by the cell site receiver. In this respect, wideband DS-CDMA that can resolve at least two paths is necessary. The effect of cell sectorization is shown in Figure 33.11. Cell sectorization

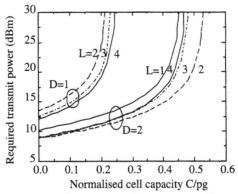

Figure 33.10 Maximum transmit power vs. cell capacity.

Table 33.2 Median-to-average conversion factor Δ

	$L = 1$	$L = 2$	$L = 3$	$L = 4$
$D = 1$		3.91 dB	2.22 dB	1.67 dB
$D = 2$	3.91 dB	3.91 dB	1.67 dB	1.02 dB

can increase the cell capacity if the mobile transmit power is limited or equivalently can reduce the transmit power if the required capacity is given.

The relationship between the cell capacity and the cell radius R is plotted in Figure 33.12 where the maximum transmit power is limited to 30 dBm. Up to $R = 4$ km, there is almost no reduction in capacity because the link is interference-limited.

33.4. CONCLUSIONS

The orthogonal multi-rate DS-CDMA access architecture that can support multimedia services and also flexible system deployment was described. To reduce the

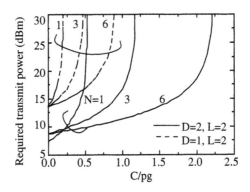

Figure 33.11 Effects of sectorization.

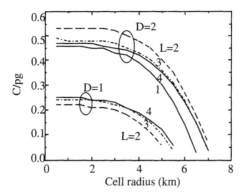

Figure 33.12 Link capacity versus cell radius.

required E_b/I_0 and accordingly increase the cell capacity, pilot-symbol aided coherent chip synchronization and rake combining are employed. User data of various rates can be transmitted by the use of a single code-channel, thus only a simple rake receiver is required at the mobile transceiver. DS-CDMA systems are interference limited systems. The proposed coherent orthogonal multi-rate DS-CDMAs periodically transmit known pilot symbols to enhance transmission performance and to increase cell capacity by the use of interference cancellation and adaptive array antenna diversity.

A field test is being conducted for the orthogonal multi-rate DS-CDMA described in this paper. Three base stations and six mobile stations are involved. Data at up to 2 Mbps was transmitted under real multipath fading environments to measure BER, and also to investigate the dependence of TPC imperfection on the bandwidth (or chip rate) and the speed of a mobile terminal [22]. The effect of chip rates (or the spreading bandwidths) was measured to show [23] that the mobile transmit power can be reduced by increasing the spreading bandwidth when TPC is applied. We are also investigating interference cancellation [24] and adaptive antenna array diversity [25].

REFERENCES

[1] Viterbi, A. J., *CDMA: Principles of Spread Spectrum Communications*, Addison-Wesley, Reading, MA, 1995.
[2] Khono, R., Meidan, R. and Milstein, L., Spread spectrum access methods for wireless communications. *IEEE Communications Magazine*, **33**, 58–67, 1995.
[3] Gilhousen, K. S., Jacobs, I. M., Viterbi, A. J., Weaver, L. A. and Wheatley, C. E., III. On the capacity of a cellular CDMA system. *IEEE Transactions on Vehicular Technology*, **VT-40**, 303–312, 1991.
[4] Padovani, R., Reverse link performance of IS-95 based cellular systems. *IEEE Personal Communications*, **1** (3), 28–34, 1994.
[5] Baier, A., Fiebig, U.-C., Granzow, W., Koch, W., Teder, P. and Thielecke, J., Design study for a CDMA-based third generation mobile radio system. *IEEE Journal on Selected Areas of Communication*, **SAC-12**, 733–743, 1994.

[6]   Andermo, P. G. and Brismark, G., CODIT, a testbed project evaluating DS-CDMA for UMTS/FPLMTS. *Proceedings of the IEEE Vehicular Technology Conference*, Stockholm, 21–25, June 1994.

[7]   McFarlane, D. A. and Allkins, M. D., Validation of advanced CDMA concepts for UMTS and FPLMTS. In *Proceedings of the IEEE Vehicular Technology Conference*, Stockholm, 36–40, June 1994.

[8]   Adachi, F., Ohno, K., Sawahashi, M. and Higashi, A., Multimedia mobile radio access based on coherent DS-CDMA. *Proceedings of the 2nd International Workshop on Mobile Multimedia Communications*, A.2.3, Bristol University, UK, 11–13, 1995.

[9]   Ohno, K., Sawahashi, M. and Adachi, F., Wideband coherent DS-CDMA. In *Proceedings of IEEE VTC*, 779–783, Chicago, U.S.A., July 1995.

[10] Adachi, F., Ohno, K., Higashi, A., Dohi, T. and Okumura, Y., Coherent multicode DS-CDMA mobile radio access. *IEICE Transactions on Communications*, **E79-B**, 1316–1325, 1996.

[11] Adachi, F., Reverse link capacity of DS-CDMA mobile radio with SIR-based transmit power control. *IEICE Transactions on Fundamentals*, 2028–2034, 1996.

[12] Dohi, T. and Sawahashi, M., Power control by employing interference power for DS/CDMA (in Japanese). IEICE Technical Report, RCS94-99, 63–68, October 1994.

[13] Adachi, F., Sawahashi, M. and Okawa, K., Tree-structured generation of orthogonal spreading codes with different lengths for forward link of DS-CDMA mobile radio. *IEE Electronics Letters*, **33**, 27–28, 1997.

[14] Lin, S. and Costello, D. J., Jr., *Error Control Coding: Fundamentals and Applications*, Prentice-Hall, Englewood Cliffs, NJ, 1983.

[15] Sampei, S. and Sunaga, T., Rayleigh fading compensation for QAM in land mobile radio communications. *IEEE Transactions on Vehicular Technology*, **42**, 137–147, 1995.

[16] Okumura, Y. and Adachi, F., Variable rate data transmission with blind rate detection for coherent DS-CDMA mobile radio. *IEE Electronics Letters*, **32**, 1865–1866, 1996.

[17] Sawahashi, M. and Adachi, F., Decision-directed coherent delay-lock PN tracking loop for DS-CDMA. In *Proceedings of the Virginia Technical Symposium on Wireless Personal Communications*, 10-1–10-8, Virginia Tech., USA, May 31–June 2, 1995.

[18] Taguchi, T., Higashi, A. and Suda, H., Decoding performance of concatenated code in DS-CDMA mobile radio (in Japanese). IEICE Technical Report, SST, March 1995.

[19] Viterbi, A. M. and Viterbi, A. J., Erlang capacity of a power controlled CDMA system. *IEEE Journal on Selected Areas of Communication*, **11**, 892–900, 1993.

[20] Ohno, K. and Adachi, F., Reverse-link capacity and transmit power in a power-controled cellular DS-CDMA system (in Japanese). *IEICE Transactions on Communications*, **J79-B-II**, 17–25, 1996.

[21] Viterbi, A. J. and Viterbi, A. M., Other-cell interference in cellular power-controlled CDMA. *IEEE Transactions in Communication*, **42**, 1501–1504, 1994.

[22] Dohi, T., Okumura, Y., Higashi, A., Ohno, K. and Adachi, F., Experiments on coherent multicode DS-CDMA. *IEICE Transactions on Communications*, **E79-B**, 1326–1332, 1996.

[23] Dohi, T., Okumura, Y. and Adachi, F., Effects of spreading chip rates on transmit power distribution in power-controlled DS-CDMA reverse link. *IEE Electronics Letters*, **33**, 447–448, 1997.

[24] Sawahashi, M., Miki, Y., Andoh, H. and Higuchi, K., Pilot symbol-aided coherent multistage interference cancellar using recursive channel estimation for DS-CDMA mobile radio. *IEICE Transactions on Communications*, **E79-B**, 1262–1270, 1996.

[25] Tanaka, S., Sawahashi, M. and Adachi, F., Pilot symbol-assisted decision-directed coherent adaptive diversity array for DS-CDMA mobile radio. *Proceedings of Wireless '97*, Calgary, Canada, 444–451, July 1997.

34

Performance and Capacity of a Voice/ Data CDMA System with Variable Bit Rate Sources

Narayan B. Mandayam, Jack Holtzman and Sergio Barberis

34.1. INTRODUCTION

As is well known, the direct sequence CDMA system capacity is interference limited [1]. In addition, a major problem in CDMA systems is the near–far effect due to which a nearby interferer can disrupt the reception (at the base station) of a highly attenuated desired signal. As a result, power control of the users in the system is required. Therefore the actual quality of service (QoS) of a CDMA system is clearly a function of the number of active users as well as the tightness of the power control in the system.

In [2], an approach was presented to study the Erlang capacity of a voice only CDMA system. For applications where multimedia services need to be provided, there is a need for better understanding of the nature of the relationships between the different QoS requirements of the services in the system and the capacity of each type of service. Voice traffic is well understood through many years of telephone experience. It is well characterized by a Poisson arrival process and a usual holding time of a few minutes. Also, the statistical characterization of the output of the variable bit rate codecs, developed recently, are starting to be available [3,4]. A particular concern is the uncertainties with the data (or actually, any non-voice) traffic. These uncertainties not only affect the capacity, but also have implications upon control, i.e., both admission control and the control of set-up calls. Note that uncertainties arise here not only because of variances of certain random variables, but also because of uncertainties of the means and variances. For example, in fax service, there is a high level of uncertainty about the information content of a page after source coding (from a few tens of kbytes for a text page only and a few hundred kbytes for a highly complex picture, depending on the compression ratio) [5]. A modeling methodology is needed which facilitates sensitivity studies of uncertain parameters. The work presented here provides a step in studying the interplay between different types of services in a DS-CDMA system. We develop a generalized Erlang capacity formulation for voice and generalized data. The calculations are reduced to using means and variances. Since data, but not voice, can be delayed, emphasis is on voice outage probability in anticipation of control strategies for data. One such data control strategy, using related analysis methodology, is studied in [6].

The analytic simplicity is achieved using an approximation from [7]. This approximation was also used in [8], in a somewhat different way, for an Erlang capacity type problem for packet CDMA. We illustrate the formulation with a CELP speech coder and fax. Section 34.2 provides the general formulation and section 34.3 gives the source models for voice and fax. Illustrative numerical results are in section 34.44 and concluding comments in section 34.5.

34.2. SYSTEM MODEL

We consider a DS-CDMA network consisting of numerous mobile subscribers communicating with one base station (corresponding to a single cell). The subscribers can request voice calls as well as data/fax calls. If the channel bandwidth is W, the approximated measure of the interference rejection capability is given by the processing gain $G_{v,i} = W/R_{v,i}$ for voice calls that are using, in that instant, the bit rate $R_{v,i}$. Similarly, the corresponding processing gain for data/fax is given as $G_{d,i} = W/R_{d,i}$ where $R_{d,i}$ is the ith bit rate allowed for data/fax. Furthermore, we will denote by K_v and K_d the number of ongoing voice calls and data calls respectively. Since we are considering an integrated voice-data/fax wireless system, typically each type of service may require a different QoS (BER or FER) and, for this reason, we will denote the received bit energy due to each bit in a packet by $E_{b,v}$ and $E_{b,d}$ for voice and data/fax respectively. This corresponds to different power control requirements for the voice and data/fax services.

Remark. Data BER requirements can be very stringent, and in practice require additional coding and ARQ. While this is not explicitly included in the numerical examples, the methodology can easily incorporate that. The effect of other-cell interference can be included in the analysis by means of an f-factor as in [2].

34.2.1. Effect of Data on Voice

We will consider a desired voice user and treat all other voice users as well as data users as interferers. The selection of a desired voice user is arbitrary since, without loss of generality, we assume all voice users to belong to the same class of service. Therefore the bit energy over total interference density ratio of any voice user reflects (on the average) that of all other voice users as well. Then, based on the QoS requirement for voice calls, we will formulate a condition for outage of voice calls where the effect of data/fax calls is also considered. With direct sequence BPSK, we obtain the ratio of the despread bit energy to noise density at the desired user as

$$\left(\frac{E_b}{\eta}\right)_v = \frac{\frac{E_{b,vo}}{N_0}}{1 + \frac{1}{G_v'} \sum_{i=1}^{K_v-1} \alpha_{vi} \frac{E_{b,vi}}{N_0} + \frac{1}{G_d'} \sum_{i=1}^{K_d} \alpha_{di} \frac{E_{b,di}}{N_0}} \qquad (34.1)$$

where N_0 is the background noise power spectral density and η is the total interference spectral density ($\eta = N_0 + I_0$). The parameters G_v' and G_d' are defined by the ratio between the spreading bandwidth and the average bit rate: $G_v' = W/E[R_v]$ and $G_d' = W/E[R_d]$. The bit rate variability is in the αs. The variables α_{vi} for voice user i and α_{dj} for data/fax user j are defined as $\alpha_{vi} = R_{vi}^{(c)}/E[R_v]$ and $\alpha_{dj} = R_{dj}^{(c)}/E[R_d]$, where $R_{vi}^{(c)}$ and $R_{dj}^{(c)}$ are the current voice and data/fax gross bit rates of user i and j, respectively. In other words, α is the current gross bit rate normalized with respect to the average gross bit rate. In section 34.3 we will give specific examples of those random variables.

The QoS requirement for voice can be specified as $(\frac{E_b}{\eta})_v \geq \gamma_v$ where γ_v is the power protection ratio for the voice calls. This translates to a specific requirement on the bit error rate for voice calls. Typically $K_v - 1$ and K_d are random variables. For the case of voice calls, K_v can be modeled as a Poisson random variable with parameter λ_v/μ_v where the voice calls arrive into the system with rate λ_v and have an exponential service time with mean $\frac{1}{\mu_v}$. It can be easily shown that $K_v - 1$ is a random variable with probability mass function given by

$$\Pr\{K_v - 1 = n \mid K_v > 0\} = \frac{e^{\frac{-\lambda_v}{\mu_v}}(\frac{\lambda_v}{\mu_v})^{n+1}}{(n+1)!} \frac{1}{1 - e^{\frac{-\lambda_v}{\mu_v}}} \qquad (34.2)$$

with the mean and the variance given by

$$E[K_v - 1 \mid K_v > 0] = \frac{\lambda_v/\mu_v}{1 - \exp(-\lambda_v/\mu_v)} - 1 \qquad (34.3)$$

$$\text{Var}[K_v - 1 \mid K_v > 0] = \frac{\frac{\lambda_v}{\mu_v}}{(1 - e^{\frac{-\lambda_v}{\mu_v}})^2} \{1 - e^{(\frac{-\lambda_v}{\mu_v})} - \frac{\lambda_v}{\mu_v} e^{(\frac{-\lambda_v}{\mu_v})}\} \qquad (34.4)$$

For a perfectly power controlled system, $E_{b,v}/N_0$ and $E_{b,d}/N_0$ are constants, while for the case of imperfections in power control, these are modeled as lognormal random variables with mean m and standard deviation σ, that reflect the inaccuracy in the power control loops [2]. Based on the QoS requirement specified above, the condition for outage of voice calls can be rewritten as

$$Z < \gamma_v \qquad (34.5)$$

where

$$Z = \frac{E_{b,vo}}{N_0} - \beta_v \sum_{i=1}^{K_v-1} \alpha_{vi} \frac{E_{b,vi}}{N_0} - \beta_d \sum_{i=1}^{K_d} \alpha_{di} \frac{E_{b,di}}{N_0} \tag{34.6}$$

with $\beta_v = \gamma_v/G'_v$, and $\beta_d = \gamma_v/G'_d$. In general, we let $K_v - 1$ and K_d be random variables and we assume $E_{b,vi}/N_0$ and $E_{b,di}/N_0$ to be independent lognormal random variables with means m_v, m_d, and variances σ_v^2, σ_d^2, respectively. We assume that the random variables $E_{b,vi}/N_0$ are identically distributed with mean m_v and variance σ_v^2, because the power is varied such that the power to current bit rate ratio $P_{vi}/R_{vi}^{(c)}$ is kept constant, and similarly for data calls of each service class as well. If we define $X = \frac{E_{b,vo}}{N_0}$, $S_v = \beta_v \sum_{i=1}^{K_v-1} \alpha_{vi} \frac{E_{b,vi}}{N_0}$, and $S_d = \beta_d \sum_{i=1}^{K_d} \alpha_{di} \frac{E_{b,di}}{N_0}$, then the outage probability for voice calls can be written as

$$\Pr\{Z < \gamma_v\} = \Pr\{X < \gamma_v + S\} \tag{34.7}$$

where $S = S_v + S_d$. The outage probability in (34.7) can be rewritten as

$$\Pr\{Z < \gamma_v\} = \int_0^\infty \Pr\{X < \gamma_v + s \mid S = s\} f_S(s)\, ds \tag{34.8}$$

where we have conditioned on, and then averaged out, the random variable S. Furthermore, since X is a lognormal random variable, we have

$$\Pr\{Z < \gamma_v\} = 1 - \int_0^\infty Q\left(\frac{\ln(\gamma_v + s) - E(Y)}{\sqrt{\text{Var}(Y)}}\right) f_S(s)\, ds \tag{34.9}$$

where $Y = \ln X$ is a Gaussian random variable with mean $E[Y]$ and variance $\text{Var}(Y)$. The relationships for evaluating $E[Y]$ and $\text{Var}(Y)$ are given as [9]

$$E[X] = \exp(E[Y] + \frac{1}{2}\text{Var}(Y)) \tag{34.10}$$

$$\text{Var}(Y) = \ln((CV(X))^2 + 1) \tag{34.11}$$

where the coefficient of variation of X is defined as $CV(X) = \sqrt{\text{Var}(X)}/E[X]$. To calculate the outage probability in (9), we use the approximation used in [7], and it is given as

$$\Pr\{Z < \gamma_v\} = 1 - \frac{2}{3}Q\left(\frac{\ln(\gamma_v + \mu_S) - E(Y)}{\sqrt{\text{Var}(Y)}}\right)$$
$$-\frac{1}{6}Q\left(\frac{\ln(\gamma_v + \mu_S + \sqrt{3}\sigma_S) - E(Y)}{\sqrt{\text{Var}(Y)}}\right) - \frac{1}{6}Q\left(\frac{\ln(\gamma_v + \mu_S - \sqrt{3}\sigma_S) - E(Y)}{\sqrt{\text{Var}(Y)}}\right) \tag{34.12}$$

where μ_S and σ_S^2 are the mean and variance of the random variable S, respectively, and can be computed as [10]

$$\mu_S = \beta_v E[K_v - 1] m_v m_{\alpha_v} + \beta_d E[K_d] m_d m_{\alpha_d} \tag{34.13}$$

$$\sigma_S^2 = \beta_v^2 \{E[K_v - 1](\text{Var}(\alpha_v)\sigma_v^2 + \text{Var}(\alpha_v)m_v^2 + \sigma_v^2 m_{\alpha_v}^2)$$
$$+ \text{Var}(K_v - 1)(m_v m_{\alpha_v})^2\} + \beta_d^2 \{E[K_d](\text{Var}(\alpha_d)\sigma_d^2$$
$$+ \text{Var}(\alpha_d)m_d^2 + \sigma_d^2 m_{\alpha_d}^2) + \text{Var}(K_d)(m_d m_{\alpha_d})^2\}. \tag{34.14}$$

where m_{α_v}, $\mathrm{Var}(\alpha_v)$, and m_{α_d}, $\mathrm{Var}(\alpha_d)$, are the means and the variances of the RVs α_v and α_d, respectively. We will use (34.12) for evaluating the outage probability for voice calls for the different cases of data traffic considered. The approximation from [7] is exact when the function is a fifth degree polynomial and the random variable S is normal and it is fairly robust to deviations from those conditions. When the active numbers of voice and data calls get large and are Poisson, the central limit theorem for filtered Poisson processes [11] can be invoked to show that S tends to a normal random variable.

Remark. Note again that we are concentrating on the effect of data upon voice QoS (in anticipation of control strategies for data).

34.3. SOURCE MODELS AND SERVICES

To use the methodology of section 2, only simple static source models are needed, that is, the fractions of time that the source is at a given rate and power are needed.

34.3.1. Voice Service

The flexibility of CDMA-based communication systems can support variable bit rate transmission [12,13]. The removal of the fixed bit rate constraint allowed the development of a class of speech codecs able to provide different output bit rates depending on the time varying local characteristics of speech. The variable bit rate codec considered here is the Codebook Excited Linear Predictive (CELP) codec designed for the CODIT project [4]. This codec has seven different modes of operations which correspond to seven different output rates. Silence periods are encoded with the lowest number of bits (so as to provide comfort noise) while the active voice periods are encoded with an increasing number of bits as the information content of the "sound" increases. The source coding process, as well as the information transmission on the air interface, is organized in fixed periods of time called "frames". The frame length is 10 ms. The instantaneous bit rate is identified by the number of bits (i.e. the physical packet length) carried by each frame. In this case the radiated power (and then, the created interference) is directly related to the instantaneous bit rate and, therefore, no power exceeding what is strictly needed in a certain instant will be transmitted. The occurrence probability of each bit rate has been estimated by means of measurements. The measurement results for the case of voice only (no background noise) are shown in Table 34.1. It should be noted that there is another advantage allowed by the use of a variable rate codec. Due to the different sensitivity to the errors of the seven possible codec outputs, it is possible to design an unequal error protection coding scheme in order to cope with the impairments introduced by the channel. Here, for simplicity of illustration, we assume to have equal error protection provided by a channel code with rate ρ (as a consequence, the gross bit rate will be $R_i = r_i/\rho$). The formulation, however, permits the inclusion of unequal error protection.

Table 34.1 Bit rate statistics for voice: $E[r] = 6185$ bps, $\sigma_r = 5346$ bps

Mode	Net bit rate	$r/E[r]$	Occurrence probability
1	$r_1 = 400$	0.0647	$p_1 = 0.40$
2	$r_2 = 3200$	0.517	$p_2 = 0.05$
3	$r_3 = 8500$	1.37	$p_3 = 0.14$
4	$r_4 = 12\,500$	2.02	$p_4 = 0.15$
5	$r_5 = 7200$	1.16	$p_5 = 0.10$
6	$r_6 = 12\,000$	1.94	$p_6 = 0.12$
7	$r_7 = 16\,000$	2.59	$p_7 = 0.04$

34.3.2. Fax Service

In this section we will briefly describe the problems related to the fax service provision in a mobile communication network [14,15] and then we will make some reasonable assumptions about the requested bit rates and their occurrence probabilities. We will consider the G3 facsimile terminals that are the most commonly adopted. A "mobile fax" service can be provided in two different ways:

1. as a transparent (or real time) service;
2. as a non-transparent (or store and forward) service.

In case (1) the facsimile information is delivered to the destination before the call between the two fax terminals is released. In case (2) the facsimile message is stored temporarily in a buffer before its transmission over the radio interface; in this case the call established between the originating terminal and "the mobile network border" is usually released before the information is delivered to the final destination. In both cases appropriate interworking units (fax adaptors – FA) must be inserted in the network in order to allow the fax terminals to communicate over the air interface. The FA are located in the mobile station (between the fax terminal and the radio interface) and in the mobile control node (MCN), between the mobile radio environment and the public switched telephone network (PSTN), as shown in Figure 34.1.

Transparent Service

In this case the FA is simpler because its main task is to convert the G3 facsimile signals into a binary information stream suitable for the transmission over the radio interface. No protocol conversion is performed. The mobile network between the two

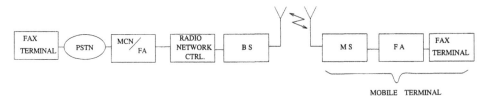

Figure 34.1 Mobile fax service.

FAs appears like an additive noise channel with a delay τ (see Figure 34.2). The only error protection available is provided by the error control coding designed for the mobile radio channel (in addition to the optional error control mode of G3 fax transmission). According to the T.30 Recommendation [16], the control messages are (usually) transmitted at 300 bit/s while the image information is transmitted at a maximum bit rate of 9600 bit/s (14 400 bit/s in recent machines and in case of very good connections). Here we assume that the transmission of the control messages and of the image information are performed at 300 bit/s and 9600 bit/s, respectively. We assume also that the mobile station is the sender and that it has to transmit a three-page document whose information content is about 60 kbytes each. On the basis of the signal sequence for the transmission of a three-page document [16,5,17] and taking into account the control message lengths, we obtain the following approximated occurrence probabilities of the two bit rates (see Table 34.2). In this probability estimation, we considered that the MS, excluding the image information transfer phase, is always transmitting at 300 bit/s. In fact, in a CDMA system, in order to maintain synchronization, enable power control and to facilitate the channel estimation, the bit rate cannot be reduced exactly to zero. At least a few hundred bit/s are always needed. As a consequence, the MS, having completed the call set-up procedure, is supposed to transmit always at 300 bit/s, excluding the periods of time related to the training sequences and image information transfer. For the same reason, when the MS is receiving a fax, on the uplink there is always a net bit rate of 300 bit/s. It should also be remarked that the probabilities of Table 34.2 have been estimated assuming that the message transmissions are always successful and neglecting the processing time at the two ends. But, taking into account the enormous variability of the information content of a page, the relative impact of the neglected parameters should be minor. The transparent approach to the fax transmission has the advantage of the simplicity and provides a "sense of immediacy" (the sender has always an immediate indication about the transmission result). The main disadvantage consists, of course, in a non-optimal efficiency in the use of the radio channel.

Non-transparent Service

In this case the fax adapter is more complicated because it has to be able to store the image information and to handle call management procedures. On the other hand, the higher complexity and cost of the FA allows a more efficient use of the radio

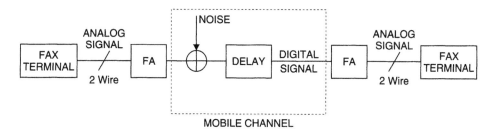

Figure 34.2 Mobile fax service (real time).

Table 34.2 Bit rate statistics for FAX: $E[r] = 8670$ bit/s, $\sigma_r = 2790$ bit/s

Net bit rate	$r/E[r]$	Occurrence probability
$r_1 = 300$	0.0346	0.1
$r_2 = 9600$	1.11	0.9

resource. In fact, once the call between the sender and the FA has been released, the fax information stored in the FA can be delivered like a constant bit rate file transfer, optimizing the use of the radio channel. It is, for example, likely to assume a constant bit rate transmission, higher or lower depending on the system load. Of course, in this case the information transmission over the radio channel is more reliable because, appropriate radio link protocols can be designed and implemented between the two FAs. The disadvantage consists in losing the "sense of immediacy" (the sender usually has no confirmation of the fax delivery to the destination before the call is cleared off). Taking account of the advantages and disadvantages of the two possible approaches to fax transmission over a mobile radio channel, a possible solution is to have available both options. In a mixed voice/fax service environment, an intelligent control entity can choose real-time transmission when the system is not heavily loaded by voice and store and forward otherwise. Therefore, by studying the variations (mean and variance) of the data/fax traffic, and its impact on voice capacity, we could develop protocols for admission, and after call set up.

34.4. NUMERICAL RESULTS

In this section we use the models previously described so as to provide some numerical examples of evaluation of outage probability for voice service in the case of an integrated voice/fax CDMA wireless system. Two main cases are considered:

1. K_d is a random variable, i.e. the number of fax call arrivals is modeled as a Poisson process with parameter λ_d/μ_d (transparent case);
2. K_d is deterministic, i.e. there is a fixed number of active calls on the uplink (non transparent or store and forward case).

For both situations the outage probability is evaluated first taking into account the source variability and then on the basis of the average source bit rates only (i.e. considering a continuous transmission performed at the average bit rate). This is made in order to compare the impact of the bit rate variance compared to the calculations based on the mean value only.

Remarks. For the illustrative results to be displayed we choose the voice power control error, σ_v, to be 2 dB. For data power control error, σ_d, we used both 2 dB and 4 dB to show the effect of possibly larger errors in controlling the data. Note that while we display results with fixed means for E_b/N_0s for varying parameters, the E_b/N_0s actually change with these parameters. This is because it is the E_b/ηs that are power controlled. The selection of the E_b/N_0s for multiple services is considered as a power optimization problem in [18,19].

1. *Real time fax.* If K_d is a Poisson random variable with parameter λ_d/μ_d the mean and the variance of the random variable S are given are given by equations (34.13) and (34.14), where $E[K_d] = \mathrm{Var}(K_d) = \lambda_d/\mu_d$. This could be a likely model when the fax is provided as transparent service. Both voice and fax call arrivals are Poisson and, depending on the average offered fax traffic, by means of (12) we can get different plots of the outage probability for voice versus λ_v/μ_v, as shown in Figure 34.3. Simulation results are also shown for reference. The approximations work extremely well when the number of voice users is large (for probability of outage 1% and larger) owing to the fact that the random variable S is approaching a normal random variable. The same kind of evaluation has been made on the basis of the average bit rates only. In this case the transmission is supposed to be continuous (i.e. $\alpha_v = \alpha_d = 1$, $m_{\alpha_v} = m_{\alpha_d}$, and $\mathrm{Var}(\alpha_v) = \mathrm{Var}(\alpha_d) = 0$) and the results are depicted in Figure 34.4. Comparing the two diagrams it turns out that, the outage probability based on the mean bit rates only is a little more optimistic than that one evaluated taking into account the bit rate variations. An interesting two-dimensional capacity diagram is illustrated in Figure 34.5. In case

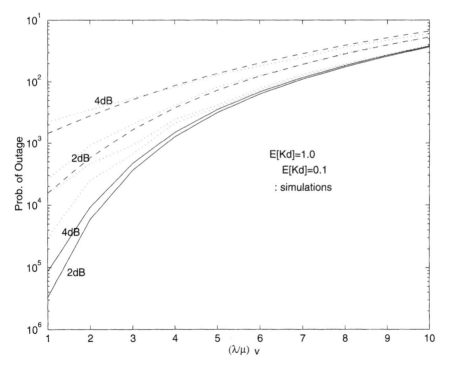

Figure 34.3 Outage probability for voice with transparent fax (VBR). The plots are shown for $\sigma_d = 2$ dB, and $\sigma_d = 4$ dB. $m_v = m_d = 20$ dB, $\sigma_v = 2$ dB, $W = 1$ MHz, and $\gamma_v = 5.0$.

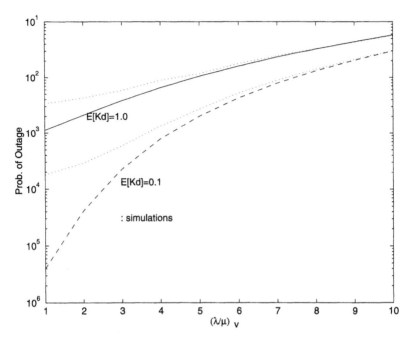

Figure 34.4 Outage probability for voice with transparent fax (average BR). The average bit rate is chosen for the voice calls and for the data for this evaluation. $m_v = m_d = 20$ dB, $\sigma_v = 2$ dB, $\sigma_d = 4$ dB, $W = 1$ MHz, and $\gamma_v = 5.0$.

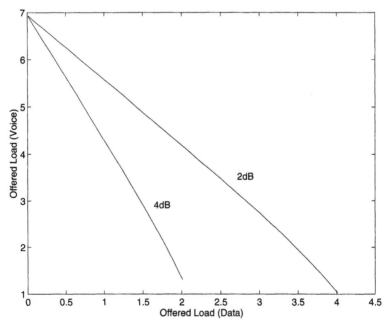

Figure 34.5 Two-dimensional capacity. The outage probability for voice calls is fixed at 0.01. The offered load of voice calls λ_v/μ_v is shown versus that for fax calls λ_d/μ_d (transparent case). The plots are shown for $\sigma_d = 2$ dB, and $\sigma_d = 4$ dB. $m_v = m_d = 20$ dB, $\sigma_v = 2$ dB, $W = 1$ MHz, and $\gamma_v = 5.0$.

of an outage probability for voice of 1%, the diagram shows the relationship between the average voice traffic and the average fax traffic.

2. *Store and forward fax*. In this case, we assume that the allowed number of active fax users is controlled by the number of voice users. Hence, K_v and K_d are deterministic. Therefore, $E[K_v] = K_v$, $E[K_d] = K_d$, and $\text{Var}(K_v) = \text{Var}(K_d) = 0$. This could be a reasonable model of the situation where the fax is provided as a non-transparent (or store and forward) service: on the basis of the current number of voice calls, if we do not want to exceed a certain outage probability value, we are not allowed to provide more than K_d contemporaneous fax calls. Also in this case, the results based on the average bit rate only are more optimistic than those obtained considering the variable bit rate sources (see Figures 34.6 and 34.7).

34.5. CONCLUSIONS AND FUTURE DIRECTIONS

In this chapter we presented an approach for the evaluation of the outage probability of voice calls in an integrated voice/data DS-CDMA wireless network with VBR sources. The variability of the sources has been described by the occurrence

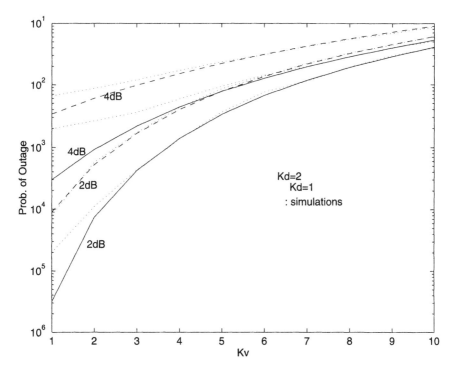

Figure 34.6 Outage probability for voice with non-transparent fax (VBR). The plots are shown for $\sigma_d = 2$ dB, and $\sigma_d = 4$ dB. $m_v = m_d = 20$ dB, $\sigma_v = 2$ dB, $W = 1$ MHz, and $\gamma_v = 5.0$.

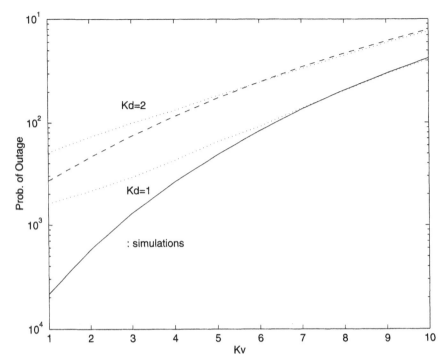

Figure 34.7 Outage probability for voice with non-transparent fax (average BR). The average
  bit rate is chosen for the voice calls and for the data for this evaluation. $m_v = m_d = 20$ dB,
  $\sigma_v = 2$ dB, $\sigma_d = 4$ dB, $W = 1$ MHz, and $\gamma_v = 5.0$.

probabilities of the various possible bit rates. Numerical examples were presented
for a CELP variable rate speech coder and fax (both transparent and non-transparent). Imperfections in the power control loop were considered in the analysis.
Although the results are presented for equal error protection and same bit energy to
background noise density ratio for all the bit rates, the methodology can be readily
extended to the unequal error protection case. Two different cases have been studied: the case of Poisson arrivals of fax calls and the case of a fixed number of fax
calls (determined on the basis of the current number of voice calls). For both cases
the results obtained with the variable bit rate sources have been compared with
those based on the average bit rates only. As expected, evaluations based on the
average bit rates provide more optimistic results. Finally, a two-dimensional capacity diagram provides the average voice traffic versus the average fax traffic for fixed
voice outage probability. Two dimensional capacity diagrams like this can be used
for algorithms responsible for the admission control of new calls in mixed service
environments. Another example of such capacity diagrams is in [20]. Extensions of
the analysis include other cell interference, unequal error protection and quality
requirements for the different service classes.

Future directions related to this work include considering situations of optimum
power control as well as as retransmission schemes for data messages. While this
work has considered the signal-to-interference ratio as a QoS measure to define

Erlang capacity, an alternate definition is possible based on total interference power (as in [2]). It is shown in [21] that the above outage condition is more stringent and hence ensures feasibility of the power control problem. An approximate analysis that takes into account imperfect power control, activity factors and different quality requirements is also presented.

Another aspect of further investigation is the guarantee of individual QoS requirements for voice and data. For delay-tolerant (but high fidelity) data services, one possible approach is to design retransmission strategies that take into account voice QoS. In [6], a protocol for retransmission of data messages in an integrated voice/data CDMA system is analyzed. Voice calls are given preemptive priority over data calls and the protocol for data calls is controlled by the quality of service (QoS) requirement for the voice calls. The QoS criterion for voice calls is specified by a target SIR (signal-to-interference ratio) that each voice call has to meet in order to avoid outage. The admission of data calls into the system is achieved by means of a "persistence state" (as in [22]) that controls the probability of each data user gaining access to the system. Imperfections in power control for both the voice and data messages are included while studying the effect of the protocol on average performance measures like throughput and blocking probability for data service. Stability conditions are derived in [23], delineating parameter regions for stable operation.

In real cellular environments, it is not the instantaneous SIR dropping below the threshold that determines outage. It is, in fact, a certain duration of time of the SIR below a threshold that determines outage. The notion of minimum duration outages has been introduced in [24] where an outage is defined as the signal-to-interference ratio (SIR) going below a threshold and staying below longer than a certain "minimum duration". Such a notion of outage includes the time correlation in the signals, which is important in real systems owing to mobility, fading and power control.

REFERENCES

[1] Gilhousen, K. S. *et al.*, On the capacity of a cellular CDMA system. *IEEE Transactions on Vehicular Technology*, **40** (2), 303–312, 1991.

[2] Viterbi, A. M. and Viterbi, A. J., Erlang capacity of a power controlled cellular CDMA system. *IEEE Journal on Selected Areas in Communications*, **11** (6), 892–900, 1993.

[3] Berruto, E. and Sereno, D., Variable-rate for the basic speech service in UMTS. In *Proceedings of VTC '93*, 520–523.

[4] Cellario, L., Sereno, D., Giani, M., Blocher, P., Hellwig, K., A VR-CELP implementation for CDMA mobile communications. In *Proceedings of ICASSP '94*, I-281–I-284.

[5] McConnell, K. R., Bodson, D. and Schaphorst, R., *FAX, Digital Facsimile Technology and Applications*, 2nd edn, Artech House, London, UK, 1992.

[6] Mandayam, N. and Holtzman, J., Analysis of a simple protocol for short message data in an integrated voice/data CDMA system. In *Proceedings of MILCOM'95*, San Diego, CA, **3**, 1160–1164, 1995.

[7] Holtzman, J. M., A simple accurate method to calculate spread-spectrum multiple access error probabilities. *IEEE Transactions on Communications*, **40** (3), 461–464, 1992.

[8] Kou, C. F., Packet CDMA performance with imperfect power control. Master of Engineering Thesis, McGill University, Montreal, Canada, November 1994.

[9] Johnson, N. and Kotz, S., *Continuous Univariate Distributions*, Vol. I, Wiley, New York, 1970.

[10]   Feller, W., *An Introduction to Probability Theory and its Applications* Vol. I, 2nd edn, Wiley, New York, 1957.

[11]   Snyder, D. and Miller, M., *Random Point Processes in Time and Space*, 2nd edn, Springer-Verlag, Berlin, 1991.

[12]   Baier, A., *et al.*, Design study for a CDMA-based third-generation mobile radio system. *IEEE Journal on Selected Areas in Communications*, **12** (4), 733–743, 1994.

[13]   TIA/EIA/IS-95. Mobile station-base station compatibility standard for dual-mode wideband spread spectrum cellular system, July 1993.

[14]   Recommendation ETSI/GSM 03.45. Technical realization of facsimile group 3 – transparent, vers. 3.0.0, January 1995.

[15]   Recommendation ETSI/GSM 03.46. Technical realization of facsimile group 3 – non-transparent, vers. 3.0.0, January 1995.

[16]   Recommendation T.30 of ITU/TSS. Procedures for document facsimile transmission in the general switched telephone network, March 1993.

[17]   Halton, K. C., The group 3 facsimile protocol. *British Telecom Technology Journal*, **12** (1), 61–69, 1994.

[18]   Yun, L. C. and Messerschmitt, D. G., Power control for variable QoS on a CDMA channel. In *Proceedings of MILCOM '94*, 178–182, 1994.

[19]   Sampath, A., Kumar, P. S. and Holtzman, J., Power control and resource management for a multimedia CDMA wireless system. In *Proceedings of PIMRC'95*, Toronto, Canada, September 1995.

[20]   Guo, N. and Morgera, S. D., The grade of service for integrated voice/data wireless DS-CDMA networks. In *Proceedings of Supercomm/ICC '94*, 1104–1110, 1994.

[21]   Sampath, A., Mandayam, N. B. and Holtzman, J., Erlang capacity of a power controlled integrated voice and data CDMA system. In *Proceedings of IEEE VTC'97*, Phoenix, Arizona, May 1997.

[22]   Mandayam, N. B., Chen, P.-C. and Holtzman, J. M., Minimum duration outage for cellular systems: a level crossing analysis. In *Proceedings of IEEE VTC*, **2**, 879–883, Atlanta, April, 1996.

[23]   Sampath, A., Mandayam, N. B. and Holtzman, J., Analysis of an access control mechanism for data traffic in an integrated voice and data CDMA system. In *Proceedings of IEEE VTC '96*, **3**, 1448–1452, Atlanta, GA, May 1996.

[24]   Viterbi, A. J., Capacity of a simple stable protocol for short message service over a CDMA Network. In Blahut, R., *Communications and Cryptography*, Kluwer Academic Publishers, Boston MA, 423–429, 1994.

35

Software Radio Design for Next Generation Radio Systems

Peter B. Kenington

35.1. INTRODUCTION

Although there is an increasing trend towards regional (and to some degree world-wide) air interface standardisation, there are still a multiplicity of current and proposed standards and consequently different handset designs. The European trend towards harmonisation, in the shape of the GSM,[1] DECT,[2] DCS1800 and TETRA[3] standards has shown subscribers the benefits of a single handset with Europe-wide

---

[1] Groupe Speciale Mobile (or Global System for Mobile Communications, as it has become known)
[2] Digital Equipment Cordless Telephone
[3] Trans-European Trunked Radio

Insights into Mobile Multimedia Communication
ISBN 0-12-140310-6

(and to some degree, worldwide) mobility. Even if these standards become dominant throughout the world (which is unlikely, given the strong penetration of US-derived systems, such as AMPS,[4] DAMPS,[5] IS-95[6] and PDC[7] – which is similar to DAMPS in many respects), there will still be many different standards for different applications. Even in Europe, there are standards for cellular (GSM and DCS1800), cordless phone (DECT) and private mobile radio (TETRA). A single handset capable of interworking with all of these systems would therefore require either three or four different circuit cards (with a common case, display and keyboard), or a *flexible architecture radio* (often referred to as a "software" radio, although there are reconfigurable hardware options available for its implementation).

The increasing convergence of mobile satellite and terrestrial cellular technologies inevitably leads to the requirement for a combined handportable terminal. At present this is envisaged as a dual-mode, or even a dual radio handset, due to the current difficulties in achieving a satisfactory performance from shared RF hardware and flexible DSP processing. Considerable research is under way in order to solve these problems and such items should be available in the medium, if not short term.

This chapter will examine some of the current and potential future configurations which could constitute a flexible architecture radio, with particular reference to the RF and baseband hardware aspects of the system design. It should be emphasised that single-band software radios have been built, including a number in which the author has been involved (and some of which are in commercial production), so the technologies described here are not "risky" or "long-term", although some multiband and broadband issues have yet to be solved in a production radio.

35.2. SOFTWARE RADIO

A software radio is a transceiver in which, ideally, all aspects of its operation are determined using reconfigurable elements. This is usually thought of in terms of baseband DSPs, but FPGAs and other techniques are also possible. It is also usually assumed to be broadband in nature, as one of its principal applications is perceived to be in replacing the numerous handsets currently required to guarantee cellular (and in the future, satellite) operation worldwide. This is strictly speaking an extension of the basic "software radio" concept into that of a broadband flexible architecture radio, since the reprogrammability and adaptability aspects of operation do not depend upon multi-band coverage. It would be possible, for example, to construct a useful software radio which operated in the 800/900 MHz area of spectrum and which could adapt between TACS, AMPS, GSM, DAMPS, CT2 and PDC. The trend is, however, for multi-frequency operation (there are emerging handsets which

---

[4]American CDMA Mobile Phone System
[5]Digital American Mobile Phone System
[6]Code Division Multiple Access
[7]Personal Digital Cellular (formerly known as Japanese Digital Cellular, JDC)

cover both GSM and DECT, generally by employing separate hardware for each, in a common case).

There are many issues which must be addressed in determining if a software radio is realistic and also to what extent it is flexible. For example, it is possible to create a single-band software radio with a narrowband channel restriction relatively easily (indeed this was done at the University of Bristol in 1989, using a linear RF transmit/ receive chain and a DSP baseband architecture constructed around two TMS320C25 processors). Coping with wider channel bandwidths and operating in multiple bands in differing parts of the spectrum is much more difficult, but nevertheless essential for a combined terrestrial/satellite telephone.

An idealised software radio is shown in Figure 35.1. It has the following features:

- The modulation scheme, protocols, equalisation etc. for transmit and receive are all determined in software within the DSP.
- The "ideal" circulator, operating with ideal (perfect) matching between it and the antenna and power amplifier impedances, results in the elimination of the requirement for a diplexer in the radio. Since the diplexer is very much a fixed-frequency component within a radio, its elimination is a key element in any multi-band or even multi-standard radio. Note that the circulator would also have to be very broadband, which most current designs are not.
- The linear power amplifier ensures an ideal transfer of the RF modulation from the DAC to high-power suitable for transmission, with low, adjacent channel emissions (ideally none).

35.2.1. Required Specifications

In considering the ideal architecture, it is worth examining the specifications required of each element, and the consequent likelihood of technological advancement over the coming years making them a realistic proposition. To do this, it is necessary to make some assumptions about the types of modulation scheme (and in some cases access scheme) which the radio is likely to need to accommodate. If this is based

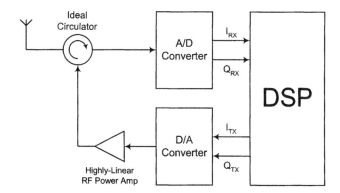

Figure 35.1 "Ideal" software radio architecture.

on current and currently proposed schemes for cellular and PMR worldwide, the specifications shown in Table 35.1 could be chosen.

The resulting specifications for the various elements of Figure 35.1 could then be derived as follows.

35.2.2. Antenna Specifications

Frequency range: 100 MHz–2.2 GHz
Gain: 0 dBi
Radiation pattern: Omnidirectional

It is clear that such an antenna has yet to be invented, particularly when the requirement for its size to be kept to a minimum is taken into account, although a number of antennae could be used (e.g. a wire antenna for VHF/UHF and a patch antenna for SHF). It may also be advantageous to steer the antenna beam away from the user, if only to prevent power being wasted in that direction.

35.2.3. Circulator

Frequency range: 100 MHz–2.2 GHz
Isolation (assuming perfect matching at each port): 47.5 dB

Table 35.1 Basic specifications for a handportable software radio

Parameter	Value	Notes
Frequency coverage	100 MHz–2.2 GHz	This would cover most PMR, cellular, PCN/PCS, mobile satellite and UMTS bands worldwide
Dynamic range	0 dBm–120 dBm*	This must not only cope with fading and in-band interferers, but *any* signals in the above frequency range.
Power output	1 W	This is decreasing as time progresses and health fears increase, but most systems still require this power level (many PMR systems require more).
Adjacent channel power	−75 dBc	This figure is slightly in excess of most known specifications in this area.
Power control range	70 dB	Determined by the GEONET FH-CDMA system.
Power ramping range	75 dB	DECT requires 71 dB and is probably the toughest requirement in this area.
Channel bandwidth	6 MHz	Based on the CDMA-only proposal from the ACTS FRAMES project (potential European UMTS standard).
Image rejection	60 dB	Based on an interpretation of the TETRA specifications.

*Based on an 18 kHz equivalent channel bandwidth – e.g. TETRA

Loss: 0 dB in both transmit and receive paths.
Power handling: 1 W

The isolation figure is calculated by assuming that the receiver intercept point is $+20$ dBm (*much* higher than current handsets), and that the stated intermodulation level (-75 dBc) is required. This results in a maximum receiver input level (for unwanted signals) of :

$$P_{Rx,\max} = \frac{IMR}{2} + IP_3 (\text{dBm}) \tag{35.1}$$

where *IMR* is the required intermodulation ratio in dB and IP_3 is the receiver intercept point in dBm.

Thus:

$$P_{Rx,\max} = \frac{-75}{2} + 20 = -17.5 \text{dBm} \tag{35.2}$$

A loss can be tolerated in the isolator if additional output power is available from the RF PA and the noise figure of the A/D is low enough to allow for the resultant added noise figure.

35.2.4. A/D Converter

Converter

Resolution: 22 or 24 bits (assuming that a 10 dB SINAD is required at maximum sensitivity – i.e. minimum signal level).
Noise figure: 1.4 dB (18 kHz equivalent modulation bandwidth with 10 dB SINAD at maximum sensitivity).
Linearity: $+20$ dBm equivalent IP_3.
Sample rate: 48 Msps (based on 4× oversampling of the channel and bandpass sampling)
Input bandwidth: DC/100 MHz–2.2 GHz

Anti-alias Filtering

Attenuation at Image frequency: 60 dB
Image frequency separation: 36 MHz (from upper frequency of wanted channel to lower frequency of alias product)
Centre frequency range (bandpass filter): 100 MHz–2.2 GHz

35.2.5. DSP

It is difficult to put precise estimates on the processing power/speed required of the DSP, as it depends heavily on the modulation format and type of receive architecture employed. Some estimates have been performed in the EC-funded ACTS FIRST[8] project, in particular for the demonstrator, and this will require multiple Analogue Devices "SHARC" processors, despite having a maximum channel bandwidth of one quarter of that indicated in Table 35.1.

The receive processing required is generally much more complex than that required on transmit; however, the peak data rate in each case is similar (assuming that "alias upconversion" can be employed). The data rate required from the transmitter can be halved if two (I/Q) channels are employed and a quadrature upconverter added (along with a local oscillator). In this case, two DAC channels are required, each operating at half of the original rate.

35.2.6. D/A Converter

Resolution: 13/14 bits (assuming that the IMD and noise floors are equal). In the case of many narrow-band modulation formats, the noise floor may prove to be a major problem, unless the conversion rate of the DAC can be altered dynamically to suit the modulation scheme. It this is not the case, then a much higher resolution may be required.

Linearity: $+20$ dBm equivalent IP_3 (assuming -1.5 dBm peak output signal power from the DAC).
Conversion rate: 48 MHz (based on $4\times$ oversampling of the channel and alias upconversion). If this is not used, the conversion rate climbs well into the gigahertz region.
Output bandwidth: DC/100 MHz–2.2 GHz

35.2.7. RF Power Amplifier

Output power: 1 W peak
Third-order intercept point: $+67.5$ dBm
Full power bandwidth: 100 MHz–2.2 GHz
Gain: 31.5 dB

Power efficiency is also an important factor in the acceptability of software radio. Current figures for quasi-linear systems, such as DAMPS and PDC, are around 40%, with traditional FM systems giving over 60%. A value in this region will therefore be expected (in terms of battery-life performance) from users of any new handset.

---

[8]A European Community funded project within the Advanced Communication Technologies and Services area. FIRST = Flexible Integrated Radio System Technology.

35.2.8. Specification Analysis

The above specifications are clearly not achievable with current technology; however, it is worth examining some of the more extreme specifications (in particular, the A/D converter) to examine what may be achievable in the future, and whether the problems highlighted will eventually be solved by technological advancement.

A/D Conversion

In a bandpass A/D sampling process, the analogue input bandwidth depends largely upon the sample-and-hold device and its operating speed. The operating speed of a sample and hold is given by

$$F_{S/H} = \frac{K_{S/H}}{L_T^2} \tag{35.3}$$

where L_T is the length of the transistor within the sample-and-hold and $K_{S/H}$ is a constant of proportionality. At present, the transistor length is decreasing by a factor of approximately 2 every four years, and this leads to an increase in operating frequency of approximately four times in that period. Based on this, the likely track-and-hold performance within the UMTS time frame (up to 2007) is:

- for a 16-bit ADC: 250 MHz;
- for a 14-bit ADC: 1000 MHz;
- for a 12-bit ADC: 5000 MHz.

This is clearly far short of the performance required in a good design, but will be appropriate for sampling at IF in most systems.

It is worth examining the minimum possible power consumption for an A/D converter to determine the lowest power requirement for the desired specification outlined above.

Generic Form of an A/D Converter

A generic form of A/D converter is shown in Figure 35.2 [1] and consists of an anti-alias filter, a sample-and-hold, a method of converting the (now constant level) analogue voltage into a digital word (the "quantiser") and finally a digital buffer. The quantiser element may be implemented in a wide variety of ways, including flash, successive approximation, sigma–delta, bandpass sigma–delta and subranging. The analysis presented below is independent of the implementation technology and is based on the requirements of the sample-and-hold element.

Note that although the anti-alias filter of Figure 35.2 is shown as a bandpass element, it may be a low-pass device in many applications.

For an A/D converter to take full advantage of the resolution available to it, the quantisation noise power must be lower than (or possibly equal to) the thermal noise power present at the converter input (within the required converter bandwidth). This noise level then sets the minimum possible peak input signal level (i.e. the minimum possible full-scale voltage) for the converter. Based on these levels, it is possible to

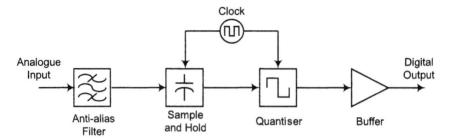

Figure 35.2 Generic form of an A/D converter for wideband digitisation at IF or RF.

calculate the converter power consumption from the minimum charging current of the converter input capacitance, which is in turn based upon the thermal noise floor (kT/C_i).

The signal to quantisation noise ratio (dynamic range) of an A/D converter is given by

$$D_C = 6n + 1.76 \text{ dB} \tag{35.4}$$

where n is the resolution (number of bits) of the converter. The thermal noise floor, on the other hand, effectively determines the limiting value of input capacitance for the converter's sample-and-hold device.

For the converter to function correctly, in terms of its accurately converting the input voltage it is presented with, this capacitance must be capable of being charged to the full-scale voltage of the converter, within its sampling interval (preferably well within). This results in an expression for the minimum power consumption for the converter of

$$P_i = \frac{kT}{t_s} 10^{(6n+1.76)/10} \text{watts} \tag{35.5}$$

For the results presented in Figure 35.3, the sampling rate of 48 MSPS was chosen as potentially representative for alias downconversion sampling of a future UMTS standard (FRAMES mode 2, with 4 times over-sampling). In addition, it was assumed that a 3 dB margin would be necessary between the quantisation noise floor and the thermal noise floor for the converter to operate over its full usable dynamic range (as defined by its resolution). Alternatively, this could be considered as a 3 dB noise figure for the converter, assuming that decimation is employed to reduce the quantisation noise floor.

Figure 35.4 shows how the situation alters with sample rate and hence what could be expected to happen if direct sampling of the RF waveform (as opposed to alias downconversion) were to be contemplated. This would have the advantage of making the anti-alias filter design more realistic and hence would be significantly closer to the ideal scenario.

Again considering the example of a 22-bit converter, but now examining the case where the sample rate is chosen to (just) Nyquist sample the complete bandwidth outlined in section 0, the resulting power consumption is around 500 W. This is clearly not realistic for a handset application and hence indicates that this

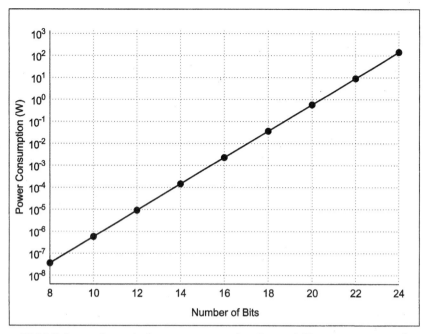

Figure 35.3 Minimum theoretical power consumption for an A/D converter, operating at 48 MSPS, for various resolution values.

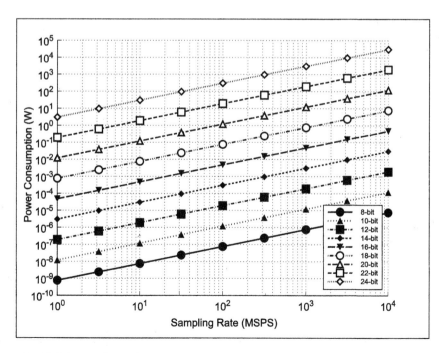

Figure 35.4 Minimum theoretical power consumption for an A/D converter over a range of sampling rates.

mechanism will not be realistic as a solution within software radio without a radical (and currently unknown) advance in technology.

Reducing the sampling rate by an order of magnitude, to accommodate, for example, a VHF software radio or wideband IF sampling for a higher frequency radio, still results in a power consumption of around 50 W. This is again clearly out of the question for a handportable unit.

It can be seen from this graph that for the 22-bit resolution chosen above, the theoretical minimum power consumption would be around 10 W when operating at 48 MSPS. This is clearly a high value for consideration to be included in a handset (in addition to the fast DSP processor(s), memory, linear transmitter etc.) and would probably be considered excessive in most designs. Any resolution above this value would obviously be far too inefficient for handset applications and would probably be considered excessive for base-station use (particularly the 24-bit device).

Anti-alias Filtering

The anti-alias filter is also a difficult component to envisage with current or forecast technologies. It is required to attenuate an "image" channel only 36 MHz away at, potentially, up to 2.2 GHz, by greater than 60 dB. This would be very difficult (possibly unachievable) with precision machined filters and even these would be unrealistically large. When the requirement to tune these filters over a 100 MHz to 2.2 GHz range is borne in mind, it clearly becomes impossible to achieve. Indeed the requirement is much greater than that of an equivalent diplexer, in terms of roll-off between pass and stop-bands.

When the small size requirements of handportable terminals are also taken into account, the result is a significant problem for this type of architecture.

Overall

Whilst the above are clearly not the only issues, they do highlight the difficulty, even in the long-term, of utilising an "ideal" software radio architecture.

It is therefore necessary to examine other architectures and/or restrictions in the specifications contained in Table 35.1, in order for software radio to become a reality in the short or medium term.

35.3. SINGLE-BAND FLEXIBLE TRANSCEIVER ARCHITECTURES

A single-band flexible transceiver is considerably easier to implement than its multi-band or broadband counterparts. This is due largely to the fact that conventional RF filtering may be employed in the receiver and diplexing between transmit and receive frequencies. Other areas are also made simpler and generally more power-efficient: for example the transmit RF power amplifier can be designed in class C with relative ease, and even class-A designs tend to be more efficient and simpler when implemented for narrow-band use.

A number of transceivers of this type have been implemented for dual-mode use in PMR/SMR applications with, for example, FM and a proprietary modulation scheme incorporated. This is usually to ensure that the product is backwards-compatible with existing FM systems.

35.4. MULTI-BAND AND BROADBAND FLEXIBLE TRANSCEIVER ARCHITECTURES

Going from single to multi-band operation introduces a number of significant difficulties, many of which are still in the "research" phase at present. It is possible, however, to place some restrictions on this general goal and simplify a number of the problem areas. An example of this is given in Figure 35.5, where a transmit/receive switch is employed to replace the diplexer, thus eliminating its inherent frequency inflexibility.

The use of a transmit/receive switch will obviously prevent the transceiver from operating with systems involving true real-time duplex (e.g. AMPS and TACS). Many of the more recent systems do not, however, intentionally employ simultaneous transmit and receive on, for example, the same timeslot (in a TDMA system). Other systems intentionally employ time-division duplex (e.g. CT2) and hence would work satisfactorily with a transmit/receive switch. It could even be a parameter considered in the design of new system standards, which would then allow an earlier adoption of flexible architecture radio systems.

35.5. CONCLUSIONS AND FUTURE DIRECTIONS

Software radio is becoming a realistic mechanism for integrating satellite and terrestrial systems into a single, cost-effective handset, although there is still much work to be done. It is clearly a better solution than attempting to house separate

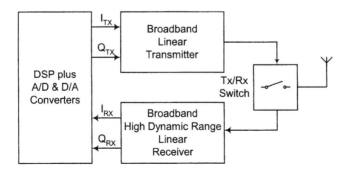

Figure 35.5 Broadband flexible transceiver architecture employing transmit/receive switching.

transceivers in a common case and much more acceptable than having separate handsets for each.

For radio equipment manufacturers, it provides a mechanism for achieving much greater economies of scale than those available at present, in handsets in particular (but also potentially in base stations); this will help to offset the perceived greater costs of software radio techniques initially. Even if a subscriber requires only a single-mode radio, this can be built as a software radio, but incorporating the software for only a single modulation format (with perhaps a lower capacity memory device, if this helps reduce costs). It also has implications for radio design itself, as teams of handset designers in each of the major markets will no longer be required; all development activity could easily be centralised at a single location in each company.

It can be seen therefore that the advent of software radio has some powerful and far-reaching consequences, for the user, the manufacturer and the network operator. It is already beginning to happen in some market niches (e.g. specialised PMR systems, telemetry radio systems), but will quickly spread to more mainstream applications.

ACKNOWLEDGEMENTS

The author is grateful to his colleagues at the Centre for Communications Research for help and advice in the preparation of this submission and in particular would like to thank Luc Astier, Ross Wilkinson, Barbara Sharp, Shixiang Chen, Paul Warr and David Bennett.

REFERENCE

[1] Wepman, J. A., Analog-to-digital converters and their applications in radio receivers, *IEEE Communications Magazine*, **33** (5), 39–45, 1995.

Part 8

Wireless LAN Technology for Future Multimedia Services

In the early 1990s, a growing number of proprietary radio LAN products appeared on the market. These products did not develop as fast as expected, partly due to a lack of standards, partly due to their limited capabilities, and partly due to poor spectrum assignment. In this section, the next generation of European, Japanese and North American wireless LAN standards are discussed. Carrier frequencies ranging from 5 GHz, 17 GHz and 60 GHz are investigated, together with the emerging use of infrared technology. The growing importance of wireless ATM is covered, both from the wireless physical layer perspective, and the underlying protocol and mobility issues.

In Chapter 36 the authors investigate the appropriate bit rates, frequency allocations and mobility issues for future multimedia applications. The chapter also includes a review of standardisation activities and a summary of the wireless LAN projects being performed under the European Union ESPRIT and ACTS research programmes. Chapter 37 focuses on Europe's emerging high-speed wireless LAN standard (HIPERLAN). A number of deployment scenarios and performance targets are discussed and the key features of the HIPERLAN standard are given. The chapter includes a summary of the future plans for the HIPERLAN standard and the possibilities for application outside of Europe.

Short range infrared data communications are a cost effective alternative to radio and are currently being standardised by Infrared Data Associated (IrDA). Chapter 38 gives an overview of the IrDA system and explains the working of the proposed protocols. As a rival to infrared, the use of the 60 GHz band has been suggested. Chapter 39 provides an analysis of the propagation characteristics (received power, r.m.s. delay spread and k-factor) as a function of the antenna types used. Coverage and outage results are produced and compared with those obtained in the HIPERLAN frequency band (5 GHz). Issues such as antenna alignment and the impact of furniture and office structures are also investigated.

To support mobility in future wireless ATM LANs, Chapter 40 investigates the design of a suitable handover protocol. The problems and challenges facing this area are discussed, and the author proposes the use of a multi-tier wireless cell clustering architecture.

36

High-performance Wireless LAN Developments for Future Multimedia Communications

Andrew R. Nix, Mark A. Beach, Cengiz Evci, M. Umehira and Masaharu Araki

36.1. INTRODUCTION

At present, wireless data transmission can be achieved via a dedicated wireless LAN or a data modem connected to the public cellular network. Cellular modems offer wide coverage but restrict the data rate to around 9.6 kb/s. The use of public cellular networks makes the modem expense to operate. The majority of wireless LAN products are currently based on the IEEE 802.11 standard and operate at 900 MHz in North America and 2.4 GHz in Europe. Relative to cellular modems, these products achieve far higher bit rates (typically 1–2 Mb/s) although they are restricted to short range indoor use. One advantage with a private indoor network is that there is no charge for the air time. The frequency at which these LANs operate is

Insights into Mobile Multimedia Communication
ISBN 0-12-140310-6

part of the industrial, scientific and medical (ISM) allocation and allows unlicensed use for products meeting certain bandwidth and power regulations. Since these wireless LANs do not use a dedicated spectrum, they must co-exist with the interference produced by other types of radio equipment. Perhaps more importantly, on-air data rates of 1–2 Mb/s are not sufficient for true multimedia and video-based applications. With wired Ethernet LANs offering up to 100 Mb/s and public ATM networks achieving up to 155 Mb/s it is clear that new dedicated, high speed, wireless access systems will be required in the near future. The market demand for wireless communications has grown rapidly and this is expected to continue into the future. This rapid growth is due to the significant advantages offered by terminal mobility.

36.2. THE TREND TOWARDS HIGH-SPEED WIRELESS NETWORKS

To address the issue of high speed public wireless networks, the Future Public Land Mobile Telecommunication Systems (IMT2000) standard has already been developed in the ITU. The standard sets the scope for the next (third) generation of wireless systems and is intended to ensure interoperability across different wireless services and mobility on a global scale. A similar specification for third-generation mobile systems has been developed in Europe under the ETSI standard for the Universal Mobile Telecommunication Systems (UMTS). However, the transmission speed of these systems is limited to 2 Mb/s, while optical fibre networks have transmission speeds as high as 155 Mb/s. It is obvious that there is still a large gap in terms of transmission speed between the wired and wireless world.

Asynchronous transfer mode (ATM) is flexible enough to support a wide bit rate ranging from voice or short text to motion pictures [1]. Flexible connection support includes asymmetric bandwidth assignments, and services based on constant bit rate (CBR), variable bit rate (VBR), available bit rate (ABR) and unspecified bit rate (UBR). ATM can transport efficiently and economically any type of user traffic. ATM technologies are also maturing rapidly and many ATM products are now available for the telecommunications market. To promote wireless access, an ATM Forum has been created with the aim of developing a set of wireless ATM specifications for both mobile ATM protocol extensions and radio access layer protocols [2].

For private networks, the field of wireless local area networks (LANs) is expanding rapidly due to the advantages of mobility and cable-free installation. There is a strong need for standardisation to allow interoperability across different wireless LANs. The first phase of wireless LAN standards was developed by the IEEE 802.11 group in the United States, the ETSI RES10 group in Europe and the Research and Development Centre for Radio systems (RCR) in Japan. These standards bodies are now discussing the next generation of high-speed wireless systems. Indeed, terminals with transmission rates of 10–20 Mb/s are already covered by these emerging standards.

At present there are no wireless ATM multimedia terminals in the marketplace. Multimedia services make the most effective use of text, voice, and image in order to communicate the information required by users. Visual information is expected to be particularly important in the next generation of wireless products. Some kind of visual display is therefore expected to be an indispensable part of the multimedia terminal. A typical wireless multimedia terminal will probably be based on a portable PC and/or a handheld personal digital assistant (PDA).

36.3. BIT RATES, FREQUENCY ALLOCATIONS AND MOBILITY ISSUES

Multimedia applications deal simultaneously with various kinds of information such as voice, data, and images. The volume of traffic required is therefore far higher than current voice or data-based applications. This increase in traffic rate is expected to become even more serious when full interactive multimedia transfers are required. As the information volume increases, so does the required instantaneous transmission rate. One of the key questions to be answered is, "What is the bit rate required for multimedia wireless access systems?". Table 36.1 shows an estimate of the bit rates required for various multimedia services. It suggests that for a single user, the peak bit rate required for a wireless multimedia system is approximately 10 Mb/s. It also implies that the aggregate radio transmission speed will need to be several tens of Mb/s since multiple users will be connected to each base station. Consequently, the 2 Mb/s supported by standards such as IMT2000 and UMTS is unlikely to be sufficient for future broadband wireless multimedia services. Furthermore, it is almost certain that the average user bit rate should be varied on demand, for example, between 1 Mb/s and 10 Mb/s.

It is important that the frequency bands chosen for wireless ATM access are suitable from a wireless design viewpoint. There are many key issues that must be considered such as the bandwidth and the propagation characteristics required for high-speed transmission and the availability of suitable devices and components for a low-cost hardware solution. As discussed above, the bandwidth must be suitable for supporting several tens of Mb/s. This implies that a bandwidth of at least 100–200 MHz is made available. Due to the need for such large bandwidth allocations, the frequency bands from 5 GHz and above appear far more promising than the 2 GHz band currently allocated for IMT2000. At present, the frequency bands below 30 GHz are desirable since RF devices and components are now becoming inexpensive for consumer use. However, the 40 GHz and 60 GHz millimetre bands are now being discussed for high-speed indoor wireless LAN application [3,4].

Table 36.1 Typical application bit rates for multimedia services

	Types of services	Bit rate
Voice/audio	CBR, low delay	8–256 kb/s
Digital data	ABR/UBR, low error	0.1–10 Mb/s
Video telephony (H261)	CBR, low error	64–384 kb/s
Motion-video (MPEG1/MPEG2)	CBR/VBR, low delay	1.5–6 Mb/s

As mentioned earlier, it is clear that the visualisation of information will be important in a future wireless multimedia system. With this in mind, users can be expected to remain still or to move slowly when accessing information. This restriction is important since it is extremely challenging to provide broadband wireless multimedia services to users moving faster than a walking pace. Consequently, broadband wireless multimedia services must be offered in certain spot locations and it may be unnecessary to hand over between these high-speed service areas.

For high-speed radio transmissions, severe multipath fading degrades the BER performance of the wireless system. There are various countermeasures that can be applied to improve the BER performance that can be achieved in a multipath fading environment. For bit rates of 20–30 Mb/s or less, techniques such as adaptive equalisation and orthogonal frequency division multiplexing (OFDM) have been considered. These methods have the advantage of being able to exploit the multipath environment and are not therefore restricted to line-of-sight locations. However, for very high bit rate systems (i.e. greater than 20–30 Mb/s), to overcome the multipath problem in a cost-effective manner some form of narrow-beam directive antenna is essential for tracking the LOS component. The problems with this approach include the difficulty in achieving omnidirectional coverage, limited mobility support and the restriction of service provision to LOS locations. One partial solution to the problem is to employ an electronic switched beam antenna at the base station. However, since beam tracking control needs a long time to evaluate the wireless link, mobility is still restricted. This trend is also true to some extent if equalisation and/or diversity is used to improve the link performance since both methods have a certain evaluation time associated with their processing. Hence, mobility must be limited during the actual communication if high-speed communications are to be achieved.

36.4. DUAL ACCESS TERMINALS

It is desirable for a future wireless terminal to be usable both indoors and outdoors, in other words, in the LAN/WAN environment. For this purpose, it is desirable to support a dual access capability; this means offering an access capability to private ATM networks in addition to public ATM backbone networks. This concept is supported in systems such as Europe's CT2 cordless telephone standard and Japan's Personal Handy Phone (PHP) system, where the cordless phone can be used at home as well as in public spaces [5]. Figure 36.1 illustrates the concept of dual access capability for private and public access. The base stations (BSs) for public access are connected to a public ATM-based backbone network and the base stations for private access, mainly in office environments, are connected to a private ATM-LAN. In this environment, a wireless ATM terminal can connect to both the public ATM network and a private ATM network via private base stations. This wireless access network will realise a flexible wireless multimedia private service in office and public spaces. To complete the dual access capability, it is desirable to

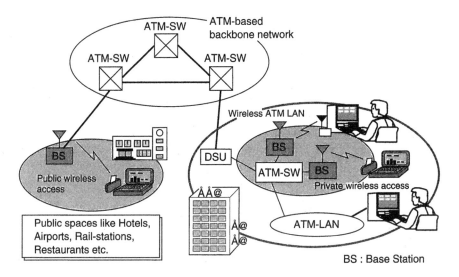

Figure 36.1 The concept of dual access capability.

use a common interface including the radio physical interface, the radio channels control interface, the signalling between the ATM node and the ATM terminal, and so on.

36.5. STANDARDISATION AND RELATED EUROPEAN HIGH-SPEED WIRELESS LAN RESEARCH PROJECTS

In 1992, as part of the European Commission's ESPRIT III programme, the Local Area User Radio Access (LAURA) project was funded to investigate high data rate transmission techniques for future indoor radio LANs. This project worked closely with the HIPERLAN ETSI RES10 subcommittee and towards the end of 1995 demonstrated communications over a prototype 15 Mb/s radio LAN demonstrator. The project focused on developing the protocols necessary for *ad hoc* radio LANs and also performed a comparative analysis of equalisation and multicarrier physical layer solutions. These results were presented to the ETSI RES10 committee and the majority of LAURA's input was later adapted into the European Standard. Details of the modulation, equalisation and synchronisation techniques proposed in LAURA can be found in [6,7].

HIPERLAN represents Europe's attempt to produce a next generation high-speed wireless LAN standard. The system was originally intended for indoor use and aimed to cover both line-of-sight (LOS) and non-LOS locations. The system supports transmissions over distances up to 100 m with optional infra-forwarding allowing multi-hop transmission beyond the radio horizon. Across Europe a dedicated   frequency allocation between 51.5 GHz and 5.3 GHz has been assigned for HIPERLAN use. The transmit power is restricted to 1 W effective isotropic radiated peak envelope power (EIRPEP) [8]. A more detailed description of the HIPERLAN standard is given in section 36.6.

The ETSI RES10 group are now focusing their attention on broadening the HIPERLAN standard into a family of radio communication subsystems intended for integration with future computer systems and products. It is now proposed that the HIPERLAN family architecture addresses four types of HIPERLANs. Type 1 represents the standard *ad hoc* radio LAN system as described above. Type 2 will be an ATM optimised version of the HIPERLAN standard with appropriate modifications being made to both the physical layer and the access protocol to more efficiently support ATM traffic. Type 3 is aimed at short hop microwave links, while type 4 is addressing bit rates up to 155 Mb/s in the 17 GHz frequency band (also allocated throughout Europe for HIPERLAN use).

Radio LAN standardisation is not restricted to Europe, for example within Japan, the Ministry of Posts and Telecommunications (MPT) has set up a study group which has dealt with technical subjects on multimedia mobile access communication systems (MMAC). This includes broadband mobile access for outdoor use and ultra-broadband radio LANs for indoor use. More than 30 organisations have participated in this committee. A field trial based on the committee's report will soon be conducted. It is planned that the Association of Radio Industries and Businesses (ARIB) in Japan will adopt national standards for high-speed wireless access systems and wireless LANs based on the results and conclusion from these field trials.

In North America, the IEEE 802.11 group have been active in the area of wireless LANs for many years. Recently the 5 GHz and 60 GHz bands have been considered for new high-speed wireless LAN products. To help develop ATM technologies, an ATM Forum has been set up by manufacturers and operators interested in the introduction of such techniques. The Forum has the objective of speeding up the standardisation of ATM and to address subjects not yet covered by the relevant standardisation bodies. The Wireless ATM (WATM) group is a relatively new subgroup of the ATM Forum. It is addressing the subject of wireless access to the ATM network. In the area of WATM, the ATM Forum is now cooperating with the ETSI BRAN group. Since ETSI BRAN is already defining the physical layer issues of WATM under the HIPERLAN type 2 standard, the ATM Forum is concentrating on the remaining functionality. In particular, the group is concentrating on the medium access control procedures (MAC), the logical link control (LLC) procedures, and mobility management in the ATM network (e.g. routeing protocols).

Under the European ACTS research programme, a wide range of projects have recently been funded in the area of wireless LANs and wireless ATM. The main ACTS projects are WAND, SAMBA, AWACS and MEDIAN. Two further projects (CABSINET and CRABS) are investigating related multimedia applications and technology.

The ATM Wireless Access Communication System (AWACS) is considering low mobility terminals operating in the 19 GHz band and aims to support full duplex user bit-rates up to 34 Mbit/s with radio transmission ranges up to a maximum of 100 m. Project partners include Alcatel CIT (F), Alcatel SEL (D), NTT (JP), University of Bristol (UK), Elektrobit (Fin) and CSELT (I). The project has the advantage of direct access to NTT's existing wireless ATM demonstrator based on 19 GHz hardware. AWACS is concentrating on the use of directional antennas as a means of overcoming the harmful effects of the multipath channel. The project will take wideband measurements of the radio channel in addition to BER and ATM cell-loss rates for a range of indoor and outdoor environments. These measurements will be repeated with a range of OMNI and directional antennas. Based on these measurements, channel models and simulations will be created to redesign and optimise the modem's performance. The demonstrator has the capability of being upgraded to support cellular like mobility in addition to spectrum and power efficient radio access technologies.

The key issues considered within AWACS include the performance evaluation of an existing 19 GHz ATM compatible modem and the generation of appropriate channel and system simulation models. Using these models, the antenna configuration at both the base station and handset will be optimised for ATM-based transmissions. The project will end in early 1998 with a field trial which aims to improve the communications between physically separated offices via telepresence technologies. This AWACS project is focused towards the new HIPERLAN type 4 standard. Figure 36.2 shows how AWACS aims to interact with other ACTS projects and also with European and worldwide standardisation bodies.

SAMBA has set its target on promoting the development of a broadband cellular radio extension (MBS) to the integrated broadband communications network, thus giving multimedia support to mobile users. The trials platform will demonstrate reliable 40 GHz radio transmission for transparent ATM up to 34 Mbit/s in addition to specifying and implementing medium access control, handover and radio resource management. Furthermore, this platform will help to develop a portable millimetre wave mobile terminal and will impact on MBS standardisation. The trial will make use of two base stations and two mobile terminals. The user applications will be related to a wireless TV camera and medical applications.

MEDIAN has a main objective of evaluating and implementing a high-speed wireless customer premises local area network (WCPN/WLAN) pilot system for multimedia applications and demonstrating the technology in a real user trial. The

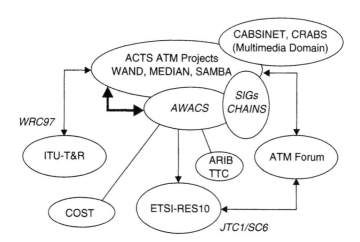

Figure 36.2 Relationship of AWACS with other ACTS Projects and Standardisation bodies.

pilot system relies on a multi-carrier (OFDM) modulation scheme which can be adapted based on the channel characteristics to vary the transmission rate. The project is also considering the development of a wireless ATM network extension. The system will be based on two mobile units and one base station which is connected to third and future generation mobile systems via the ATM interface. MEDIAN will operate at 60 GHz and offers broadband wireless coverage in confined areas. The project aims to demonstrate multimedia ATM applications, voice and video over the MEDIAN network with data rates up to 155 Mbit/s.

WAND aims to develop and evaluate, in realistic user environments, a wireless ATM transmission facility at both 5 GHz and 17 GHz (system studies will be performed at the higher frequency band) with the hardware demonstrator operating at 5 GHz. Wireless ATM will allow users to transmit and receive data at rates over 20 Mbit/s and controlled service levels that match those of the wired ATM world. The project will provide a wireless ATM access network demonstration system and contribute to the necessary standardisation bodies. WAND, like AWACS, is also contributing to both the HIPERLAN type 2 and type 4 standards.

CABSINET is investigating the exploitation of interactive broadband wireless end-user access. It aims to design, build and demonstrate a working interactive cellular TV architecture based on the 40 GHz frequency band. This demonstrator will be capable of delivering both broadcast and ATM-derived services. The trials will encompass a complete macrocell/microcell deployment, including an example of interworking with a trans-European broadband fibre network over two national host platforms.

CRABS is dealing with a 42 GHz stationary cellular radio system for interactive TV and broadband services. The communication is extremely unsymmetrical, with tens of Mbit/s on the downlink (for broadcast) but only a maximum of 2 Mb/s on the return channel.

Under the continuing ESPRIT programme, at least two new HIPERLAN related development projects began at the end of 1997. One of these projects, Wireless INnovation in the HOME (WINHOME), includes partners from the University of Bristol (UK), Dassault (Fr), Grundig (UK/GER), SCT (UK) and Eutelsat (Fr). This projects aims to integrate HIPERLAN 1 technology into currently emerging cable and satellite digital set-top box products. The resulting demonstrator will allow the wireless transmission of MPEG 2 transmission at 5 GHz to portable computers and televisions throughout the home. In addition, this technology will enable services such as interactive television and true multimedia applications. High-speed wireless Internet access will also be made available throughout the home via the use of high-speed cable modems and satellite transmissions.

36.6. AN OVERVIEW OF THE HIPERLAN STANDARD

This section describes some of the technical aspects of the HIPERLAN type 1 standard [8]; however, Chapter 37 gives a more complete description of the emerging wireless LAN standard. HIPERLAN represents a wireless LAN standard in which all nodes communicate using a single shared communication channel. The standard supports pre-arranged (i.e. star and hub networks) as well as *ad hoc*-based configurations. *Ad hoc* networks allow any two or more HIPERLAN nodes to communicate without the necessity for any additional local infrastructure (i.e. the presence of a dedicated base station). Nodes are allowed a degree of mobility (up to 1.4 m/s). Importantly, HIPERLAN compatible hardware will be licence exempt and, due to its standardisation, rival manufacturers' equipment will be interoperable. The system is packet-based and uses the frame structure shown in Figure 36.3.

Before each packet can be sent, a process of channel acquisition must be performed. This consists of listening to the channel and, if the channel remains quiet, transmitting an assertion pulse to reserve the medium. The length of time spent listening and asserting is related to the data packet priority level. The standard can support both asynchronous and time bounded traffic at various priority levels (this allows interactive and time-critical services such as remote computing and real-time voice and video transmission to be supported). One drawback with the type 1 system is its inability to guarantee bandwidth (this is being addressed in the ATM specific type 2 standard). The concept of power conservation is supported to preserve battery life. Confidentiality can be achieved through the MAC protocol by the use of an optional encryption–decryption scheme.

The user data is modulated using Gaussian minimum shift keying (GMSK) with a bandwidth–bit period product (BT) of 0.3. The high-speed "on-air" bit-rate is 23.5294 Mb/s and this occupies a bandwidth of 23.5294 MHz. Within each European member state, a minimum of three carrier frequencies will be available with a further two being available on a national basis. The system includes a low bit rate data modem for sending packet address information. This low bit rate modem uses frequency shift keying (FSK) at a bit rate of 1.4706 Mb/s. The low bit rate data is not intended to be processed by the equaliser but is used to determine whether the following high-speed packet is intended for the current node.

To overcome the harmful effects of the radio channel [9], a training sequence is inserted at the beginning of each packet to allow adaptive equalisation to be performed at the receiving node. Equalisation was chosen since it offers a compact all-digital solution that is compatible with PCMCIA implementation. The high throughput rates required by the equaliser (which must train in real time at almost 24 Mb/s) represents a considerable implementation challenge. While the standard does not specify the type of equaliser, an LMS-based decision feedback approach is generally accepted to offer the best compromise between complexity, performance and power consumption [10]. The performance of various equalisation structures and a number of methods for reducing implementation complexity are reported in [11,12].

			Frame 0	Frame m-2	Frame m-1
450 BIT SYNC	496 BIT MPDU	. . .	496 BIT MPDU	496 BIT MPDU

Figure 36.3 Simplified HIPERLAN packet structure.

Each data packet consists of H high rate bits and L low rate bits. After sending a low bit rate identification sequence, the data packet begins with a fixed 450-bit sequence which is used to synchronise and train the receiver's equaliser. Following this sequence there are a number of 496-bit data blocks ($m < 48$). Each data frame is protected by the use of a BCH (31,26) block code. This channel coding scheme is used to significantly improve the overall packet outage probability. The low bit rate header is used to conserve power by ensuring that only the intended node turns on its equaliser to receive the data sequence. H and L are integers with H being given by the expression, $H = 450 + 496m$, where m is an integer representing the number of 496 bit data blocks. The maximum value of m is 47, this value being chosen based on a mobile speed of 1.4 m/s. Note: for higher mobile speeds a lower value of m is recommended.

36.7. CONCLUSIONS

A number of new high-speed wireless LAN standards are now being developed around the world that will allow maximum combined data rates of between 24 and 155 Mb/s to be achieved over distances up to 100 m. These radio tails are also being optimised to integrate with fixed, high-speed, ATM networks. In Europe, the HIPERLAN type 1 standard has already been defined and dedicated spectrum is available. The European Commission has funded a wide range of new projects under the ACTS and ESPRIT initiatives to support the development of a number of new wireless LAN and ATM demonstrators at frequencies ranging from 5, 17, 19, 40 and 60 GHz. ETSI are continuing to develop a number of new HIPERLAN standards and through cooperation with bodies such as the Wireless ATM Forum, there is hope that the HIPERLAN physical layers will also be suitable for use in the North American and Japanese markets. Research projects such as ACTS AWACS are also bridging the standardisation gap between Europe and Japan by disseminating results into both the European and Japanese standardisation bodies.

ACKNOWLEDGEMENTS

Elements of the work described in this chapter were carried out under ACTS project no AC228 "AWACS". Dr Andrew Nix and Dr Mark Beach are particularly grateful to their colleagues at the Centre for Communications Research, University of Bristol. The authors would like to acknowledge the support offered by the AWACS project partners in the preparation of this chapter. In particular, thanks are given to NTT research labs (Japan) for the loan of the pre-prototype AWA demonstrator hardware.

REFERENCES

[1]   Umehira, M. *et al.*, An ATM wireless access system for tetherless multimedia services. In *Proceeding of ICUP '95*, 1995.

[2]   Dellaverson, L. P., Wireless ATM standardisation in ATM Forum. *Wireless ATM Workshop*, Espoo, Finland, 1996.

[3]   Fernandes, L., Developing a systems concept and technologies for mobile broadband communications. *IEEE Personal Communication*, **2** (1), 54–59, 1995.

[4]   Driessen, P. F. and Greenstein, L. J., Modulation techniques for high-speed wireless indoor systems using narrowbeam antenna. *IEEE Transactions in Communication*, **43** (10), 2605–2612, 1995.

[5]   Hattori, T. *et al.*, Personal Communication – Concept and Architecture. *Proceeding of the IEEE ICC '90 Conference*, 1990.

[6]   Nix, A. R. *et al.*, Modulation and equalisation considerations for high performance radio LANs (HIPERLAN). *IEEE PIMRC*, 964–968, September 1994.

[7]   Sun, Y., Li, M. and Nix, A., Modulation and equalisation considerations for high performance radio LANs. In *Wireless Personal Communications*, Kluwer Academic Press, New York, 1997.

[8]   ETSI Radio Equipment and Systems. HIgh PErformance Radio Local Area Network (HIPERLAN). Functional Specification Version 1.1 (Draft), January 1995.

[9]   Athanasiadou, G. E., Nix, A. R. and McGeehan, J. P., An efficient "image-based" propagation model for LOS and non-LOS applications. In *Proceedings of the IEE Colloquium on Propagation in Buildings*, 1995/134, June 1995.

[10] Nix, A. R., Athanasiadou, G. E. and McGeehan, J. P., Predicted HIPERLAN coverage and outage performance at 5.2 and 17 GHz using indoor 3-D ray-tracing techniques. In *Wireless Personal Communications*, Kluwer Academic Publishers, New York, December 1996.

[11] Sun, Y., Nix, A. R. and McGeehan, J. P., HIPERLAN performance analysis with dual antenna diversity and decision feedback equalisation. In *Proceedings of the IEEE VTC'96 Conference*, Atlanta, GA, April 1996.

[12] Perry, R., Bull, D. R. and Nix, A. R., An adaptive DFE for high data rate applications. In *Proceedings of the IEEE Vehicular Technology Conference*, Atlanta, GA, 686–690, April 28–May 1, 1996.

37

HIPERLAN - An Air Interface Designed for Multimedia

Tim Wilkinson

37.1. INTRODUCTION

The HIgh PErformance Radio Local Area Network (HIPERLAN) is a new standard for high-speed multimedia wireless networking that has recently been developed by the European Telecommunications Standards Institute (ETSI) in Europe. HIPERLAN is a common air interface standard, which means that it is essentially a specification of the lower two layers of the ISO OSI model. What follows is an outline of this standard's past, present and future.

37.2. BACKGROUND

In the early 1990s there was a growing number of proprietary radio LAN products appearing on the market in the US. Most of these products operated in the 902–928 MHz industrial, scientific and medical (ISM) band, which only exists in the US. Soon versions of these products started to appear in other countries in the 2.4–2.5 GHz ISM band, which exists in slightly different forms world-wide. The market for these products did not develop as fast as expected and three reasons were frequently cited for this.

1. The ISM bands present a hostile interference environment, as they are essentially a free-for-all, which means that there is potential for considerable mutual interference between these and other ISM band systems such as microwave ovens and hence performance could not easily be guaranteed.
2. The FCC in the US, the CEPT in Europe and other spectrum regulatory bodies specified the use of spread spectrum techniques for communication systems operating in these bands to limit the effect of this interference which essentially limited the transmission rates to 1–2 Mbps for commercially viable systems.
3. There were no standards and hence the market was confused. To fix this the IEEE started the formulation of IEEE802.11 but this was taking time to define and it was subject to both of the above constraints.

It was apparent that to enable this market to develop there should be a standard defined which would operate in a dedicated spectrum to enable reliable communication with sufficient bandwidth for acceptable transmission rates. In early 1991 ETSI started work on the HIPERLAN standard to meet this need. The general goal was to develop a wireless networking standard that would be equivalent in performance to wired networking standards such as Ethernet. In late 1996 the functional specification [1] was published. The conformance testing specification [2] followed in early 1997 allowing immediate manufacture of HIPERLAN products.

37.2.1. Spectrum Allocation

Spectrum is a scarce resource in most of the populated areas of the world. There are few vacant and usable parts of the radio spectrum. Typically if there are, they are reserved for experimental purposes such as the ISM bands, they are absorption bands such as the oxygen absorption at 60 GHz or they have been allocated but not used. The latter was the case in the HIPERLAN allocation. The band 5.15–5.25 GHz was originally allocated to the civil aviation authorities as part of a bigger allocation 5.00–5.25 GHz for the Microwave Landing System (MLS). However, this part of the allocation was left unused. The reason often cited for this was to leave a guard band between this safety critical system and potentially hostile high-power military radars, which were located in spectrum immediately above this band. HIPERLAN was seen to be a system benign enough and robust enough that it could sit in a spectrum between these two types of system. Hence in 1992 the

CEPT recommended [3] that the band 5.15–5.25 GHz should be allocated to HIPERLAN in Europe with an extension 5.25–5.30 GHz in most countries as shown in Figure 37.1. The maximum transmit power was specified as 1 W or 30 dBm equivalent isotropic peak envelope power (EIRPEP). This band overlaps with the Mobile Satellite Service (MSS) band and there has been much debate as to whether or not these systems can co-exist. At the time of writing the MSS interests are still campaigning for a reduction in the HIPERLAN power limit. There is good reason for concentrating on the detail of this allocation. The point is that this allocation was in vacant spectrum linked to an international spectrum allocation. Consequently there was the potential for an international spectrum allocation for HIPERLAN or HIPERLAN-like systems, which was ultimately the case, as we shall see later.

37.2.2. Deployment Scenarios

At conception two quite different deployment scenarios were perceived for HIPERLAN: that with wired infrastructure and that without. Extreme examples of these are shown in Figures 37.2 and 37.3. In Figure 37.2 the value of the wireless element is in mobility and the elimination of cable management problems around an individual's work area. Local re-wiring costs are also eliminated but the initial installation of the infrastructure could be more expensive than the standard wired network if the system is not carefully designed. In Figure 37.3 the value of this type of wired network is the ability to set up a LAN anywhere without infrastructure. There is obviously no infrastructure present in the environment shown in Figure 37.2 however, more typical environments are in a small office or at home where the installation of wired infrastructure is undesirable.

These two deployment scenarios have quite different requirements and constraints and these attributes are compared in Table 37.1.

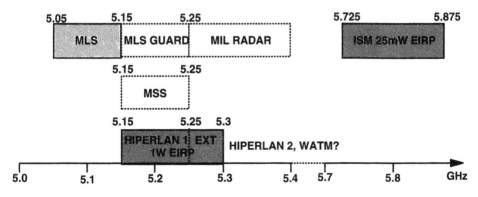

Figure 37.1 The HIPERLAN spectrum allocation.

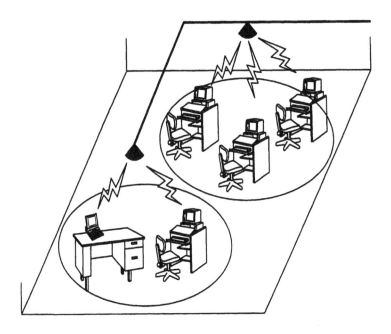

Figure 37.2 WLAN with wired infrastructure.

Figure 37.3 WLAN without wired infrastructure.

Table 37.1 Comparison of attributes of deployment scenarios

Deployment scenario	With wired infrastructure	Without wired infrastructure
Natural communication topology	Star	Mesh
Ideal medium access control	Centralised	Distributed
Range requirement	Short	Long
Line-of-sight	Likely	Unlikely
Fading statistics	Rician	Rayleigh
Dispersion or delay spread	High	Low
Capacity requirement	High	Low
Spectral re-use time/frequency	Planned or self-managing	Must be self-managing

The communication topology and the medium access control are closely related and are driven by the application. It is clear from the examples of these deployment scenarios which topology is appropriate in either case. A star communication topology is more appropriate with wired infrastructure, where all communication passes through the wired access points, but a mesh topology is more appropriate without wired infrastructure, where it would be a great burden to give a device the responsibility of acting as a hub. This assertion is by no means absolute. The network with infrastructure could easily use a mesh communication topology within an access point coverage area and the network without infrastructure may have a device in the network that is wired to external infrastructure and is mains powered and thus would be capable of taking on the burden of acting as a hub. An obvious compromise is to have a device co-ordinating communication without all local communication passing through this as a hub. This debate will continue to rage and the correct choice is always application dependent; however, the compromise decision in the HIPERLAN standardisation was that the mesh topology was the most general supporting both types of network with and without infrastructure.

It is apparent from these examples of deployment scenarios that the network with wired infrastructure will have a reduced range requirement over that without infrastructure where the extent of the network could be limited by the maximum communication range.

It is also apparent that in a network with infrastructure the access points could be placed in elevated positions and hence a line-of-sight radio propagation path is likely, whereas in a network without infrastructure the antennas will be buried in the room clutter and a line-of-sight is unlikely.

The range and likelihood of a line-of-sight affect the degree of fading and dispersion. If the range is short and a line-of-sight is likely the fading will be Ricean rather than Rayleigh and the delay spread will be low. Hence the channel is much more benign for the network with infrastructure than that without. The shorter-range communication in the case of the network with wired infrastructure enables higher capacity but the typical scenario demands this. The capacity of the network without infrastructure is reduced if the maximum range is long but the capacity requirement is also low so this is not an issue.

In both cases the medium re-use should be self-managing so that these networks are easy to deploy, increasing their usability and reducing their cost.

37.3. PERFORMANCE TARGETS

The fundamental aim in the creation of the HIPERLAN standard was to create a wireless networking standard that would be equivalent in performance to wired networking standards such as Ethernet. This essentially means that it must have a user data rate of around 10 Mbps shared between the users of the network. It must be able to do this reliably over a radio channel. It was assumed that this channel would be a multipath Rayleigh fading channel with an RMS delay spread of less than 100 ns. At this time few propagation measurements had been made at 5 GHz in the target environments but subsequent measurement campaigns in commercial and domestic environments [4,5] confirmed this assumption. It must be able to support the deployment scenarios described previously in an in-building environment. Consequently it would have to have a range of 50 m and support mobility of 10 m/s and 360°/s. The target packet failure rate was 0.01.

There was also an obvious necessity to support multimedia traffic or a mix of asynchronous and isochronous traffic. It was assumed that this meant the support of asynchronous traffic at a rate of 10 Mbps (equivalent to Ethernet) and isochronous data from 32 kbps (audio) with a delivery delay of < 10 ms to 2048 kbps (video) with a delivery delay of 100 ms. The following section describes the design of the standard to satisfy these requirements. The rationale behind the decisions that were made is given in [6] and a good outline of the features of both HIPERLAN and IEEE802.11 is given in [7].

37.4. THE HIPERLAN STANDARD

The HIPERLAN standard was designed with two transmission rates: the primary or high bit rate of 23.5 Mbps for the transmission of data and the secondary or low bit rate of 1.5 Mbps (1/16 of the high bit rate) for the transmission of packet headers and acknowledgement packets.

The modulation for the high bit rate is pre-coded non-differential Gaussian minimum shift keying (GMSK) with a *BT* (bandwidth bit–period product) of 0.3. This is designed to be demodulated coherently. At this rate it is envisaged that the demodulation would require adaptive equalisation to compensate for the multipath distortion imposed on the signal by the channel. Although equalisation is not specified in the standard, the standard is designed with this in mind. The channel bandwidth is 23.5 MHz and there are five channels specified in the 150 MHz system bandwidth allowing for two guard bands at either end of the allocation.

The modulation for the low bit rate is wideband frequency shift keying (FSK) with a frequency deviation of 368 kHz. This is designed to be demodulated noncoherently and without equalisation so that it can be done at a significantly reduced power to facilitate power saving.

The HIPERLAN data packet and acknowledgement packet are shown in Figure 37.4. The data packet consists of three parts: the low bit rate header, the synchronisation and training sequence and the data payload.

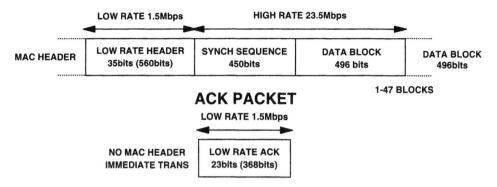

Figure 37.4 The HIPERLAN packets.

The low bit rate header contains a hashed version of the destination 48-bit MAC address. As this can be demodulated with a low power demodulator the high bit rate and high power demodulator is only activated if the hashed version of the MAC address in this header is recognised.

The synchronisation and training sequence consists of three repetitions each of five different 31-bit m-sequences, the last of which is truncated giving a 450-bit sequence. This sequence can be used for equaliser training, channel estimation and antenna selection. This sequence may seem unnecessarily long for these functions but this is to allow for the least sophisticated equaliser training algorithm and it is still a small fraction of the total packet overhead.

The data consists of m 496-bit data blocks, where m is from 1 to 47. The maximum packet length of 47 blocks is specified such that the channel will be stationary over this period with a transmitter or receiver relative velocity of 1.4 m/s. For faster relative velocities smaller packets should be used. This packet size is also compatible with Ethernet packet sizes.

The user data is divided into 416-bit blocks. These are segmented into 16 segments that are block interleaved and then FEC encoded with a (31,26) BCH code to give a 496-bit block. This code can correct single-bit errors. However, an equaliser is likely to produce a burst of errors so interleaving is employed to break up these bursts. The interleaver is a 16×16 block interleaver.

There are three classes of transmitter power, 30 dBm, 20 dBm and 10 dBm equivalent isotropic radiated peak envelope power (EIRPEP) and three classes of receiver sensitivity, −50 dBm, −60 dBm and −70 dBm. Only certain combinations of transmitter power and receiver sensitivity are allowed to avoid powerful and insensitive links. Antenna diversity is an option and directional antennas can be used but the transmit power must be scaled accordingly.

The data link layer in the standard ISO OSI is typically supplemented with a medium access control (MAC) sublayer in radio systems. In HIPERLAN it is supplemented by a medium access control layer and a channel access control layer as shown in Figure 37.5 [1]. The purpose of this split is to separate

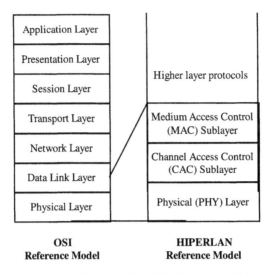

Figure 37.5 The HIPERLAN reference model.

functions that are time critical. However, as this split is unique to HIPERLAN we will discuss these parts of the system under the more general heading of MAC to avoid confusion.

The only workable way to implement a distributed MAC protocol that is the only natural choice for this type of network was to use a variant of carrier sense multiple access (CSMA) or listen-before-talk, which is used to great effect in Ethernet. The difficulty with implementing this over a radio channel is that the large dynamic range between transmit and receive signal strengths makes collision detection (CD) or listen-while-talk very difficult. Hence, most radio versions of CSMA use collision avoidance (CA), which is a randomisation of the collision probability rather than actual avoidance. The classic problem with this type of multiple access scheme is the hidden node problem. This arises because in a transmission the interference is assessed at the transmitter but the effect of this interference occurs at the receiver where the interference environment could be different. The classic solution to this problem is either to execute a handshake with short packets to inhibit other transmissions at either end of the link or to ensure that a transmission preamble is heard well beyond the data transmission range of the transmitter to the same effect. This later solution is employed in HIPERLAN. An adaptive threshold is used in the CCA (clear channel assessment) to allow a HIPERLAN device to adjust to the background interference and compensate for component tolerances.

The HIPERLAN MAC protocol uses a form of listen-before-talk with a random listen-talk sequence prior to data packet transmission. The most interesting feature of this protocol is the way in which it handles multimedia asynchronous and isochronous traffic. The MAC protocol mechanism is described below.

If no activity has been sensed on the channel for 85.00 ms (2000 bits) the channel-free condition, transmission can take place after 0–3 time slots of 8.50 ms (200 bits) where the probability of the period being n slots is 0.25.

Otherwise, it is assumed that all stations wishing to transmit will synchronise to the end of the last transmission, the synchronised channel condition and execute three phases in channel access. These are the prioritisation phase, the contention phase and the transmission phase, all of which are of variable duration.

The prioritisation phase exists to prioritise transmissions. The prioritisation phase consists of 1–5 time slots of 7.14 ms (168 bits). Priority is asserted by starting transmitting in a slot, 0 to 4, 0 for highest priority, 4 for lowest priority. Stations listen until they transmit. The priority phase ends when one or more stations assert their priority and listening stations with lower priority hear this and defer.

Packet priority is related to the normalised residual lifetime, which is a function of residual packet lifetime and number of hops to final destination, and the user defined priority as shown in Table 37.2. So asynchronous traffic such as file transfer, which has an infinite lifetime, has low priority, whereas isochronous traffic such as audio or video, which has a finite lifetime, has high priority. The priority for a packet increases until the lifetime expires and then the packet is discarded in the MAC. This has interesting implications for multimedia applications. Although this priority cannot give guarantees it does give a best effort attempt at either latency or integrity, latency for isochronous traffic with finite lifetime which may be discarded if this expires and integrity for asynchronous traffic because infinite lifetime packets will never be discarded and will ultimately be transmitted.

The contention phase exists to resolve contention between stations with the same priority. The contention phase consists of an elimination phase and a yield phase. The elimination phase consists of 0–12 time slots of 10.88 ms (212 bits) and one time slot of 9.01 ms (256 bits). Stations entering the elimination phase after the prioritisation phase will continue to transmit in successive slots with probability 0.5 then listen for a single time slot. If they hear nothing, they then enter the yield phase. The yield phase consists of 0–9 time slots of 7.14 ms (168 bits), where the probability of the period being n slots is 0.1. Stations continue listening during this period, if they hear nothing they then enter the transmission phase.

The entire MAC header is 0–219 ms (0–5152 bits) in duration. This requires a transmit-to-receive turn-around time of 6 ms. The MAC protocol should ensure that with 256 contending stations there is a 3.5% collision probability. The transmission phase simply consists of packet transmission. Packet reception is acknowledged by immediate transmission of an acknowledgement.

Figure 37.6 shows four stations contending for the channel. Station four has a lower priority packet than the other stations and is eliminated in the prioritisation

Table 37.2 HIPERLAN packet priorities

Normalised residual lifetime	High user defined priority	Low user defined priority
NRL < 10 ms	0	1
10 ms < NRL < 20 ms	1	2
20 ms < NRL < 40 ms	2	3
40 ms < NRL < 80 ms	3	4
NRL > 80 ms	4	4

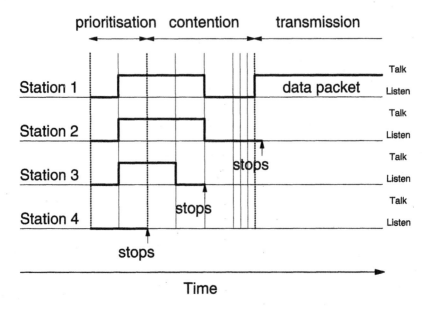

Figure 37.6 HIPERLAN MAC protocol mechanism.

phase. Station three is eliminated in the elimination part of the contention phase. Station two is eliminated in the yield part of contention phase and station one gets the channel.

The HIPERLAN standard has other elegant features. HIPERLAN has multi-hop relaying capability, which effectively extends a network coverage area without infrastructure. HIPERLAN also has power saving involving sleep/wake cycles where traffic for temporarily sleeping devices (p-savers) which could be battery powered is buffered and then forwarded by always awake devices (p-supporters) which should be mains powered.

37.5. HIPERLAN IMPLEMENTATIONS

At the time of writing there are no HIPERLAN products on the market so it could be said that take up is slow but the truth is perhaps that the standard was somewhat ahead of its time. There are, however, several companies who are actively commercialising HIPERLAN technology.

There have been a number of HIPERLAN demonstrators. In the early days of the standard there were two European collaborative ESPRIT projects, LAURA and HIPERION, both of which served to validate the standard. Several simulation studies indicated that HIPERLAN could achieve the target packet failure rates of 0.01 in the envisaged channels [8,9]. More recently results from HIPERLAN demonstrator trials [10] illustrated the capability of the standard by showing that an office area approximately 18 m × 35 m with 1 mW transmit power.

Initially there was some scepticism as to the commercial viability of the standard. This centred on the complexity of the digital signal processing required to implement the equaliser and associated functions. Much of this scepticism was unjustified as it assumed implementation of these functions in software running on a general purpose processor core whereas in reality these functions would be implemented in hardware on an ASIC with around three orders of magnitude lower power consumption and easily manageable silicon area. However, this scepticism did motivate investigations into reduced complexity equaliser implementations yielding novel modular architectures employing non-linear number representation to simplify the computation [11]. Other less conventional equalisation was proposed in [12].

More recently it has become apparent that it is not the equaliser complexity that is the challenge but the associated analogue RF/IF circuitry required to support a DFE the preferred equalisation technique. This requires a linear front end with complex I/Q down-conversion along with high stability local oscillators and/or local oscillator frequency offset compensation. Local oscillators are required in the standard to have a frequency offset of < 10 p.p.m., which results in a maximum frequency offset of 100 kHz, occurring at either end of the link. This problem can be solved either by measuring the frequency offset in the training sequence and using feed-forward frequency offset compensation or by somehow incorporating the compensation in the equaliser architecture and making this part of the equaliser continually adaptive [13]. Neither technique is ideal and the effect of local oscillator frequency offset can result in high packet failure rates if it is not adequately controlled. This prompted research into differential receiver architectures that have an inherent resilience to local oscillator frequency offset [9] and [14]. The difficulty with such architectures is that the differential detection places a nonlinearly between the equaliser and the channel, which compromises the performance of the equaliser, which can only compensate for linear distortion [9]. Consequently, alternative equalisation techniques, such as analogue equalisation, have been proposed. There is an inherent problem with differential detection of the HIPERLAN modulation in that as the modulation is non-differential and errors in differential detection can result in catastrophic error propagation. However, any errors in a packet that cannot be corrected with the FEC will cause a packet failure so this may not be a serious problem.

The research into the optimum receiver architecture will continue and as none of these details are specified in the standard this will be an area where there will be considerable competition.

There has been less published about the performance and implementation of the HIPERLAN MAC. The main reason for this is that it is very difficult to simulate how such a MAC will perform in a realistic scenario carrying typical traffic. Early simulations by the ETSI project team indicated that its performance was adequate. Regarding implementation it is clear that the HIPERLAN MAC is simpler than the IEEE802.11 MAC in that it has a single solution to support all types of traffic but it is also obvious that the HIPERLAN MAC may have more time critical elements. The management of the buffers for example to facilitate the prioritisation of the traffic in the MAC will not be trivial.

37.6. BEYOND HIPERLAN

Since completion of the HIPERLAN standard in 1996 there has been growing interest in air interfaces optimised for ATM traffic. This requires efficient transport of small packets, a single ATM cell, and some degree of call admission control to be able to guarantee quality of service an important part of the ATM philosophy. In 1997 ETSI formed Broadband Radio Access Networks (BRAN) to meet this need. The aim of BRAN is to specify other versions of HIPERLAN designed to carry ATM traffic at different rates and for different applications. These are HIPERLAN types 2, 3 and 4 as shown in Figure 37.7 [15]. There is even talk of HIPERLAN type 5 at 60 GHz.

These standards are being defined in collaboration with the ATM Forum. BRAN will define the air interfaces and the ATM Forum will specify ATM inter-working and support for the wireless physical layer. At present this activity has only reached the requirements phase and these have been published in [15].

There is most aggressive activity in the specification of HIPERLAN type 2 where there is most interest. HIPERLAN type 3 may use the same air interface as HIPERLAN type 2 but in a different scenario. It is clear that there are many applications of these two standards envisaged beyond the wireless LANs for public and private systems. At present neither of these standards has a spectral allocation at 5 GHz and although BRAN is lobbying for additional spectrum for these standards it is not clear at this time whether they will get this. The target user data rate for these standards is ~ 20 Mbps, which implies a transmission rate of much higher than this, but the assumption is that the channels will be of similar bandwidth to HIPERLAN type 1. The aim is to have the air interface defined by mid-1998. Currently there are a number of proposals on the table, most of which have gravitated around coded orthogonal frequency division multiplexing (COFDM) modulation with a conservative number of carriers and a centralised MAC for call admission control.

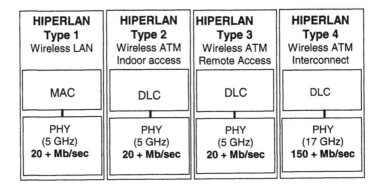

Figure 37.7 The HIPERLAN family.

37.7. HIPERLAN IN THE REST OF THE WORLD

The most important aspect about the HIPERLAN European spectrum allocation was that it was in vacant spectrum linked to an international spectrum allocation. Consequently there was the potential for an international spectrum allocation for HIPERLAN or HIPERLAN-like systems. In 1996 several companies in the US involved in the HIPERLAN standardisation in Europe petitioned for this spectrum to be allocated for similar systems in the US. This culminated in the FCC issuing a Notice of Proposed Rule Making (NPRM) in 1996 and then finally a Report and Order (R&O) in 1997 allocating 5.15–5.35 GHz and 5.725–5.825 GHz for Unlicensed National Information Infrastructure or UNII devices. These rules were an addition to the part 15 FCC rules (part 15.407) for unlicensed devices. Their definition of UNII devices was "Intentional radiators operating in the frequency bands 5.15–5.35 GHz and 5.725–5.825 GHz that provide a wide array of wide-band high data rate digital mobile and fixed communications for individuals and businesses and institutions". It is clear from this statement that the bands were allocated for a much wider scope of devices than just HIPERLAN but it is also apparent from the NPRM that HIPERLAN is the reference system. To allow a variety of systems in the band the FCC have regulated the spectrum to a minimal degree specifying only maximum transmit power and maximum power spectral density for three sections of the band as shown in Figure 37.8. The lower and mid-bands are more attractive than the higher band as this shares the spectrum with the ISM band (Part 15.249) and the spread spectrum band (Part 15.247). The lower band overlaps with the MSS band, hence the lower power limit.

The UNII regulations were designed so that a system must have a bandwidth in excess of 20 MHz in order to be allowed the maximum transmit power. This encourages the use of wide-band systems in the band but does not discourage the use of modulation-inefficient systems. However, as there is no direct link between modulation efficiency and spectral efficiency, which is the more important measure

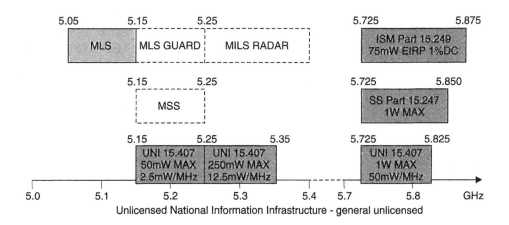

Figure 37.8 The UNII spectrum allocation.

of how effectively the spectrum is being used, the FCC decided not to restrict modulation efficiency in the regulations. The most important thing about the band is that whilst there are few restrictions to what sort of systems can use the spectrum there is no compulsory etiquette to manage interference between these systems. An etiquette is being developed by WINForum and although the FCC have said that they would endorse such an etiquette it is likely that this will be voluntary rather than mandatory. This is because systems are already designed to operate under the current set of rules and these could not possibly comply with a future etiquette.

The important implication of these rules is that if systems are to be reliable they will have to have some degree of interference avoidance such as dynamic channel assignment or interference randomisation such as spread spectrum. At the time of writing HIPERLAN type I could be operated in the UNII bands. However, in order for it work reliably, there would need to be modifications to the standard to enable a HIPERLAN to sense interference and if it is adversely affected move to another channel to avoid this. This is quite important as HIPERLAN was only ever designed to co-exist with itself. This would also be true of any other common air interface standard designed for dedicated spectrum such as HIPERLAN type 2. Very recently there has been an indication of interest in these bands in Japan so the possibility of a near global allocation is real.

37.8. CONCLUSIONS AND THE FUTURE OF 5 GHz SYSTEMS

The HIPERLAN standard was one of the first air interfaces designed to transport multimedia traffic in unlicensed spectra. Whilst its original focus was LANs this standard should see application for the interconnection of all sorts of computing, communications and entertainment appliances. This should enable high-speed in-building wireless networking for offices and homes. The newer versions of HIPERLAN should enable even more wireless networking applications, public and private, from wireless Internet service provision to community networking. The importance of these standards is further enhanced by the fortuitous nature of the spectrum allocation, making global spectrum availability a reality. Although these standards may not become globally ubiquitous the know-how and component volume advantages that HIPERLAN has spawned will propagate into proprietary standards that proliferate outside of Europe.

REFERENCES

[1] ETSI Radio Equipment and Systems (RES). High performance radio local area network (HIPERLAN) type 1. Functional Specification ETS 300 652.
[2] ETSI Radio Equipment and Systems (RES). High performance radio local area network (HIPERLAN) type 1. Conformance Testing Specification ETS 300 836.

[3] CEPT Recommendation T/R 22-06. Relating to the harmonised radio frequency bands for high performance radio local area networks (HIPERLANs) in the 5GHz and 17GHz frequency range.

[4] Airs, J., Comparative indoor RF channel soundings at 2, 5 and 17GHz. *Wireless Personal Communications*, **3** (4), 353–363, 1996.

[5] Hafezi, P., Wedge, D., Beach, M. A. and Lawton, M., Propagation measurements at 5.2GHz in commercial and domestic environments. In *PIMRC* 509–513, 1997.

[6] Wilkinson, T., Phipps, T. G. C. and Barton, S. K., A report on HIPERLAN Standardisation. *International Journal of Wireless Information Networks*, **2**, 99–120.

[7] LaMaire, R. O., Krishna, A., Bhagwat, P. and Panian, J., Wireless LANs and Mobile Networking: Standards and Future Directions. *IEEE Communications Magazine*, 86–94, 1996.

[8] Sun, Y., Nix, A. and McGeehan, J. P., HIPERLAN performance analysis with dual antenna diversity and decision feedback equalisation. In *Proceedings of IEEE VTC*, 1549–1553, 1996.

[9] Tellado, J., Khayata, E. and Cioffi, J. M., 'Equalising GMSK for high data rate wireless LAN. In *PIMRC*, 16–20, 1995.

[10] Wittneben, A. and Liu, W., The European wireless LAN standard HIPERLAN: key concepts and testbed results. In *Proceedings of IEEE VTC*, 1317–1321, 1997.

[11] Perry, R., Bull, D. R. and Nix, A., An adaptive DFE for high data rate applications. In *Proceedings of IEEE VTC*, 686–690, 1996.

[12] Liu, W. and Fleischmann, M., Advanced low-complexity HIPERLAN receiver using combined antenna switching diversity and simple equaliser. In *Proceedings of IEEE VTC*, 2037–2041, 1997.

[13] Tellambura, C., Johnson, I. R., Guo, Y. J. and Barton, S. K., Equalisation and frequency offset correction for HIPERLAN. In *PIMRC*, 796–800, 1997.

[14] Wittneben, A. and Dettmar, U., A low cost non-coherent receiver with adaptive antenna combining for high speed wireless LANs. In *Proceedings of IEEE VTC*, 173–177, 1997.

[15] ETSI Radio Equipment and Systems (RES). High performance radio local area networks (HIPERLANs) requirements and architectures for wireless ATM access and interconnection.

[16] FCC 97-5 Report and Order In the Matter of Amendment of the commission's Rules to Provide for Operation of Unlicensed NII Devices in the 5GHz Frequency Range, January 1997.

38

The IrDA Platform

Stuart Williams and Iain Millar

38.1. INTRODUCTION

Since its formation in June 1993 the Infrared Data Association (IrDA) has been working to establish an open standard for short range infrared data communications. At the time of its formation there were a number of vertical, non-interoperable infrared communications technologies. Today IrDA is a strong contender for anyone considering adding infrared data communications to their product. Indeed, whilst supporting their own legacies, vendors who have been offering infrared solutions for years are embarked on the transition to an IrDA-based solution.

The key goals for the IrDA are interoperability, low cost, and ease of use. Interoperability is addressed through the creation of an open standard with widespread, multi-vendor support.[1] Low cost refers to the marginal cost of adding an IrDA interface to products in high volume manufacture. For the most part the cost of adding the digital logic required to provide an IrDA interface is regarded as

---

[1] Including several major manufacturers of computers, PDAs, printers, modem and mobile phones; computer software companies; PTTs and component vendors.

negligible. The few thousand gates that it takes to implement even the recent higher speed proposals are regarded as coming free in an environment where ASIC functionality is limited largely by pin-count rather than gate utilisation. This leaves the marginal cost of adding the optoelectronic transceiver which is estimated as $2–$3 and is set to fall further in future with the availability of transceiver modules from optoelectronic suppliers.

Lastly there is ease of use. The IrDA usage model is for short range directed communication link that supports *ad hoc* point-and-shoot and place-and-play communications. The nominal operating envelope is a 1 m cone with 15° half-angles. One of the IrDA frequently asked questions over the past year has been "How do I aim my printer?" The point is that it is all very well to be able to point a PDA at a printer, but it is not really tenable for a printer to sprout legs and point back. Whilst the term "directed" is used to describe an IrDA system, it would be unfair to suggest that it requires highly accurate alignment. Indeed the physical specification allows for more omnidirectional behaviour at ranges of less than 1 m.

The IrDA system design, which is the focus of the bulk of this paper, is also a significant factor in establishing IrDA platforms as easy to use. Users of conventional communications applications have had to deal with having the correct cables to connect a computer or terminal to a peripheral such as a printer or a modem. They have had to do battle with baud rates, and bits per character and parity. They have also had the responsibility of ensuring that the correct software was loaded at opposite ends of the communications channel.

Whilst the IrDA aims to replace the serial cable for *ad hoc* peripheral connection, it also aims to add ease of use features that enable applications to identify peer entities with which they can communicate. Thus a printing subsystem; a file sharing client; a calendar management application; a business card exchange utility... can all identify and locate matching peer entities in order to make use of their services.

The IrDA chose to base its initial standards on a 115 kbit/s UART-based physical layer that had been developed by Hewlett-Packard (HP-SIR) [1,2], and an HDLC-based Link Access Protocol (IrLAP) originally proposed by IBM [3,4,5].

During the course of its first year the need to multiplex multiple application-to-application streams over a single IrLAP connection was identified and with it the need to provide a means for locating and identifying the function of application entities offerings services over an IrDA interface. These needs led to the development of the IrDA Link Management Protocol (IrLMP) [6].

This paper provides an introduction to the services provided by a IrDA platform. These are services upon which new families of infrared-aware applications will be built. End users will not be tied to either applications or platforms from a single vendor.

38.2. IrDA SYSTEM OVERVIEW

The IrDA Architecture is shown in Figure 38.1.

There are now three components to the physical IrDA layer:

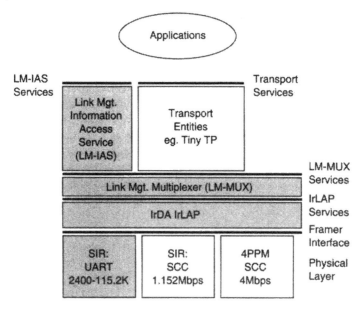

Figure 38.1 IrDA architecture.

1. the original 2400 bps–115.2 kbps HP-SIR-based [1] scheme using a conventional UART with character-stuffed packet framing;
2. a 1.152 Mbps scheme that retains the same modulation scheme, but uses a synchronous communications controller and conventional HDLC bit stuffing [7];
3. a 4 Mbps scheme that uses a 4PPM modulation scheme and frames packets with a sequence of code violations [7].

From the point of view of the Link Access Protocol (IrLAP) [5], the 1.152 Mbps and 4 Mbps extensions are regarded merely as extra speeds that may be negotiated when a device-to-device connection is established. All three physical layer schemes are designed to have a range of 1 m at off-axis angles of up to ±15°. In practice, due to component tolerancing, on-axis ranges can be substantially greater, and satisfactory operation can be achieved at off-axis angles of 30° or more.

The Link Access Protocol (IrLAP) is a variation of multi-drop HDLC [3]. It provides facilities for:

1. controlling hidden terminal problems;
2. device discovery;
3. device-to-device connect/disconnect and QoS negotiation;
4. data transfer.

IrLAP is an asymmetric protocol and uses HDLC in its normal response mode (NRM). This means that once an IrLAP connection has been established, one station becomes a primary whilst the other becomes a secondary. In the context of a point-to-point connection there is very little difference between the behaviour of primary and secondary stations. However, as we shall see, IrLAP has the potential to be extended to provide point-to-multipoint device-to-device connectivity. In this case a single primary device would be able to communicate with several secondary devices, but the secondary devices will not be able to communicate directly with each other.

The Link Management Protocol (IrLMP) [6] consists of two parts, a connection-oriented multiplexer (LM-MUX) and a directory service (LM-IAS). With the exception of the directory service itself, there are no fixed addresses within the IrDA architecture. Device addresses are chosen at random and exchanged during IrLAP discovery. Address space collisions are resolved by the device that initiates discovery. Likewise "port" space above LM-MUX is dynamically assigned. The LM-IAS directory service then serves as a means to identify the application services present within a device and the addressing information required to establish contact between application peers.

38.2.1. Addressing

Within the basic IrLAP/IrLMP IrDA platform there are three levels of addressing:

1. *Device addresses.* 32-bit randomly chosen identifiers exchanged between devices during IrLAP/IrLMP device discovery.
2. *IrLAP connection addresses.* 7-bit HDLC secondary addresses assigned to a secondary device by the primary during IrLAP connection establishment and used for the duration of that connection.
3. *IrLMP multiplexer connection addresses.* Logically an LM-MUX service access point is addressed by the concatenation of a 32-bit device address and an 8-bit multiplexer port selector. Once an IrLAP connection is established the IrLAP connection address serves as a synonym for the device address. A multiplexer connection is labelled by the addresses of the LM-MUX service access points at either end of the connection.[2]

The relationship between IrLAP connections, LM-MUX connections/connection end-points and LM-MUX service access points is shown in Figure 38.2

38.2.2. Link Access Protocol (IrLAP)

IrLAP, the IrDA Link Access Protocol [3], is based on HDLC operating in the normal response mode (NRM) [3]. Typically this mode of operation has been used on multi-drop serial lines between say a terminal controller and a group of terminals sharing the serial line. The terminal controller acts as a primary and regularly polls each of the attached terminals. There are two attractive artefacts to this behaviour in the context of directed short-range infrared communications:

---

[2]This is similar to a TCP/IP connection being labelled by the concatenation of IP address and port number at each end of the connection. Also this leads to the restriction that there may be at most one TCP/IP connection between the same pair of TCP ports. A similar restriction applies to LM-MUX connections.

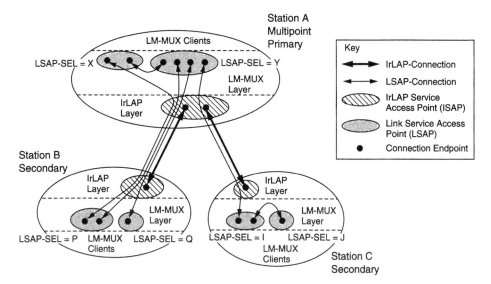

Figure 38.2 IrLAP and IrLMP connections.

1. Once the device-to-device connection has been established then in the absence of aberrant behaviour, access to the shared media is contention free.
2. The constant reversal of the "line" due to the polling mechanism acts as a beacon to indicate to other devices that approach an active link that the media is in use.

In today's world of peer-to-peer communication a master/slave protocol may seem something of an odd choice. However, it is reasoned that for directed communication the majority of real life scenarios can be addressed by the provision of a single point-to-point link between a pair of devices. In this context the difference between primaries (masters) and secondaries (slaves) becomes moot. Indeed, all differences between primary and secondary are masked by the IrLMP layer above IrLAP so that applications need not be aware of this minor asymmetry.

IrLAP operates in two main modes:

1. *Contention mode*. Procedures that occur in contention mode are device discovery, address conflict resolution and connection establishment. All contention mode traffic occurs at 9600 bps over the HP-SIR/UART physical layer.
2. *Connection mode* is entered at the point that an IrLAP connection is established. The communication speed is changed to the rate negotiated in the connection setup messages.

Connection mode traffic has priority over contention mode traffic.

Contention Mode MAC rules

Once the IrDA stack in a device has been enabled it must sense the media for a minimum of 500 ms. Also, a device must sense the media as idle for at least a further

500 ms prior to repeating a contention mode procedure. 500 ms is the absolute upper bound on the time that either a primary or secondary station may retain the right to transmit frames. Shorter intervals may be negotiated during connection establishment.

Connection Mode MAC rules

Once an IrLAP connection has been established access to the media is mediated by the exchange of a token (the P/F bit in the HDLC control field). Both primary and secondary stations monitor the exchange of this token and provide status indications upward in the event of deteriorating link quality or loss of connection. A station may transmit a number of frames, up to a limit bounded by the negotiated window size, maximum data packet size and the overriding turnaround time for returning the token to its peer.

Hidden Terminal Management

We have already touched on the hidden terminal management capabilities of IrLAP. By placing an absolute upper bound on the link turn-around time it is possible to ensure that each end of an IrLAP connection makes regular transmissions that act as a beacon to indicate that there is an established IrLAP connection in the vicinity. Hidden terminals (hidden from one end of the connection) remain silent in the presence of an active IrLAP connection.

Device Discovery and Address Conflict Resolution

An IrDA device address is a 32-bit identifier that a station randomly assigns to itself. Within the relatively small extent of the "reachspace" the probability of two or more devices choosing the same address is relatively small. IrLAP provides the facility for its client (IrLMP) to instruct devices with colliding addresses to select new device addresses. IrLMP drives the address resolution process by making a single attempt to resolve each address conflict.

Device discovery takes place in contention mode. Device discovery is used to retrieve $< DeviceAddress > < DeviceInfo >$ tuples from devices in the vicinity.

$$< DeviceInfo > = < ServiceHints > < DeviceNickName >$$

Service hints is an extensible bit map that provides for a very coarse characterisation of the services offered by the device: currently defined hints bits can specify a device as offering the services of a PDA, a computer, a printer, a modem, a fax, LAN access, telephony, or a file server. It is incumbent on the designers of an application service to state what hints bits will be set if an instance of that service is available with a station.

The Device Nickname is a short name that may be presented to the user in order to select between two otherwise identical devices.

A slotted discipline is used for discovery. The station initiating discovery issues a request that specifies the use of 1, 6, 8 or 16 slots. Stations receiving this request randomly select a slot between 0 and the specified upper bound minus 1 in which to make their response. The initiating station then "calls" out each slot in turn, marking its start with a packet that encodes the slot number slot being polled. Finally the initiating station marks the end of discovery with a final packet that includes the station's own discovery information. Device discovery is illustrated in Figure 38.3.

Connection Establishment

An IrLAP connection is initiated by the transmission of a set normal response mode (SNRM) frame, using contention mode MAC rules, by the station that will initially become the primary station. The SNRM frame contains a number of negotiable QoS parameters, including data rate capabilities; turnaround requirements negotiable from 500 ms down to 50 ms; maximum data packet size; receiver window size, 1 through 7; link disconnect and threshold times to deal with packet loss. The SNRM also contains a device address (retrieved by a recent discovery operation) and assigns a 7-bit connection address for use during the connection.

The station addressed by the SNRM responds to the SNRM with an unnumbered acknowledge (UA) frame, also at the 9600 bps contention rate, that contains the results of the negotiation process. At this point both stations apply the newly negotiated parameters. The primary immediately sends a receiver ready (RR) frame to indicate to the secondary that it is now using the negotiated communication parameters.[3]

Connection establishment is illustrated in Figure 38.4.

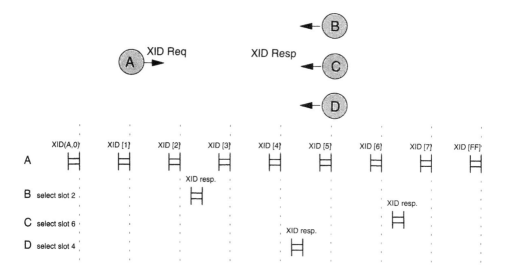

Figure 38.3 Eight-slot device discovery.

---

[3]Careful timing is needed here to ensure that both the primary and secondary have applied the new parameters prior to the transmission of the RR.

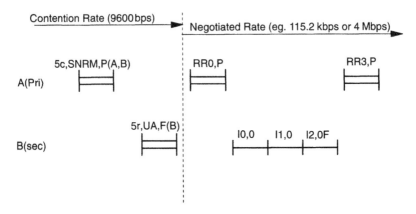

Figure 38.4 IrLAP connection establishment.

IrLAP Data Transfer Services

Once an IrLAP connection has been established then the data transfer service between devices is similar to that provided by HDLC operating in normal response mode. It provides for the reliable sequenced exchange of packets. Provision is also made for the transmission of unnumbered information (UI) frames. These frames are sent at the negotiated data rate and have priority over the exchange of sequenced information (I) frames, but they are also subject to loss without recovery.

38.2.3. Link Management Protocol (IrLMP)

The IrDA Link Management Protocol (IrLMP) [6] provides two distinctly different types of services. Firstly it provides a level of connection oriented multiplexing (LM-MUX) on top of IrLAP. Secondly it provides an information base that holds details of the application entities present in the local station that are currently offering services to other IrDA devices. Objects in this information base carry the essential addressing information necessary to establish communication with the corresponding application entities. Access to this information base is provided by an information access server and a corresponding client. Collectively the information base, the server and the client provide an information access service (LM-IAS). Both LM-IAS client and server entities are LM-MUX clients.

The Multiplexer

The LM-MUX provides a simple level of switching over the top of an IrLAP connection. It also hides the master/slave nature of IrLAP from the application and provides a symmetrical set of services to IrLMP clients.

The key goal here is to allow multiple independent sets of application entities to share access to the underlying IrLAP connection. This is increasingly important as the ability of portable platforms to multi-task improves. Also the mix of applications running on a portable platform tends to be chosen by the end-user from a potentially varied list of vendors. With this degree of "end-user" integration it is simply not tenable to offer a solution that does not allow applications to share access to the IR media. For example: consider a file sharing application that allows portable device to access files on a desktop machine. This may result in a relatively long lived connection between the devices and the end user may not really be conscious of its existence. It would not be acceptable to have to shut down the file-sharing software in order to then gain access to say an E-mail, printing or fax services.

The functionality provided by LM-MUX has already been presented in Section 38.2.1 and Figure 38.1. LM-MUX also provides a device discovery operation that combines IrLAP device discovery and address conflict resolution into a single operation.

The introduction of a simple multiplexing function above the reliable device-to-device data transfer service of IrLAP does introduce one problem. In general, multiplexing LM-MUX channels over a single IrLAP connection can lead to either data loss or deadlock. To illustrate consider a pair of peer application entities **A** and **B** connected by two LM-MUX connections. One LM-MUX connection is used to exchange data, while the other is used to send control. **A** sends on both connections and **B** receives on both connections.

Consider what happens if **A** sends a large amount of data to **B** while **B** attempts to read first from the control channel and then from the data channel. The following code fragments illustrate this behaviour:[4]

```
/* Behaviour of sender A */
for(...) {
    .
    res = send(data_fd, data, sizeof(data));
    .
}
res = send(control_fd, ctrl,sizeof(ctrl);
```

```
/* Behaviour of receiver B */
    recv(control_fd, ctrl,sizeof(ctrl));
    recv(data_fd, data, sizeof(data));
```

Data inbound for **B** is not being read. At some point buffer space holding inbound data for **B** becomes exhausted. We could allow the IrLAP flow-control mechanism to pause the sending of data by **A**. However, if we do that there is

---

[4]Assuming blocking send and receive operations.

then no way that the control information from **A** to **B** can now be sent to allow **B** to progress to reading the data. The system is deadlocked. This is a general problem for any system that multiplexes channels over a reliable channel.

Alternatively, we avoid this deadlock situation by allowing IrLMP LM-MUX to discard inbound data that it cannot deliver. However, this destroys the reliable delivery property that IrLAP gives us.

Allowing the deadlock possibility is by far the greater of these two evils, so the designers of IrLMP took the view that LM-MUX may discard inbound data that it is unable to deliver. The LM-MUX data transfer service is therefore best effort rather than reliable. The only possible cause of packet loss in these circumstances is inbound congestion within an LM-MUX channel. This inbound congestion may be completely avoided by the inclusion of a flow-control mechanism within the channel between application entities.

The IrDA has addressed this problem by defining a simple credit-based flow-control scheme dubbed Tiny TP [8].

As an alternative to flow control it should also be noted that designers of LM-MUX clients may choose instead to implement a retransmission scheme to recover from data lost due to inbound congestion; and in some circumstances the loss of data could simply be tolerated (e.g. playback of audio data).

Since these three alternatives exist, the IrDA has not mandated the use of any one particular.

There is one last facet of the multiplexer that is worthy of note. The negotiation present within IrLAP means that there is a deterministic upper bound to the time between the reversals of the underlying IrLAP connection. Some application designers may wish to exploit this deterministic behaviour; however, the presence of multiplexed streams provided by IrLMP obscures any guarantees provided by IrLAP. LM-MUX provides a mode of operation that grants exclusive access to the underlying IrLAP connection to just one LM-MUX connection.

Information Access Services

So far we have described a relatively straightforward communication mechanism that supports multiplexed communication channels between a pair of devices. A key goal for the IrDA has been ease of use. Previous IR solutions have had a tendency to disappoint users because the burden of ensuring that compatible peers application entities are active at each end of the link has fallen on the end-user. The LM-IAS within IrLMP provides the means for an application entity to identify and locate a compatible peer entity.

The LM-IAS information base contains a number of simple objects. Each object is an instance of a given class and contains a number of named attributes.

The class of an object implies the nature of the application entity that it represents, the data transfer method(s), the semantics of the information stream exchanged between peer entities etc. It also scopes the semantics of the attributes contained within instances of the given object class.

Both class names and attribute names may be up to 60 octets long. Since the meaning of an attribute is scoped by the class of the enclosing object there is no strict requirement to administer the attribute name space. Object class names do need to be administered; however, it is intended that with such a large name space some sensible conventions will ensure that class name collisions do not occur. For example, object class definitions defined by the IrDA all start with the root "IrDA:".

Whilst in general attributes are scoped by the class of the enclosing object some attributes are of such general utility that they may be regarded as having global scope. In a formal sense this requires that they are identically defined in all object classes that adopt the use of such global attributes. Typically global attributes arise in order to express an address within the IrDA environment. For example, IrLMP defines the attribute "IrDA:IrLMP:LsapSel" to identify the LM-MUX service access point of a directly attached application entity. Likewise Tiny TP defines the attribute "IrDA:TinyTp:LsapSel". Application entities advertise their accessibility via these mechanisms by the inclusion of the corresponding attribute.

There are three attribute value types:

- *Integer*. A 32-bit integer.
- *User strings*. Intended for presentation via a user interface. Up to 255 octets in length with multilingual support.
- *Octet sequence*. An opaque sequence of up to 1024 octets of information. The attribute may impose further structure on the contents of the sequence. This is a good way to cluster a body of information under one attribute.

The IrDA requires that every IrDA compliant device provides a "Device" object that carries a long form of the device name and an indication of the version of IrLMP implemented on the device and the optional features that have been implemented. The long device name is useful as it allows names of up to 255 octets in length whereas the nickname exchanged during device discovery is restricted to 19 octets (< 10 characters if Unicode is used to encode the nickname).

Access to a remote IAS information base is provided by a local IAS client entity that communicates with an IAS server entity on the remote device. The IAS server is statically bound to LSAP 0×00 on the multiplexer. This is the ONLY fixed address in the IrDA environment. All other application services are located by inspection of the information base. The IAS client and IAS server entities provide a number of querying operations on the information base. Get Value By Class is the only mandatory operation that both must support. This provides a "shot-in-the-dark" mode of retrieving attributes from objects. The notion is that a client application entity knows what application service it seeks to make use of. For example a file-sharing entity would be looking to make contact with a matching file-sharing entity. It therefore knows the object class name of information base objects that represent such an entity and implicitly it knows the name and semantics of attributes attached to such objects. There is therefore little value in the application entity browsing the information base, it merely needs to attempt to retrieve known attributes from an instance of a known object class. This is precisely what Get Value By Class does.

The remaining optional IAS operations: Get Information Base Details; Get Objects; Get Value; Get Object Info and Get Attribute Names provide for richer interactions with the information base including the ability to browse the information base.

IRLMP Client Example

Consider a fax modem that offers independent data and control channels. The fax modem advertises its data service by installing the following object in its local information base:

```
object 1 class FaxModemData {
  attribute IrDA:TinyTP:LsapSel =
      Integer(0x05);
}
```

The FaxModemData service makes use of Tiny TP and is accessible with an LM-MUX service access point selector of 5.

A client application entity that wishes to make use of the FaxModemData service performs the following operations:

```
IasValue *lsapSel;
DiscoverList *dl;
DeviceAddress *da;
FILE *fp;

/* Device Discovery */
dl = LM_DiscoverDevices(slots);

fp = NULL;

while (fp == NULL) {

/* Search Hints for Fax Device */
  while (dl != NULL) {
    if(dl->deviceInfo.hints & FAX_MASK){
       da = dl->deviceAddress;
       break;
    }
       dl = dl->next;
  }

/* Check for end of Discovery List */
  if (dl == NULL)
    break;

/* Read the LM SAP Selector */
  lsapSel = LM_GetValueByClass(
              da,"FaxModemData",
              "IrDA:TinyTP:LsapSel");

/* Connect if we got an LM-MUX SAP Sel */
  if(lsapSel != NULL) {
    fp = TinyTP_Connect(lsapSel,...)
  }
  dl = dl->next;
}
```

38.2.4. Upper Layers

We have already made mention of Tiny TP which may also be regarded as part of the plumbing as it forms part of the conduit between peer application entities. Tiny TP [8] defines a credit-based flow-control scheme and relies on LM-MUX for multiplexing that provides: per transport connection flow control; segmentation and reassembly of arbitrary sized PDUs.

A key application of IrDA is its use as a means to access local area networks. IrDA itself is not a LAN; however, it may be used as a LAN access technology. The IrLAN Specification [9] describes the protocols agreed by several IrDA LAN access developers for this purpose.

The IrDA has recently adopted protocol for application level object exchange, OBEX [10]. OBEX is based on HTTP from the World-Wide Web community. OBEX objects may be records from personal information managers, diary entries, business cards etc.; word processor, spreadsheet other traditional types of file; or other parcel of information e.g. an information hunting robot launched into a network from say a PDA attached via IR to a payphone in an airport lounge.

38.2.5. Application Interfaces

Widespread implementation of the IrDA platform services described previously and the availability of consistent application programming interfaces (APIs) is the key to creating a market for IrDA aware applications.

The IrDA community is also currently working on two types of application programming interfaces (APIs) for IrDA: legacy communications APIs and native IrDA APIs.

Legacy APIs

There is a general perception of infrared as merely a cable replacement technology. From the preceding discussion it should be apparent that the inclusion of IrLMP, particularly LM-IAS, makes it much more. Nevertheless this perception persists and there is a desire to be able to run legacy serial and parallel port communications applications over infrared in much the same way that terminal emulation applications were transitioned onto LANs from RS232 cables, see Figure 38.5. These needs are addressed by the IrDA IrCOMM specification [11].

Native APIs

New applications will take full advantage of the potential that IrLMP offers to enable compatible application peers to identify and locate each other. This requires a native API for IrDA that exposes the full functionality of IrLMP and Tiny TP to the application programmer.

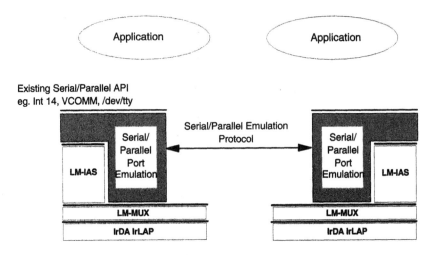

Figure 38.5 Legacy communications API support.

IrDA members are interested in defining a Winsock 2 Service Provider and API semantics [12,13]. Mapping the LM-IAS services into Winsock 2 is likely to prove a particular challenge! Figure 38.6 shows the sockets-based approach to the API.

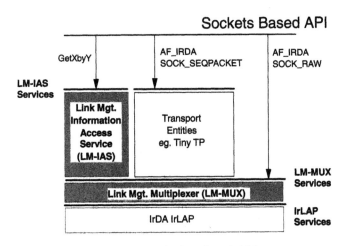

Figure 38.6 Sockets-based API.

38.3. APPLICATION SERVICES

With a stable platform on which to build, application and application service designers can rapidly populate the space above the base platform. Some services will come about as the result of either open or closed collaborations, others will be the work of an individual or a single organisation. It is incumbent on application service designers to specify:

1. the IrLMP service hints that will be set if an instance of the service is being advertised;
2. an object class to carry the parameters essential to establish communication between peers;
3. data transfer methods: i.e. raw, over Tiny TP or some other method specified in the service definition;
4. the application level protocol.

The degree to which the application service designer makes this information open is a matter of judgement for them. They may choose to seek endorsement from the IrDA; they may simply publish the specification of their service. Alternatively it may be regarded as proprietary and closed.

38.4. CONCLUSIONS

The future of short range directed infrared data communications looks bright!

With some 130 member companies the technology developed within the IrDA is available on just about every significant mobile computing platform. End users will be able to casually exchange, print and share information from whole documents to snippets from a diary or a business card. They will be able to do this without the hassle of needing to have the right cables and without having to do combat with pages of configuration and setup dialogues commonplace with serial and networked communication.

The ease of use that is characteristic of IrDA applications is largely due to the device discovery and QoS negotiation facilities in IrLAP and the information access services specified in IrLMP.

REFERENCES

[1] Brown, P. D., Moore, L. S. and York, D. C., Low power optical transceiver for portable computing devices. US Patent No. 5,075,792. Assignee: Hewlett-Packard Company, December 24 1991.
[2] Infrared Data Association Serial infrared (SIR) physical layer link specification. Version 1.0, April 27 1994.
[3] International Organisation for Standardisation (ISO). High level data link control (HDLC) procedures – elements of procedures. ISO/IEC 4335, September 15, 1991.
[4] Williams, T. F., Hortensius, P. D. and Novak, F., Proposal for: infrared data association serial infrared link protocol specification. Version 1.0, IBM Corporation, August 27, 1993.
[5] Infrared Data Association. Serial infrared link access protocol (IrLAP). Version 1.1, June 16 1996.
[6] Infrared Data Association. Link management protocol. Version 1.1, January 23, 1996.
[7] Infrared Data Association. Serial infrared physical layer specification. Version 1.1, October 17 1995
[8] Infrared Data Association, TinyTP: A Flow-Control Mechanism for use with IrLMP. Version 1.1, October 20, 1996.

[9]   Infrared Data Association. LAN access extensions for link management protocol IrLAN. Version 1.1, January 1, 1997.

[10] Infrared Data Association, IrCOMM: serial and parallel port emulation over IR (Wire replacement). Version 1.0, November 7, 1995

[11] Infrared Data Association IrDA Object Exchange Protocol. Version 1.0, January 22 1997.

[12] Winsock 2. Windows sockets 2 application programming interface: an interface for transparent network programming under microsoft windows. Revision 2.0.6, February 1, 1995.

[13] Winsock 2. Windows sockets 2 service provider interface: a service provider interface for transparent network programming under microsoft windows. Revision 2.0.6, February 1, 1995.

39

Analysis of Antennas and Propagation for 60 GHz Indoor Wireless Networks

M. Williamson, G. Athanasiadou, A. Nix and T. Wilkinson

39.1. INTRODUCTION TO WIRELESS LOCAL AREA NETWORKS

In the forthcoming years, the market for indoor wireless networking facilities is expected to grow considerably in both commercial and domestic sectors. A wireless local area network (LAN) can be easily installed, maintained and extended, and has the potential to provide flexible, high data-rate access to network facilities. The basic principle of a wireless LAN is a radio coverage cell within which two-way communication is supported between a fixed base station (BS) and a number of mobile stations (MS). Typically the base station is mounted on a wall or ceiling and provides a data link between the mobile stations and the existing fixed network. In addition, a number of BS hubs can be interconnected to provide increased capacity or coverage. Mobile stations can potentially include personal digital assistants (PDAs) and laptops as well as fixed PCs and workstations (Figure 39.1).

39.2. BROADBAND WIRELESS ACCESS

In North America, Europe and Japan, there is a great deal of research focused on the development of future wireless access systems. In particular, there is much interest in the design of wireless LANs which support data rates in excess of 100 Mbps and thus allow transfer of broadband voice, video and data between wireless terminals. The UHF bands conventionally used for mobile communication systems are not suitable for broadband applications due to the lack of available bandwidth. In this respect, the millimetre-wave frequencies above 20 GHz are considered most suitble [1,2]. In December 1995, the US Federal Communications Commission (FCC) allocated spectrum in the 59–64 GHz millimetre-wave band for general purpose unlicensed operations [3], and consequently there is significant interest in developing systems which operate at these frequencies [4,5,6].

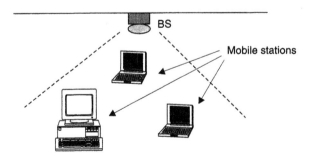

Figure 39.1 Wireless LAN coverage cell.

39.3. CHARACTERISTICS OF OPERATION AT 60 GHz

39.3.1. Compact Directional Antennas

In any high data rate, wireless application, delay distortion due to multipath propagation causes severe impairment of performance. For the purposes of current wireless LAN systems, adaptive equalisation and multi-carrier techniques are regarded as the most suitable means of combating multipath effects. However, in terms of cost and computational complexity, both of these techniques are somewhat undesirable. In the millimetre-wave region, directional antennas of relatively compact size can be realised due to the wavelengths involved. Advances in antenna miniaturisation and monolithic millimetre-wave IC technology mean that it is now feasible to employ electronically-switched, multiple sector antennas to track the LOS, or significant signal components, and thus eliminate many unwanted multipath components. This use of antenna directionality may remove the need for more complex anti-multipath techniques, but operation is most likely to be restricted to locations where there is a line-of-sight (LOS) or dominant reflecting path between BS and MS [7].

39.3.2. Severe Attenuation

60 GHz signals are heavily attenuated by transmission through concrete and other standard building materials, as well as many other objects in typical office or domestic environments [8]. The exceptions to this are materials such as plasterboard and glass. Typically this limits the likely operating range to around 10–20 m, but enables good spectrum efficiency and higher user density to be achieved since frequency re-use between neighbouring rooms is viable.

39.3.3. Oxygen Absorption

In the 60 GHz range signals suffer additional attenuation due to atmospheric oxygen of around 15 dB/km [9]. This is insignificant in relation to the attenuation due to transmission through walls and local objects, but means that in comparison with operation at other frequencies, there is less interference from external signals.

39.3.4. Safety Aspects and Power Limitations

For millimetre waves the maximum safe power density for continuous exposure is 1 mW/cm^2 [10]. The report of the millimetre-wave communications working group to the FCC [11] recommends that in the 59–64 GHz band, at a distance of 3 m, the average power density of any emission should not exceed 9 μW/cm^2, and the peak power density should not exceed 18 μW/cm^2. In practice, for a typical transceiver unit used at a mobile station in a 60 GHz wireless LAN, a transmitter output power of around 10–20 mW may be achieved. It should be noted that at 60 GHz there is around 68 dB path loss at 1 m from the transmitter.

39.4. MODELLING THE 60 GHz CHANNEL

It is essential to characterise the nature of the indoor propagation channel before designing a suitable system architecture. In recent years ray-tracing has emergd as a viable technique for producing deterministic channel models. For the purposes of this analysis, the three-dimensional, image-based ray-tracing algorithm presented in [12,13] was used to model propagation in a single-floor indoor environment. The model considers each ray separately, with reflections and transmissions computed using vector mathematics, and each object characterised by its permittivity, conductivity and thickness.

The ray-tracing prediction model is used to study the variation in received power, RMS delay spread and Rician k-factor in a typical office environment. The benefits provided by using directional antennas are investigated in both LOS and non-LOS locations and the performances of different BS and MS antenna configurations are compared. The effects of non-optimal antenna alignment are observed, and the area of operational coverage is determined for transmission at 10 Mbps and 100 Mbps. In addition, the dependence of coverage on wall material characteristics, and the effects of adding dividing partitions to the environment are investigated.

Figure 39.2 shows the simulated environment used for this study. It consists of various rooms and corridors (with 30 cm-thick concrete walls), containing windows, doors and partitions. The ceiling height is 4.5 m and the base station (BS) height is 3 m. The BS output power is 10 mW (10 dBm) and the channel impulse response is analysed over a 2 m resolution grid of mobile station (MS) locations at a height of 1 m.

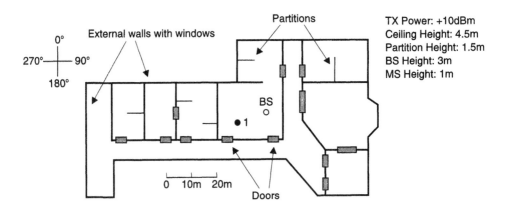

Figure 39.2 Simulated Indoor Environment.

39.5. SIGNAL PROPAGATION AT 60 GHz

Figure 39.3 shows the variation in average received power in the study environment for transmission at 5.1 GHz, a frequency used for current wireless LAN standards such as HIPERLAN [14]. Figure 39.4 shows the corresponding variation in received power at 60 GHz. In each case the reflection and transmission parameters used in the ray-model are based on values that have been measured at these particular frequencies. The characteristics of the materials used in the ray-tracing model are shown in Table 39.1 for both 60 GHz [8] and 5.1 GHz [15].

A comparison of Figures 39.3 and 39.4 illustrates the significant difference in the coverage provided at the two frequencies. For 5 GHz systems, operation can potentially be supported when the BS and MS are located in adjacent rooms, and a number of wall boundaries are required before frequency re-use is viable. At 60 GHz, there is considerably more attenuation with distance, and the coverage cell is well defined by the perimeter of the room, due to heavy attenuation by walls. This reduces problems due to interference and potentially allows higher capacity due to increased spectrum re-use. However the area of coverage is limited mainly to single rooms. From the power levels indicated, operation in the study environment is likely to be restricted to locations where there is either a LOS path between BS and MS, or a reflecting path that does not undergo wall transmission. In addition, some significant power levels are observed in the adjacent corridors, due to signals escaping through doors.

If frequencies are to be re-used between neighbouring rooms then this may cause co-channel interference problems. Figure 39.5 shows the variation in received power at the two frequencies over a cross-section of locations within the environment, as indicated by the dotted lines in Figure 39.3 and 39.4. This highlights the additional attenuation through walls at 60 GHz. It should be noted that the peak power is received when the MS is located directly beneath the BS such that the distance between them is 2m.

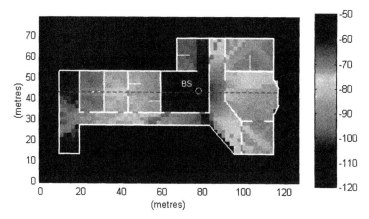

Figure 39.3 Variation in received power at 5.1 GHz (10 mW transmit power). (For a colour version of this figure see the plate section which appears between pp. 614 and 615.)

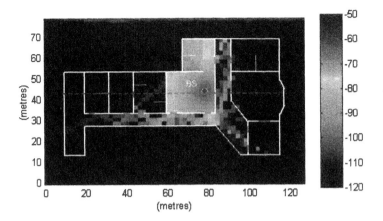

Figure 39.4 Variation in received power at 60 GHz (10 mW transmit power). (For a colour version of this figure see the plate section which appears between pp. 614 and 615.)

Table 39.1 Characteristics of materials used in ray-tracing model [8,15]

Object	Material	Thickness (m)	Height (m)	Frequency (GHz)	Relative permittivity ε_r	Conductivity σ (S/m)	Attenuation coefficient α (dB/cm)
Wall	Concrete	0.3	4.5	5.1	5	0.1368	1
				60.0	6.14	1.006	6.67
Door	Wood	0.05	2.5	5.1	2	0.0006	0.007
				60.0	1.57	0.3218	4.22
Partition	Chipboard	0.04	1.5	5.1	—	—	—
				60.0	2.86	0.5308	5.15
Window	Glass	0.01	2.5	5.1	6	10^{-12}	6.7×10^{-12}
				60.0	5.29	0.8476	6.03

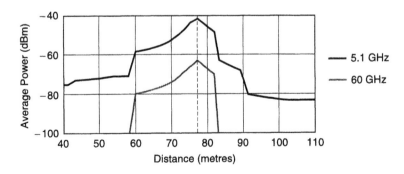

Figure 39.5 Comparing power variation at 5 GHz and 60 GHz.

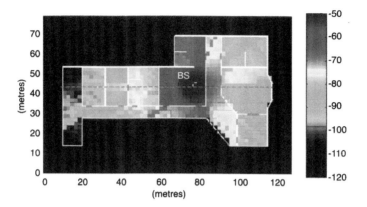

Figure 39.3 Variation in received power at 5.1 GHz (10 mW transmit power).

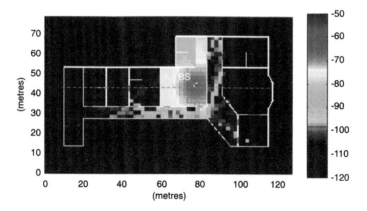

Figure 39.4 Variation in received power at 60 GHz (10 mW transmit power).

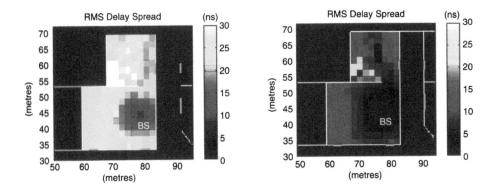

Figure 39.6 Variation in RMS Delay Spread: (a) omni:omni; (b) 60°:60°.

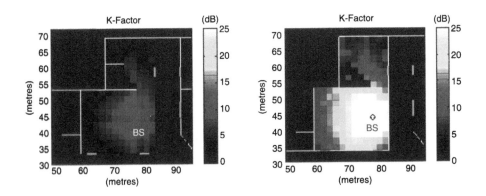

Figure 39.7 Variation in k-factor: (a) omni:omni; (b) 60°60°.

39.6. MEASUREMENT PARAMETERS

In order to investigate the nature of the 60 GHz indoor radio channel, the variations in received power, RMS delay spread and Rician k-factor are studied throughout the simulated office environment. The RMS delay spread, R, is given by equations (39.1) and (39.2), where a_k and τ_k are the amplitude and time of the kth ray respectively, and m is the total number of rays within a 30 dB window relative to the peak of the delay profile. Analysis of RMS delay spread is of benefit since channels with suitably low delay spreads do not suffer problems due to inter-symbol interference (ISI) at high bit rates.

$$R = \sqrt{\frac{\sum\limits_{k=1}^{m}(\tau_k - t_a)^2 a_k^2}{\sum\limits_{k=1}^{m} a_k^2}} \tag{39.1}$$

$$t_a = \frac{\sum\limits_{k=1}^{m}(\tau_k a_k^2)}{\sum\limits_{k=1}^{m} a_k^2} \tag{39.2}$$

The k-factor, K, of the channel impulse response, is taken as the ratio of the power in the LOS or dominant signal component, a_{max}, to the sum of that in the $m-1$ random multipath components. This is shown in equation (39.3). The k-factor has a direct impact on the quality of the "eye-diagram" at the receiver, and consequently high bit-rate operation can only be suppored when suitably high k-factors are achieved.

$$K = \frac{a_{max}^2}{\left(\sum\limits_{k=1}^{m} a_k^2\right) - a_{max}^2} \tag{39.3}$$

39.7. EFFECTS OF ANTENNA DIRECTIVITY

Figure 39.6 shows the spatial variation in RMS delay spread in the room where the BS is located, for two different antenna combinations. For the left-hand plot omni-directional antennas are used at both BS and MS, with the half-wave dipole radiation pattern assumed for both antennas. For the right-hand plot, directional antennas are used at both the BS and MS. The directional antennas have ideal Gaussian-shaped main beams with a 3 dB beamwidth of $60°$ in both azimuth and elevation planes, a constant sidelobe level of -30 dB, and a gain of 12 dBd in the maximum direction. For each grid location the antennas are optimally aligned such

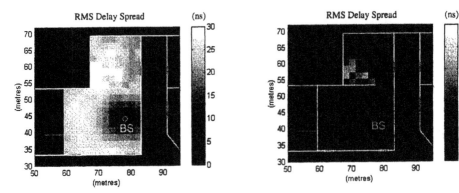

Figure 39.6 Variation in RMS Delay Spread: (a) omni:omni; (b) 60°:60°. (For a colour version of this figure see the plate section which appears between pp. 614 and 615.)

that the peak of the radiation pattern points directly towards the angle of arrival, or angle of departure, of the most significant ray. Using omnidirectional antennas, the average delay spread in the room is 22.7 ns and using the directional antennas this is reduced to 8.2 ns, Figure 39.7 shows the spatial variation in Rician k-factor for the same two antenna configurations. For the left-hand plot, using omnidirectional antennas, the average k-factor is 4.5 dB; for the right-hand plot, using 60° directional antennas, the k-factor is increased to 15.4 dB.

Figures 39.6 and 39.7 demonstrate that the use of directional antennas has considerable effect on both RMS delay spread and k-factor. Performance is improved by the use of directional antennas for all LOS locations and those non-LOS locations where there is a first-order reflecting path between MS and BS. However, the benefits are less noticeable in the areas immediately behind the concrete wall that divides the two areas of the main room. This is due to the fact that even the most dominant signals arriving at the MS have undergone two or more reflections.

Figure 39.8 shows the impulse response experienced at a typical LOS location, position 1 (Figure 39.2), using omnidirectional antennas at both BS and MS. In order to illustrate the degree of multipath suppression provided by using 60° direc-

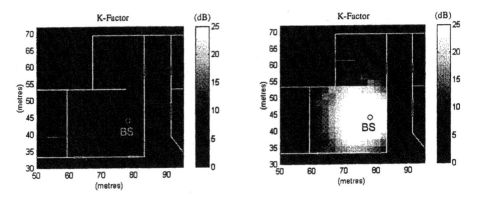

Figure 39.7 Variation in k-factor: (a) omni:omni; (b) 60°:60°. (For a colour version of this figure see the plate section which appears between pp. 614 and 615.)

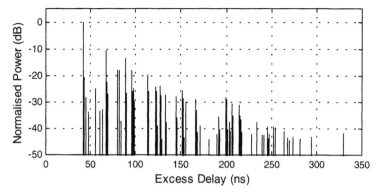

Figure 39.8 Impulse response (position 1, omni:omni).

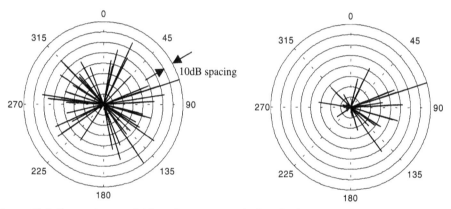

Figure 39.9 Comparing spatial impulse responses (azimuth plane) (position 1, omni:omni and 60°:60°).

tional antennas. Figure 39.9 compares the spatial impulse responses in the azimuth plane, for omni:omni and 60°:60° configurations. The directional antennas are optimally aligned as previously.

For the omni:omni configuration, although the LOS path is clearly dominant, first-order reflected rays of relatively high power (10–20 dB below LOS) arrive from various directions in the azimuth plane, giving an RMS delay spread of 24.8 ns and k-factor of 5.6 dB. Using 60° directional antennas provides suppression of most major rays outside of the 3-dB beamwidth, giving an RMS delay spread of 12.1 ns and k-factor of 18.3 dB.

Figure 39.10 shows the difference between impulse responses for the two antenna combinations. Using directional antennas, the power in the LOS path is increased due to the additional gain whilst the majority of random components are suppressed. The other components that show an increase in power correspond to signal paths that undergo various orders of reflection and arrive at the MS within the 60° 3 dB beamwidth. The LOS component can be seen to have gained around 24 dB, which corresponds to twice the maximum gain of the 60° antenna.

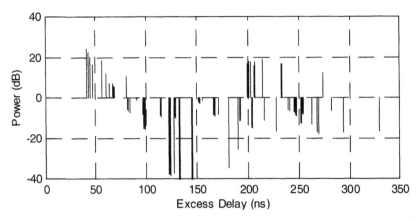

Figure 39.10 Impulse response difference plot (position 1, 60°:60° – omni:omni).

39.8. COMPARING ANTENNA COMBINATIONS

For the design of future 60 GHz wireless systems, it is necessary to compare the performance of a number of antenna combinations. This will enable decisions to be made as to whether directional antennas should be used at the central, ceiling-mounted BS, at the MS, or at both. In addition, suitable antenna beamwidths must be chosen. This is dependent on the cost, complexity and availability of suitable antenna technology and will also determine the number of sectors required if a switched-beam configuration is to be employed. In theory, at 60 GHz very small beamwidths are possible due to the wavelengths involved, and these are realised in practice using either dielectric lens antennas or arrays of patches. For a square antenna aperture the relationship between 3 dB beamwidth, B, in degrees, and aperture diameter, D, in wavelengths, is given approximately by $B = 51\lambda/D$ [16]. So for example, given that the wavelength at 60 GHz is 5 mm, for a 3 dB beamwidth of 60° an aperture diameter of 4.25 mm is required. For a typical MS, such as a laptop PC, antennas are likely to be mounted on or within the casing above the screen, or else mounted on the visible end of the PCMCIA card which is slotted into the PC.

For the purposes of this investigation, the performances of omni:omni, 60°:omni, 30°:omni, 60°:60°, and 30°:30° (BS:MS) antenna configurations are compared. Figure 39.11 shows the cumulative distribution of RMS delay spread for LOS and non-LOS locations in the room where the BS is located. Figure 39.12 shows the corresponding distribution of k-factor. In each case the directional antennas are optimally aligned.

It can be seen that the 30°:30° configuration gives rise to the lowest delay spreads and highest k-factors. For 80% of the locations studied the received impulse response has a delay spread of less than 7 ns, or k-factor of more than 11 dB (indicated by arrows in Figures 39.11 and 39.12). This is a considerable improvement over the omni:omni configuration, for which 80% of the locations have a delay spread greater than 18 ns, or k-factor of less than 8 dB. It has been found that

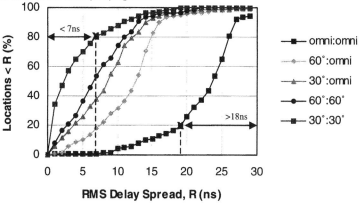

Figure 39.11 CDFs of RMS delay spread.

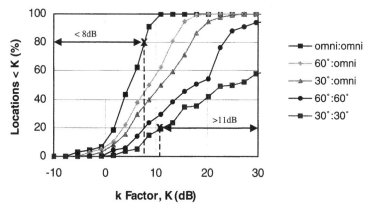

Figure 39.12 CDF of k-factor.

relative to omnidirectional antennas, the use of optimally aligned directional antennas gives rise to a reduction in delay spread and an increase in k-factor for all LOS locations and many obstructed ones.

39.9. ANGULAR VARIATION OF DELAY SPREAD AND k-FACTOR

39.9.1. Comparing Different Antenna Beamwidths

Figure 39.13 shows the variation in RMS delay spread at position 1 (Figure 39.2) for both 30° and 60° directional antennas as they are rotated through 360°. An omnidirectional BS antenna is assumed. Figure 39.14 shows the angular variation in k-factor for the same location. The angle from position 1 to the BS is 72°, and both antennas exhibit minimum delay spread and maximum k-factor for this orientation,

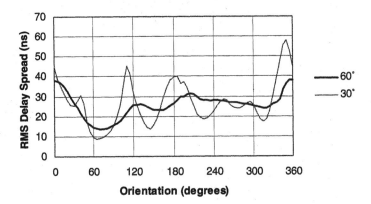

Figure 39.13 Angular variation in RMS delay spread (position 1).

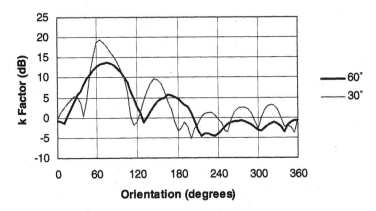

Figure 39.14 Angular variation in k-factor (position 1).

with the 30° degree antenna achieving the more favourable values (9 ns, 19 dB) due to increased multipath suppression. However, moving either side of this pointing angle gives rise to large values of delay spread, up to a maximum of 58 ns, and k-factor values as low as −5 dB.

39.9.2. Comparing LOS and non-LOS Locations

Figure 39.15 shows the variation in average power, RMS delay spread and k-factor as a directional antenna with a 20° 3 dB beamwidth is swept through 360° at the LOS mobile station position indicated in Figure 39.16, where there is a direct signal path from the base station. The MS antenna has a gain of 15 dBi and an omnidirectional antenna is used at the base station, with the transmit power set at 1 mW (0 dBm).

At an angle of approximately 80°, when the antenna is pointing directly at the base station, there is a peak in the received power level, with a corresponding minimum delay spread value, and maximum k-factor. This is clearly the ideal orientation

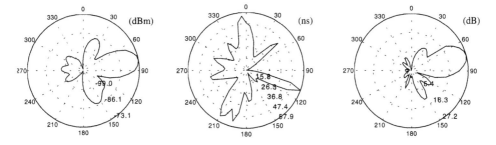

Figure 39.15 Angular variation in power, RMS delay spread and k-factor at LOS location.

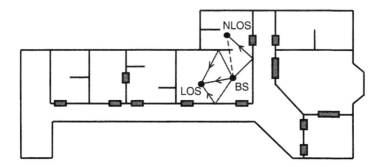

Figure 39.16 Studied LOS and non-LOS Mobile Station Locations.

for a directional mobile station antenna, but there are also two other positions with relatively high power, low delay spread and high k-factor. These are at angles that correspond to the first order reflections off the walls either side of the base station as indicated by rays shown in Figure 39.16. This demonstrates that operation can potentially be supported even if the LOS path is blocked, for example by someone temporarily standing in the way. However, either side of the ideal pointing angle, we encounter values of delay spread as high as 60 ns. This is far worse than the corresponding delay spread using only omnidirectional antennas and gives an indication of how received signal quality can be degraded if directional antennas are incorrectly aligned.

Figure 39.17 shows the variation in the same parameters for the non-LOS mobile station location shown in Figure 39.15. Here the received power is at a maximum when the antenna is orientated at approximately 135°, rather than the angle where the mobile station points directly towards the base station. This is the direction of arrival of a first-order reflecting ray as indicated in Figure 39.16. This angle also corresponds to a high k-factor and low RMS delay spread, and in this case is the ideal orientation.

For most LOS locations, as well as the direct orientation, there are other potentially suitable orientations for data transmission between BS and MS, providing sufficient signal power is available. However, these characteristics are highly dependent on the wall materials used in the environment, and in this case particularly

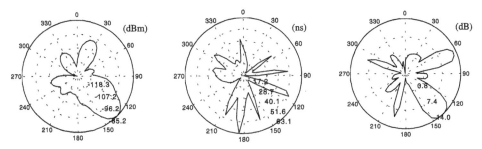

Figure 39.17 Angular variation in power, RMS delay spread and k-factor at non-LOS location.

strong reflections occur since the walls are 30 cm thick and made of concrete (see section 39.13). For non-LOS locations the optimum antenna orientations are at angles corresponding to a dominant first- or second-order reflecting path rather than the direct (transmission) path between BS and MS.

39.10. NON-OPTIMAL ANTENNA ALIGNMENT

The results presented in section 39.9 demonstrate that for both LOS and non-LOS locations there are optimum MS antenna orientations where low delay spread and high k-factor can be achieved. However it is also evident that incorrect antenna alignment can give rise to delay spreads greater than 50 ns and k-factors of less than −5 dB. It is of interest to further investigate the performance statistics for directional antennas that are not optimally aligned. Figure 39.18 shows the cumulative distribution of RMS delay spread when the BS antenna is offset by different

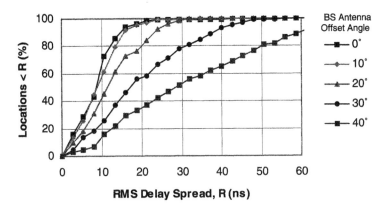

Figure 39.18 CDF of RMS delay spread for different angle offsets.

angles relative to the optimum alignment. A 30° directional antenna is used at the BS with an omnidirectional antenna at the MS. For small angle offsets ($< 10°$) there is no noticeable effect on the delay spread and the same was found to be true for k-factor. This means that some inaccuracy in antenna alignment can be tolerated before operation is affected. However, once the MS antenna is offset by more than 20°, performance is significantly degraded.

In a further study comparing the effects of non-optimal alignment for a range of different antenna combinations, it has been shown that offsets of less than 30% of the 3 dB beamwidth generally have no noticeable effect on either delay spread or k-factor. However, if a directional antenna is offset by an angle equal to its 3 dB beamwidth, the RMS delay spread is typically doubled.

39.11. RELATIONSHIP BETWEEN TRANSMISSION RATE, DELAY SPREAD AND k-FACTOR

The normalised delay spread of a channel is given by $R.T_r$, where R is the RMS delay spread and T_r is the transmission rate. In a Rayleigh-type channel, multipath components arrive at the mobile station with a random nature and with no single dominant (e.g. LOS) component. For this type of channel it is normally assumed that the maximum tolerable normalised delay spread is 0.1, in order to achieve a BER floor of 10^{-3}. Table 39.2 shows the maximum tolerable delay spread for operation at 10 Mbps and 100 Mbs using this assumption.

In a Rician-type channel, one dominant signal component arrives at the MS with significantly higher power than the other multipath components. In this case if the k-factor of the channel is sufficiently high then a higher (i.e. worse) RMS delay spread can be tolerated for a given data rate. Results presented in [17] demonstrate this relationship and a selection of these are shown in Table 39.3. Here a BER floor of 10^{-3} is assumed for QPSK operation.

The results in Table 39.3 show that transmission at 100 Mbps can be supported with RMS delay spreads of more than 5 ns provided the channel has a k-factor of 10 dB. The ray-tracing simulation has shown that such delay spreads and k-factors are feasible when correctly oriented directional antennas are used, and this demonstrates the fact that wireless LANs can operate at high data rates without the need for equalisation or multi-carrier techniques.

Table 39.2 Maximum tolerable delay spread (Rayleigh channel)

Transmission rate (Mbps)	Max. tolerable delay spread (ns)
10	10
100	1

Table 39.3 Maximum tolerable delay spread for given k-factor (Rician channel) [17]

Transmission rate (Mbps)	Rician k-factor (dB)	Max. tolerable delay spread (ns)
10	6	20
	10	70
100	6	2
	10	7

39.12. COVERAGE AREA FOR 10 MBPS AND 100 MBPS OPERATION

Using the data in Tables 39.2 and 39.3 we can determine which locations in the study environment will be covered for operation at 10 Mbps and 100 Mbps. Figures 39.19–39.22 show the area of coverage for four different BS and MS antenna combinations, using a transmit power of + 10 dBm. In each case it is assumed that a minimum received power of −75 dBm is required.

1. Simplest case. Figure 39.19 shows the area of coverage using omnidirectional antennas at both BS and MS. Operation at 10 Mbps is only supported for 19% of locations in the room where the BS is located, all within a radius of about 10 m around the BS. Operation at 100 Mbps is not possible at all due to high delay spreads.

2. Ideal case. Figure 39.20 shows the area of coverage using 30° degree directional antennas at both BS and MS. For each MS location the directional antennas are optimally aligned. Operation at 10 MBps is supported in 96% of locations in the room where the BS is located, including all those where there is a LOS path; 100 Mbps operation is supported in 84%. In addition, at both data rates operation is possible in some locations in the adjacent corridors, due to signals escaping through the doors.

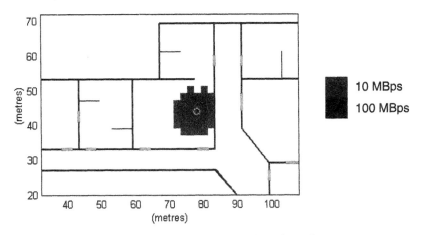

Figure 39.19 Area of coverage (omni:omni).

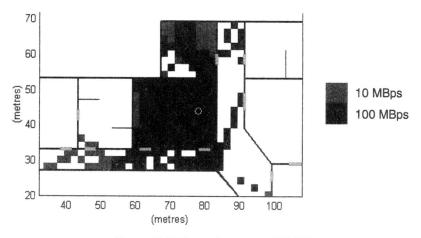

Figure 39.20 Area of coverage (30°:30°)

At both data rates, outage in non-LOS locations is due to a lack of received power since the dominant ray undergoes two or more orders of reflection. In this particular environment the concrete dividing wall causes problems in those MS locations which it directly obscures from the BS. In any indoor environment the location of the BS will have a significant effect on the area of coverage and in this case it is possible that these problems could be overcome by relocating the BS nearer to the gap between the two sections of the room.

3. Practical implementation 1 – Antenna complexity at BS. It may be desirable for a wireless LAN system to incorporate the majority of antenna complexity at the central, ceiling-mounted or wall-mounted BS, whilst using simple omnidirectional or wide-beam directional antennas at the MS. This minimises the cost and complexity of the portable equipment whilst ensuring that the user does not have to consciously re-align the MS antenna towards the BS in order to gain network access. To obtain the benefits provided by directivity, multiple antenna sectors can be implemented at the central BS; alternatively, adaptive antenna arrays can be used, but this is likely to be less cost-effective. The probability of a favourable transmission channel can also be increased by using spaced antenna diversity. This can be implemented at the MS, since the spacing required to ensure uncorrelated paths is small at 60 GHz.

Figure 39.21 shows the area of coverage using 12 fixed antenna sectors of 30° beamwidth at the BS and an omnidirectional antenna at the MS; the orientation of each antenna sector is indicated. For each MS location the best BS antenna sector is chosen based on the highest received power level. For the majority of locations (both LOS and non-LOS) this has been found to correspond with minimising the RMS delay spread and maximising the k-factor. The coverage is clearly not as good as that obtained using 30° directional antennas at both terminals (Figure 39.20). At 10 Mbps operation is supported in 85% of locations in the room in which the BS is located, nearly all of which have a LOS path between BS and MS. In most non-LOS locations operation is not possible due to a lack of received power. At 100 Mbps operation is possible in 38% of locations, limited to a radius of about 15 m around the BS. In other locations there is insufficient k-factor or too large a delay spread

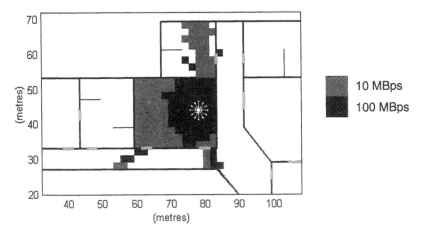

Figure 39.21 Area of coverage (12 sectors of 30°:omni).

since the omnidirectional antenna does not provide suppression of multipath components.

4. Practical implementation II: multiple sectors at both BS and MS. In a future 60 GHz wireless LAN system it may be impractical to implement a multiple-sector antenna with as many as 12 sectors, due to excessive antenna switching losses and the fact that the required aperture size increases as the beamwidth is reduced. For these reasons the antenna configuration described in section 39.12.3 may not be feasible. Another possibility is the use of multiple sectors at both BS and MS, but using a wider beamwidth and correspondingly fewer sectors. For example possibilities include 60° (six sectors), 72° (five sectors), 90° (four sectors), or 120° (three sectors).

Figure 39.22 shows the area of coverage using four fixed antenna sectors of 90° beamwidth at both BS and MS. The orientation of each BS antenna sector is shown and the orientation of the MS is randomly assigned for each location. In each case the best BS and MS antenna sectors are chosen based on maximising the received power. Operation at 10 Mbps is supported in 74% of locations in the room where the BS is located and this does not include any non-LOS locations, since the wider-beamwidth antenna sectors provide less effective multipath suppression. At 100 Mbps, 28% of locations are covered, limited to a radius of about 12 m around the BS.

The antenna parameters and percentage coverage at 10 Mbps and 100 Mbps for section 39.12 are summarised in Table 39.4.

39.13. DEPENDENCE ON MATERIAL PARAMETERS

The heavy attenuation of 60 GHz signals by certain materials means that propagation is heavily dependent on the nature of the building materials encountered, and RMS delay spread and k-factor values can vary considerably as a result. Figure

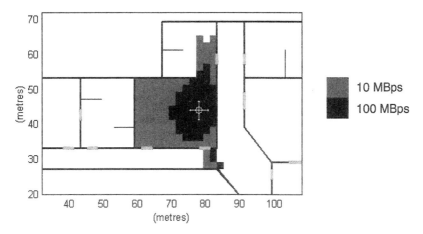

Figure 39.22 Area of coverage (four sectors of 90°: four sectors of 90°).

Table 39.4 Summary of antenna characteristics and areas of coverage

BS beamwidth (degrees)	BS gain (dBi)	MS beamwidth (degrees)	MS gain (dBi)	Coverage at 10 Mbps (%)	Coverage at 100 Mbps (%)
Omnidirectional	2.1	Omnidirectional	2.1	19	0
30 (optimally aligned)	16.6	30 (optimally aligned)	16.6	96	84
30 (12 sectors)	16.6	Omnidirectional	2.1	85	38
90 (four sectors)	7.1	90 (four sectors)	7.1	74	28

39.23 shows the cumulative distribution of RMS delay spread for the same room in the environment shown in Figure 39.2; however, this time the walls are made of aerated concrete rather than concrete [8]. Figure 39.24 shows the corresponding cumulative distribution of k-factor values. The electrical parameters of the two materials are compared in Table 39.5.

Comparing Figures 39.23(a) and 39.23(b) with Figures 39.11 and 39.12 demonstrates the extent to which a variation in wall composition can alter propagation within the environment. The RMS delay spread values are considerably reduced for each of the five different antenna configurations, in many locations dropping to ¼ of the value previously calculated. Similarly the k-factor is noticeably higher than previously, particularly in LOS locations near to the BS.

Table 39.5 Electrical characteristics at 60 GHz [8]

Material	Relative permittivity, ε_r	Conductivity σ (S/m)	Attenuation coefficient α (dB/cm)
Aerated concrete	2.26	0.339	3.70
Concrete	6.14	1.006	6.67

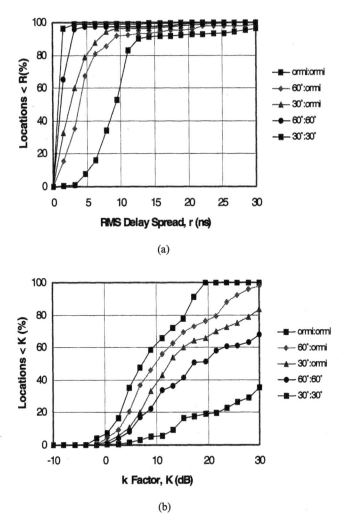

Figure 39.23 (a) CDF of RMS delay spread; (b) CDF of k-factor (using aerated concrete walls).

This variation can be attributed principally to the fact that the aerated concrete is less reflective than concrete. Consequently, in LOS locations there is an improvement in performance, since multipath (reflected) components are less prominent in comparison with the LOS component, thus reducing the delay spread and increasing k-factor. However, in non-LOS locations, where operation relies on reflected signals, this means that the performance is worse, although this is offset somewhat by the fact that signals are less heavily attenuated by transmission through the aerated concrete than through concrete.

These results highlight the need for careful analysis of building structure and layout when installing a 60 GHz wireless LAN. In this respect the environment studied in Sections 39.4–39.12 represents a "worst-case" scenario, since the thick concrete walls are particularly reflective and provide such a high attenuation that they are effectively impenetrable to 60 GHz signals.

39.14. EFFECTS OF OFFICE PARTITIONS

In many modern open-plan office environments, partitions are used to divide up a large area into a number of office "cubicles". These partitions are normally 2–5 cm thick and vary in height typically from 1 m to 2 m. In some cases partitions are made of lightweight wood or plastic, in other cases they may contain metal. This has serious implications for 60 GHz wireless LAN systems, due to the high attenuation experienced when transmitting through certain materials. For example if partitions do contain metal and are reasonably high relative to the ceiling, then the majority of potential MS locations will be non-LOS and thus rely on reflections off nearby partitions in order for operation to be supported. In addition many partitions have metal strips along their edges which may act as strong, undesirable reflectors at a short range from the MS. This is likely to cause additional multipath problems and may lead to an intolerably high delay spread.

Figure 39.24 shows the same office study environment used in the previous sections, with a number of dividing partitions added in the main room. The partitions have a height of 1.5 m and their characteristics are also described in Table 39.1. The walls are modelled as aerated concrete as in section 39.13.

Figure 39.25 shows the cumulative distribution of RMS delay spread using omnidirectional antennas at both BS and MS, for four different environment scenarios:

1. the original "empty" environment, as shown in Figure 39.2;
2. the environment with dividing partitions, as shown in Figure 39.24;
3. as for (2) but with strips of metal (width 5 cm) along the top of each partition;
4. as for (2) but with 30% of the partition surface covered with random metal reflectors.

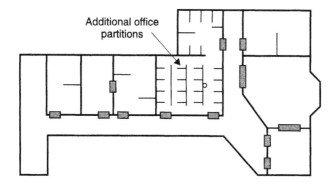

Figure 39.24 Office environment with partitions.

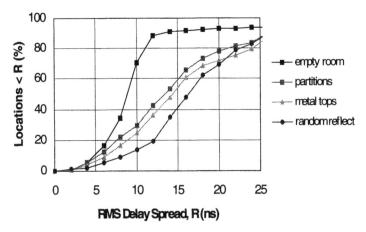

Figure 39.25 CDF of RMS delay spread (omni:omni).

The last of these scenarios is designed to represent a worst cast office environment where a number of objects can act as strong reflectors close to the MS, thus contributing to multipath problems. Figure 39.26 compares the CDF of delay spread for the same four configurations, using 60° directional antennas at both BS and MS.

Studying Figures 39.25 and 39.26 we see that the addition of partitions in the ray-tracing simulation does have a significant effect on the RMS delay spread encountered. For both of the antenna configurations considered, the delay spread is typically increased by a factor of 1.5 to 2 times relative to that encountered in the "empty" environment. There is also a corresponding but less noticeable reduction in k-factor (not shown). This is an important result and demonstrates that an increase in the complexity of the environment leads, on average, to a reduction in the quality of the channel between BS and MS, and thus a reduction in the area in which operation can be supported at a given data rate. It is also clear that the

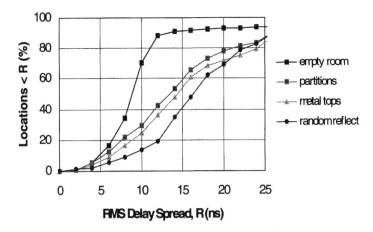

Figure 39.26 CDF of RMS delay spread (60°:60°).

addition of metal strips to the top of partitions or random metal reflectors to the surface does have a detrimental effect, further increasing the delay spreads encountered. However, this is not as significant as the difference between the empty environment and the environment with partitions added.

Figure 39.27 shows the area of coverage provided using four fixed antenna sectors of 90 beamwidth at both BS and MS, for the "empty" environment of Figure 39.2. It should be noted that this is not the same as Figure 39.22 since the walls are now modelled using aerated concrete, rather than standard concrete. In the room where the BS is located, operation is supported in 78% of locations at 10 Mbps and 63% of locations at 100 Mbps. (This is in fact a larger coverage area than that shown in Figure 39.22, since lower delay spreads are encountered due to the change in wall material.) Figure 39.28 shows the area of coverage using the same antenna configuration, for the environment with partitions added. Operation is now supported in 61% of locations at 10 Mbps and 43% of locations at 100 Mbps.

As is to be expected, the locations which are most heavily affected by the addition of partitions are those where previously there was an LOS or dominant path which has now been obstructed by a partition. In locations where the LOS path remains unobstructed there is little change in performance, and in a few instances delay spread and k-factor are improved since significant reflected (multipath) components have been blocked by a partition. Table 39.6 summarises the areas of coverage for the two different environment scenarios. In a further study of a number of different antenna configurations, the addition of partitions has been found to reduce the area of coverage at both data rates by 10–30%.

39.15. WIRELESS LAN NETWORK DESIGN ISSUES

The ability of 60 GHz wireless LANs to provide wideband transmission capability in offices and other working environments means that the number of potential applications is considerable, ranging from file transfer to full multimedia interaction,

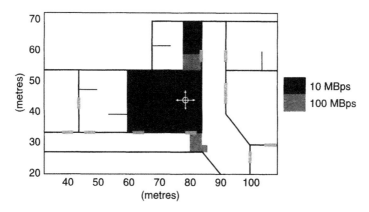

Figure 39.27 Area of coverage (no partitions).

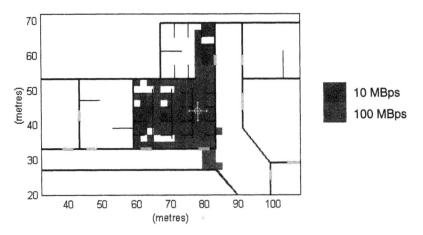

Figure 39.28 Area of coverage (with partitions).

Table 39.6 Summary of areas of coverage

Environment	Coverage at 10 Mbps (%)	Coverage at 100 Mbps (%)
Empty	78	63
With partitions	61	43

requiring real-time audio and video. For a given environment, transmission is limited to data rates which are less than those obtainable on the existing fixed network, since additional FEC coding schemes will be required in order to reduce the BER to an acceptable level. Currently the maximum data rates of fixed networks range from about 10 Mbps (10BaseT Ethernet) to 155 Mbps (ATM). Section 39.12 has highlighted the difference in coverage area for operation at two different data rates. It is likely to be advantageous for 60 GHz wireless LANs to support operation at a range of different data rates. This will ensure that users can derive the most benefit from a strong LOS link when close to the BS, but still be able to perform a low-rate transfer when there is significant movement in the local environment, or if the MS is located further from the BS.

From the propagation analysis in sections 39.8–39.12 it appears possible that a single well-located BS can provide all the networking needs of an MS within a 15 m range, so long as an appropriate antenna configuration is chosen. If multiple sector antennas are used in a configuration similar to those described in sections 39.12.3 and 39.12.4 then even if the LOS path is obstructed, an alternative sector (or sectors) can be chosen and operation can continue via a strong (normally single-order) reflecting path. In order to allow more than one MS to communicate with a single BS, some form of multiple-access scheme is required. Time division multiple access (TDMA) is probably the most viable technique, with time slots allocated to each MS depending on the rate of data transfer required and the number of terminals requiring network services at any given time.

For increased capacity multiple BSs can be used with an FDMA scheme allowing each to operate at a different frequency or using a different set of frequencies. The location of each BS and the number required can be tailored to particular network requirements, such the likely number of users, the maximum required data rate and the types of services needed (fixed bit rate, variable bit rate, non-time-critical etc.). One of the principle advantages of 60 GHz operation is the large amount of bandwidth available, and this has the potential to allow multiple broadband frequency channels in the same operating environment. If we assume that the available spectrum (59–64 GHz) is divided into high data-rate channels of, say 125 MHz bandwidth, there is a maximum of 40 channels available. However, full data-rate transfer is unlikely to be required by all users all the time, and so the available spectrum can be divided into variable bandwidth channels as required. If a large area is to be covered by a single wireless LAN then channels can be divided into sets and a simple cellular-based frequency re-use plan implemented. As highlighted in Section 39.5, the materials from which the building is made, and the layout of rooms will govern the frequency re-use distance.

In any particular loction the phase of the multipath components may add destructively such that a local fade in the received power level is encountered. In order to overcome such problems a separate antenna sector may be chosen, or else operation can be switched to a separate frequency channel. This technique requires some form of dynamic channel allocation and thus a central coordination of the BSs within an operating environment. For two-way communication between any given BS and MS, either a time-division duplex (TDD) or frequency-division duplex scheme (FDD) scheme can be implemented. Of these the TDD approach may be favourable, since the up-link and down-link will share a reciprocal channel, allowing antenna/channel selection to be carried out at the BS. However, an FDD scheme allows higher aggregate rates of data transfer between terminals, although this will require twice the bandwidth.

The results in section 39.14 suggest that the more complex an environment is in terms of partitions, the shorter the range in which operation can be reliably supported at a given data rate. Consequently a higher density of BS locations may be required to serve a partition-based open plan office environment. Based on the assumption (from Figure 39.28) that operation can be adequately supported within a radius of 10 m around the BS, a spacing of less than 20 m between base stations is required in order to provide contiguous coverage.

39.16. CONCLUSIONS

The characteristics of indoor propagation at 60 GHz have been investigated using an image-based ray-tracing prediction model. It has been demonstrated that 60 GHz signals are heavily attenuated by transmission through walls and consequently operation is likely to be restricted to locations where there is either an LOS path between base station and mobile station, or a reflecting path that does not undergo wall transmission.

For LOS and non-LOS locations the use of directional antennas provides significant suppression of multipath components giving a reduction in RMS delay spread, and a corresponding increase in Rician k-factor. However, incorrect alignment of directional antennas can give rise to high delay spreads and low k-factors. In general reducing the beamwidth of either the BS or MS antenna reduces the delay spread and increases the k-factor, and offsets of < 30% of the antenna 3 dB beamwidth can be tolerated without noticeable degradation in performance. It has also been found that for many LOS locations, as well as the direct path between BS and MS, there are a number of antenna orientations for which a reasonable power level, low delay spread and high k-factor are encountered. This suggests that if the LOS path is blocked, for example by someone standing in the way, operation may be continued by switching to an alternative antenna sector.

The relationship between transmission rate, RMS delay spread and Rician k-factor has been discussed and it has been shown in principle that, using directional antennas, high data-rate transmission can be supported without the need for equalisation or multi-carrier techniques. Two practical implementations have been suggested; the first uses 12 fixed antenna sectors at the BS and an omnidirectional antenna at the MS; the second uses four fixed sectors at both BS and MS. These configurations provide coverage at 100 Mbps for MS locations within 10–15 m of the BS and coverage at 10 Mbps for the majority of LOS locations.

The characteristics of 60 GHz signal propagation vary considerably with the construction materials used. As an example, it has been shown that much lower delay spreads and higher k-factors are encountered if walls are made of aerated concrete rather than concrete. This demonstrates that the construction and layout of a particular environment must be carefully analysed before a wireless LAN system is installed. The addition of partitions to the study environment has been shown to reduce the areas of coverage at 10 Mbps and 100 Mbps. However, MS locations near the BS are less noticeably affected. This implies that as the complexity of an environment is increased, a higher density of BS locations will be required in order to provide contiguous coverage.

REFERENCES

[1] Manabe, T., Miura, Y. and Ihara, T., Effects of antenna directivity on indoor multipath propagation characteristics at 60GHz. In *Proceedings of IEEE PIMRC 1995*, Toronto, 1035–1039, 1995.

[2] Guérin, S., Pradal., C. and Khalfa, P., Indoor propagation narrowband and wideband measurements around 60GHz using a network analyser. In *Proceedings of IEEE Vehicular Technology Conference*, Atlanta, 160–164, 1995.

[3] Rules to permit use of radio frequences above 40GHz for new radio applications. FCC ET Docket No. 94-124, first report and order/second notice of proposed rulemaking, December 1995.

[4] Falconer, D., A system architecture for broadband millimetre – wave access to an ATM LAN. *IEEE Personal Communication Journal*, **3** (4), 36–41, 1996.

[5] Skellern, D. J. *et al.*, A mm-wave high speed wireless LAN for mobile computing – architecture and prototype modem codec implementation. In *Proceedings of IEEE Hot Interconnects*, Stanford, CA, 1996.

[6] Williamson, M., Athanasiadou, G. E. and Nix, A. R., Investigating the effects of antenna directivity on wireless indoor communication at 60GHz. In *Proceedings of IEEE PIMRC*, Helsinski, 635–639, 1990.

[7] Tharek, A. R., Propagation and bit-error rate measurements in the millimetre-wave band around 60GHz. Ph.D. thesis, University of Bristol, 1988.

[8] Correia, L. M. and Frances, P. O., Transmission and isolation of signals in buildings at 60GHz. In *Proceedings of IEEE PIMRC*, 1031–1034, 1995.

[9] Smulders, P. F. M. and Correia, L. M., Characterisation of propagation in 60GHz radio channels. In *Proceedings of IEE Electronic Communication*, 73–80, April 1997.

[10] Smulders, P. F. M., Broadband wireless LANs: a feasibility study. Ph.D. thesis, Eindhoven University of Technology, 1995.

[11] Report and recommendations of the millimetre wave communications working group to the FCC. December 1996.

[12] Athanasiadou, G. E., Nix, A. R. and McGeehan, J. P., A new 3-D indoor ray tracing model with particular reference to predictions of power and RMS delay spread In *Proceedings of IEEE PIMRC*, Toronto, Canada, 1161–1165, September 1995.

[13] Athanasiadou, G. E., Nix, A. R. and McGeehan, J. P. Indoor 3D ray tracing predictions and their comparison with high resolution wideband measurements. In *Proceedings IEEE Vehicular Technology Conference*, Atlanta, GA, 36–39, 1996.

[14] Nix, A. R., Athanasiadou, G. E. and McGeehan, J. P., Predicted HIPERLAN coverage and outage performance at 5.2 and 17 GHz using indoor 3-D ray-tracing techniques. *Wireless Personal Communication Journal HIPERLAN special edition*, Kluwer, New York, 1996.

[15] Von Hippel, A. R., *Dielectric Material and Applications*, The Technology Press of MIT of John Wiley & Sons Inc., New York, 1954.

[16] Kraus, J. D., *Antennas*, McGraw-Hill, New York, 1988.

[17] Karasawa, Y., Kuroda, T. and Iwai, H., The equivalent transmission-path model – a tool for analyzing error floor characteristics due to intersymbol interference in Nakagami–Rice fading environments. *IEEE Transactions on Vehicular Technology*, **46** (1), 194–202, 1997.

40

The Design of a Handover Protocol for Multimedia Wireless ATM LANs

C.-K. Toh

Insights into Mobile Multimedia Communication
ISBN 0-12-140310-6

40.1. INTRODUCTION

40.1.1. Problems Associated With Mobile Handovers

Mobile handover is the process of passing the responsibility of communication connectivity from one base station (BS) to another. The resulting handover traffic is a function of many factors, such as the number of mobile hosts (MHs), the size of wireless cells, the mobility profile of MHs, the desired channel capacity and the MH's migration speed and direction.

As the size of the wireless cells gets smaller, the signalling traffic as a result of handovers can be substantial. Hence, there is a need to reduce the handover traffic, so that more bandwidth can be devoted to data transfer. Handovers can also cause connectivity interruption if it is not performed properly and quickly. Since video and audio data are intolerant to delay and jitter, traffic interruption as a result of the handover can be damaging. Therefore, there is also a need to provide continuous handover. The decreasing wireless cell size also results in a decrease in the wireless cell overlapping width. With the MH migrating at a certain speed and direction, there will come a point when the handover mechanism cannot be completed in time. Hence, a fast handover is also necessary. The above, therefore, summarises the need for a *fast, efficient and continuous* handover mechanism.

40.1.2. Current Handover Schemes

A lot of work has been done on mobility support for connectionless WLANs. An example is the IETF[1] Mobile IP[2] group. One must not, however, rule out the possibility of a connection-oriented WLAN. If multimedia services are to be provided to MHs, one of the possibilities is to have a local wireless ATM (LWATM) [1] network.

Handovers can be initiated by the MH, the network or a combination of both. Several methods have been suggested for MH handovers for wireless ATM networks. They can be broadly classified under: (a) full re-establishment, (b) incremental re-establishment, (c) multicast re-establishment and (d) connection extension-based schemes.

The full re-establishment scheme requires a completely new connection to be set up from the source MH to the destination host. This is slow and inefficient. The incremental re-establishment scheme, however, requires only the establishment of a new partial path (which connects to a portion of the original connection) and it also allows circuit reuse. Hence this scheme is fast and efficient. Examples of such schemes are found in [2] and [3]. The multicast re-establishment scheme is fast but not necessarily efficient as the problems of ATM cell duplicates and the abundant pre-reservation of network bandwidth exist. Examples of this scheme are mentioned in [4], [5] and [2]. Finally, the connection extension scheme [6] extends an existing

---

[1] Internet Engineering Task Force.
[2] Internet Protocol.

mobile connection from the old access point to the new access point. This results in fast handovers but the lengthen path increases end-to-end delay and route loops can occur if the MH migrates in a zig-zag fashion. The protocol proposed in this chapter will be based on the incremental re-establishment scheme.

40.2. WIRELESS CELL CLUSTERING ARCHITECTURE

40.2.1. Principle

To exploit the advantage of locality, a multi-tier wireless cells clustering architecture is proposed. Figure 40.1 shows that a cluster is a collection of base stations (BSs) which are connected to a cluster switch. MH migrations occurring under the coverage of the cluster switch are known as intra-cluster migration, while those occurring between clusters are called inter-cluster migration. The clustering of wireless cells

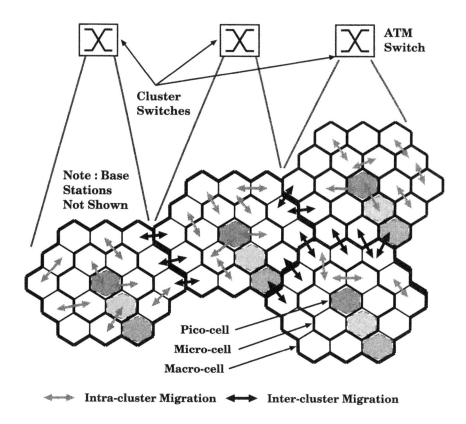

Figure 40.1 The concept of intra- and inter-cluster migration.

and the grouping of BSs to cluster switches result in a distributed handover management, location management and call admissions control architecture.

Intra-cluster migrations result in fast handovers since the handover management entity resides in the cluster switch, which is only a single hop away from the BS. Hence, a cluster of wireless cells is the region where fast handovers can be achieved and where handovers can occur most frequently. Inter-cluster handovers (as a result of the MH migrations between the macrocells), however, give rise to slower handovers, since the handover management entity is now performed by an ATM switch in the backbone network.

Crossover ATM Switch Discovery

As stated earlier, the incremental re-establishment scheme allows the BS to set up a new partial path instead of a completely new path. As shown in Figure 40.2, the node where the convergence occurs (i.e. the switch where the new partial path meets the old connection path) is called the crossover switch (CX). The purpose of the CX discovery mechanism is to derive a minimum-hop new partial path with reasonable circuit reuse efficiency[3] by locating a suitable CX. Algorithms that govern the selection of a CX and the effects of network topology on CX discovery are mentioned in [7].

CX discovery is only invoked for inter-cluster handovers, where the CX is a switch in the backbone network. For intra-cluster handovers, the cluster switch is the crossover switch, and hence no CX discovery needs to be invoked. The need

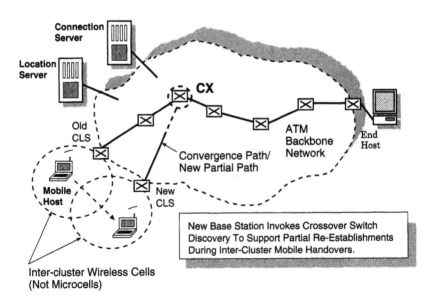

Figure 40.2 The concept of crossover switch discovery.

---

[3]Circuit reuse efficiency is defined as the ratio of the reuse path to the old source-to-destination connection path.

to cater for both intra- and inter-cluster handovers results in a hybrid handover protocol.

40.3. HANDOVER PROTOCOL DESIGN ISSUES

Exploitation of Locality

Many handover schemes have exploited the advantages of locality. In an environment where bandwidth is always abundant, the "footprint" scheme [5] can be used. This method is, however, not efficient in terms of network resource consumption. Alternatively, one can use the concept of clustering. By grouping a set of BSs to a cluster switch, fast handovers can be achieved by performing ATM cell re-routeing operation at the cluster switch. This handover operation is considered "local" and the cluster switch need not set up any connection segments beyond a single hop, except during inter-cluster MH migration.

Exploitation of Radio Hint

A suggestion [2] to achieve a seamless handover is through the usage of a hint in terms of radio signal strength. To keep track of the locations of MHs, beacons are generated periodically by the BSs. Due to the degradation of signal strength with increasing propagation distance and the fact that the MH can listen to the BSs beacons simultaneously, this hint can be used as an advance signal to trigger the handover earlier, even before the MH fully enters the new wireless cell.

Having a radio hint, however, means that the new BS has to derive information about the MH's existing connections from some other sources. This can be obtained from the old BS via BS-to-BS signalling or by having a scheme whereby each BS will maintain a cache containing its peer BSs' MH connection information. The former method can be established easily, but the latter requires more memory per connection per MH and the resulting traffic for updating the connection tables can be substantial.

Resource Reuse (Efficiency)

As mentioned earlier, an inefficient approach is that of the "footprint" method, where all the surrounding wireless cells have pre-established backbone network connections prior to any pending handovers. However, if multicast connections exist for MHs, the "footprint" method will be complicated and will consume even more network resources. An alternative scheme using partial re-establishment serves to set up only a small portion of the new circuit path, which results in lower resource consumption.

Scalability

Scalability here refers to the ability of the handover protocol to service as many handover requests as possible. If the handover protocol supports independent source and destination mobile handovers and it does not consume excessive resources (bandwidth allocation, buffers per stream per connection, etc.), then the handover protocol can scale well to a large number of MHs.

Service Disruption Time (Continuity)

Traffic disruption relates to how long the data path between the source MH to the destination MH is being held up while the handover operation is taking place. While signalling occurs once in a connection set-up, handovers cause signalling to occur anytime and many times. "Soft" handovers rely on buffering at the MH and BSs to ensure that data are not lost and traffic continuity can be sustained. However, depending on the handover protocol, the service disruption time can be undesirably significant. Through a more comprehensive handover protocol design, this service disruption time can be reduced to an acceptable level.

ATM Cell Loss and Sequencing

After a handover, the connection re-routeing operation at the crossover switch causes ATM cells to be transported over the new path. However, data that remain transported over the old path may arrive late or be transmitted (over wireless media) late due to congestion at the old BS. Under such circumstances, for a system based on wireless cell reuse (for example, FDMA), switching frequency to a new wireless cell will result in ATM cell loss. For a system not based on wireless cell reuse (for example, CDMA), however, ATM cells will arrive out of sequence.

Exploitation of Personality

A mobile user may exhibit certain patterns of repetitive movement. If a prediction of the next movement can be done accurately, then advance handover procedures can be completed well before the MH enters the wireless cell overlapping zone. However, to date, the accuracy of such mobility profiles remains questionable. While some users seldom move at all once they reach certain locations (this mean that the hand-over scheme can be more relaxed), other users may have unpredictable movements. Existing efforts employ artificial intelligence, where mobility profiles are fed to a suitable self-learning neutral net to predict a user's next movement.

Consideration For Data Looping

As MHs' movement can be highly dynamic and sometimes abrupt, looping can occur when an MH enters a new wireless cell and suddenly withdraws and re-enters the old wireless cell. This results in "zig-zag" handovers. If the data forwarding methodology is employed in the handover scheme, then a data looping cycle can arise. Consequently, the design of a handover protocol must consider this factor.

Traffic Disruption Symmetry

Both the upstream and downstream data flows are both affected during the handover of a mobile connection. For interactive multimedia applications, asymmetric upstream and downstream data disruption time can adversely affect the application. Since the re-routeing of a mobile connection affects both the upstream and downstream data flows, the handover protocol must be designed to handle both data flows simultaneously.

Robustness of Protocol

The handover protocol must be robust in order to handle situations like link failures or high reception error rates. Mechanisms such as time-outs, retries and aborts are useful in such cases.

Handover Requests Queueing

From the connectivity point of view, blocking an on-going call is significantly less desirable than blocking a new call attempt [8]. Hence, one may increase the probability of call blocking while achieving a lower probability of forced termination. It is sensible to allow queueing of new originating calls as they are less sensitive to delays than handover requests for an on-going call. This is a trade-off between an increase in the service quality and a decrease in the total carried traffic. Handover requests should be serviced in the order of *urgency*, not on a first-come–first-served basis [9].

Service Traffic Characteristics

The services provided by the public cellular networks (such as GSM) are mainly telephony and facsimile. While data streams are highly tolerant to delays, congestion and unreliable transmissions, continuous media traffic in a multimedia WLAN environment demands more stringent traffic characteristics, such as timeliness, jitter, etc. Hence, the types of traffic to be supported also govern the design of a handover protocol. Other than audio and the conventional data services, we envisage that video applications will be present for mobile users, but these video images will appear at a smaller picture size and operate at a lower frame rate.

40.3.1. Handover Quality of Service

QoS Before and After Handover

The handover protocol must be capable of supporting QoS requirements. The QoS issue can be two-sided, i.e. one can provide hard QoS guarantees (where QoS is always assured despite MH migration) or soft QoS guarantees (where QoS degradation is possible during MH migration between wireless cells having different available bandwidth). As MHs enter and leave the wireless cells freely, the wireless network capacities can vary greatly. Some wireless cells can be crowded with many live MHs (i.e. having on-going communication) and any new incoming MHs with high QoS requirements can result in the following: (a) its handover request be forced to terminate (undesirable), (b) its QoS requirements are degraded prior to handoff (possible), (c) some existing MHs in the new wireless cell might need to have their QoS degraded to fulfil this incoming MH's QoS requirements (questionable), and (d) suspend or reject queued originating calls (i.e. blocking of new calls) by the MHs in the new wireless cell to fulfil this incoming MH's QoS requirements (possible).

QoS Consistency Problem

There are four factors that have to be considered for a mobile QoS. Firstly, should QoS renegotiation be allowed during handovers? Given the time constraint imposed by the MH's migration speed, direction and the wireless cells overlapping width, QoS renegotiation may not be feasible. The second factor concerns the "upgrading" and "downgrading" of QoS during MH migration from one wireless cell to another. Should the original mobile QoS requirements specified during connection set-up be used as the basis of reference for the consideration of QoS "upgrade" or "downgrade"? Thirdly, the main governing factor for QoS adjustments is the wireless channel quality. Finally, QoS consistency throughout the wired and wireless segments of a mobile connection should be maintained after each handover.

40.3.2. Envisaged Operating Environment

When designing a handover protocol, one should also be concerned with its scalability in terms of: (a) the number of mobile users to be supported, (b) the number of BSs within a Wireless ATM LAN, (c) the number of nodes in the backbone network, etc. Our Wireless ATM LAN model is one with several hundreds (not thousands) of BSs, providing wireless cells of 10 m in diameter. Several hundred of mobile users are expected. Hence, a bigger network may be partitioned into multiple wireless ATM LANs, with support provided for roaming between these LANs.

40.3.3. Support For Uniform and Unified Handovers

Handovers of Heterogeneous Mobile Connections

In designing a handover protocol for connection-oriented WLANs, the handover procedures must be applicable to handovers of both *unicast* and *multicast* connections. While existing research work has been concentrated on supporting unicast connection handovers, handovers of multicast connections should not be neglected, especially in a wireless ATM network, where multicasting is an expected feature of the ATM.

Support For Roaming Between Wireless ATM LANs

Network engineers often partition a big network into smaller subnetworks. With multiple wireless ATM LANs existing on a site, mechanisms to support roaming between such LANs are desirable. The proposed handover protocol and CX discovery scheme can support such handovers in a consistent manner. Further details can be found in [10].

40.4. A NEW HANDOVER PROTOCOL

40.4.1. Handover With Radio Hint

Control Flow

Referring to Figure 40.3, the with-hint handover protocol allows advance setup to be made. When the MH receives the beacons from the new BS, after a certain signal threshold level, the MH sends a handover $hint_{(1)}$ message to the current BS (BS_{OLD}). This results in BS_{OLD} sending a handover $invoke_{(2)}$ message, which contains the MH existing list of connections, to the target BS (BS_{NEW}).

The BS_{NEW} then determines if the handover is "intra" or "inter". For intra-cluster handovers, the cluster switch is the crossover switch (CX). Hence, no CX discovery is required. However, if the next-cell beacon flag signifies an inter-cluster handover, the CX discovery process is invoked to compute the location of the CX. New partial paths are then established from the BS_{NEW} to the CX. This is denoted by $set\text{-}up_{(3)}$ and $setup\text{-}ack_{(4)}$. Hence, by the time the MH enters well into the overlap region, the new partial paths have already been established. However, if the new partial paths cannot be established and the BS_{NEW} does not receive the $setup\text{-}ack_{(4)}$ message, the BS_{NEW} will time-out and invoke another CX discovery procedure to locate a different CX.[4] This is illustrated by the BS_{NEW} protocol state machine diagram in the appendix.

The MH will now command the BS_{NEW} to initiate a handover via the $greet_{(5)}$ message. This causes the BS_{NEW} to send a connection $redirect_{(6)}$ message to the CX, so that the ATM cells can now be re-routed over the new path. This mechanism avoids the "sudden withdrawal" problem of the MH, i.e. if the MH suddenly withdraws from entering the new cell, time-outs will abolish all the newly created partial paths, as shown by the BS_{OLD}, BS_{NEW} and CX protocol state machine diagrams in the Appendices. Once the CX has performed the redirection, it informs the BS_{OLD} to disconnect the old partial paths. If BS_{OLD} does not receive the $tear\text{-}down_{(8a)}$ message from the CX, it will time-out. Likewise, if the BS_{NEW} does not receive the $greet_{(5)}$ message from the MH, it will time-out and the handover process will be aborted.

---

[4]Alternatively, one may employ the alternate routeing approach where the second best partial path previously computed can be used for the next attempt.

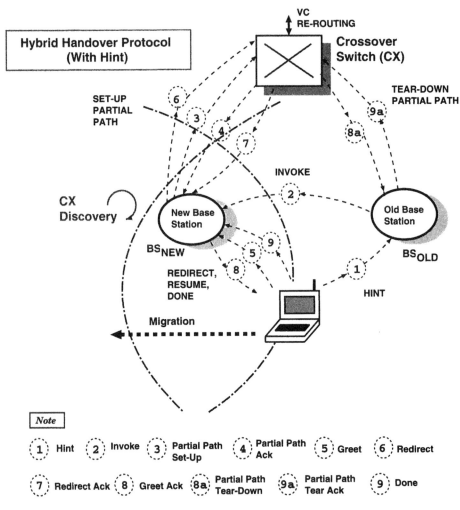

Messages (8) and (8a), (9) and (9a) can be performed in parallel.
Other than messages (1), (5), (8), and (9), all other messages occur
over the wired ATM network.

Figure 40.3 The with-hint handover protocol control flow diagram.

Finally, the MH returns a $done_{(9)}$ message to the BS_{NEW}, indicating that it is now ready to receive and transmit data via the new BS. Due to the advance set-up and the use of buffering at the BSs and CX, handovers can now be performed quickly and seamlessly. To maintain ATM cell arrival sequence, one possibility is to return the $done_{(9)}$ message only after the *"end-of-cell"*[5] marker is received by the MH. If, for some reason, this indicator cell does not arrive within a time limit, the $done_{(9)}$

---

[5]This *end-of-cell* marker can be inserted using in-band signalling or out-of-band. It is inserted by the CX prior to re-routeing the mobile connection.

message will still be sent to BS_{NEW} to complete the handover process and cell loss or resequencing has to be handled by the higher layer protocols.

A possible problem can exist when after a MH has sent the $hint_{(1)}$ message, this message is somehow not received by BS_{OLD} and the MH proceeds to issue a $greet_{(5)}$ message to BS_{NEW}. Since no new partial paths have been set up, handovers cannot be performed and no $greet\text{-}ack_{(8)}$ message will be returned to the MH. However, from the MH's protocol state machine diagram, the MH will eventually time-out and enter a state where it will issue an explicit handover request to BS_{NEW}. This request is similar to the one issued by the without-hint handover protocol, which will be explained in the next subsection.

Data Flow

A connection can exist as a bidirectional path for the transfer of information between two end communicating entities. During a handover, both the upstream and downstream data flows are affected. The data flow is analysed below, with reference to Figure 40.4.

Considering the upstream data flow, before a handover, the MH transmits data via the BS_{OLD} and CX to the destination host. Upon a handover, the $greet_{(5)}$ message is sent and the MH starts to queue the transmitted data in its upstream buffer, until the BS_{NEW} responds with a $greet\text{-}ack_{(8)}$ message. This is the upstream traffic disruption period. Thereafter, the upstream data flow resumes over the new wired and wireless segments.[6]

For the downstream case, prior to the handover, the downstream data flows from the destination host through the CX and BS_{OLD} to the MH. During a handover, the downstream data flow is disrupted at the point when the $redirect_{(6)}$ message is received at the CX. Once the cell redirection is successful, the downstream data flow resumes through the new wired segment and is buffered at the BS_{NEW}[7] until it receives a $done_{(9)}$ message from the MH (this is the downstream disruption experienced by the MH). Thereafter, the BS_{NEW} flushes its downstream buffer and the data now flows through the new wireless segment.

40.4.2. Handover Without Radio Hint

There will be cases where radio hint is not available, such as when the radio link fails to function. In such circumstances, the advantage of radio hint cannot be realised and some ATM cells will be lost. Hence, there is a need for a mechanism to: (a) perform handover when an existing wireless link fails, such as when a BS fails and (b) when, for some reasons, BS_{OLD} does not received the $hint_{(1)}$ message and a $greet_{(5)}$

---

[6]The mobile application need not re-create a socket during handover. The new wired segment mentioned here still supports the original source-to-destination connection.

[7]The downstream buffer at the BS_{NEW} is allocated in advance, during the reception of the $greet_{(5)}$ message, since once the cell redirection is performed, downstream data will flow immediately to the BS_{NEW}. If the downstream buffer is only allocated after having received the $redirect\text{-}ack_{(7)}$ message, cells will be lost if the data stream arrives at the BS_{NEW} earlier than the $redirect\text{-}ack_{(7)}$ message.

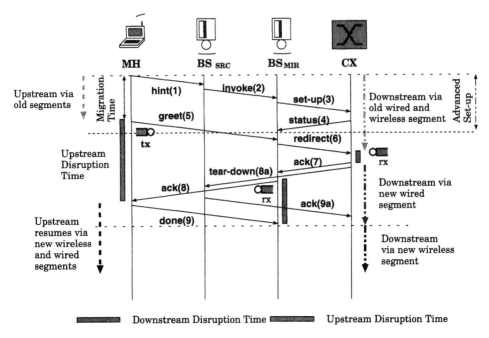

Figure 40.4 The with-hint handover protocol data flow diagram.

message has been sent by the MH to the BS_{NEW}. A without-hint handover protocol which enables the MH to continue to perform a handover under such circumstances is proposed below.

Control Flow

Figure 40.5 shows that once a wireless link failure is detected, the migrating MH sends an explicit handover request, i.e. *explicit-greet*$_{(1)}$ message to the BS_{NEW}. Unlike the earlier *greet*$_{(5)}$, this explicit greet message contains sufficient information[8] about the MH's current connections. Hence, this unavoidably incurs a penalty for using the wireless interface to convey the MH connections information. Once the BS_{NEW} receives the explicit greet message, it invokes the CX discovery (for inter-cluster handovers) and sends a partial path *setup*$_{(2)}$ request message to the CX. After receiving the *setup-ack*$_{(3)}$ message, the BS_{NEW} immediately sends the connection *redirect*$_{(4)}$ request to the CX, since the advance set-up is no longer present. When the BS_{NEW} has received the *redirect-ack*$_{(5)}$ message from the CX, a *greet-ack*$_{(6)}$ message (similar to *greet-ack*$_{(8)}$ in the with-hint handover case) is then returned to the MH to inform it about the re-connection. Lastly, the MH sends a *done*$_{(7)}$ message to the BS_{NEW} to signify that it is ready to receive and transmit data.

---

[8]This includes the connections' identifiers and their respective QoS requirements.

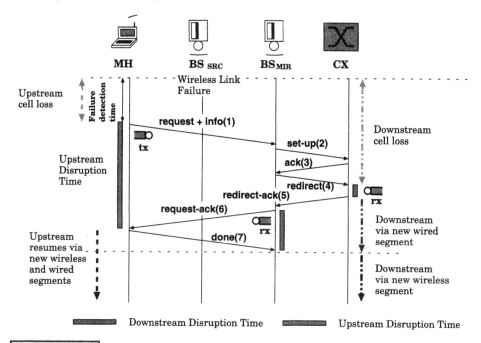

On CX failure, re-invoke crossover switch discovery to re-select a new CX.
On MIR failure, re-select a new MIR based on beacons received.

Figure 40.5 The without-hint handover protocol control flow diagram.

When a BS fails, it could be: (a) only its radio section fails to work or, (b) the whole BS fails. In the former case, the old partial path tear-down process can still proceed. This tear-down process can occur in parallel with messages *greet-ack*$_{(6)}$ and *done*$_{(7)}$. In the latter case, however, the BS_{OLD} will no longer respond to the *tear-down*$_{(6a)}$ message. From the CX protocol state machine diagram in the appendix, CX will time-out and return to its initial state.[9]

Data Flow

The MH normally takes some time to detect the failure of a wireless link, hence data will be lost during this period. Once a handover request is initiated, the MH begins to accumulate its upstream data in its transmit buffer. This buffer will not be flushed until the new partial path is established and the cell re-routeing operation is completed. For the downstream data, they will be buffered at the CX and later at the

---

[9]Depending on the underlying ATM connection management scheme employed, when an intermediate stage of a connection fails, the subsequent stages may be automatically informed to remove the associated VCs and release resources reserved for the connection.

BS_{NEW} until the MH acknowledges that it is ready to receive data. Figure 40.6 illustrates both the upstream and downstream data flows.

40.5 HANDOVER PROTOCOL SUMMARY

40.5.1. Protocol Robustness

The previous sections reveal that the handover protocol has some features of fault tolerance. A with-hint handover may revert to one without hint when the BS_{OLD} fails to received the $hint_{(1)}$ message sent by the MH. A without-hint handover is triggered by the MH when it realises that the radio link associated with its current BS has gone away. The MH will then "scan" its beacon channel to select a new BS to perform the handover. Consequently, we have seen how a protocol can complement the other during a radio link failure.

40.5.2. Handover Process

Figure 40.7 summarises the handover process for both with and without hint handovers. The process comprises of intra- and inter-cluster handovers. For the former case, no CX discovery is required as the cluster switch is the CX. However, there is still the need for mobile QoS validation to be performed at the BS_{NEW}, so that decision can be made to fulfil or degrade the requested QoS requirements.

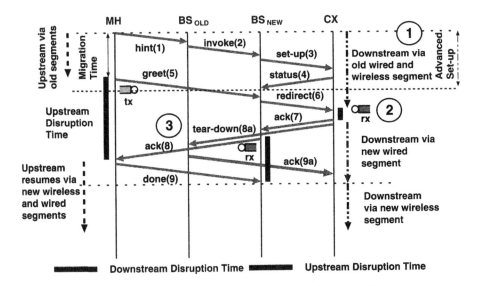

Figure 40.6 The without-hint handover protocol data flow diagram.

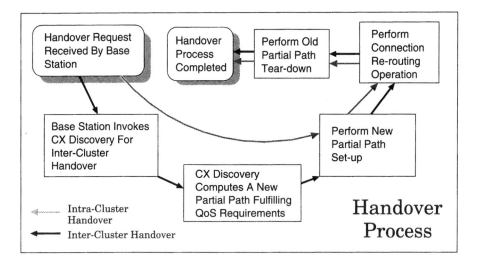

Figure 40.7 The resulting handover process – from handover initiation till completion.

For inter-cluster handovers, the CX discovery process can be invoked by the BS_{OLD} or BS_{NEW}. Once invoked, CX discovery will utilise the existing network topology, routing and QoS information to derive a new partial path from BS_{NEW} to CX . When the new partial path is established and the MH is ready to perform a handover, connection re-routeing is then executed by the CX. The handover process is completed when the old partial path from CX to BS_{OLD} is torn down.

40.5.3. Summary of Handover Types

Handovers can occur for a single MH or multiple MHs with single or multiple on-going connections. For each of these cases, the mobility functions required to support the handovers are different and they are summarised in Table 40.1.

While Type 1 handovers are single-connection (unicast) handovers, Type 2→4 are concerned with handovers of multiple and multipoint connections. Because Type 2 handovers are related to multiple unicast connections (over the same route) between the same source and destination MH pair, the CX is common, i.e. multiple VCs are handed over at a common CX. However, if multiple connections exist over different routes, then multiple redirects, cell re-routeing operations and CXs exist. Type 3 handovers are concerned with point-to-multipoint connections. While source-rooted multicast connections have multiple CXs, shared-based tree connections require only a single CX. The handovers of source and receivers of point-to-multipoint connections are described in [11]. Finally, Type 4 handovers are related to multiple MHs having multiple on-going connections to multiple destinations, i.e. multipoint-to-multipoint. Note that an MH can have heterogeneous connections (i.e. a combination of any of the above-mentioned connection types) prior to handovers. As seen from Table 40.1, due to the fact that different CXs are selected to collaboratively

Table 40.1 Types of handovers and how they are being supported by our proposed handover protocol

Handover (HO) Types	No. of HO REQs/ACKs	No. of CXs	No. of redirects	No. of Cell re-routeing operations
1. Single MHsrc, single MHdest, single connection	Single	Single (common)	Single	Single
2. Single MHsrc, single MHdest, many connections	Single	Single (common)	Single	Many (separate)
3. Single MHsrc, multiple MHdest, many connections	Single	Many (separate)	Many (separate)	Many (separate)
4. Multiple MHsrc, multiple MHdest, many connections	Many (separate)	Many (separate)	Many (separate)	Many (separate)

support the handovers of multiple connections (especially for MHs migrating in groups), the possibility of overloading a particular CX is greatly reduced. More importantly, the presence of distributed handover management has shortened the overall handover time for multi-connection handovers.

40.6. CONCLUSION

In this chapter, the problems and challenges faced in a wireless ATM LAN environment are presented. In particular, there is a need for a fast, efficient and continuous handover mechanism. To cater for both high and low rate of mobility and to exploit the advantages of locality, a multi-tier wireless cells clustering architecture is proposed. To support fast, efficient and continuous handover, a hybrid handover protocol based on the incremental re-establishment scheme and the concept of crossover switch discovery is introduced. The protocol has the provision for QoS and can function in a with or without radio hint situation. More importantly, the protocol can support handovers of heterogeneous mobile connections.

REFERENCES

[1] Leslie, I. and McAuley, D. R., ATM everywhere. *IEEE Network Magazine*, 7 (2), 40–46, 1993.
[2] Keeton, K., Providing connection oriented network services to mobile hosts. In *Proceedings of the USENIX Symposium on Mobile and Location Independent Computing*, 83–103, August 1993.

[3] Acampora, A., Control and quality-of-service provisioning in high speed microcelluar networks. *IEEE Personal Communications Magazine*, Second Quarter, 1994.

[4] Ghai, R. and Singh, S., A protocol for seamless communication in a picocellular network. In *Proceedings of IEEE Supercomm/ICC*, 192–196, May 1994.

[5] Earnshaw, R., Footprints for mobile communications. In *Proceedings of the 8th IEE UK Teletraffic Symposium*, April 1991.

[6] Mishra, P. P. and Srivasatava, M. B., Programmable ATM switches for mobility support. In *Proceedings of OPENSIG (Open Signalling For Middleware and Services Creation)*, April 1996.

[7] Toh, C.-K., Crossover switch discovery schemes for fast handovers in wireless ATM LANs. *ACM Journal on Mobile Networks and Applications – Special Issue on "Routing In Mobile Communication Networks"*, **1** (2), 1996.

[8] Tekinay, S., Handover and channel assignment in mobile cellular networks. *IEEE Communications Magazine*, **29** (11), 1991.

[9] Jun, Y., A novel priority queue scheme for handoff procedure. In *Proceedings of the IEEE Supercomm/ICC*, 182–186, May 1994.

[10] Toh, C.-K., Supporting roaming across wireless ATM LANs. In *IEE Colloquium on Network Aspects of Mobile Radio Systems*, 1–6 February 1996.

[11] Toh, C.-K., A handover paradigm for wireless ATM LANs. In *Proceedings of the ACM Symposium on Applied Computing (SAC'96) – Special Track on Mobile Computing Applications and Systems*, February 1996.

APPENDIX A
WITH-HINT HANDOVER PROTOCOL STATE MACHINES

A.1. Mobile Host (MH) and Crossover Switch (CX) FSM

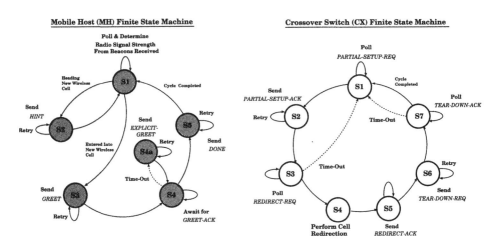

A.2. Old (BS_OLD) and New (BS_NEW) Base Stations FSM

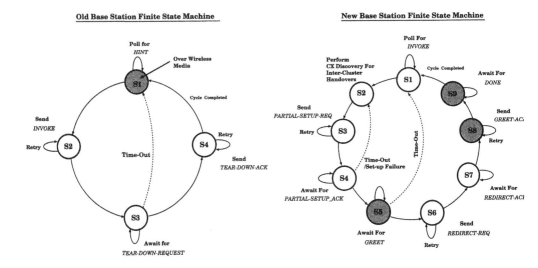

APPENDIX B
WITHOUT-HINT HANDOVER PROTOCOL STATE MACHINES

B.1. Mobile Host (MH) and Crossover Switch (CX) FSM

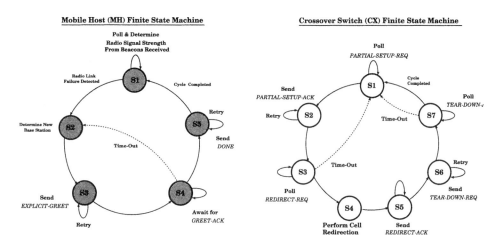

B.2. New (*BS_NEW*) Base Station FSM

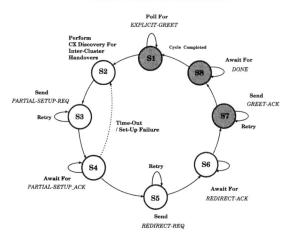

41

A Vision of the Future of Multimedia Technology

Murat Kunt

41.1. INTRODUCTION

A *passive* prediction of the future can be made via logical deductions and extrapolations based on a subjective observation of the past and present. In contrast, an *active vision* is as defined by Walt Disney when he said "If you can dream it you can make it". This chapter is written with this latter view in mind, describing some of the dreams and efforts to "*make them*". The framework is defined by two present day tendencies : globalisation and independence of time and space, both placed in the context of continuous change.

The second half of the 20th century has been rich in technical developments. These arrive at such a rate that it is often difficult for the layman to keep track of them. They are no longer secondary to the essence of society but play an increasing role in

Insights into Mobile Multimedia Communication
ISBN 0-12-140310-6

dictating patterns of behaviour and means of interaction; lifestyle changes are often made in reaction to them as much as dictating their innovation. Such points open a wide debate which is well beyond the scope of this chapter. However, it is the author's belief that the developers of new technology should pay attention to these issues in order to avoid a two speed society in the future.

The current convergence of broadcast, computing and communications technologies heralds a new era, bringing with it large numbers of unexpected and unforeseen changes in our daily life. The relationship of the exploitation of man by man is being reversed into a relationship of seduction. Initially focused on accumulating knowledge, education must turn more to exploring talents, giving more emphasis to know-how than to knowledge. In addition, our relationship to information will cease to impose passivity (such as with the classical TV set or an expert) and instead will encourage inspired and active behaviour. The most appreciated wealth will become the expression of human depth and the enlarged conscience. Beyond the good or bad, the creative dream gains value.

This chapter concentrates first on a summary of key past developments. A subjective view of future requirements is then offered, discussing multimedia technology and standards. A typical example based on visual information compression, problems related to security issues and a possible educational framework are also discussed. Other promising and useful applications of multimedia systems such as telemedicine are not included due to limitations on space.

41.2. TECHNICAL DEVELOPMENTS

Traditional radio broadcasting was the first generally accessible electronic medium. It is an analogue one-way communication system offering point to multipoint communication of audio information. It is a frozen system with few modalities (basically at the modulation and bandwidth levels). It is unable to take full advantage of the digital audio revolution introduced through the compact disc since high quality digital signals must be converted back to analogue form before transmission. The recently introduced higher bandwidth digital audio broadcast system (DAB) has however gone some way to rectifying this.

Television is by far the most successful multimedia system ever designed. Various estimates agree that there are around a billion TV receivers around the world. Its basic principles, introduced half a century ago, are still valid and operational. However, its success and long life cannot solely be attributed to its design. It is an established *frozen* infrastructure that has been difficult to change. The failed introduction of high definition television in Europe, despite the enormous expenditure, exemplifies this. The first major change, since the introduction of colour transmission in the 1960s, in the form of digital video broadcasting (DVB) based on the MPEG-2 coding standard, is now about to be launched. Due to its availability on cable and satellite as well as terrestrial services it is likely to rapidly gain market penetration. Despite the additional functionalities and some quality improvements that digital TV brings, it still remains a patchwork which needs to accommodate the old analogue components of the global system. This is a consequence of the need for "backward compatibility" imposed by the frozen and closed nature of the initial system. In addition to these features, TV is, like radio, currently a one way communication system with a single source and numerous destinations. With the emergence of interactive services, such as banking and shopping and in the longer term, full interactive television, the issue of the return channel needs serious attention.

Older than the television and less widespread, the telephone system was also analogue in nature. This is another example of a closed and frozen system that has progressively and successfully moved into a digital framework with backward compatibility (rotary and tone devices operate on the same network). Despite the flexibility that digital telephony offers, it cannot be called a truly open system. The recent emergence of wireless cellular phones has improved this situation. Although GSM networks are currently only capable of handling voice and data traffic, future generations such as UMTS and Hiperlan will enable wireless computing and provide the basis of an infrastructure for future truly flexible mobile multimedia communications. It should not be forgotten, however, that related technology, for example related to extending battery life, must progress in parallel with communications and computing developments if such systems are to be truly flexible in their use.

Computers, in contrast with TV and telephone, had a relatively short analogue life and rapidly evolved into a digital form, unfrozen and open from the early days. A given hardware platform can handle a plethora of application programs for various problems. When the program becomes too complex and the hardware too slow or limited, upgrading is simple and affordable. In addition, the picture (screen) format for a computer is flexible whereas it is frozen for the TV system. Computers have been interconnected for some time using existing wired networks and are capable of exchanging all types of data. The rudimentary electronic mail (e-mail) of the 1970s and 1980s has evolved into the Internet and World Wide Web (WWW) services that are commonplace today.

41.3. MULTIMEDIA

Multimedia communications has, in the past, been likened to teenage sex:

- everyone thinks about it;
- everyone thinks everyone else is doing it;
- everyone talks about doing it, but almost no one is really doing it;
- the few who are doing it aren't doing it very well;
- everyone hopes it will be great when they finally do it.

Despite the enormous efforts being invested in the area, this analogy still rings true. The difficulty perhaps lies in attempting to accommodate too many conflicting interests, existing systems, partial and incomplete solutions, storage formats and a diversity of communication methods with mutually incompatible characteristics.

A further difficulty is related to the explosion of information, not necessarily always useful. In the past, information was produced by authors – a small group who had something to communicate in written, spoken or visual form. Written or unwritten rules often existed to enable quality control of the information prior to release. Nowadays, anyone who can use a keyboard or manipulate a mouse can copious amounts of information and become an author. In many situations he or she is his or her only quality controller and our only hope is that a spell check is performed before release! Because a long list of *publications* of any sort is a measure of achievement, everyone writes and few people read; we run the risk of creating a write-only society. This is clearly demonstrated by the present levels of congestion on the Internet. If everyone attempts to become a "Fellini" or a "Beethoven" the difference between the real author and the rest, or more generally, between professionals and amateurs, must still be measured in terms of talent rather than access to tools. Today's information search engines should therefore evolve into talent search engines measuring the quality as well as the extent of coverage.

In the past, it was a common belief that the environment could absorb any garbage thrown into it. Only when the amount thrown exceeded the "absorption rate", did we became pollution conscious. As far as one can see, we are repeating the same mistake with the Internet. Bandwidth and resources will always remain finite, especially over mobile channels, and increased pollution consciousness is urgently needed.

A taste of what multimedia services will offer is provided today by the Internet, DVD and CD-ROM devices. However, their value is much more in "multi" than in "media". The quality of each medium and the performance of information manipulating software, hardware and networks still need considerable improvement. Multimedia systems will definitely fail if they remain limited to the simple duplication of old media into electronic form. They will succeed proportionally to the new functionalities they offer. The question to ask is not how to transpose the old material on the new medium but how to most efficiently exploit all the possibilities offered by the new medium to create new material.

The ultimate goal should be to define a system in which a plethora of information types (audio, video, text, data) can be communicated efficiently over a large number of possible channels using a wide variety of equipment. The research and innovations presented in earlier chapters of this book are hopefully creating a framework for this. Let's dream it and ... let's make it!

41.4. STANDARDS

Standards are necessary when there are many potential solutions to any problem that involves large scale information exchange. Almost all existing standards are designed for limited scope and are based on frozen systems. As a consequence, a plethora of standards exist which relate to equipment and devices that are often incompatible and which rely on patchwork engineering to achieve even minimal compatibility. The pace of globalisation suffers from localised standards as anyone who has attempted to use a British razor in a German hotel room will have found to their cost. Another problem arises when one travels with autonomous devices. In a train or aeroplane when travelling long distances, one needs to carry a large number of spare batteries since it is almost impossible to charge them despite the numerous power lines running under the seats. All that is needed is a simple universal plug!

Also, despite the increased compatibilities between cellular phone systems across the world, it is still at the time of writing not possible to use a European mobile phone in the USA.

As shown above, the key problems with standards are not technical. National or continental pride, particular economic interests, successful or lousy lobbying, subjective ambitions and lack of global vision all play a much more significant role. Furthermore, any standardisation group, by the virtue of its democratic rules, is often forced to work from the bottom up rather than from the top down. This should not stop us seeking the most global, open and flexible solutions since the pace of new standards activities often lags well behind that of technical progress.

The need for a standard for any means of communication is so obvious that it is not necessary to emphasise it further in the context of multimedia information exchange. The main problem is to establish what kind of standard(s) are needed and how open they should be.

Digital technology has introduced a new concept which is not sufficiently well accepted and used outside research laboratories; that of an open system. This can be viewed in its most general sense, as a system that does more than simply enable a given machine to be connected to others. A truly open system can evolve without questioning its existing structure. At the product level the need for an open system is very evident. Customers no longer follow the fashion of buying a new car radio or a CD player every month because the new model has a few more buttons or functions than its predecessor. The pace of the progress is much higher than the acceptance of the users. Open systems are the obvious solution with which all the new functions can easily be integrated to the existing system.

41.5. POSSIBLE REQUIREMENTS OF FUTURE SYSTEMS

Future illiterates will be those who are not capable of using terminals for the four basic operations: typing a text for text processing, searching for information in a data base, reading and writing accounting type information in a preprogrammed spreadsheet and communicating with another person electronically.

The list of multimedia functions and devices we can imagine for our entertainment, education and work place is still finite but growing rapidly. These include hideable flat screens of various sizes for the house or office, with integrated microphones, loudspeakers, headsets and cameras accompanied by printers and portable terminals/keyboards. Portable terminals should be the size and the weight of a typical book with easy to use input–output facilities (e.g. reliable voice to text conversion). Whereas a movie theatre atmosphere may be desirable in the family room, a smaller display is acceptable for the electronic book offering the same visual comfort as that of a book page. Autonomous operation over a long working day (say 15 hours) dictates a minimum power specification for the portable "book" which should offer information exchange capabilities anywhere on Earth and in the air.

The above list can be extended with enhanced machine interfaces such as multiple camera systems, 3-D displays, augmented reality and virtual reality devices.

However, as is the case with today's DVD devices, it is likely that the equipment will be available before the content. If content production does not closely parallel these developments, they may fail to gain widespread acceptance. Access to games, movies, news, educational courses, video telephony or conferencing, information search and computing facilities will all be required in future systems. These will need to operate over fixed and mobile networks offering an appropropriate quality of service on flexible plug and play platforms.

Technical solutions for all these requirements are almost in place, and those that are missing will come sooner than we can expect. The question to be asked is: Is a single standard possible to handle all these functions on all these devices? Only if the standard is an open one. Such a standard will however bear greater resemblance to a language than a system block diagram with frozen components. The standard becomes a toolbox and an intelligent system will even be capable of checking a remote configuration, updating any missing components.

This brings us to the topic of the network computer, the re-invented French Minitel at a global scale. Clearly it is more efficient to download a piece of software from a server than to copy it to diskette, pack it, ship it and distribute it to a retailer. However, for this to become reality, we need secure, robust and uncongested networks, with a delivery quality similar to the steady supply of water and electricity to our homes.

The above views transpose the uniqueness of the standard to the uniqueness of the tool. There may well be several competing tools for the same function and the best tool for a given scenario will need to be selected. To some extent, sub-optimum tools can be used, although in some cases this can introduce new problems. A good example is information compression. We are accustomed to using the same compression algorithm for almost all types of information, for example, the discrete cosine transform (DCT) for all visual information. A number of standards, such as H.261, H.263, JPEG, MPEG1 and MPEG2 employ this method today. The drawback is that it is far from optimum in terms of efficient bandwidth use in all cases and does not facilitate many of the functionalities possible with a digital representation of the information. A possible solution, referred to as dynamic coding is discussed later in this chapter. MPEG-4 is the first standard that is following this argument. It is hoped that successive compromises will not detract from this goal.

Another important requirement is the adaptation of the information both to the device and to the user. Let us assume that the original information (video, audio and/or data) is generated within a specific format, translated and compressed into a bit-stream. A high definition screen in the family room must decode more bits than a regular size display in the kitchen or a small screen in the garage. This may be viewed as a basic form of scalability. However, to make for efficient storage and transmission, a top-down scaling from higher to lower resolution (or from lossless to lossy) is not necessarily the best solution. The receiving device should interact with the source to avoid receiving more bits than necessary. Since digital representations allow us to segment images and image sequences into objects, the end user should be able to select his or her preferred objects within a subjective list of priorities, precision and functionalities. Preliminary results in this area and possible directions to follow are discussed further in section 41.7.

Medium to medium transposition or translation has been mainly applied in areas such as text to speech or speech to text conversion. Although many software packages are available today for both of these, their performance is often limited. In the future, it should be possible to translate from any medium to any other. If an image is worth thousands of words, we should have a method that can analyse any given image and produce these words. Such a method would, for example, be of great help to the visually impaired, to car owners who cannot drive and read GPS maps simultaneously and also in the case of extracting metadata for video data base preparation. In reverse, it may also be useful to translate a verbal or textual description into an image. Interactive software packages that do this for police photo-fits are typical of attempts in this direction. Without elaborating on each possible medium to medium translation, let us end with image to image sequence transformation whereby a few words or a list of guidelines can animate an initial picture (cartoon production). An exhaustive analysis of the translation problems reveals a large number of challenging problems to be solved. More generally, these translation methods will be useful when, at a given time and location, a user doesn't have an appropriate input–output device for a given medium. Because of the global tendency toward independence from place and time, this situation will occur increasingly often.

41.6. DYNAMIC CODING

Similar to a piece of furniture that comes in kit form with a set of tools, dynamic coding is an attempt to append the representation model or a compression algorithm to each piece of visual information that can be of any shape, colour, texture and motion. To achieve global efficiency, individual compression techniques are dynamically activated or discarded according to the environment and/or content. Within a given application, the set possible solutions is determined first. Then, the best method is selected for a given image segment. Provided that the receiver or the decoder is aware of the possible methods available, this scheme requires only a few bits more per segment to indicate the label of the method used. If this is not the case, then the code of the algorithm needs to be transmitted as well as the data. Further details of the above mentioned principles can be found in [1]. These results showed that black and white video can be compressed down to 4 kbit/s with quality comparable to that of an H.261 based codec. This is significantly less than the voice bandwidth of a current GSM phone. In order to choose the best method for coding an image segment, a selection criterion is needed. This could be based on compression efficiency, complexity, delay or any combination thereof. The results described above were obtained using a cost function involving rate for a given distortion or distortion for a given rate. The latter case produces varying quality output in a bufferless system.

Research into objective image quality measurement was almost abandoned recently due to a lack of progress over the last two decades. Fortunately, the topic has regained momentum thanks to exchanges between vision researchers and the image processing community. Even though it is well known that mean-square related measures do not accurately follow subjective assessments of image quality, the coding community has been addicted to them due to the lack of any meaningful alternative. Serious effort is needed in the future to further develop initial attempts based on VdBs (visual dBs) [2]. An easily computable objective measure that agrees with our subjective assessments is desperately needed.

41.7. SCALABILITY

Scalability is an extremely attractive feature for multimedia information. The representation of visual information in digital form naturally offers the possibility of scaling various parameters. The most trivial of these is spatial scalability, whereby the size of an image is changed in terms of the number of pixels per width or height. For a given size of image, the quality may be scaled in terms of pixels per inch or VdBs. Both are applicable, not only to the picture or sequence itself but also to the individual objects making up the picture. In the case of an image sequence, temporal scalability is another possibility which alters the number of pictures transmitted or displayed per second. If a required decoder is too complex to decode at the rate of the initial time scale, complexity scalability can also be related to the reduction of the time scale. The soon to be finalised MPEG-4 standard will incorporate such scalable modes of operation.

Scalability, like dynamic coding, relies heavily on data structures. If the right data structure is used, implementing any scalability or dynamic coding becomes an easy task. However, the highly non-stationary nature of the visual data implies that structure must be data-driven for the highest possible efficiency.

As described in Chapter 23, subband decomposition methods can be made hierarchical by iteratively applying a basic two-band structure to the bands produced in the previous levels. The number of levels may be viewed in terms of a quantised quality scale. Linear filters are commonly used for selecting different bands. However, non-linear filters can offer other functionalities such as size dependent filtering to select objects of a given size. Parent–children relationships can also be defined across levels permitting prediction from level to level. Combined with the so-called successive quantisation of the coefficients in each band, this can lead to a fully embedded bit stream for easy quality scalability. It is for example sufficient to decode a small number of bits to yield a poor quality picture which then can be successively refined by decoding the following bits in a more or less continuous way. Such a scheme is very suitable for rapid data-base search for compressed pictures. It is also appropriate for use over heterogeneous networks of variable capacity where display devices of different resolution and quality are simultaneously connected.

Such schemes can be generalised to image sequences by incorporating successively quantised motion vectors in the embedded bit stream to allow scale variations between the original quality of the raw data and the poorest acceptable one. It is hoped that model based representations and interactivity will, in the future, facilitate scaling on both sides of the raw data quality.

41.8. SECURITY

With the increasingly large number of people using wired and wireless services (such as the Internet) for the exchange of widely different types of source material, security issues will play an even more prominent role in the future. This problem has been addressed in Chapter 11 in the case of wireless services. The exponential increase in the problems and the solutions witnessed over the last few years will certainly continue at higher rates. One fundamental problem is how to electronically sign a document, using either a public or a private key, so that the origin of the document can be identified, the integrity of its content be guaranteed and its author authenticated. Such a signature should be invisible, robust to all types of possible attack and should protect not only the entire document but also any "valuable" components within it. Furthermore, signatures should be accessible for control, not only in electronic form but also in the printed version of the images or the analogue version of the signals. A number of possible solutions to this problem have already been proposed. However they are still some way from achieving the ideal requirements listed above.

Intensive work on cryptography has resulted in highly sophisticated and robust ways of preserving or checking the integrity of a bit stream. However, in the context of multimedia traffic (for example a bit stream resulting from the digitisation of a black and white page), these solutions need significant adaptation. A single bit switching from 0 to 1 (or vice versa) perhaps only corresponds to a tiny white dot becoming black or the reverse. Such changes do not affect the integrity of the semantic content but are detected as attacks. So-called "hashing" approaches, appropriate for single bit reversals, should be made softer such that they are sensitive only to semantic changes, ignoring bit reversals. Most multimedia information will be stored or transmitted in a form that is not directly usable, for example compressed or transposed to another representation. Security techniques should also be applicable with the same features to these indirect forms.

Since digital representation enables the segmentation of an image into objects, photomontage becomes a simple operation that anyone with basic equipment can perform. This will open the doors to blackmail or misuse unless a simple analysis can show that the picture or the signal has been manipulated. Tools for such an analysis are not yet available but they must be in place before the use of the technology becomes widespread. Since it is difficult to control access to the raw data, to prevent misuse, either camera manufacturers or material producers must globally agree to introduce signatures or watermarks onto every single picture taken to identify photomontage, or the final result must be analysed to verify the conformity of the basic physics (light sources, reflections, shadows, etc). As with medium to medium translation, there are in this area a number of very challenging problems yet to be solved.

Applications such as labelling, copyright protection, author authentication and proof of data integrity must be developed together with "fingerprinting" methods (a log file to store the identity of those who have accessed the information), to track and monitor multimedia information.

41.9. EDUCATION

Computers have been employed for more than two decades to support teaching in one way or another. Many computer aided teaching (CAT) programs have been developed. Reflecting on these efforts, it is possible to observe more unfulfilled expectations than success. Their introduction has often been accompanied by a fear on the part of the teacher, of being replaced by a machine; something that most teachers, with good reason, consider inappropriate. The main reason for the failure of computer aided teaching is, in the author's opinion, the misplaced goal of replacing conventional teaching tools with their electronic equivalent. For example a slide designed with the latest version of Powerpoint doesn't fundamentally improve the quality of the teaching, compared to handwritten text on a blackboard. This may be viewed as a particular case of electronic multimedia documents in which the content has been copied from the old medium. To obtain maximum advantage from digital electronic media and multimedia communications, the content (be it educational or entertaining) must be designed with the user and the medium in mind.

Each teacher has a specific way of teaching, and likewise each student has his or her own way of understanding the material taught. Those of us who teach know very well that only 20–30% of the class follow the material when delivered. A similar percentage of those remaining may however very easily understand another teacher covering the same topic. Pushing this to its limit, we may argue that different students must be taught differently. It is clear that high quality teaching relies on committed teachers, involved parents, creative administrators, diligent students and not solely on the use of computers and remote access.

Multimedia systems should allow teachers and students (and also parents and administrators) to share worldwide experience and use the best possible teaching materials in each case. The goal should not be to suppress schools and teachers. Students should still continue to attend classes, listen to their teachers, ask questions, experiment in laboratories and do homework. In addition they could enrich their education with new tools for better efficiency.

At the Swiss Federal Institute of Technology, Lausanne such a tool has been designed that will be trialled in the near future. The tool is targeted at technical topics at the university level and will be used initially for regular courses on signal and image processing. The basic question asked was how do we transmit the educators' knowledge to the student other than via linear methods such as textbooks, chalk, blackboards and transparencies. The aim was to design the material with the electronic medium in mind. The general tool will also facilitate distance teaching. Fully illustrated description of concepts, exercises, laboratory sessions, controls and examinations will also be possible. Through highly user-friendly interactivity, students can perform a number of experiments using the acquired concepts and check a plethora of configurations that will allow them to obtain a deeper knowledge and expertise on the subject.

The main means of supporting classical teaching, the textbook, has several drawbacks: it is intrinsically static (it only supports text and still figures), it induces a sequential mode of reading and does not allow dynamic evolution of its contents according to new items and readers' feedback. All these problems can be overcome using new technologies. These allow any kind of access to the existing information and can contain any sort of medium (text, still images, video, audio, interactive material and live classes), that can be added to or modified at any time.

The basic notion of the system proposed is called a brick. This is the smallest unit that contains a basic concept. It corresponds to the classical notion of a paragraph. Bricks are grouped to form walls (like sections in a book) and these are also grouped to form rooms (chapters). Finally, several rooms make a house (book).

Inside a brick as shown Figure 41.1, the student who accesses it for the first time must follow a given path. As a first step, he or she can choose either a mathematical or an intuitive explanation of the concept. The intuitive approach explains the subject in a comprehensive way, without equations, whereas the mathematical approach gives a detailed description using rigorous equations if necessary. As a second step the user could access examples to illustrate the concept applied to both an academic and to a real case. Finally, there are a number of proposed exercises and, if the concept is broad enough (or at a terminal brick) an examination.

For every house, room and wall there is a series of guided paths recommended to the user. Figure 41.2 shows a typical case. Inside a brick sequential access is straightforward, and for each brick there are a number of prerequisite bricks that provide the knowledge required to understand its content. There is also a connection to other bricks, to define concepts that can be learned after the present one has been acquired. The present brick becomes thus a prerequisite for the new one.

At higher levels, the walls and rooms also have interconnections, to define a suggested hierarchy of concepts or groups of concepts that are required to access the others. All the visited items are managed by a personal log file, which contains

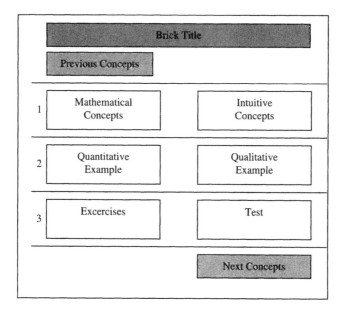

Figure 41.1 Internal structure of a brick.

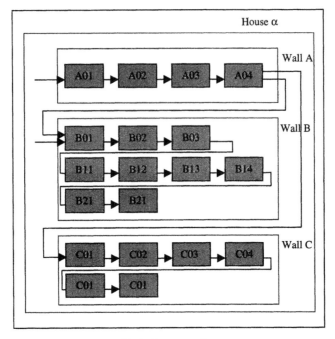

Figure 41.2 Links between the components.

the visited bricks together with other useful information, such as grades (if exercises or examinations have been completed).

For a global view of a course, a condensed look can be inspected at any time. This is the equivalent of a table of contents. The difference lies in the structure: in a book the table of contents is presented as a sequence of chapters and sections, whereas in the new system all the components can be viewed at a glance, having a general and clear idea about the building modules and their interrelationships. Moreover, searching capabilities are enhanced. For instance, intelligent search can be performed through all the bricks and those that are directly related to the desired information are highlighted.

This leads finally to the key problem of information access and exchange. It is apparent that research into multimedia content preparation must be run in parallel with the development of campus networks or intranets which provide sufficient quality of service and openess on both wired and wireless platforms.

41.10. CONCLUSIONS

In this chapter a subjective view of the future has been given. A short summary of the technical developments that led to the multimedia era are discussed. The rather anarchic situation of multimedia systems and functions are outlined, stressing the need for talent search engines and pollution conscientiousness. Problems related to standards and the urgent need for a single but open standard are emphasised. A non-exhaustive list of requirements are discussed as a possible framework for such a standard and the corresponding systems which would operate within it. The key idea of appending tools to information has been developed and exemplified in the dynamic coding of image sequences. Media conversion and unconstrained scalability issues are discussed. Finally, issues related to security and education are developed. These require new design approaches at both the system and content level so that multimedia research and development can progress in an innovative way, not simply copying old contents or functions from existing systems.

ACKNOWLEDGEMENTS

The author would like to express his sincere thanks to his colleagues and Ph.D. students who contributed to this text in various ways, namely to Drs T. Ebrahimi, J.M. Vesin, O. Egger and to MM. F. Pinol and M. Kutter.

REFERENCES

[1] Reusens, E. *et al.*, Dynamic approach to visual data compression. *IEEE Trans. Circuits and Systems for Video Technology*, **7** (1), 197–211, 1997.
[2] van den Branden Lambrecht, C., Characterization of the human visual sensitivity for video. *Signal Processing*, **67** (3), August 1998.

Index